EXTRATERRESTRIAL SEISMOLOGY

Seismology, the study of waves traveling through a medium, is a highly effective tool for investigating the internal structure of the Earth. Similar techniques have also successfully been used to study other planetary bodies (planetary seismology), the Sun (helioseismology), and other stars (asteroseismology). Despite obvious differences between stars and planetary bodies, these disciplines share many similarities and together form a coherent field of scientific research.

This unique book takes a transdisciplinary approach to seismology and seismic imaging, reviewing the most recent developments in these extraterrestrial contexts. With contributions from leading scientists, this timely volume systematically outlines the techniques used in observation, data processing, and modeling for asteroseismology, helioseismology, and planetary seismology, drawing comparisons with seismic methods used in geophysics. Important recent discoveries in each discipline are presented along with the interdisciplinary work that underpins them.

With an emphasis on transcending the traditional boundaries of astronomy, solar, planetary, and Earth sciences, this novel book is an invaluable resource and reference for undergraduates, postgraduates, and academics.

VINCENT C. H. TONG is Senior Lecturer in Geophysics at Birkbeck, University of London. He is also one of the principal investigators in an international interdisciplinary research consortium involving solid Earth science and oceanography. In addition to his research in geophysical imaging, he has published cross-disciplinary studies in solar and planetary seismology, and is a strong advocate for teaching and outreach enhanced by interdisciplinary scientific research.

RAFAEL A. GARCÍA is Director of Research in Solar and Stellar Physics at the Service d'Astrophysique, French Alternative Energies and Atomic Energy Commission in Saclay, France. His research focuses on global helioseismology and more recently on asteroseismology of Sun-like stars and red giants. He has participated in the development of various helio- and asteroseismic space missions of CNES, NASA, and ESA.

EXTRATERRESTRIAL SEISMOLOGY

Edited by

VINCENT C. H. TONG
Birkbeck, University of London

RAFAEL A. GARCÍA
Commissariat à l'énergie atomique et aux énergies alternatives (CEA)

CAMBRIDGE
UNIVERSITY PRESS

CAMBRIDGE
UNIVERSITY PRESS

University Printing House, Cambridge CB2 8BS, United Kingdom

Cambridge University Press is a part of the University of Cambridge.

It furthers the University's mission by disseminating knowledge in the pursuit of education, learning and research at the highest international levels of excellence.

www.cambridge.org
Information on this title: www.cambridge.org/9781107041721

First published 2015

Printed in the United Kingdom by TJ International Ltd. Padstow Cornwall

A catalogue record for this publication is available from the British Library

Library of Congress Cataloguing in Publication data
Extraterrestrial seismology / edited by Vincent C.H. Tong, Birkbeck, University of London, Rafael A. García, Commissariat à l'énergie atomique (CEA, Saclay).
pages cm
Includes bibliographical references and index.
ISBN 978-1-107-04172-1 (hardback : alk. paper) 1. Extraterrestrial seismology.
I. Tong, Vincent C. H., editor. II. Garcia, Rafael A., editor.
QB455.5.E98 2015
551.220999 – dc23 2015008918

ISBN 978-1-107-04172-1 Hardback

Additional resources for this publication at www.cambridge.org/tong

Contents

Color plates are located between pp. 232 and 233.

Contributors

Conny Aerts
Institute of Astronomy, KU Leuven, Celestijnenlaan 200D, B-3001 Leuven, Belgium and Department of Astrophysics, IMAPP, Radboud University Nijmegen, PO Box 9010, 6500 GL Nijmegen, The Netherlands

Thierry Appourchaux
Institut d'Astrophysique Spatiale, UMR8617, Université Paris-Sud, Bâtiment 121, 91405 Orsay Cedex, France

Jérôme Ballot
CNRS, Institut de Recherche en Astrophysique et Planétologie, 14 Avenue Edouard Belin, 31400 Toulouse, France and Université de Toulouse, UPS-OMP, IRAP, 31400 Toulouse, France

Sarbani Basu
Department of Astronomy, Yale University, PO Box 208101, New Haven, CT 06520-8101, USA

Rory P. Bigger
Southwest Research Institute, 6220 Culebra Road, San Antonio, TX 78238, USA

Douglas C. Braun
NorthWest Research Associates, 3380 Mitchell Ln, Boulder, CO, USA

Allan Sacha Brun
Laboratoire AIM, CEA/DSM-CNRS-Université Paris Diderot, IRFU/SAp, Centre de Saclay, 91191, Gif-sur-Yvette Cedex, France

William J. Chaplin
School of Physics and Astronomy, University of Birmingham, Edgbaston, Birmingham B15 2TT, UK

Sylvain Chaty
Laboratoire AIM, CEA/DSM-CNRS-Université Paris Diderot, IRFU/SAp, Centre de Saclay, 91191, Gif-sur-Yvette Cedex, France, and Institut Universitaire de France, 103, Boulevard Saint-Michel, 75005 Paris, France

Sidney Chocron
Southwest Research Institute, 6220 Culebra Road, San Antonio, TX 78238, USA

Jørgen Christensen-Dalsgaard
Stellar Astrophysics Centre, Department of Physics and Astronomy, Aarhus University, Ny Munkegade 120, DK-8000 Aarhus C, Denmark

Laura J. Cobden
Department of Earth Sciences, Utrecht University, Budapestlaan 4, 3584CD Utrecht, The Netherlands

Andreas Fichtner
Department of Earth Sciences, Swiss Federal Institute of Technology (ETH), Sonneggstrasse 5, 8092 Zurich, Switzerland and Department of Earth Sciences, Utrecht University, Budapestlaan 4, 3584CD Utrecht, The Netherlands

Rafael A. García
Laboratoire AIM, CEA/DSM-CNRS-Université Paris Diderot, IRFU/SAp, Centre de Saclay, 91191, Gif-sur-Yvette Cedex, France

Patrick Gaulme
Department of Astronomy, New Mexico State University, PO Box 30001, MSC 4500, Las Cruces, NM 88003-8001, USA and Apache Point Observatory, 2001 Apache Point Road, PO Box 59, Sunspot, NM 88349, USA

Frank Grundahl
Stellar Astrophysics Centre, Department of Physics and Astronomy, Aarhus University, Ny Munkegade 120, DK-8000 Aarhus C, Denmark

Tristan Guillot
Laboratoire Lagrange, Université de Nice Sophia Antipolis, UMR 7293 Observatoire de la Cote d'Azur (OCA), Nice, France

Shravan Hanasoge
Department of Astronomy and Astrophysics, Tata Institute of Fundamental Research, Mumbai 400-005, India

Rachel Howe
School of Physics and Astronomy, University of Birmingham, Edgbaston, Birmingham B15 2TT, UK

Walter F. Huebner
Southwest Research Institute, 6220 Culebra Road, San Antonio, TX 78238, USA

Steven D. Kawaler
Department of Physics and Astronomy, 12 Physics, Iowa State University, Ames, IA 50011, USA

Taichi Kawamura
Université Paris Diderot – Sorbonne Paris Cité, Institut de Physique du Globe de Paris, 35 Rue Hélène Brion, Case 7071, 75205 Paris Cedex 13, France

Trent T. Kirchdoerfer
Southwest Research Institute, 6220 Culebra Road, San Antonio, TX 78238, USA

Irina N. Kitiashvili
Stanford University, 488 Escondido Mall, Stanford, 94305, CA, USA

Martin Knapmeyer
DLR Institute of Planetary Research, Rutherfordstr. 2, 12489 Berlin, Germany

Alexander G. Kosovichev
Big Bear Solar Observatory, New Jersey Institute of Technology, 40386 North Shore Lane, Big Bear City, CA 92314-9672, USA

Philippe Lognonné
Université Paris Diderot – Sorbonne Paris Cité, Institut de Physique du Globe de Paris, 35 Rue Hélène Brion, Case 7071, 75205 Paris Cedex 13, France

Stéphane Mathis
Laboratoire AIM, CEA/DSM-CNRS-Université Paris Diderot, IRFU/SAp, Centre de Saclay, 91191, Gif-sur-Yvette Cedex, France

Savita Mathur
Space Science Institute, 4750 Walnut Street, Suite#205, Boulder CO 80301, USA

Benoît Mosser
LESIA, CNRS, Université Pierre et Marie Curie, Université Denis Diderot, Observatoire de Paris, 92195 Meudon Cedex, France

Yosio Nakamura
Institute for Geophysics, Jackson School of Geosciences, The University of Texas at Austin, J. J. Pickle Research Campus, Bldg. 196 (ROC), 10100 Burnet Road (R2200), Austin, TX 78758-4445, USA

Guust Nolet
Geoazur, Université de Nice/Sophia Antipolis, France and Department of Geosciences, Princeton University, NJ, USA

Pere L. Pallé
Instituto de Astrofísica de Canarias, Observatorio del Teide, Vía Láctea s/n, 38200 La Laguna, Tenerife, Spain

Mark Panning
University of Florida, Department of Geological Sciences, 241 Williamson Hall, PO Box 112120, Gainesville, FL 32611, USA

W. Thomas Pike
Department of Electrical and Electronic Engineering, Imperial College London, Exhibition Road, London SW7 2AZ, UK

S. P. Rajaguru
Indian Institute of Astrophysics, Koramangala II Block, Bangalore – 560034, India

Nicholas Schmerr
University of Maryland, Department of Geology, College Park, MD 20742-4211, USA

François-Xavier Schmider
Laboratoire Lagrange, Université de Nice Sophia Antipolis, UMR 7293 Observatoire de la Cote d'Azur (OCA), Nice, France

Takashi Sekii
Solar and Plasma Astrophysics Division, National Astronomical Observatory of Japan, Mitaka, Tokyo 181-8588, Japan

Hiromoto Shibahashi
Department of Astronomy, School of Science, University of Tokyo, Bunkyo-ku, Tokyo 113-0033, Japan

Dennis Stello
Sydney Institute for Astronomy, School of Physics, University of Sydney, NSW 2006, Australia

Michael J. Thompson
National Center for Atmospheric Research, PO Box 3000, Boulder, CO 80307-3000, USA

Vincent C. H. Tong
Department of Earth and Planetary Sciences, Birkbeck, University of London, Malet Street, London WC1E 7HX, UK

James D. Walker
Southwest Research Institute, 6220 Culebra Road, San Antonio, TX 78238, USA

Renee C. Weber
NASA Marshall Space Flight Center, 320 Sparkman Drive, Huntsville, AL 35805, USA

Preface

Stars and planetary bodies have a special place in our study of the universe. Identifying new planets and stars has a long history in astronomy, and studying their complex surface phenomena continues to be an important scientific endeavor in planetary science and astrophysics. With the advent of modern observational techniques and approaches to data analysis, it is now possible to probe their internal structure at increasingly high resolution and to monitor subtle changes in the deep interiors.

Indeed, having more accurate and detailed information about the stellar and planetary interiors helps us understand the evolution of stars and planetary bodies. Such information is invaluable for studying the links between subsurface structures and surface processes, too. Seismology is therefore central to our quests for understanding their interiors *and* surfaces. It provides a set of powerful and versatile tools capable of revealing the hidden phenomena inside the Earth, the Moon, and the Sun, amongst other stars, asteroids, planets, and their satellites.

Why extraterrestrial seismology?

In seismic studies, it is critically important to be able to relate any measurable or scientifically significant effects associated with seismic waves to the spatially varying physical properties in the subsurface. Research into the modeling of seismic sources and wave propagation in different stellar and planetary media has allowed scientists to build this foundation for imaging the interiors of stars and planetary bodies. Advances in the acquisition, processing, and modeling of seismic data as a whole have drastically improved the quality of the seismic images and information we have about the subsurface.

Encompassing both the theoretical studies of seismic phenomena and technical applications of seismic imaging developed over the past few decades, seismology is both a pure and an applied science. The pure and applied nature of seismology underpins many major breakthroughs across modern solid-earth planetary, and astrophysical sciences: the discovery of the Earth's and lunar cores, the presence of complex time-dependent differential rotation patterns of the solar interior, and the characterization of the inner radiative cores of red giants, to name but a few.

The development of seismic methods for studying this vast range of astrophysical bodies other than our own planet has naturally taken different paths. The approaches vary according to the physics pertinent to the targeted bodies, responding to and overcoming specific limitations in data acquisition as well as data quality and availability. Given the existence of gaseous and rocky planetary bodies of different sizes in the Solar System, it is perhaps not surprising that both the study of the seismic phenomena and the application of seismic imaging vary considerably within planetary seismology.

Solar seismology, or helioseismology, is different from planetary seismology in the sense that helioseismic analyses are based on the detection and analysis of changes in the electromagnetic radiation that the Sun emits. With the availability of ultra-high-resolution images of the visible solar surface, opportunities and challenges associated with the joint analysis of surface observations and seismic data in the solar context are as plentiful as they are unique.

Stellar seismology, or asteroseismology, builds on our understanding of the internal structure of the Sun and techniques developed by helioseismologists. However, the spatial resolution of the stellar phenomena that can be investigated in reality is vastly different from the solar ones, given the much greater distances between these stars and the Earth. Interestingly, the limited spatial resolution of seismic models of distant stars is compensated by the availability of seismic data from a large number of stars at different stages of their life cycles, resulting in the development of the statistical study of "ensemble" asteroseismology.

In other words, the development of stellar, solar, and planetary seismology is intricately linked to the nature of seismic sources, composition, and physical conditions of the stellar and planetary bodies. The distance between the Earth and the targeted body also affects the volume and quality of the seismic data available and hence the way to analyze and interpret them. Given these diverse flavors of planetary, solar, and stellar seismology, what is our rationale for grouping them under the umbrella of extraterrestrial seismology?

Despite the obvious physical differences between the targeted bodies, the methodologies used in the data processing and modeling share many similarities across the seismology spectrum. Challenges in the detection of seismic wave propagation and complex seismic resonance patterns are present in all seismic investigations. So are those related to the processing and inversion of seismic data with constraints from theoretical models and surface observations.

In order to acquire seismic data of the highest possible quality and with the optimal spatial coverage, extraterrestrial seismic studies require not only the development of state-of-the-art networks of ground-based telescopes but also pushing the frontiers of space science. From the development of nanosatellites to the launches of major multinational space missions, and from the incorporation of next-generation optical detector systems to the deployment of robotic seismologists to other planetary bodies, it is clear that scientists and engineers working in extraterrestrial seismology play a key part in driving the development of a wide range of subdisciplines in space science.

The very idea of extraterrestrial seismology does not imply that this collection of seismic disciplines bears few similarities with its terrestrial counterpart. In fact, earth seismologists also have to overcome difficulties and make technical advances in the acquisition, processing, and modeling of seismic data. Like extraterrestrial seismology, novel ways of analyzing seismic data are being developed for studying the Earth, with monitoring based on ambient seismic noise and high-resolution seismic imaging of ocean currents being two recent examples. Even space science has become part of the recipe in terrestrial seismology, with satellite data being used to monitor strong ground motions and to study deformations caused by earthquakes.

However, seismic investigations of other planetary and stellar bodies are still relatively young compared with terrestrial seismology, which has long been an established part of earth science. Generally speaking, planetary, solar, and stellar seismology are not featured regularly as a subject area in astrophysics or planetary science at undergraduate levels. In contrast, seismology has long been taught and prominently featured in undergraduate programs in geology and earth sciences. Furthermore, extraterrestrial seismic investigations do not (yet) have the immediate societal impact that many terrestrial seismic studies have in terms of natural hazards and commercial exploration of natural resources.

A cross-disciplinary subject space for research and education

We therefore believe that there is a strong case for creating a coherent "subject space" for research and education by drawing planetary seismology, helioseismology, and asteroseismology together. Sharing similarities in terms of methodologies, such as acquisition, processing, and modeling of seismic data, allows critical comparisons across disciplines that have been shaped by their own scientific priorities and traditions. This commonality in methodologies serves as an excellent context or even as a catalyst for both interdisciplinary research and education.

However, unfamiliar terminologies, conventions, and scientific contexts can present formidable barriers impeding possible cross-disciplinary exchanges between researchers, educators, and students, who may have different academic training and backgrounds. Moreover, as terrestrial, planetary, solar, and stellar seismologists are likely to be affiliated to different university departments and research institutes with different disciplinary focuses, the formation of a coherent research and education area of extraterrestrial seismology has hitherto been particularly challenging.

This volume is an important part of our efforts to create a new and accessible subject space for promoting cross-disciplinary research and education based on extraterrestrial seismology. This subject space has three cross-disciplinary dimensions:

1. The first part of the book highlights their commonalities and differences in research methodologies (Parts I to III). The aim is to encourage an understanding of *research methodologies in multidisciplinary contexts* by drawing attention to these key aspects of the research methods used across planetary, solar, and stellar seismology. These sections also serve as overviews to those who are new to the respective research approaches.

2. The part on scientific discoveries across planetary science and astronomy is aimed at highlighting *transdisciplinary findings* based on the use of extraterrestrial seismic methods. This part discusses how subsurface phenomena have been discovered and studied (Part IV), showcasing the immense contribution of extraterrestrial seismology to our study of the physical universe.
3. *Interdisciplinary research* involving other branches of planetary and astrophysical sciences (Part V) and studies incorporating terrestrial seismic methods (Part VI) are featured in the final parts of the book.

It is our wish to demonstrate the three cross-disciplinary aspects of this new exciting subject space, in order to facilitate exchanges of scientific ideas in explicit, helpful, and systematic ways. With an emphasis on the cross-disciplinary aspects, we do not intend to impose an overarching "discipline" of extraterrestrial seismology per se. In fact, we believe that thespectrum of seismology with inherently different conventions and scientific priorities should enrich this subject space for research and education.

The book comprises 26 chapters written by leading stellar, solar, planetary, and terrestrial seismologists currently based in 12 countries. In addition, following this preface, there are two short introductory articles with some personal thoughts on the development of planetary seismology (Mosser) and stellar-solar seismology (Christensen-Dalsgaard). The chapters are grouped under six parts, which are designed to provide a systematic survey of the multidisciplinary research methodologies, transdisciplinary findings, and interdisciplinary research.

The first part, entitled "Observation and space missions," is about the acquisition of extraterrestrial seismic data. The chapters together showcase how state-of-the-art space science and ground-based observations lie at the heart of the data acquisition. The chapter on asteroseismology (Appourchaux and Grundahl) and the one on helioseismology (Pallé, Appourchaux, Christensen-Dalsgaard, and García) together show similarities and differences between observational techniques used for studying stars. In contrast to the remote sensing techniques used in stellar and solar seismology, Lognonné and Pike discuss the design and deployment of seismometers on the Moon, Mars, and other planetary bodies.

Part II is on the relationship between "data and physical parameters," and the theme of the chapters in this part is to explore the crucial link between observations and information about the interiors of stellar and planetary bodies. These are explored in the stellar context (Kawaler) and are compared and contrasted between the Sun and solar-like stars (Chaplin and Howe). For the study of more local solar structure, Braun discusses in his chapter how helioseismic data can be analyzed. Nakamura reviews the links between data and models based on the landmark Apollo lunar experiments and the Viking experiments on Mars.

Building on the first two parts, Part III explores another multidisciplinary aspect of research methodologies – seismic data modeling. This is examined systematically in Aerts' chapter on asteroseismic modeling, Rajaguru's chapter in helioseismology with more mathematical details, and a comprehensive review that covers both lunar and martian seismology (Weber, Knapmeyer, Panning, and Schmerr).

Part IV comprises five chapters reviewing recent transdisciplinary discoveries in a diverse range of planetary and stellar contexts. They together underline the transdisciplinary power of applying seismic techniques to the study of a wide range of astrophysical bodies. The chapter by García and Stello shows how our knowledge of red giants has been dramatically enhanced by asteroseismology. Basu and Thompson review our understanding of the solar internal structure and dynamics based on the application of helioseismology. The local solar phenomena found relatively close to the Sun's visible surface are discussed in Sekii and Hiromoto's chapter. There are two chapters on planetary seismology, with a chapter on gaseous planets focusing on Jupiter's seismology (Gaulme, Mosser, Schmider, and Guillot) and one on our current understanding of the internal structure of the Moon (Knapmeyer and Weber).

The final parts of the book are on interdisciplinary research methodologies. Part V consists of eight chapters, showing exciting interdisciplinary developments in planetary and astrophysical sciences. Chaty's chapter on the new field of diskoseismology explains how matter accumulation under gravity can be exploited in seismic analyses. Kitiashvili reviews the use of numerical simulation in studying complex convection patterns in the solar interior and its links with solar seismology. Linking planetary seismology for both terrestrial and gaseous planets, a chapter by Lognonné and Kawamura introduces the field of impact seismology.

The next four chapters in Part V focus on solar and stellar phenomena and their links to seismology. The transport of angular momentum in stars during their evolution is explored in Brun and Mathis's chapter. By drawing on research progress in astrophysics, Christensen-Dalsgaard presents his suggestions for future research directions in stellar and solar seismology. The relationships between magnetic activity cycles and rotation patterns in stars are discussed in Mathur, Ballot, and García's chapter. Kosovichev's contribution investigates an important link between the solar interior and solar atmosphere: solar flares and helioseismic phenomena. Last but not least, the part ends with the contrasting branch of seismology: the seismic study of small bodies such as asteroids (Walker, Huebner, Bigger, Chocron, and Kirchdoerfer).

Part VI, the final one, is on interdisciplinary research linking terrestrial seismic methods and extraterrestrial seismology. The three chapters showcase the close relationships in research methodologies. The chapter on how models can be obtained by data inversion in solar and terrestrial contexts (Nolet) is complemented by Hanasoge's review on the use of cross-correlation in analyzing solar and terrestrial seismic data. The similarities and differences between seismic tomography based on studying seismic waveforms are compared and contrasted in the final chapter (Cobden, Fichter and Tong).

Interseismology: The book and beyond

Apart from creating this cross-disciplinary subject space, we endeavor to enhance the accessibility of the content to a wider readership beyond the science communities of both extraterrestrial and terrestrial seismology. This volume is not a textbook or a conventional edited research volume. Nor is it a collection of review articles written in journal styles. Instead, all the book chapters are relatively short overview articles featuring the chapter authors' perspectives on well-defined topics, contributing to the overall cross-disciplinary aims of the volume.

Most of the book chapters are written with minimum use of mathematical equations in this otherwise highly quantitative research area and do not assume specialist knowledge in the respective subjects. Equations are used to illustrate key scientific concepts and methods only, and the equations should therefore be useful to specialists and researchers who are interested in the mathematical description of the methodologies. Most chapters can therefore be readily used in undergraduate programs and courses in astrophysics, and planetary and earth sciences. Chapters on transdisciplinary findings (Part IV) are also suitable for use in non-seismology courses in physical sciences, including those at introductory levels.

Although there is some cross-referencing between book chapters, all chapters are essentially self-contained and can be read independently and in any order. This is particularly important for encouraging comparisons with different combinations of research methodologies and in different branches of extraterrestrial seismology. This format also allows the book to be used in university courses with different disciplinary (or even interdisciplinary) focuses and in flexible ways.

Inevitably there are overlapping topics and concepts that are relevant to a number of chapters. These commonly encountered concepts, such as "normal modes" and "tomographic inversion," are introduced by chapter authors in a way and to the level that are appropriate to the respective chapters. Comparing how these common concepts in seismology are introduced in different chapters may indeed be part of enhancing multidisciplinary understanding of the research methodologies.

Linked content is available on the publisher's dedicated webpage for the book. As part of our efforts to create the cross-disciplinary subject space, there is an associated project with multimedia research content and educational material about extraterrestrial seismology and its links to terrestrial seismology. Here is the web address for the Interseismology Project:

interseismology.org

It is a place for researchers and educators, including the chapter authors, to promote interdisciplinary dialogues and understanding in seismology. The website offers a dedicated platform to reach out to other branches of seismology and the wider physical sciences community about their work in a coherent subject space.

We hope that this book, together with other resources in the Interseismology Project, will be useful to you as well as to our colleagues and students in the earth, planetary, and astrophysical science communities. We would like to thank all our chapter authors for their immense contribution to this book project with their invaluable insights in a fast-developing subject, as well as their efforts into reviewing other book chapters.We are enormously grateful to Dr Emma Kiddle, Rosina Piovani, Zoë Pruce, Bronte Rawlings, Emily Trebilcock, and Martin Tynan at Cambridge University Press for their extremely generous support, help, and advice throughout this book project. We would like to thank

Margaret Patterson for her wonderful copy-editing work. We are grateful to Gabriel Pérez SMM/IAC for his help with the book cover. RAG would like to thank his family for their patience and support. VCHT would like to thank Professor Mike Thompson and Professor Mike Warner for introducing him to the fascinating world of cross-disciplinary studies involving solar and terrestrial seismology. He is grateful to Professor Sir Peter Knight for his infectious enthusiasm for both scientific research and teaching, and would like to thank Professor Alan Jenkins for his encouragement and unyielding guidance on research-enhanced education.

Acknowledgements

The author of Chapter 6 (Braun) would like to thank Charles Lindsey, Martin Woodard, and Thomas Duvall, Jr. and acknowledges support from the NASA Heliophysics Division and the Solar Terrestrial Program of the US National Science Foundation.

The author of Chapter 7 (Nakamura) would like to thank all of his colleagues who contributed to the successful execution of the Apollo and Viking seismic experiments, although they are too numerous to be listed here. Every one of them played a significant role in what is described in the chapter. The experiments and data analyses were supported by grants from the National Aeronautics and Space Administration of the United States of America.

The authors of Chapter 10 (Weber, Knapmeyer, Panning, and Schmerr) would like to thank the NASA Jet Propulsion Laboratory for partially supporting R. Weber and M. Panning through the Mars InSight mission, the Helmholtz Association for supporting M. Knapmeyer through the research alliance "ROBEX – Robotic Exploration in Extreme Environments," the NASA Planetary Geology and Geophysics Program (grant NNH10ZDA001N-PGG) for supporting R. C. Weber, and the NASA Lunar Advanced Science and Exploration Research Program (grant NNX09AO80G) and the NASA Post-doctoral Program for supporting N. Schmerr. Figure 10.2b was reprinted from *Journal of Geophysical Research*, Vol. 111, E11006, Knapmeyer, M.; Oberst, J.; Wählisch, Deuchler, C.; Wagner, R.; Working models for spatial distribution and level of Mars' seismicity, doi:10.1029/2006JE002708, Copyright (2006), with permission from Wiley. Figures 10.2c and 10.2d were reprinted from *Icarus*, Vol. 220 No. 2, Hempel, S.; Knapmeyer, M.; Jonkers, A. R. T.; Oberst, J.; Uncertainty of Apollo deep moonquake locations and implications for future network designs, pp. 971–980, Copyright (2012), with permission from Elsevier.

The authors of Chapter 11 (García and Stello) would like to acknowledge Tim Bedding for comments on the manuscript, the Australian Research Council for financial support, and the European Community Seventh Framework Programme ([FP7/2007-2013]) under grant agreements no. 269194 (IRSES/ASK) and no. 312844 (SPACEINN).

An author of Chapter 12 (Thompson) acknowledges that the National Center for Atmospheric Research is sponsored by the National Science Foundation.

The authors of Chapter 15 (Knapmeyer and Weber) would like to thank the Helmholtz Association for supporting M. Knapmeyer through the research alliance "ROBEX – Robotic

Exploration in Extreme Environments," and the NASA Planetary Geology and Geophysics Program for supporting R. C. Weber through grant NNH10ZDA001N-PGG. The authors also wish to thank Elsevier for permission to reprint Fig. 5 from *Physics of the Earth and Planetary Interiors*, Vol. 182, Weber, R. C., Bills, B. G., Johnson, C. L., A simple physical model for deep moonquake occurrence times, pp. 152–160, © 2010, and Fig. 7 from *Earth and Planetary Science Letters*, Vol. 243, Chenet, H.; Lognonné, Ph.; Wieczorek, M.; Mizutani, H., Lateral variations of lunar crustal thickness from the Apollo Seismic Data set, pp. 1–14, © 2006.

The authors of Chapter 19 (Brun and Mathis) acknowledge the support received from the European Community Seventh Framework Programme ([FP7/2007-2013]) under grant agreements no. 312844 (SPACEINN), no. 269194 (IRSES/ASK), and from the CNES GOLF/SOHO and CoRoT grants at CEA. Brun acknowledges the support received from the ERC-STARS2 207430 grant, ANR Toupies grant, INSU/PNST, and the CNES via the Solar Orbiter grant. He is also grateful to both the University of Kyoto, RIMS and Professors M. Yamada, K. Shibata, H. Isobe, and S. Takahiro for their invitation in Fall 2013 and to the University of California at Santa Barbara, KITP, and the organizers of the waves and flow program in Spring 2014. Mathis acknowledges the support received from CNRS/INP/PEPS "Physique Théorique et ses Interfaces."

The author of Chapter 20 (Christensen-Dalsgaard) would like to acknowledge the Danish National Research Foundation for funding for the Stellar Astrophysics Centre (Grant DNRF106), and the European Research Council for support of the ASTERISK project (ASTERoseismic Investigations with SONG and Kepler; Grant agreement no. 267864).

An author of Chapter 21 (Mathur) would like to acknowledge support from the NASA grant NNX12AE17G, while another author (García) acknowledges the ANR (Agence Nationale de la Recherche, France) program IDEE(ANR-12-BS05-0008) "Interaction Des Etoiles et des Exoplanetes" and the CNES for the GOLF/SOHO and CoRoT grants at CEA.

The author of Chapter 22 (Kosovichev) would like to acknowledge support from NASA grants (NNX14AB68G and NNX14AB70G). The observational data are provided by the SDO/HMI and SOHO/MDI teams. SOHO is a project of international cooperation between ESA and NASA.

The authors of Chapter 23 (Walker *et al.*) thank Walter Huebner for his vision and tenacity and Southwest Research Institute for its continued support.

The author of Chapter 24 (Nolet) would like to acknowledge support from the European Research Council (Advanced Grant #226837).

An author of Chapter 26 (Cobden) acknowledges funding support from the Dutch Science Foundation (NWO) on grant number NWO: VIDI864.11.008.

Abbreviations

ACRIM	active cavity radiometer
AMP	Asteroseismic Modeling Portal
ASE	Active Seismic Experiment
ASLEP	Apollo Lunar Surface Experiment Package
asteroFLAG	Astero Fitting at Low Angular degree Group
ATHENA	Advanced Telescope for High ENergy Astrophysics
BiSON	Birmingham Solar Oscillation Network
CCD	charge-coupled device
CFHT	Canada France Hawaii Telescope
CME	coronal mass ejection
CNES	Centre National d'Etudes Spatiales
CoRoT	Convection Rotation and planetary Transits
CV	cataclysmic variable
CZ	convection zone
EASEP	Early Apollo Surface Experiment Package
EIS	Extreme-ultraviolet Imaging Spectrometer
EOS	equation of state
ESA	European Space Agency
ESO	European Southern Observatory
EVRIS	Etude de la Variabilité, de la Rotation et des Intérieurs Stellaires
FD	finite difference
FFT	fast Fourier transform
FT	Fourier transform
FTS	Fourier transform spectrometry
GEMINI	Green function of the Earth by MINor Integration
GOES	Geostationary Operational Environmental Satellite
GOLF	Global Oscillations at Low Frequencies
GONG	Global Oscillation Network Group
GRB	gamma-ray burst
HARPS	High Accuracy Radial velocity Planetary Search
HFT	high-frequency teleseism

HMI	Helioseismic and Magnetic Imager
HMXRB	high-mass X-ray binary
HR	Hertzsprung–Russell
IRIS	International Research on the Interior of the Sun
IRTF	Infrared Telescope Facility
ISCO	innermost stable circular orbit
JUICE	JUpiter ICy moon Explorer
KASC	Kepler Asteroseismic Science Consortium
LES	large eddy simulation
LM	Lunar Module
LMXRB	low-mass X-ray binary
LOFT	Large Observatory For X-ray Timing
LOI	Luminosity Oscillations Imager
LP	long period
LRO	Lunar Reconnaissance Orbiter
LSG	Lunar Surface Gravimeter
LSPE	Lunar Seismic Profiling Experiment
MCMC	Markov chain Monte Carlo
MDI	Michelson Doppler Imager
MHD	magnetohydrodynamic
MOF	magneto-optical filter
MOST	Microvariability and Oscillations of Stars Telescope
NASA	National Aeronautics and Space Administration
OGLE	Optical Gravitational Lensing Experiment
OHP	Observatoire de Haute Provence
OLA	optimally localized averages
OPD	optical path difference
PDS	power density spectrum
PLATO	PLAnetary Transits and Oscillations of stars
PRISMA	Probing Rotation and Interiors of Stars: Microvariability and Activity
PRNU	pixel-to-pixel response non-uniformity
PSE	Passive Seismic Experiment
QPO	quasi-periodic oscillations
RHESSI	Reuven Ramaty High Energy Solar Spectroscopic Imager
RLS	regularized least squares
RXTE	Rossi X-ray Timing Explorer
SCLERA	Santa Catalina Laboratory for Experimental Relativity by Astrometry
SDO	Solar Dynamics Observatory
SEIS	Seismic Experiment for Interior Structure
SEM	spectral element method
SL9	Shoemaker–Levy 9 comet
SMBH	supermassive black hole

SOHO	Solar and Heliospheric Observatory
SOI	Solar Oscillation Investigation
SOLA	subtractive optimally localized averages
SP	short period
SPICE	Seismic wave Propagation and Imaging in Complex media: a European network
SSM	standard solar/stellar model
TESS	Transiting Exoplanet Survey Satellite
VBB	very broad band
VIRGO	Variability of Irradiance and Gravity Oscillations
WIRE	Wide Field Infrared Explorer
WPP	Wave Propagation Program
XRB	X-ray binary
YSO	young stellar object
ZAMS	zero-age main sequence

Planetary seismology: High risk, high return

BENOÎT MOSSER

Except for specialists, planetary seismology refers first to Earth seismology, and stimulates thoughts about earthquakes, tsunamis, and all such rare and brutal events that everyone knows by experience and/or fears. According to etymology, seismology shakes and shocks our bodies and our minds. According to semantics, all languages make the difference between "superficial" and "deep" things, so that everyone turns out to be a seismologist, encouraged to prefer a deep to a superficial analysis, to get to the bottom of things, to look at an issue in more depth, to delve into something, to get to the root of the matter, to seek out the underlying issues...

We have basically a 2D topological experience of the Earth since we live on it and are attached to its surface. Of course, in tunnels and by plane, we are invited to explore the third dimension, a thin envelope in fact, so shallow with respect to the Earth radius. Even this Earth radius is an ethereal concept (sorry for the hiatus), since we lack normal circumstances where we can simply feel that the Earth is not as flat as a pancake. When the Earth was a pancake, there was a great debate about the nature of its edge. In a 3D spherical view, defining and measuring the Earth radius implies necessarily prospecting the medium between the surface and the center.

Hidden signatures

The favorite activity of a planetary seismologist consists in seeking to look well below the planetary surface, as done by Athanasius Kircher, a German Jesuit (1601–1680), in his model of the Earth's interior. Volcanoes were seen as outlets of the Earth's internal fires. Similarly, any trace on a planetary surface can be related to properties of the interior structure. On the Moon, the absence of surface faulting indicates tectonic inactivity. On Mars, the presence of huge volcanoes has the same consequence. However, this does not imply that the objects are seismologically inactive. On Jupiter, Saturn, and Neptune, the net flux radiating from the planet indicates a significant inner energy source (approximately equivalent to the solar flux), hence convection for extracting this energy, then possibly

Extraterrestrial Seismology, eds. Vincent C. H. Tong and Rafael A. García. Published by Cambridge University Press.

non-negligible seismic activity. These surface signatures are, however, not enough to scan the interior in detail. To do so, we need seismology.

The aim of seismology is, in essence, hidden. As a consequence, the paradigm of instrumentation for planetary seismology is demanding: Complex and sensitive devices are required, for mostly unknown results to be recorded in complex conditions. When an instrument is designed and built for, e.g., measuring various components in a planetary atmosphere with an unprecedented sensitivity, much information is available on this atmosphere. First of all, the existence of this atmosphere is certain, and many molecules just wait to be detected. When you aim to observe the Moon, Mars, or Jupiter with a seismometer, you need the most sensitive instrument, since you seek for tiny signals, and you pray for an efficient source of excitation, and the absence of noise.

Instrument and instrumental noise

The noise level of an efficient instrument must be as low as possible. As usual, this implies contradictory specifications. The white noise level of an *in-situ* seismometer with a suspended proof mass is inversely proportional to the square root of the product of the proof mass, the period, and the quality factor of the suspension (see Chapter 3 by Lognonné and Pike). Such seismometers are, basically, equivalent to instruments used on Earth. However, for use in space, mass is an issue. Mid-term solutions must be found for the quality factor too, since a high quality factor in favor of a noise-free instrument implies a narrow bandpass, hence an instrument efficient in only very limited cases.

Similarity with helioseismology is useful for defining the instrumental principle for seismic observations of a planet with a fluid surface. In order to retrieve a tiny seismic Doppler signal embedded in multiple much stronger Doppler-like signatures (differential rotation, convective movements), such an instrument has to simultaneously achieve imaging and high-resolution spectrometry. This requires a large number of photons to be detected, hence a large photon collector. The quality factor of the spectrometer has to be as large as possible, which goes against the first specification.

Having a single instrument for seismology is, as one can imagine, not appropriate. With one single instrument, the information to be recovered is highly degenerate. In the opposite case, reconstructing a precise tomographic view of an inhomogeneous body requires a large number of seismometers adequately distributed. For cases in between, the more instruments you have, the more informative the results can be. The single seismometer deposited on Mars by the Viking mission could not work in favorable conditions. For Jovian seismology, the first observations made without spatial resolution could only retrieve integrated signals. Current projects for Jovian seismology all consider an imaging instrument able to deliver seismic information on the whole visible hemisphere.

Excitation and signal

The question of excitation plays a central role. As seen above, for other bodies a tectonic origin as on Earth is unlikely. Identifying a phenomenon able to excite waves is tough;

quantifying the excitation is awfully complex. Apart from the Moon, which benefitted from multiple observations that led to the detection of various types of events (see Chapter 15 by Knapmeyer and Weber), the situation is unclear since we lack information. Obtaining this information would require seismology. . .

The examples of helio- and asteroseismology can be usefully exploited for understanding the possible excitation of oscillations in the fluid giant planets. Oscillations in these objects are called normal modes (according to the terminology used for solid planets) or global oscillations (according to the terminology used in helio- or asteroseismology). In practice, long-lived modes constructed by resonant pressure waves (P waves) are observed, not only transient waves. In such fluid media, shear waves (S waves), with transverse displacement, do not exist. Solar-like oscillations are seen in all stars showing an external convection envelope. Convective turbulent cells breaking just below the photosphere, responsible for the granulation seen on the Sun, stochastically excite waves. The levels of excitation are low, and the mechanism implies that the oscillations are embedded in a high background noise. Fortunately, the coherence of the long-lived oscillations allows their detection. A similar excitation is expected for the giant planets, except Uranus.

Results and prospects

So far, unique results have been obtained for the seismic Moon, benefitting from the multiple manned missions on the Moon from 1969 through 1977. Passive and even active seismometric observations were performed, and yielded various results, with the detection of waves generated by meteoroid impact, thermal waves, or deep Moonquakes. Mars' seismicity was monitored by the Viking mission, from 1976 through 1978. Only one seismic station was operative. The seismometer could not be deposited on the ground but was mounted on the lander to minimize the weight and complexity. It failed to produce any positive seismic detection, with all events certainly due to wind shifts. For Jovian planets, the best result, limited to the detection of seismic global parameters, was obtained with a dedicated ground-based instrument.

Even if seismology is a powerful tool for inferring information on the planetary (and stellar) interior structure, any other way to provide such information is good and should not be ignored. The gravitational moments of Jupiter, quickly measured by the NASA space probe Juno during its nearby flybys of the planet, can be used to relate the geometrical properties of the external gravitational field of the planet to its mass density distribution. However, such information corresponds to quantities integrated from the center to the surface of the planet and has no precision compared to seismic observations, which provide us with a differential view of the body. At the time of writing, we expect unique results from the ESA mission Rosetta. The lander Philae will carry much information on the structure of a comet core. Radar will perform tomography of the nucleus by measuring electromagnetic wave propagation between the Philae lander and the Rosetta orbiter through the comet nucleus. Other instruments are aimed at measuring physical properties of the outer layers of the comet.

Despite the high cadence of planetary probes traveling to Mars and landing on it, the seismic analysis of Mars is still to come. The situation will fortunately change with the NASA Discovery Program mission InSight (Interior Exploration using Seismic Investigations, Geodesy and Heat Transport). InSight aims to measure the size, composition, and physical state of the martian core. It is designed to determine the thickness and structure of the martian crust, the composition and structure of the martian mantle, the thermal state of Mars' interior, to measure the magnitude, rate and geographical distribution of Mars' internal seismic activity, and to measure the rate of meteorite impacts on the surface of Mars. The only way currently possible to provide more rapid results in planetary seismology should come from the NASA *Kepler* 2.0 mission. With photometry, this mission could lead to a major breakthrough in planetology by providing the first seismic observations of Neptune.

A bright outlook for helio- and asteroseismology

JØRGEN CHRISTENSEN-DALSGAARD

As documented in the following, very substantial results have already been obtained from helio- and asteroseismology, based on huge observational projects, such as the GONG project, the helioseismic instruments on the Solar and Heliospheric Observatory (SOHO) and Solar Dynamics Observatory (SDO) satellites, and asteroseismology with the CoRoT and *Kepler* missions. But this is just the beginning. New asteroseismic projects are on the way, and there are also plans for new developments in observational helioseismology, including multi-wavelength observations (Hill *et al.*, 2013). For local helioseismology a potential breakthrough will come from ESA's Solar Orbiter mission (Gandorfer *et al.*, 2011), which will for the first time allow detailed investigations of the regions near the solar poles.

However, there is much to be done with the existing data, and much to be done further to develop the analysis techniques. The helioseismic data have been far from fully exploited, with the inverse analyses being typically based just on a small fraction of the available observations. In particular, the high-resolution observations from the Helioseismic and Magnetic Imager (HMI) instrument on SDO may provide reliable frequency data on high-degree modes, which would allow more detailed investigations of the thermodynamics of the region of helium ionization (e.g., Rabello-Soares *et al.*, 2000). Also, the base of the solar convective envelope certainly deserves further investigation, including an updated search for possible latitude variations (Monteiro and Thompson, 1998) and a follow-up to the hint of solar-cycle variations found by Baldner and Basu (2008). The helioseismic data on frequency splittings have been extensively used to study variations of solar internal rotation during the solar cycle (for a recent example, see Howe *et al.*, 2013). However, further work is certainly required on constraining the rotation of the solar core, where improvements in the inversion techniques may allow a better localization and balance between resolution and error. I note that Vorontsov and Jefferies (2013) are developing new techniques for the analysis and interpretation of helioseismic data which offer substantial promise for improved inferences on structure and rotation.

Extraterrestrial Seismology, eds. Vincent C. H. Tong and Rafael A. García. Published by Cambridge University Press.

Local helioseismology has the potential to provide information about the rapidly evolving subsurface structure and dynamics of the Sun. At present, and in the foreseeable future, the most detailed results are obtained from the HMI on the SDO satellite (Scherrer *et al.*, 2012). These include daily images of active regions on the solar backside obtained with helioseismic holography (e.g., Lindsey *et al.*, 2011, see http://jsoc.stanford.edu/data/farside/), the potential, not yet fully realized, of providing generally available maps of flow-field wave-speed perturbations several times a day, as well as synoptic maps of these quantities for each Carrington rotation (Zhao *et al.*, 2012c). Such results clearly have great value for the study and potential understanding of solar activity, and for space weather analysis and prediction, and their availability and user friendliness should be further improved.

An extensive discussion of the state of helioseismology and its future development, including the very many open issues, was provided by Gough (2013).

For distant stars, inverse analysis, in the manner successfully applied to the Sun, is severely limited owing to the restriction of the observed modes to those of low degree (see also Thompson and Christensen-Dalsgaard, 2002). Analysis of low-degree solar data (Gough and Kosovichev, 1993) and simulated stellar data (Basu *et al.*, 2002) indicates that some resolution of the structure may be possible in the core of solar-like stars. These techniques are based on linearizing the frequency differences between the star and a reference model; it has been found that such linearization may be inaccurate unless the reference model is already quite close to the star. A conceptually very different approach to structure inversion, which does not require linearization, has been proposed by Roxburgh and Vorontsov, based on phase matching (for a brief overview, see Roxburgh, 2010). This has yielded very promising results when applied to artificial data and was recently applied to observed data with encouraging results (Roxburgh, 2015). Inversion for stellar rotation has been attempted in a few cases, to obtain localized information about the core rotation in unevolved red giants, while inferences aimed at determining the near-surface rotation had some success in suppressing the unwanted contributions from the core. Further tests of these techniques, including use on artificial data, are also required.

As a complement to these detailed investigations of individual stars, further development is required of efficient and reliable techniques to analyse observations of large numbers of stars. An extensive analysis of main-sequence and subgiant stars was carried out by Chaplin *et al.* (2014) based on *Kepler* observations of global seismic parameters. However, the determination and fitting of individual frequencies still require substantial care and human intervention in the analysis. Even techniques that in principle provide an unbiassed best estimate of the fit and its statistical properties depend on appropriate assignment of weight to the frequencies, frequency combinations, and other observables which is difficult to make automatic. Also, work is still required on the proper implementation of error correlation in these fits. In the case of *Kepler*, detailed data are potentially available for a few hundred main-sequence and subgiant stars, and hence it is possible to invest considerable effort in the analysis of individual stars. For future missions we can expect such data for thousands of stars, and hence automated procedures become mandatory.

Given the huge number of stars observed with present and coming projects we also need to consider how best to make use of the data. Extensive modeling efforts will probably be confined to a relatively small number of "best" cases, with data of exceptional quality or otherwise presenting particularly interesting features. Thus we need to develop methods to identify such cases. For the bulk of the stars, ensemble studies will be carried out, to determine the dependence on global stellar parameters of the properties of the stars, including specific aspects of their internal structure. This also has the potential to provide a wide-ranging test of stellar modeling. Cases of clear discrepancies between the models and the observations will be obvious targets for more detailed investigation, to identify the reasons for the failure of the models. It is evident that reliable statistical information is required to ascertain whether the inferred discrepancies are indeed significant.

A key goal of these investigations is of course to improve our understanding of stellar structure and evolution, with emphasis on the physical processes controlling stellar properties. Stellar modeling is undergoing a substantial evolution, with the inclusion of increasingly realistic descriptions of effects of rotation and other instabilities, based in part on insights obtained from large-scale hydrodynamical simulations. We can expect a strong interaction between this improved modeling and the continuing improvements in the asteroseismic inferences.

Looking beyond the likely horizon of the present author, it is clear that a major limitation of present asteroseismology is the restriction to low-degree modes. For example, this essentially precludes any search for stellar analogues of the solar tachocline, which is expected to play a major role in the generation of the solar magnetic activity. Thus, observations able to detect oscillations of degree at least as high as 60 are highly desirable. This is one of the goals of the Stellar Imager (Carpenter *et al.*, 2009), a mission concept for interferometric stellar observations with rather high spatial resolution. Such a mission will require very considerable technical development and likely precursor missions to document technical feasibility. Should the full Stellar Imager be carried out, which I certainly hope, we would get an even closer relation between the seismology of the Sun and the distant stars.

Part I

Observation and space missions

1

Asteroseismology observations and space missions

THIERRY APPOURCHAUX AND FRANK GRUNDAHL

1.1 Introduction

Variable stars have been observed since the discovery of the variability of Mira by David Fabricius at the end of the sixteenth century (Olbers, 1850) with a variation of 10 stellar magnitudes! The first known photometric measurement of variability was done by Goodricke (1783) for Algol, comparing with stars differing by two magnitudes only. Using the same technique, Goodricke and Englefield (1785) discovered that δ Cephei was a variable star showing a periodical non-sinusoidal variation of about one magnitude.

Roberts (1889) suggested the use of photography for studying variable stars. The technique for detecting variable stars was quickly improved by photographically recording variations in the stellar spectra (Fleming, 1895), which would lead to visual identification of these variables (Reed, 1893).

The use of the selenium photometer by Stebbins and Brown (1907) and the development of the photoelectric cell, as anticipated by Stebbins (1915), introduced the electric determination of variability of stars (Stebbins, 1911). This was the start of an era: the detection of photons using the photoelectric effect. Improvement in the instrumentation led to lower detectable levels of variability that led to the discovery of rapidly oscillating Ap stars by Kurtz (1978). These stars oscillate with periodicity of about 6 to 10 minutes with an amplitude of about a thousandth of a magnitude (mmag). Kurtz and Shibahashi (1986) were even able to reach a noise limit of about 0.02 mmag on HR3831 (Kurtz and Martinez, 2000).

From space, lower levels of variabilities had already been detected on the Sun using the ACRIM (Active Cavity Radiometer Irradiance Monitor) instrument aboard the Solar Maximum Mission (Woodard and Hudson, 1983); the noise level was about one part per million (ppm, close to one μmag). From the ground, clear detection of solar-like oscillations (pressure modes or p modes) in intensity and with spatial resolution was made possible using a property of the atmosphere combined with the significant angular diameter of the Sun (Appourchaux *et al.*, 1995) (see also Chapter 2 of this book). Unfortunately, due to

Extraterrestrial Seismology, eds. Vincent C. H. Tong and Rafael A. García. Published by Cambridge University Press.

the minute angular diameter of the stars, their intensity fluctuations are not limited by transparency fluctuations but by scintillation (Young, 1967).

The observation of variable stars possesses some astrophysical importance for the understanding of the physics in our universe. The variety of variability observed since the sixteenth century led to the discovery of various classes of variable stars such as Cepheids, δ Scuti, Mira type, rapidly oscillating Ap stars, white dwarfs, solar-like oscillations; each having different interesting fields of physics at play. This list of variable stars is a subset of what is available across the Hertzsprung–Russell diagram (Eyer and Mowlavi, 2008).

With so many variable stars, the detection of such periodicities led to the development of a new type of science: asteroseismology. This new field of science was defined by Christensen-Dalsgaard (1984) as "The science of using stellar p modes for the study of the properties of stars, including their internal structure and dynamics." In this seminal paper, Christensen-Dalsgaard (1984) introduced what could be done for stars similar to the Sun. The feasibility of detecting stellar p modes similar to those of the Sun was theoretically given by Christensen-Dalsgaard and Frandsen (1983). They provided predicted amplitudes of about 10 cm s^{-1} for radial velocity and a few tens of ppm for intensity fluctuations. The detection of the first solar-like oscillations in another star by Bouchy and Carrier (2002) was the real start of asteroseismology of solar-like stars. A few years later, several other stars were found to exhibit solar-like oscillations, as shown by Kjeldsen *et al.* (2008b), with amplitudes up to 70 cm s^{-1}. The long story of the start of asteroseismology of solar-like stars is provided later in this chapter.

The goal of this chapter is to provide a review of how one can detect and observe such solar-like oscillations. The first section focuses on the scientific requirements derived from the properties of the oscillations. The second section treats the history of the first detection attempts that eventually led to the implementation of the more ambitious projects described in the third section. In the last section, we conclude with a focus on the next generation of ground- and space-based projects.

1.2 What are the requirements?

1.2.1 How can we recover the signal?

The design of the instrumentation capable of detecting solar-like oscillations is driven by three parameters: the highest mode frequency, the mode amplitude (in radial velocity or intensity), and the mode lifetime. The first parameter provides the sampling time of the time series. The second and third parameters provide the level to which the instrument should be stable and the duration of the observations, respectively.

Sampling time

This is closely related to the acoustic cut-off of the frequencies. Since most of the excited modes in solar-like stars are pressure modes, there is a frequency at which the modes are no longer reflected at the surface of the star (i.e., there are no eigenmodes); above the acoustic cut-off frequency, the pseudo-modes observed in the Sun by García *et al.* (1998) in radial velocities are due to standing waves, as modeled by Kumar and Lu (1991). The acoustic cut-off ν_{ac} was given by Brown *et al.* (1991) as

$$\nu_{ac} \propto \frac{M}{R^2\sqrt{T_{eff}}} \propto \frac{M}{L}T_{eff}^{3.5}, \tag{1.1}$$

where M, R, and T_{eff} are the stellar mass and radius in solar units, and the effective temperature. The highest cut-off frequency is obtained for stars smaller than the Sun having a larger surface gravity. For stars with mass of 0.8 $M_\odot$, a stellar luminosity of 0.4 $L_\odot$ and an effective temperature of 5650 K, the acoustic cut-off frequency scaled to that of the Sun is about 10 200 μHz or 98 s. The application of Shannon's theorem implies that the maximum sampling time should be less than 49 s, i.e., the Nyquist frequency is the previous acoustic cut-off frequency. In order to minimize the high-frequency noise, it is also advisable to integrate the signal over the sampling time, i.e., to have a duty cycle close to 100%. (See Appourchaux, 2014, for an application of various digital-world approaches to asteroseismology.)

Signal-to-noise ratio

In order to detect the modes at the maximum of power, it is reasonable to put an upper limit on the background noise of three times lower (target) and ten times lower (goal) than the mode height to be detected. The background noise is composed of two sources of noise: the instrumental noise, which can be controlled and reduced, and the stellar noise, which cannot be controlled. The goal of a signal-to-noise ratio of 10 is derived from the theoretical expression of the mode frequency precision provided by Libbrecht (1992) as

$$\sigma_\nu = \sqrt{f(\beta)\frac{\Gamma}{4\pi T}}, \tag{1.2}$$

where β is the inverse of the signal-to-noise ratio ($= H/B$ where H is the mode height and B is the background noise), Γ is the linewidth of the mode profile assumed to be Lorentzian, T is the observation time, and $f(\beta) = \sqrt{1+\beta}(\sqrt{1+\beta}+\sqrt{\beta})^3$. A typical frequency resolution of 0.2 μHz is required in order to have meaningful structure inversions and for comparison with theoretical models (Basu, 2003). With a signal-to-noise ratio better than 10, a frequency precision smaller than 0.2 μHz can be obtained for observation times longer than 2 months, and with signal-to-noise ratio better than 3 for time longer than 4 months (see Figure 1.1). These signal-to-noise ratio figures of 3 and 10 are typical arbitrary upper limits that can be tweaked according to various engineering and scientific strategies. The tweaking can certainly be done for even lower noise values.

Background noise

The required background noise is then simply derived from the theoretical mode amplitude and mode linewidth, using a signal-to-noise ratio set to 10. The mode amplitude was theoretically derived by Christensen-Dalsgaard and Frandsen (1983) and Houdek *et al.* (1999). The latter theoretically confirmed the scaling relation provided

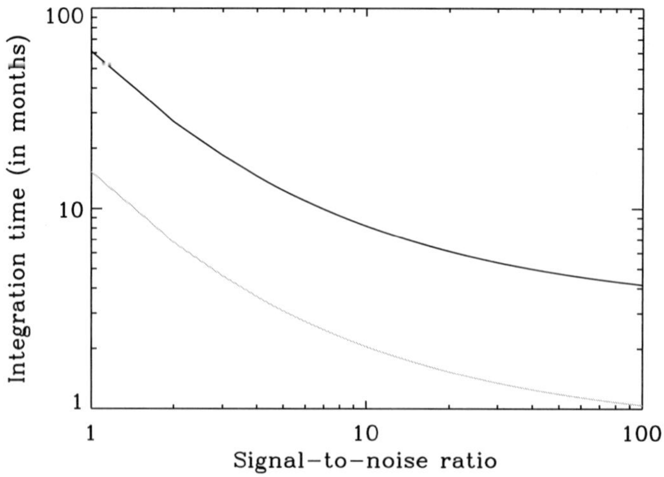

Figure 1.1 Observation time as a function of signal-to-noise ratio in the power spectrum for a mode frequency precision of 0.1 μHz (black line) and of 0.2 μHz (gray line) for a 1 μHz linewidth. The mode is a singlet.

by Kjeldsen and Bedding (1995):

$$\frac{A}{A_\odot} \approx \frac{L}{M}\left(\frac{T_\odot}{T_{\rm eff}}\right)^s, \tag{1.3}$$

where A is the maximum stellar mode amplitude, $A_\odot$ is the maximum solar mode amplitude and L is the stellar luminosity in units of the solar values, $T_\odot$ is the effective temperature of the Sun, and where $s = 0$ for radial velocity and $s = 2$ for intensity fluctuations. The units of A are either cm s^{-1} or ppm. The maximum of the mode amplitude is obtained for a frequency $\nu_{\rm max}$, which follows the same scaling relation as Equation (1.1) (Brown *et al.*, 1991). Since the stellar luminosity scales as M^4 for main-sequence stars, the lowest mode amplitude is obtained for stars less massive than the Sun. The maximum mode height in the spectral density is directly linked to the maximum mode amplitude and the mode linewidth as

$$H = \frac{2A^2}{\pi\Gamma}, \tag{1.4}$$

where H is the mode height and Γ is the mode linewidth. The relation is derived assuming that the mode profile is Lorentzian. Combining Equations (1.1), (1.3), and (1.4), one can obtain a relation between the mode height and the frequency of maximum power as

$$H = \frac{2A_\odot^2}{\pi\Gamma_{\rm max}}\left(\frac{T_{\rm eff}}{T_\odot}\right)^{7-2s}\left(\frac{\nu_\odot}{\nu_{\rm max}}\right)^2. \tag{1.5}$$

Similar relations were derived by Chaplin *et al.* (2009). The theoretical mode linewidths were derived by Houdek *et al.* (1999). They typically gave mode linewidths of about 2 μHz at the location of the maximum power for solar-like stars. It was shown by Baudin *et al.* (2011) that the mode linewidth at maximum mode height was a function of effective

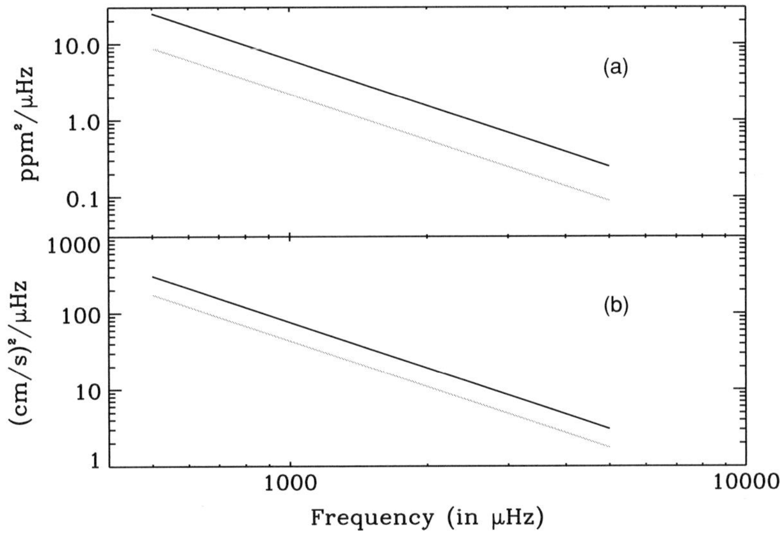

Figure 1.2 Required background noise as a function of the frequency of the maximum mode power required for detecting stellar p modes with a signal-to-noise ratio of 10, panel (a) for intensity, panel (b) for stellar radial velocity; for the solar effective temperature (black) and for an effective temperature of 6500 K (gray), the mode linewidths being 1 μHz and 4 μHz, respectively. The highest frequency corresponds to a 0.8 $M_\odot$ star, while the lowest frequency corresponds to a 2 $M_\odot$ star.

temperature, which was not predicted by any theoretical calculations. Therefore we used a mode linewidth of 1 μHz for a sun-like star ($T_{\rm eff} = 5777$ K) and of 4 μHz for hotter stars ($T_{\rm eff} = 6500$ K). It was not completely fair to use posterior knowledge for that review but the following figure provides a more accurate current picture of the required background noise. Figure 1.2 shows the value of the background noise (goal) for stellar radial velocity and intensity, assuming that $A_\odot$ is about 3 ppm and 10 cm s^{-1} for the Sun. In Figure 1.2, the highest frequency corresponds to a 0.8 $M_\odot$ star, while the lowest frequency corresponds to a 2 $M_\odot$ star; both stars being on the main sequence.

For the 0.8 $M_\odot$ star, the lowest values are 1.5 cm^2 s^{-2} μHz^{-1} and 0.08 ppm^{2}μHz^{-1}. The conservative approach for noise design is to assume white noise. Therefore, assuming that the noise power spectrum is flat from frequency 0 to the Nyquist frequency, we can deduce that the maximum rms noise is 1.5 m s^{-1} and 34 ppm in radial velocity and intensity, respectively. For the 2 $M_\odot$ star, the corresponding levels are 17.3 m s^{-1} and 360 ppm. If one assumes a more realistic, but less conservative, $1/\nu$ noise, then these design levels can be somewhat increased.

Observation duration

Finally, the duration of the observation will be driven by the mode linewidth or lifetime. The mode lifetime is related to the mode linewidth by the following relation:

$$\tau = \frac{1}{\pi \Gamma}. \tag{1.6}$$

In order to resolve the mode linewidth, the observation time should be such as to have a resolution three times better than this linewidth. This implies, using the previous equation, that the observation time is given by

$$T = 3\pi\tau. \tag{1.7}$$

Hence we need to observe at least 10 lifetimes of the mode to resolve the linewidth. For a lifetime of 3.7 days (= 1 μHz), the observation duration should be at least 37 days. The modes at the low-frequency end have long lifetimes (Houdek *et al.*, 1999), such that observations typically 1 year long would give access to a linewidth of 0.1 μHz.

1.2.2 How can we measure the oscillations?

The angular diameter of a star is so small that it is much easier to measure the stellar radial velocity or intensity without any resolution, i.e., integrated over the stellar disk. The effect of the integration over the stellar disk is known to filter out all modes above $l = 4$ in velocity (Dziembowski, 1977; Christensen-Dalsgaard and Gough, 1982), and above $l = 3$ in intensity (Dziembowski, 1977; Toutain and Gouttebroze, 1993), where l is the degree of the mode.

The measurement of radial velocity for detecting p modes was performed on the Sun using a resonance cell (Claverie *et al.*, 1979; Grec *et al.*, 1983). For intensity measurement, irradiance measurements led to the detection of p modes using the ACRIM radiometer (Woodard and Hudson, 1983). For stars, these two techniques needed to be adapted due to the lack of photons.

Stellar radial velocity

The measurement of stellar radial velocity is performed using many absorption lines. The difficulty in the design of such spectrophotometers is to make an image of the stellar spectrum over a wide wavelength band without losing photons and at high spectral resolution. The seminal work that led to the development of these spectrometers is due to Connes (1985), which predicted the limit provided by a 1 m diameter telescope on a 5th-magnitude star similar to the Sun; the limit was 0.36 $\text{cm}^2\ \text{s}^2\ \mu\text{Hz}^{-1}$, which roughly corresponds to using 3000 absorption lines instead of one line (Harvey, 1988). The working principle of the spectrometer of Connes (1985) was to use an echelle spectrograph for measuring the displacement of many absorption lines, coupled with a low finesse Fabry–Perot and stabilized laser. The measurement of the radial velocity was not done using correlation but using an optimal weighting scheme. Given the sheer complexity of these spectrophotometers, only one star can be studied at any one time, making the measurement a mono-object technique.

The precision reached by such a technique depends primarily on the number of photons gathered (telescope diameter, optic efficiency), the spectral coverage, the resolving power of the spectrometer and the star rotational velocity. Bouchy *et al.* (2001) studied the impact of all these parameters on the velocity noise induced by photon noise. They concluded that a spectrometer with a spectral resolution better than 5×10^4, with a spectral coverage of 300 nm, and a telescope diameter of one meter, would give a velocity spectral density induced by photon noise of typically 0.9 $\text{cm}^2\ \text{s}^2\ \mu\text{Hz}^{-1}$ for a slowly rotating star like the Sun, sufficient for detecting stellar p modes according to Figure 1.2.

The concept described by Connes (1985) was implemented in the EMILIE spectrograph (Schmitt *et al.*, 2000) and then used in a simplified version by his former student for detecting p modes on α Cen (Bouchy and Carrier, 2001) with the CORALIE instrument (Queloz *et al.*, 2000). In comparison, the HARPS (High Accuracy Radial velocity Planetary Search) instrument is designed for high spectral stability using a spectral calibration lamp for wavelength reference, high spectral resolution ($R = 120\,000$), high efficiency and vacuum operation (Pepe *et al.*, 2000).

Stellar intensity

In contrast to stellar radial velocity, the main advantage of doing photometry with charge-coupled device (CCD) arrays is the possibility to simultaneously observe many objects. Several attempts were made by Gilliland and Brown (1988, 1992) and Gilliland *et al.* (1991, 1993) at asteroseismology from the ground using such arrays. These trials were not successful because the measurement of stellar intensity from the ground for bright stars is limited mainly by the effect of scintillation, not by photon noise. The angular diameter of stars being typically of the order of a few milliarcsec, combined with a seeing of 0.4 to 1.5 arcsec, produces large intensity fluctuations, as shown by Young (1967), known as scintillation noise. We must point out that no amount of telescope defocusing will reduce the scintillation noise.

The original result of Young (1967) was calibrated on several observatory sites, providing scintillation noise higher by a factor 1.4 for telescope diameters less than 10 cm (Kornilov *et al.*, 2012). For apertures larger than 8 m, it was theoretically shown by Kornilov (2012) that the scintillation noise would drop faster than that re-calibrated figure. The reduction is due to the role of the outer scale of optical turbulence, which was neglected by Young (1967). Figure 1.3 shows the scintillation noise due to the calibrated result of Kornilov *et al.* (2012), together with the more recent theoretical values of Kornilov (2012). It shows that there is a potential reduction in telescope diameter by a factor of 3. A much lower scintillation noise has been extrapolated by Kenyon *et al.* (2006) for a 4 m diameter telescope using data from Dome C (0.17 $\text{ppm}^2/\mu\text{Hz}$).

It was thought to be clear back in the 1990s, during the preparation of an ESA (European Space Agency) asteroseismic space project called PRISMA (Probing Rotation and Interiors of Stars: Microvariability and Activity), that a telescope diameter as large as 50 m was not a feasible option for reducing the scintillation noise (Appourchaux *et al.*, 1991, 1993). The alternative developed in those days was to make space-based observations instead of ground-based observations. The results of Kenyon *et al.* (2006) and of Kornilov (2012)

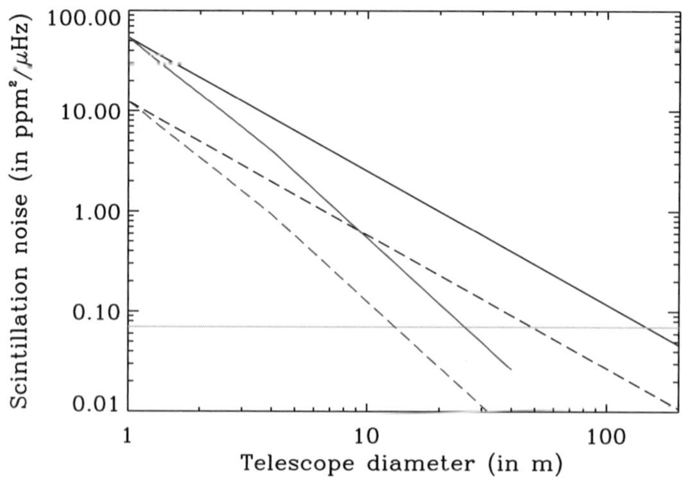

Figure 1.3 Scintillation noise as a function of telescope diameter calibrated by Kornilov *et al.* (2012) for an airmass of $X = 1.5$ (solid line) and of $X = 1.0$ (dotted line), for an observatory at an altitude of 4000 m. The gray line is the goal set as 0.07 ppm^{2}μHz^{-1} for a 0.8 $M_\odot$ star. The corresponding gray lines are the theoretical corrections due to Kornilov (2012) for the respective scintillation noise.

may well resuscitate such a debate. For the time being, we will assume that the result of the debate would still be in favor of going to space.

For space-based observations, the main source of noise besides photon noise is the combined effect of pointing jitter and CCD pixel-to-pixel response non-uniformity (PRNU). The fact that the stars can be considered as point sources serves to reduce the effect of pointing jitter by defocusing the image of the stars on the detector. Gruneisen (1998) provided the photometric noise induced by pointing jitter and by the PRNU as

$$\sigma_{\text{jitter}} = \sqrt{2}\frac{\text{PRNU}\delta x}{N^{3/4}}, \tag{1.8}$$

where σ_{jitter} is the relative photometric variation, δx is the jitter in units of CCD pixels, PRNU is in relative units and N is the equivalent number of pixels making the defocused image of the star. For a typical PRNU of 1%, a pointing jitter of 0.1 pixel rms and a typical number of pixels of 300, the pointing jitter is about 20 ppm, which imposes typical pointing stability better than 1 arcsec peak-to-peak. Such levels have been obtained with the CoRoT (Convection and Rotation of Stars) mission using the science detector as a stellar attitude sensor, and pointing jitter corrections have been devised for compensating larger pointing effects (Drummond *et al.*, 2006; De Oliveira Fialho, 2007). As we will see in Section 1.4, the major source of noise for intensity measurements when the jitter noise is under control is solely the photon noise.

Ground vs. space

Most astronomical observations performed in the visible can be done from the ground. The justification for going to space is usually based upon wavelengths being blocked by the atmosphere (UV, EUV, X-rays), the fullness of the sky coverage, the absence of day/night interruptions, and thermal requirements (operations close to 0 K). As for stellar radial velocity and intensity measurement, the sky coverage and the absence of interruptions would be enough to justify having a spectrophotometer put in space. In addition, we noted above that the absence of scintillation noise is an added benefit, even though large telescopes might be able to alleviate this source of noise.

What are the real advantages/disadvantages of space observation vs. ground-based observations? The argument that is often used for preferring ground to space is that it is less expensive. Ground-based observations are less immune to a single-point failure such as a failed launch! There is also the benefit of being flexible in allowing the instrumentation to be improved if needed. Another benefit is the possibility of having very long observations (decades) usually not possible for space missions due to the cost penalty of equipment lifetime. The benefit of space missions really lies in the scientific needs and the so-called "need to go to space." Space missions can only be justified on the real gain compared to what cannot be done from the ground. Such a need to go to space required hard work at the time of the design of the PRISMA mission study (Appourchaux *et al.*, 1991, 1993). In that case, scintillation noise was the key driver at the time for going to space, combined with the absence of interruption (as a collateral benefit).

1.3 Historical perspective of detection

1.3.1 Ground-based efforts

The search for solar-like oscillations of bright stars started in the mid 1980s using stellar radial velocity measurements (Gelly *et al.*, 1986; Frandsen, 1987; Ando *et al.*, 1988).

Several attempts were made at detecting the modes on Procyon using spectrophotometers: the first credible detection of the mode envelope by Brown *et al.* (1991) and the first measurement of the large separation (Mosser *et al.*, 1998) and of individual modes by Martić *et al.* (2004). Finally, a coordinated campaign, including eight sites distributed around the globe and 11 different instruments, provided the definite answer for the p modes in Procyon (Arentoft *et al.*, 2008). The campaign provided individual mode frequencies, mode lifetime, and a clear échelle diagram (Bedding *et al.*, 2010a). The noise achieved by that campaign could be compared with the requirement we set above for the noise level in stellar radial velocity. Arentoft *et al.* (2008) provided values of $20\,\mathrm{cm}^2\ \mathrm{s}^{-2}\mu\mathrm{Hz}^{-1}$ at $1000\,\mu\mathrm{Hz}$, and of $4\ \mathrm{cm}^2\ \mathrm{s}^{-2}\mu\mathrm{Hz}^{-1}$ above $2000\,\mu\mathrm{Hz}$ (flat), which are about two to three times lower than the required noise level of Figure 1.2 for a 6500 K star.

The detection of modes on η Boo was also claimed by Kjeldsen *et al.* (1995) by measuring the area of an absorption line (or equivalent linewidth), a measurement insensitive to scintillation noise. This claim could not be confirmed in stellar radial velocity (Brown *et al.*, 1997). Nevertheless, Kjeldsen *et al.* (2003) confirmed their earlier finding using an equivalent linewidth in combination with stellar radial velocity. The levels quoted in Kjeldsen *et al.* (2003) are consistent with a signal-to-noise ratio of 1 when compared with Figure 1.2. The undisputed detection of modes in η Boo came from Carrier *et al.* (2005), who reported several modes using a two-station network at a signal-to-noise level higher than 3; the background noise was about 270 cm^2 s^{-2}μHz^{-1} at 750 μHz, about 2.7 higher than the required noise level of Figure 1.2.

α Cen A was also extensively studied by Brown and Christensen-Dalsgaard (1990), Pottasch *et al.* (1992) and Edmonds and Cram (1995) using stellar radial velocity measurements. They all reported noise level above 160 cm^2 s^{-2}μHz^{-1}, which is 16 times above the required noise level of Figure 1.2 at 2400 μHz for the Sun, i.e., with an effective signal-to-noise ratio of 0.6. Finally, the detection of p modes in α Cen A was confirmed by Bouchy and Carrier (2001) with a noise level of 8 cm^2 s^{-2}μHz^{-1} at 2400 μHz commensurate with the required noise level of Figure 1.2. This work was the real start of asteroseismology on solar-like stars since it provided a clear and indisputable detection of p modes in a star other than the Sun.

Frandsen *et al.* (2002) reported the detection of solar-like oscillation in a red giant. This was also the start of a renewed interest in red giants. Detection of solar-like modes was also reported in a few other stars using stellar radial velocities (see Bedding, 2014, for details).

In comparison, the detection of p modes in intensity was not so successful. There were several attempts by Belmonte *et al.* (1990a,b,c), none of which could be confirmed since the amount of noise was above 1700 ppm^{2}μHz^{-1}, a level at least 100 times larger than the highest values of Figure 1.2. The use of CCDs was introduced by Gilliland and Brown (1988). They provided a noise level of about 87 ppm^{2}μHz^{-1}, lower than the previous figure but still too high for a credible mode detection. They improved the technique by using a larger telescope area to reach a value of 25 ppm^{2}μHz^{-1} at 1000 μHz (Gilliland *et al.*, 1993), resulting in a signal-to-noise ratio of about 1 to 3 according to the levels of Figure 1.2. Nevertheless, although the detection of modes in solar-like stars in intensity is difficult from the ground, the technique can be easily applied to red giants having a frequency of maximum power below 200 μHz (Stello *et al.*, 2007), implying that the mode amplitudes are also much larger. Obviously the detection of p modes in intensity would be far easier from space due to the absence of scintillation.

1.3.2 Space-based efforts

As we outlined before, the combination of multi-object observations in intensity combined with space observations was an attractive perspective for doing asteroseismology on thousands of stars. The design of such a space mission started in the 1980s with the submission of the EVRIS (Etude de la Variabilité, de la Rotation et des Intérieurs Stellaires) mission to the French space agency CNES (Centre National d'Etudes Spatiales) (Baglin *et al.*, 1993). It eventually led to a series of studies performed at ESA for various missions such as PRISMA, STARS, and Eddington (Appourchaux *et al.*, 1993; Fridlund *et al.*, 1995; Favata, 2004). A full account of the history of space missions can be found in Roxburgh (2002) and references therein. Table 1.1 provides a comparison of the various asteroseismic missions planned, launched, or to be launched.

Table 1.1 *Main characteristics of past, present and future space missions enabling asteroseismology. D is the effective diameter of the telescope, FOV is the field of view, m_V is the limiting magnitude. The sixth column provides the number of stars brighter than the limiting magnitude. The seventh column gives the photometric performance at the limiting magnitude.*

Project	Launch	D (cm)	FOV (deg × deg)	m_V	Number of stars	Noise ($\mathrm{ppm^2\,\mu Hz^{-1}}$)
MOST	2003	15	0.4 × 0.4	<6	<6	5.7
CoRoT	2006	25	1 × 1	<7	10	1.7
Kepler	2009	95	10.5 × 10.5	<12	1300	8.8
Brite	2013	3	24 × 24	<4	500	60.0
TESS	2017	10	23 × 90	<12	5×10^5	7.6
PLATO	2024	67	47 × 47	<13	10^6	18.1

None of the series of ESA studies became missions, though Eddington came very close to such a status. The precursor to all asteroseismic missions, EVRIS, was launched in November 1996 aboard the MARS96 mission, which failed to reach orbit. This event put forward on the scene the CoRoT mission, which had been submitted to CNES back in 1993 (Catala *et al.*, 1995; Roxburgh, 2002), and was launched in December 2006.

The CoRoT mission was preceded by a simple mission akin to the EVRIS mission: the MOST (Microvariability and Oscillations of Stars) mission of the Canadian Space Agency, launched in June 2003 (Matthews *et al.*, 2000). The report by Matthews *et al.* (2004) of the non-detection of p modes in Procyon by MOST started an interesting controversy. The photometric level observed at 1000 μHz was about 30 $\mathrm{ppm^2\,\mu Hz^{-1}}$, which is a factor of 10 higher than the required noise level of Figure 1.2, i.e., a potential signal-to-noise ratio of 1. This is the high level of noise (due to Earth scattered light) that was responsible for the lack of p-mode detection in Procyon as shown by Bedding *et al.* (2005).

A special mention goes to the WIRE (Wide Field Infrared Explorer) mission, whose failure was the *real* unplanned start of asteroseismology from space. WIRE was launched in March 1999 and lost all of its cryogenic capabilities shortly after launch. Buzasi (2000) then saw the opportunity of doing asteroseismology using the CCD of the 5 cm diameter star tracker of the spacecraft; tens of stars were measured by the star tracker (Buzasi *et al.*, 2004). Using WIRE data, Schou and Buzasi (2001) claimed the detection of modes in α Cen A without providing a properly normalized power spectrum, hence making difficult any comparison with theoretical expectations. Bruntt *et al.* (2005) provided a value of 8 $\mathrm{ppm^2\,\mu Hz^{-1}}$ at 1000 μHz for Procyon, making their detection of the hump of mode power credible; the required noise level of Figure 1.2 is about three times lower.

Asteroseismic space missions finally came to fruition with the launch of CoRoT in December 2006 and of *Kepler* in March 2009. The next section is devoted to their contributions to this two-stage revolution.

1.4 The revolution

1.4.1 The CoRoT mission

The first revolution was French, with the launch of the CoRoT mission in 2006. To date there have been 666 refereed papers published since January 2007 bearing CoRoT in their title. The highlights of the main asteroseismic discoveries of CoRoT are available in Baglin *et al.* (2012), such as the so-called F-star problem (Appourchaux *et al.*, 2008), the detection of mixed modes (Deheuvels *et al.*, 2010) and the study of red giants (De Ridder *et al.*, 2009).

The first clear detection of p modes by CoRoT was reported by Appourchaux *et al.* (2008), who provided a noise level of 0.3 $\text{ppm}^2\mu\text{Hz}^{-1}$ at 1800 μHz and a typical signal-to-noise ratio of 10 for the highest amplitude. This measured noise level is a factor of 10 lower than the highest required noise level of Figure 1.2, therefore with such a requirement the modes should have had a signal-to-noise ratio of 1. The reason for the discrepancy is that the assumed linewidth of 1 μHz in panel (a) of Figure 1.2 was eight times lower than measured by Appourchaux *et al.* (2008), hence with such a measured linewidth the *required noise level* should have been more than eight times lower, thereby recovering a signal-to-noise ratio close to 10. As mentioned previously, larger mode linewidths were observed in several stars hotter than the Sun (Baudin *et al.*, 2011); a dependence that was unexpected at the time of the launch of CoRoT. The noise level encountered in other solar-like stars ranged from 0.3 to 1 $\text{ppm}^2\mu\text{Hz}^{-1}$ at 1800 μHz, depending on the magnitude of the star (Michel *et al.*, 2008); these levels are all lower than the specification of panel (a) of Figure 1.2.

The detection of solar-like oscillations in hundreds of red giants is also a landmark of the CoRoT legacy (De Ridder *et al.*, 2009). The maximum mode heights as observed by Hekker *et al.* (2010) for several red giants at around 25 μHz range from 2×10^4 to 14×10^4 $\text{ppm}^2\mu\text{Hz}^{-1}$. This is consistent within a factor of 2 with Equation (1.5) using a linewidth of 0.2 μHz and an effective temperature of 4000 K. Therefore, for red giants, the required background noise level also scales with ν_{max}^{-2}, as given for solar-like stars by Figure 1.2.

1.4.2 The Kepler mission

The second revolution was American, with the launch of the *Kepler* mission in 2009. To date there have been more than 1200 refereed papers published since March 2009 bearing *Kepler* in their title. There have been ample papers describing the detection of solar-like p modes across the Hertzsprung–Russell diagram with *Kepler*, but the reader can refer to Chaplin and Miglio (2013) for a review of the main findings. The noise as observed by *Kepler* is limited by photon noise as shown by Chaplin *et al.* (2011b). In comparison, CoRoT has a higher instrumental noise mainly due to its effective telescope diameter being smaller by a factor of 3.8. The noise level of 0.15 $\text{ppm}^2\mu\text{Hz}^{-1}$ obtained for a 5.77 magnitude star (Appourchaux *et al.*, 2008) would have been obtained for a magnitude of 8.7 with the theoretical *Kepler* telescope diameter. For the true *Kepler* telescope, this previous noise

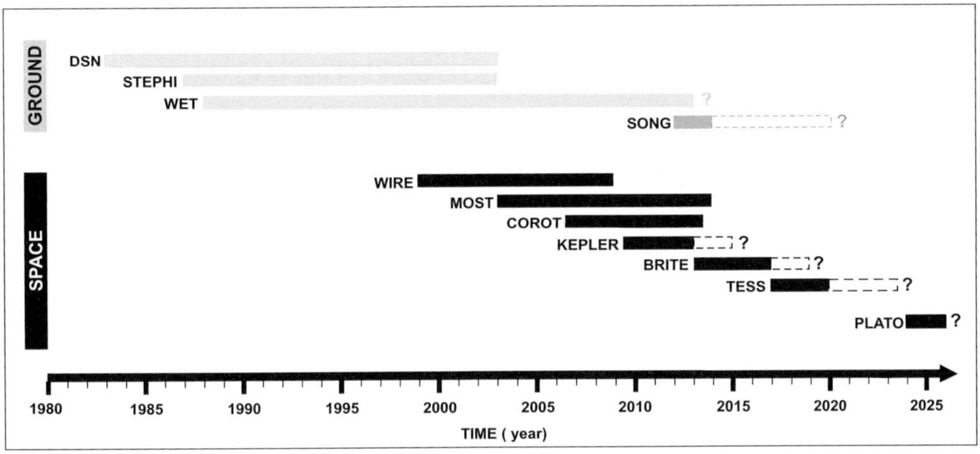

Figure 1.4 Timeline of the space- and ground-based facilities for asteroseismology. The ground-based ones are in gray (light gray for intensity measurements and dark gray for Doppler velocity measurements). Space telescopes are in black and they all correspond to intensity measurements. (Credit: Ramon Castro, SMM/IAC.)

figure is only reached for stars brighter than 7.5 magnitude. Nevertheless, it is now clear that for any star brighter than 11th magnitude, the *Kepler* performances are better than the specifications of Figure 1.2.

1.5 How to push the envelope

With about 30 years of development, *modern* asteroseismology has come a long way from the age of non-detection to the age of revolution summarized in Figure 1.4. The question the reader may ask is the following: what is left to do that would provide viable science returns?

History has taught us that many unexpected results were obtained, such as the detection of solar-like oscillation modes in red giants. From this point of view, there is a renewed interest in red giants and subgiants because these stars exhibit mixed modes with high amplitudes and long lifetimes sampling their very cores. The sounding of these stellar cores has already provided very useful information about their internal rotation (Deheuvels *et al.*, 2012; Mosser *et al.*, 2012b).

Beyond the use of field stars that are useful for doing ensemble asteroseismology (Chaplin *et al.*, 2011a), binary and cluster stars are the next targets because these stars have a common age, chemical composition, and distance. Several tens of red-giant binaries have been detected by Gaulme *et al.* (2013). Only a few main-sequence binary systems have been studied, such as α Cen A and B (Bouchy and Carrier, 2002; Carrier and Bourban, 2003) and Cyg A and B (Metcalfe *et al.*, 2012). According to Miglio *et al.* (2014), we can expect about 1% of binaries to have both stars showing detectable seismic signals.

The PLATO (PLAnetary Transits and Oscillations of stars) Mission selected in February 2014 as the next M3 mission of ESA will be able to detect solar-like oscillations in solar-like stars in a handful of open clusters (Catala, 2009). Stellar distances from the GAIA (Global Astrometric Interferometer for Astrophysics) mission and PLATO asteroseismic measurements will improve our knowledge of stellar physics.

With the end of the CoRoT mission and the recent news that the *Kepler* mission is now using a special pointing mode to compensate for the loss of one of its reaction wheels, the prospects of having high-quality photometric data for asteroseismology might be somewhat reduced (Chaplin *et al.*, 2013b). With the launch of PLATO in 2024, there is now a mere 10 years to wait for fresh data. The coming decade could be a good opportunity to re-measure the classical stellar parameters that are needed for stellar modeling using asteroseismic data (Mészáros *et al.*, 2013).

From these prospects, a project such as SONG (song.au.dk) is worth pursuing. SONG is a network of eight spectrophotometers located at existing observatories. There will be four of these in each hemisphere equipped with identical telescopes and instrumentation. SONG will fill a gap in the quality of the detected modes compared to what is currently being achieved in intensity. Furthermore, by directing attention to the nearest and brightest stars, it will be possible to obtain highly precise angular diameters, distances, and temperatures from astrometry (Hipparcos, GAIA) and interferometry.

Finally, let us dream about the instrument of the future: a hyper-telescope able to image the stars for higher degree modes than currently possible (say $l > 50$). With that capability, we will be able to study and understand the internal rotation of stars as a function of mass and age. This will be a great asset for improving our understanding of stellar physics. A hyper-telescope, likely to be a space mission, would image the star using a set of small telescopes distributed over an equivalent diameter of the order of 1 km. The focal plane instrument could either measure stellar radial velocity using a Mach–Zehnder type of instrument, such as the Echoes instrument developed for doing Jovian seismology aboard ESA's JUICE mission (Soulat *et al.*, 2012), or measure intensity using a CCD array. Such a mission, called the Stellar Imager, was proposed for the US decadal survey in 2010 by Carpenter *et al.* (2010). This kind of mission requires that the small telescopes, a few kilometers distant from each other, interferometrically combine their optical beams at the focal plane. This challenging mission has been studied by Labeyrie *et al.* (2009) and a road map has even been laid out by Labeyrie *et al.* (2008). Time will tell whether this mission is the ultimate asteroseismic mission.

2

Helioseismology: Observations and space missions

PERE L. PALLÉ, THIERRY APPOURCHAUX, JØRGEN CHRISTENSEN-DALSGAARD, AND RAFAEL A. GARCÍA

2.1 Introduction

The great success of helioseismology resides in the remarkable progress achieved in the understanding of the structure and dynamics of the solar interior. This success mainly relies on the ability to conceive, implement, and operate specific instrumentation with enough sensitivity to detect and measure small fluctuations (in velocity and/or intensity) on the solar surface that are well below one meter per second or a few parts per million. Furthermore, the limitation of ground observations imposed by the day–night cycle (thus a periodic discontinuity in the observations) was overcome with the deployment of ground-based networks – properly placed at different longitudes all over the Earth – allowing longer and continuous observations of the Sun and consequently increasing their duty cycle.

In this chapter, we start with a short historical overview of helioseismology. Then we describe the different techniques used for helioseismic analyses along with a description of the main instrumental concepts. We particularly focus on the instruments that have been operating long enough to study the solar magnetic activity. Finally, we highlight the main results obtained with such high-duty cycle observations (>80%) over the past few decades.

2.2 A historical overview

The detection of solar oscillations goes back to more than 50 years ago when Leighton *et al.* (1962) discovered the five-minute oscillations in the solar photosphere. However, it was only with the observations of Deubner (1975) that their identity as standing acoustic waves, i.e., normal modes of the Sun of high spherical-harmonic degree, was established, confirming previous theoretical inferences by Ulrich (1970) and Leibacher and Stein (1971), thus setting the scene for the development of helioseismology. Coincidentally, a strong inspiration was the announcement in 1975 by the SCLERA (Santa Catalina Laboratory for Experimental Relativity by Astrometry) group of evidence for oscillations in the solar

Extraterrestrial Seismology, eds. Vincent C. H. Tong and Rafael A. García. Published by Cambridge University Press.

diameter (see Hill *et al.*, 1976), of apparently truly global nature and hence containing information about the entire Sun. Although these detections were later found not to be of solar origin, they contributed substantially to the early investigations of the diagnostic potential of helioseismology (e.g., Christensen-Dalsgaard and Gough, 1976). Low-degree solar five-minute oscillations were announced by Claverie *et al.* (1979), quickly followed by the spectacular observations by Grec *et al.* (1980) from the geographic South Pole. With the development of observations of intermediate-degree modes (Duvall and Harvey, 1983; Brown, 1985; Libbrecht and Kaufman, 1988) the full spectrum of solar oscillations could be identified, and early helioseismic inverse analyses yielded the first inferences of solar internal rotation (Duvall *et al.*, 1984) and structure (Christensen-Dalsgaard *et al.*, 1985).

The early determinations of solar internal rotation gave no evidence for a rapidly rotating core with a significant contribution to the oblateness of the solar gravitational field, thus contributing to the confirmation through observations of Mercury's orbit of Einstein's theory of general relativity. Also, observations by Libbrecht (1989) of rotational splittings allowed inversion for the solar internal rotation as a function of latitude and distance to the centre (Christensen-Dalsgaard and Schou, 1988; Dziembowski *et al.*, 1989). The results showed that the surface differential rotation with latitude persists through the convection zone, while the radiative interior appears to rotate essentially as a solid body. This gives rise to a region of rotational shear at the base of the convection zone, the so-called *tachocline*.

Careful analyses of low-degree oscillations (e.g., Elsworth *et al.*, 1990), which are sensitive to the solar core, effectively eliminated modifications to solar models proposed to account for the low observed solar neutrino flux, thus strengthening the case for neutrino oscillations. This was confirmed by later investigations based on full helioseismic inversions (Bahcall *et al.*, 1997).

These early results both demonstrated the potential of helioseismology and indicated the importance of securing continuous observations over extended periods of time. Radial-velocity observations in integrated light were achieved from the ground by the International Research on the Interior of the Sun (IRIS, Fossat, 1991) and the Birmingham Solar Oscillation Network (BiSON, Chaplin *et al.*, 1997a, still in operation). Continuous long-term spatially resolved observations were made possible by the establishment of the Global Oscillation Network Group (GONG, Harvey *et al.*, 1996). Moreover, the Solar and Heliospheric Observatory (SOHO) satellite (Domingo *et al.*, 1995) carries three instruments for helioseismology: the Global Oscillations at Low Frequencies (GOLF) instrument for integrated-light radial-velocity observations (Gabriel *et al.*, 1995, 1997), the Variability of Irradiance and Gravity Oscillations (VIRGO) instrument for irradiance and photometric observations, also mostly in integrated light (Fröhlich *et al.*, 1997), and the Solar Oscillation Investigation/Michelson Doppler Imager (SOI/MDI) instrument for spatially resolved velocity observations (Scherrer *et al.*, 1995). Of these, GOLF and VIRGO are still in operation, whereas MDI was switched off after the start of operations with the Helioseismic and Magnetic Imager (HMI) instrument (Scherrer *et al.*, 2012) on the Solar Dynamics Observatory (SDO) mission (Pesnell *et al.*, 2012).

Although early stability calculations by Ando and Osaki (1975) indicated that solar oscillations were unstable in the observed frequency range, later calculations taking perturbations due to convection into account convincingly demonstrated their stability (Balmforth, 1992b), implying that the modes are instead excited by the acoustic noise generated by the near-sonic speed of near-surface convection. Early estimates of such stochastic excitation were made by Goldreich and Keeley (1977); they concluded that this mechanism failed to predict the oscillations claimed by Hill *et al.* (1976) by a substantial margin. However, later analyses (e.g., Balmforth, 1992a) have shown that stochastic excitation can account for the observed properties of the oscillations in the five-minute region, while statistical analysis of the observed low-degree modes was consistent with the stochastic nature of the mode amplitudes (Chaplin *et al.*, 1997a; Foglizzo *et al.*, 1998).

From this evolution has emerged a new and in a true sense more profound branch of solar physics. The understanding of the origin of the solar oscillations has in return inspired the application of similar studies to other stars, with *asteroseismology*.

2.3 Techniques to measure solar oscillations

In the Sun, acoustic waves are responsible for the compression and expansion of the gas in the solar atmosphere. An intuitive and straightforward way to detect them is to measure the shifts in the absorption lines – a velocity signal – in the solar spectrum. In addition to this effect – because of the radiative damping that takes place as a consequence of the compression – the brightness increases while it decreases during the expansion. Therefore the solar irradiance is modulated by the effect of the oscillations. For individual acoustic modes, the magnitude of such variations is really small: below 15 cm/s for velocity shifts and 3×10^{-6} for brightness variation (10^{-3} K in temperature). Given the very small amplitudes and the required high signal-to-noise ratio, observation of solar oscillations has required the development of sophisticated technologies, although it is aided by the high solar brightness. Depending on the scientific objectives behind the desired observations, different strategies and instrumentation are possible, so the best-suited ones could be adopted for each particular case.

The main differences between solar radial velocity and intensity measurements are summarized in Figure 2.1. This figure shows the comparison between power spectra obtained from Sun-as-a-star velocity observations with the GOLF instrument (García *et al.*, 2005) and power spectra computed with the observations of the Luminosity Oscillations Imager (LOI, Appourchaux *et al.*, 1997) of the VIRGO instrument. We clearly see that the main difference lies in the convective background, i.e., the continuous signal on top of which the discrete spectrum of the acoustic oscillations lies. Indeed, the granulation is much smaller in solar radial velocity compared to intensity. As a result, the maximum signal-to-noise ratio of the modes is around 30 in intensity and as high as 300 for solar radial velocity. By doing radial velocity measurements we are also able to detect more modes at low frequency (typically twice as many) and modes with longer lifetimes up to a year compared to intensity (Toutain *et al.*, 1997; Salabert *et al.*, 2009a).

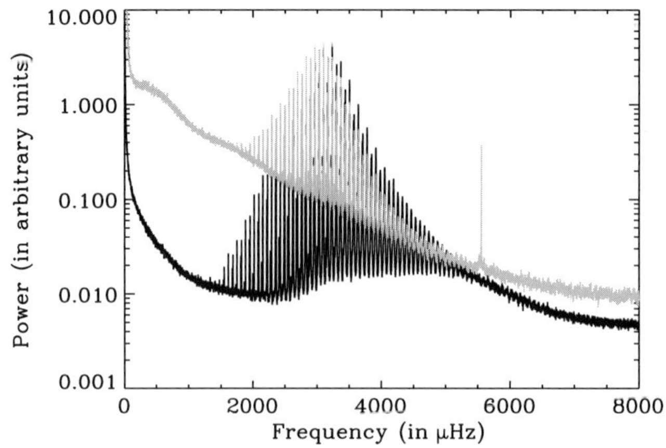

Figure 2.1 Power spectra as a function of frequency for the Sun smoothed over 1 μHz: solar radial velocity measured by GOLF (black line), and intensity measured by the LOI (gray line). The GOLF time series is 12.75 years long while the LOI is 16.9 years long. Power units are arbitrary. The large spike at 5555 μHz is an artefact of the basic cadence of the data acquisition system of VIRGO, which is 3 minutes.

2.3.1 Ground- vs. space-based observations

The day/night systematic interruptions generate daily aliases located at harmonics of 11.57 μHz (24 hours), making the mode identification difficult when the mode separation is close to that value. Such interruptions can be overcome by having a network of instruments located all around the Earth, as demonstrated in the mid 1980s by Hill and Newkirk (1985). The typical reduction in the amplitude of the first alias is about $(1 - D)^2$, where D is the data fill, i.e., for a 90% data fill, the alias amplitude is 1% of the main peak. The need to minimize interruptions gave rise in the 1990s to two types of approaches and thus the subsequent types of projects in helioseismology: ground-based networks (GONG, BiSON, IRIS, TON, etc.) and the space-borne mission SOHO with three helioseismic instruments aboard (SOI/MDI, GOLF, and VIRGO). The filling factor D achieved by such projects has been, in some cases, remarkably high (>85%) and maintained for years and even decades. In addition, avoiding the regularity in the distribution of the gaps clearly reduces the spurious effect of the aliases as proved, first using simulated data (Hill, 1984) and later with real space observations (Domingo *et al.*, 1995).

2.3.2 Velocity measurements

By measuring this physical parameter, a very high signal-to-noise ratio can be achieved because of the low level of the solar background noise. However, we should take into account the non-uniform velocity sensitivity across the solar disk arising from the projection on the line of sight, as well as possible variations in sensitivity induced by the effects of solar

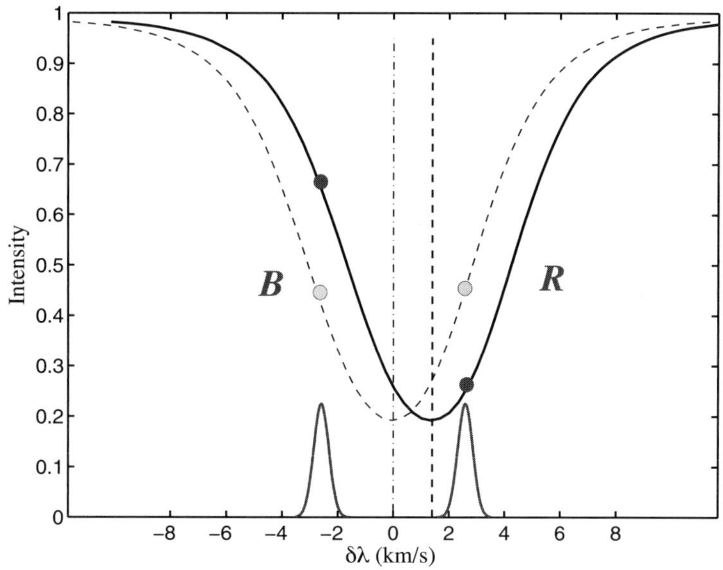

Figure 2.2 Principle of operation of the Doppler shift measured by using narrow-band filters. The relative displacement of the solar line (solid) with respect to the laboratory one (dashed) can be calculated by measuring at both wings (B and R) and making an appropriate ratio between them.

rotation, corresponding to a velocity gradient of roughly $2\,\mathrm{m\,s^{-1}\,arcsec^{-1}}$ (Christensen-Dalsgaard, 1989).

The Doppler velocity shifts are obtained by using narrow-band filters acting on the red (R) and/or blue (B) wings of a selected solar absorption line. As shown in Figure 2.2, any global shift would increase the transmitted light in one of the wing filters and decrease it in the other wing due to the solar line shape. At first order, the difference between the transmitted light in both wings ($B - R$) is a measure of the displacement of the solar line relative to its nominal (rest) position.

This approach has several constraints and disadvantages: some kind of normalization, i.e. ($R + B$), is necessary to reduce transmission changes due to the atmosphere and/or the instrument; variations in the slope of the line profile at the working points of the filter would change the sensitivity; the lack of stability of the bandpasses will produce spurious (fictitious) Doppler velocity signals, etc.

The most representative techniques (and associated instruments) of this group are the following.

Atomic resonance cell

This technique has been proven to solve (to a large extent) the issue of stability and to maximize the sensitivity of the velocity measurements, by exploiting the natural stability of resonating atoms. Briefly, sunlight around a given solar absorption line (mainly corresponding to alkaline metals such as potassium or sodium) is properly filtered and circularly polarized. A stable vapor cell of the same chemical element (K or Na), placed in a longitudinal magnetic field, causes the two Zeeman components to sample the line profile. By alternately measuring the intensities on either side of the line (B and R) the relative position of the solar and laboratory lines may be found and the Doppler velocity shift determined. The first solar spectrophotometers based on this principle were successfully developed and operated in the mid 1970s by British and French teams (Brookes *et al.*, 1978; Grec *et al.*, 1976). They discovered almost simultaneously the five-minute solar oscillations and concluded that they were the normal modes of oscillation of the Sun (Claverie *et al.*, 1979; Grec *et al.*, 1980). The high sensitivity and robustness of this kind of spectrophotometer resulted in the deployment and operation of two observational ground-based networks: IRIS (Fossat, 1991) and BiSON (Chaplin *et al.*, 1996), the latter being operational and providing high-quality data since 1976 (more than three solar activity cycles) with an outstanding long-term stability: $\sim$1 part in 10^{-3} over the last 40 years (Roca Cortés and Pallé, 2014). In the top panel of Figure 2.3, a schematic of one of the BiSON nodes instruments, Mark-I, is shown.

Thanks to the success of this technique and in order to exploit the potential of helioseismic tools, the GOLF instrument (Gabriel *et al.*, 1995) was chosen as a payload of the ESA/NASA cornerstone mission SOHO. Surrounded by 11 other instruments, GOLF is still providing continuous and useful data since the beginning of operations back in 1996.

The high stability of the Sun-as-a-star observations performed with these instruments allowed us to make a precise determination of the characteristics of low-degree (high spatial scale) modes, in particular those probing the deep layers of the Sun and thus providing structural and dynamical information of these regions.

Magneto-optical filter (MOF)

This technique is also based on the atomic resonance concept with an imaging capability. By an adequate combination of a double-cell magnet system and polarizing elements (see the bottom panel of Figure 2.3) global blue and red intensity images of the solar surface are obtained. From these observations, a solar surface velocity map is derived (see for example Cacciani and Fofi, 1978; Tomczyk *et al.*, 1995).

This type of instrument provides the so-called medium-degree modes that cannot be observed by Sun-as-a-star observations while being very stable at low frequencies.

The data and the derived results were really impressive but never crystalized on a robust instrument network capable of achieving high-duty cycles for long-term observations (years). Efforts to deploy networks of MOF instruments took place by the end of the nineties (HDHN – High Degree Helioseismology Network – Rhodes *et al.*, 1995, and ECHO/LOWL – Experiment for Coordinated Helioseismic Observations – Tomczyk *et al.*, 2000) but only a few years of operations were performed.

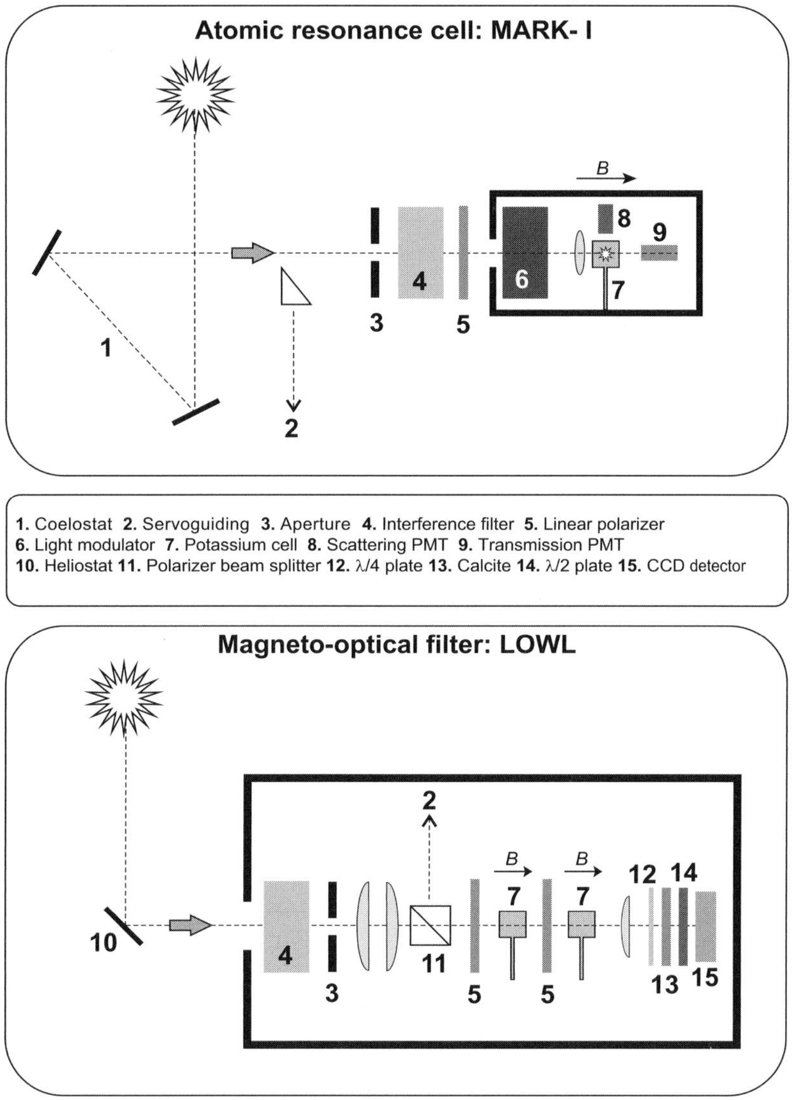

Figure 2.3 Schematic representation of the two types of resonant scattering spectrophotometer. The legend box in the middle applies to both schemes. (Credit: Ramon Castro SMM/IAC.)

A new revised concept is under development (MOTH – Magneto-Optical filter at Two Heights), which would include various atomic cells with different chemical elements, allowing us to simultaneously sample different heights in the solar atmosphere (Cacciani *et al.*, 2003).

Fourier tachometer and Michelson interferometer

These two technical approaches have many commonalities and represent a powerful way to produce extremely narrow transmission bandpasses on a solar line profile with higher flexibility compared to atomic cells, because they can be conveniently tuned in wavelength. However, they have a much higher technical complexity. The Fourier tachometer – a special purpose Fourier transform spectrograph – provides measurements of the solar line wavelength unaffected by the non-linearity of the line profile and of its variations (Brown, 1981). Briefly, the incoming solar light is appropriately filtered around the solar absorption line and passes through a single tunable Michelson interferometer. The analysis of its transmission function (with the interferometer properly tuned) provides three independent measurements directly related to the central wavelength, its strength, and the brightness of the continuum near the solar absorption line. As a consequence, we can measure velocity, modulation and intensity. This is the Doppler analyzer selected for the GONG helioseismology ground-based network (Harvey *et al.*, 1998): the only ground-based network (with six stations) with imaging capability currently operating, successfully collecting high-quality data since 1996, and achieving nearly continuous observations of the Sun's pulsations. In order to increase spectral purity of the solar spectral line region, an improved concept was developed where the largest path-difference elements in a Lyot birefringent filter are replaced with equivalent elements involving polarizing Michelson interferometers. This combination provides stable and reproducible narrow passbands, well below 100 mÅ width. This approach has been used in the MDI instrument (Scherrer *et al.*, 1995) aboard SOHO and, more recently, by its successor, the HMI instrument (Scherrer *et al.*, 2012) aboard the Solar Dynamics Observatory launched in 2010. A schematic layout of the HMI instrument is shown in Figure 2.4. In addition to standard Doppler measurements (with four points of measurement across the line profile) with high spatial resolution on the solar surface, further refinements allowed simultaneous estimates of other quantities such as continuum intensity, line-center depth, and magnetograms. The MDI instrument was switched off after the operation of its successor, HMI, started in 2011, which is providing a nice temporal continuity and extending the capabilities of the SOHO MDI instrument. HMI is measuring with higher spatial resolution and data cadence as well as making high-resolution measurements of the longitudinal and vector magnetic field over the entire visible solar disk.

2.3.3 Intensity measurements

The brightness fluctuations of the solar surface due to the presence of the pressure waves (p modes) are also a physical measurement that carries helioseismic information. Although it may appear rather simple conceptually, the amplitude of many oscillations is close (or even lower) than the intrinsic noise limits (solar background), therefore a precise detection technique is required. However, by measuring the brightness fluctuations instead of line-of-sight velocity, the sensitivity to fluctuations is almost uniform over the entire

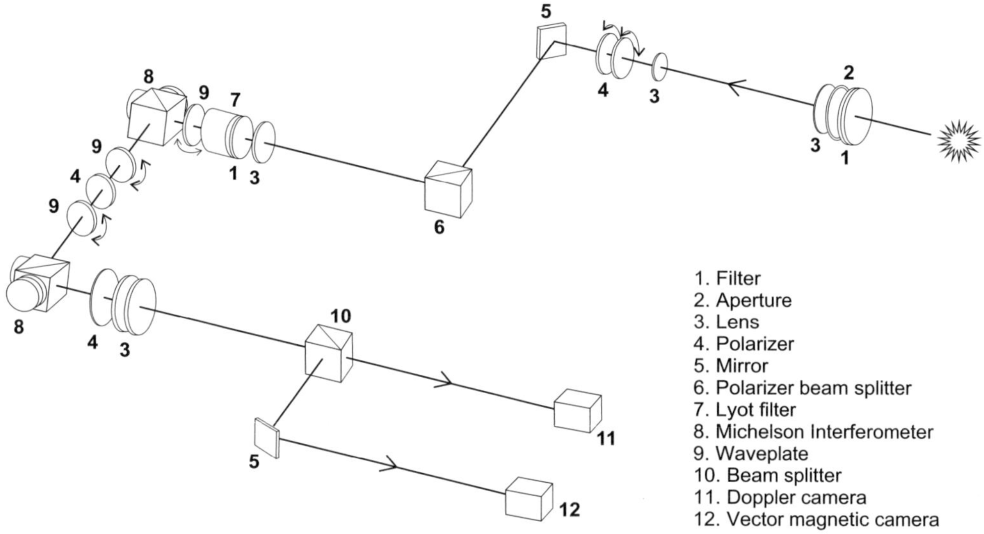

Figure 2.4 Schematic layout of the HMI Michelson interferometer. The combination of extremely narrow and stable bandpasses on the solar spectral line (Fe I 6173 Å), by means of a Lyot filter and two Michelson interferometers plus the various polarizing elements, allows recording of simultaneous filtergrams (images of the Sun at specific wavelengths and polarization) from which accurate Doppler velocity and vector magnetic field solar maps are obtained. (Adapted from Couvidat *et al.*, 2012b. Credit: Ramon Castro SMM/IAC.)

solar disk, including limb fluctuations where line-of-sight velocity measurements are almost insensitive.

Broad-band intensity

Observations of the continuum intensity of the Sun-as-a-star have been carried out mainly outside the Earth's atmosphere in order to avoid the fluctuations of the atmospheric transmission that, ultimately, limit the detection of solar oscillations. Intensity fluctuations were first measured with radiometers, the most famous being the active cavity radiometer, ACRIM, type (Willson, 1984) placed in several spacecraft over the past decades: SMM, IPHIR, etc. Although the much higher solar background noise in intensity limits the total number of individual modes detected in the spectrum, these observations should help in the detection of solar gravity modes. This is the reason why the SOHO spacecraft included the VIRGO package (Fröhlich *et al.*, 1995). It is composed of radiometers, broad-band photometers, and the LOI instrument with a specially designed detector to allow the identification of solar modes up to the degree of 9 (instead of degree 4 for classical disk-integrated measurements).

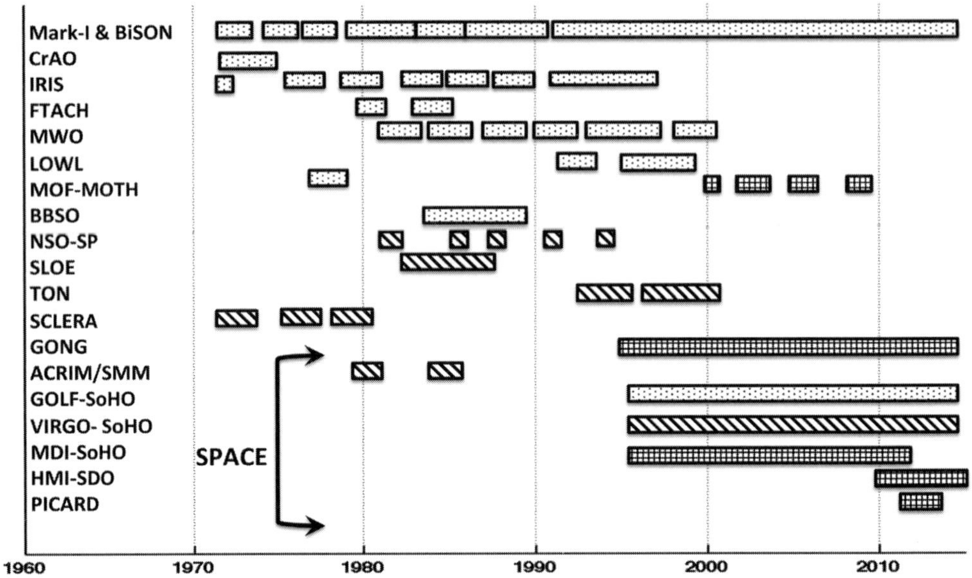

Figure 2.5 The main observation initiatives in helioseismology covering the last 50 years. The different shades identify the physical magnitude observed in each project: Doppler velocity (dotted area), intensity (downward-sloping lines), or both simultaneously (gridded area). Ground- and space-based programs are clearly identified. (Credit: Ramon Castro SMM/IAC.)

Line-core intensity

The limiting factors in the case of broad-band photometry (small amplitudes of the fluctuations and a large background contribution from convection) can be minimized by observing at a high altitude in the solar atmosphere (core of the spectral line), thus increasing substantially the signal-to-noise level. This kind of observation has been performed mainly in the Ca II K line by using simple and robustly designed instrumentation: a telescope and a suitably designed filter. The pioneer instrument was located and operated at the South Pole during austral summers (Duvall *et al.*, 1986); it demonstrated the potentiality of this technique and of the science behind it. It was followed by a number of similar single-site instruments (Bachmann *et al.*, 1995; Ronan and Labonte, 1994) as well as ground-based networks such as the Taiwan Oscillation Network (Chou *et al.*, 1995).

2.4 Summary and future prospects

During the last 50 years, a large number and variety of instrumentation has been conceived and has been in operation. This indicates the wealth of helioseismology that represents the key tool to obtain detailed information on the dynamics and structure of the Sun, all the way from the solar atmosphere down to the core. In Figure 2.5, recognizing the effort of many instrument builders and observers, a summary of past and present instrumental initiatives in the field of helioseismology is shown.

The science objectives for future helioseismic observations are: full disk Doppler velocity, vector magnetic field images, and intensity images, at a variety of wavelengths, high temporal cadence (<60 s), high duty cycle ($>90\%$), and at least a 25-year lifetime to cover a full solar activity cycle. For the technical requirements, such an instrument would need a number of conditions: precise determination of the spectral lines, a resolution of at least 4k by 4k, an aperture of at least 0.5 m, adaptive optics, high-speed image post processing, and high stability.

3

Planetary seismometry

PHILIPPE LOGNONNÉ AND W. THOMAS PIKE

3.1 The technical challenges of planetary seismometry

3.1.1 Basics of seismometry

Seismology is based on the recording, analysis, and inversion of seismic waves. It therefore requires appropriate instrumentation to measure these waves. Although first proposed and attempted in the early days of space exploration with the Ranger program (see Section 3.2.1), it was not until July 1969 that the first successful measurements were made in planetary exploration, with the installation of the Apollo 11 seismometer (Figure 3.1).

This instrument, like all seismometers, is an inertial system detecting the ground acceleration generated by the seismic waves. Strictly speaking, these inertial systems are not only detecting the ground acceleration, but the sum of all temporal changes of the gravity, which include those related to the ground relative acceleration plus the local gravity change due to the displacement of the sensor (horizontal tilt and free air anomaly) and the gravitational change due to global mass redistribution generated by the seismic event and associated waves.

To achieve the detection of inertial acceleration, most seismic instruments measure the motion of a mass suspended by a spring with either velocity transducers based on a coil/magnet system generating a voltage, for example in geophones, or with a displacement transducer based on capacitive or linear variable differential transformer outputs, as used in modern seismometers. Ground acceleration or velocity is then recovered through the instrument transfer function.

3.1.2 Instrumental noise limitations

The smallest seismic signal that can be detected by a seismometer is determined by the ability to resolve it above the aseismic background, which will inevitably also contribute to the recorded data. The lower this background, the better the performance of the instrument. This background, most often called the instrumental noise, can be divided into

Extraterrestrial Seismology, eds. Vincent C. H. Tong and Rafael A. García. Published by Cambridge University Press.

Figure 3.1 Installation of the Apollo 11 seismometer by Buzz Aldrin. This first experiment was powered by a solar panel power supply, while later ones were by radio thermal generator. The solar panel worked during the day time of the first lunation, restarted at sunrise of the second lunation and stopped 9 days later, with a total of 21 days of operation (Latham *et al.*, 1969a,b). (Credit: NASA.)

two contributions: the inherent noise generated by the instrument in the absence of any outside influences, known as the self noise, and the transduction of the aseismic signal that is produced by the unavoidable response of the instrument to non-vibrational influences.

Instrument self noise

The self-noise sets the ultimate performance of the instrument if a perfect deployment eliminating all aseismic influences is possible. Planetary seismology is pushing instruments to this limit, as both the Apollo seismometers, deployed on the Moon, and the Viking seismometer, deployed on Mars, have shown time periods during which no signal detection was made with an amplitude larger than the instrument resolution. This is a major difference from Earth, which is a relatively seismically noisy object due to the oceanic, atmospheric, and human activity.

The thermodynamics of the system sets the largest contributions to the self noise, with two unavoidable contributions coming from the primary mechanical and, for all but direct-write seismographs, the subsequent electronic transduction of the mechanical signal. Taking first the mechanical element, the external seismic vibration is transduced to a relative displacement or velocity of elements within the seismometer by a mass on a suspension. As the suspension will produce some finite energy losses, by the fluctuation-dissipation theorem (Callen and Welton, 1951), this will also produce random motion of the proof mass, known as the suspension noise. This white noise, with equal power density at all frequencies, is inversely proportional to the square root of the product of the proof mass, period, and quality factor of the suspension. This immediately presents a challenge for any compact low-noise planetary seismometer, which will have to comply with restrictions in mass. Reducing the damping requires decisions about the materials of the suspension as well as gas damping of the motion of the proof mass.

The motion of the suspension is transduced electronically mainly through a capacitive displacement measurement (used for all planetary long-period and the short-period components of the InSight/Mars2016 NASA mission), or with an electromagnetic velocity measurement (used in all past short-period instruments), or for higher frequencies piezoelectric measurements (used for the SESAME onboard the Rosetta/Philae lander; Kochan *et al.*, 2000). Other displacement measurements have not yet been space qualified, such as STS1-LVDT displacement, superconducting or optical sensing.

The second, electronic, contribution to the self-noise is most often set by the noise performance of the first amplifier used in the readout circuitry that makes this measurement. This amplifier noise is set in turn by its internal transistor technology, and as well as a broad-band component also has a noise that increases at low frequencies, known as flicker noise. The latter can be avoided by modulating the readout circuitry at a frequency above where the flicker noise is dominant. However, the broad-band noise remains and its amplitude increases with the differential base resistance and output loading of the first input transistor pairs of the amplifier, fundamentally as a product of the fluctuation-dissipation theorem.

Other contributions to the self-noise include surface electric-field fluctuations on capacitive sensing plates (the "patch effect"; Speake, 1996), electromagnetic damping by eddy currents in conductive moving elements due to the instrument's own magnetic field, feedback noise, especially when integrators are used for correcting the diurnal large temperature variation, and the readout noise mainly due to analog-to-digital conversion. All these sources of noise must be designed within the constraints of available space-qualified components to minimize the total instrument noise, which is inevitably above the best terrestrial instrumentation.

Environmental noise limitations

There are two types of noise from the environment: non-vibrational noise generated directly in the sensor from unwanted transduction of aseismic signals, such as temperature variation, and vibrational noise generated by aseismic sources that deform the ground supporting the sensor.

The largest contribution from the first type is associated with temperature fluctuations, either originating from external temperature variations or from variations in the sensor heating or power dissipation. The temperature sensor sensitivity, typically between 10^{-5} and 10^{-4} m s^{-2} °C^{-1} is mainly related to the temperature dilatation of the instruments and, for the vertical or oblique seismometer counterbalancing the gravity at equilibrium, to the temperature sensitivity of the balancing forces. These balancing forces are directly related to the dependence on the Young modulus of the material of any suspension. Other temperature sources of noise will be injected from the electronics. In both cases, the best mitigation of this noise is through thermal shielding and thermal compensation, the latter based on self-compensated materials for the spring or by offsetting one thermal response by another of the opposite sign.

The second type of noise is manifested by the very large difference in background signals seen on Earth at different locations due to variability in atmospheric and oceanic contributions: this variability is characterized by the high and low noise models (Peterson, 1993), with a 40–50 dB difference between the best and worst seismic stations. Even at the quietest sites, the background noise is above the instrumental noise floor of current seismometers. In contrast, the noise recorded by the Apollo seismometers on the Moon, without atmosphere or ocean, is extremely low compared to Earth, except during sunset or sunrise. On Earth, lower background noise than the Moon can only be found at the quietest locations below 0.01 Hz and above 1 Hz.

While Mars has no noise contribution from oceans, it does see atmospheric contributions. The resulting buoyancy pressure forces on the proof mass can be avoided by using an evacuated or hermetically sealed enclosure. Other pressure-induced signals will be, on Mars, related to the static loading of the surface by the atmosphere and even gravitational attraction of the atmosphere (Figure 3.2), and require pressure decorrelation techniques, as already performed on Earth (e.g., Beauduin *et al.*, 1996). The amplitudes of these noises are estimated to a few 10^{-9} m s^{-2} $Hz^{-1/2}$ (Lognonné and Mosser, 1993; Lognonné *et al.*, 2012) depending on the wind amplitude. Unlike Earth, where seismic measurements are generally made in underground vaults, and the airless environment of the Moon, there will also be a wind contribution to the environmental noise. There will be wind drag acting on the lander, if the seismic sensors are not separately deployed. If they are, these forces will act on the enclosure for the instruments as well as on any tether connecting the instrumentation to the main body of the lander. In either case, the wind-induced aseismic signal will be exacerbated by the expected low rigidity of the Martian surface.

Finally, magnetic field variations generate noise on seismometers with springs with non-zero magnetostriction effects or with magnetic sensitivity related to motor magnets, or coils, as observed on the Apollo short-period seismometer during magnetic storms. Mitigation can be done either by mu-metal shielding, active compensation with Helmoltz coils, or decorrelation (Forbriger *et al.*, 2010). In the last case, this requires the additional measurement of magnetic field, as planned for the InSight mission.

3.2 The Lunar golden age

3.2.1 From Rangers to Surveyor

Planetary seismology began in 1959 with the Ranger program. While the Ranger probes were designed for a hard landing on the Moon, a "survival sphere" designed to support a 3000 g impact contained a vertical-axis 3.4 kg seismometer with a natural frequency of 1 Hz designed by the California Institute of Technology (Figure 3.3a; Lehner *et al.*, 1962). All three Ranger probes carrying this package failed, but the Earth version of the seismometer is still commercially available (SS-1 seismometer from Kinemetrics).

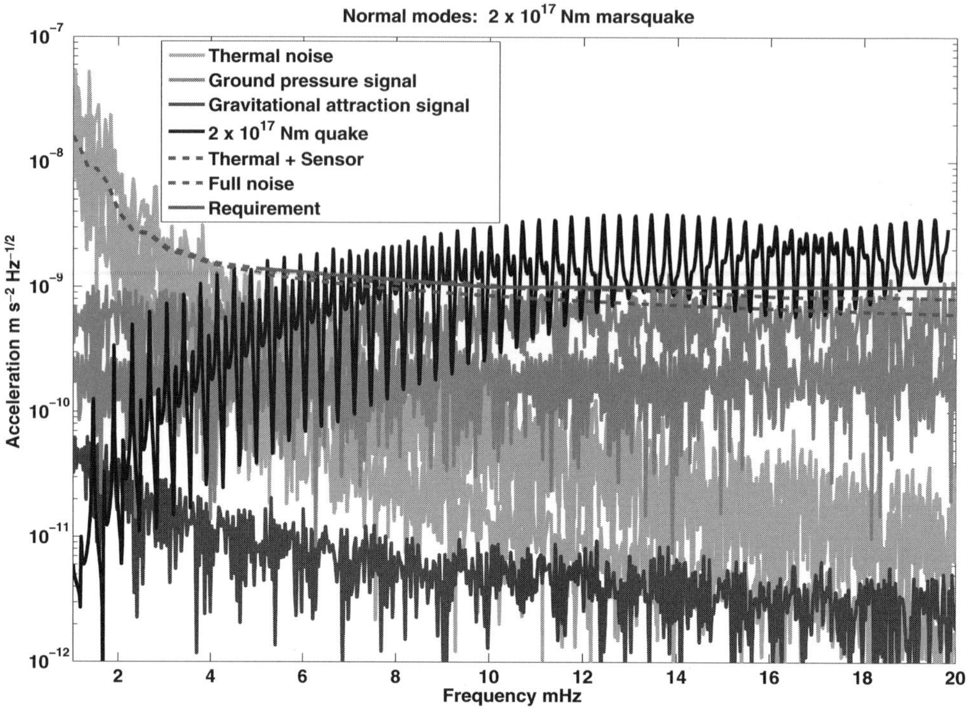

Figure 3.2 The detectability of normal modes for large quakes on Mars is shown in this plot of the calculated acceleration amplitude spectrum (black) for a quake with seismic moment 2×10^{17} Nm compared to SEIS/InSight required (red) and expected (red dashed) sensitivity. Epicentral distance is 90 degrees. Also shown are modeled environmental noise sources included in the red dashed curves: thermal (cyan); atmospheric pressure (green); atmospheric gravity (blue). A black and white version of this figure will appear in some formats. For the color version, please refer to the plate section.

A few years later, a three-axis long-period (LP) seismometer and a short-period (SP) vertical-axis seismometer, with a total mass of 11.5 kg, were built in the Lamont Doherty Geological Laboratory (Sutton and Latham, 1964) for the Surveyor mission. Because of weight and power restrictions, this instrument was first descoped to a single SP seismometer and then cancelled. Both the LP and SP seismometers had to await the Apollo missions, but the groundwork had been done to provide high-quality seismic instrumentation.

3.2.2 The Apollo instruments

The Passive Seismic Experiment

The first successful installation of a seismometer on the Moon was achieved in July 1969, during the Apollo 11 mission (Figure 3.1). The Passive Seismic Experiment (PSE), based on the 11.5 kg Surveyor payload, consisted of a triaxial LP seismometer, with a natural

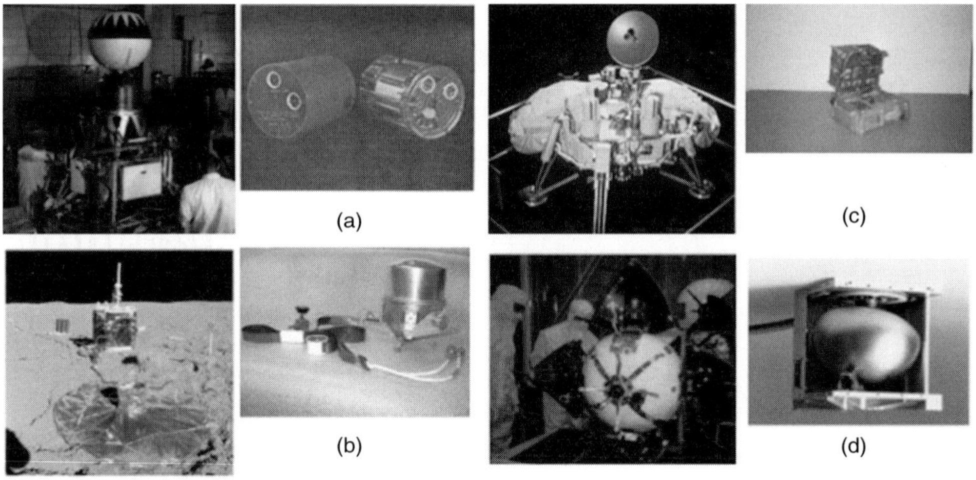

Figure 3.3 Launched missions with the corresponding number of instruments (inst.) deployed in different locations: (a) Rangers (3 inst.), (b) Apollo PSE (5 inst.), (c) Viking (2 inst.), (d) Mars96/Small Surface Stations (2 inst., mounted by the top on the lander structure). Numbers in parentheses indicate the number of instruments deployed during the program or by the mission. Not shown are the Apollo Active Seismic experiments, the Mars 96 penetrators and the Rosetta Lander Philae, all with short-period or ultra-short-period instruments and the Apollo 17 gravimeter at the other frequency end. Note the different strategies for the instrument installation on the robotic missions: Ejection of a seismometer "ball" prior the crash for Ranger, lander deck mounting for the Viking, or inside lander mounting for Mars96. The Apollo seismometer (here a PSE unit in the collection of National Air and Space Museum) was installed by astronauts and covered by a thermal shield after proper leveling and orientation. (Credit: IPGP-SODERN-CNES for (d); Allan Needell, National Air and Space Museum for PSE units (b); AGU for Ranger unit (a). All other NASA.)

period of 15 s, and an SP seismometer with a natural period of 1 s. The PSE had a volume of 12 liters and a power consumption of between 4.3 and 7.4 W (Figure 3.3b; Latham *et al.*, 1969a,b). It is notable that most of the structural elements were machined out of beryllium, with a much lower density (1850 kgm^{-3}) than other metals with comparable structural properties (e.g., titanium, 4400 kgm^{-3}) but this came with a very significant cost impact. These seismometers were extremely sensitive, capable of detecting a minimum signal of 300 pm for the LP in flat mode, 50 pm for the LP in peaked mode at 0.45 Hz, and 50 pm for the SP seismometer at 8 Hz. The nominal response curves are shown in Figure 3.4.

A few months later, Apollo 12 installed a new PSE station. This time, a radioactive thermal generator allowed continuous operation through the long lunar night (with thermal regulation keeping the temperature stable day and night) and the deployment was performed more carefully (Figure 3.3). Other stations were deployed by Apollo 14, 15, and 16 to form a seismic network of four stations. All but the Apollo 12 SP seismometer and Apollo 14 vertical LP seismometer operated until the end of September 1977, when all were shut down from the Earth.

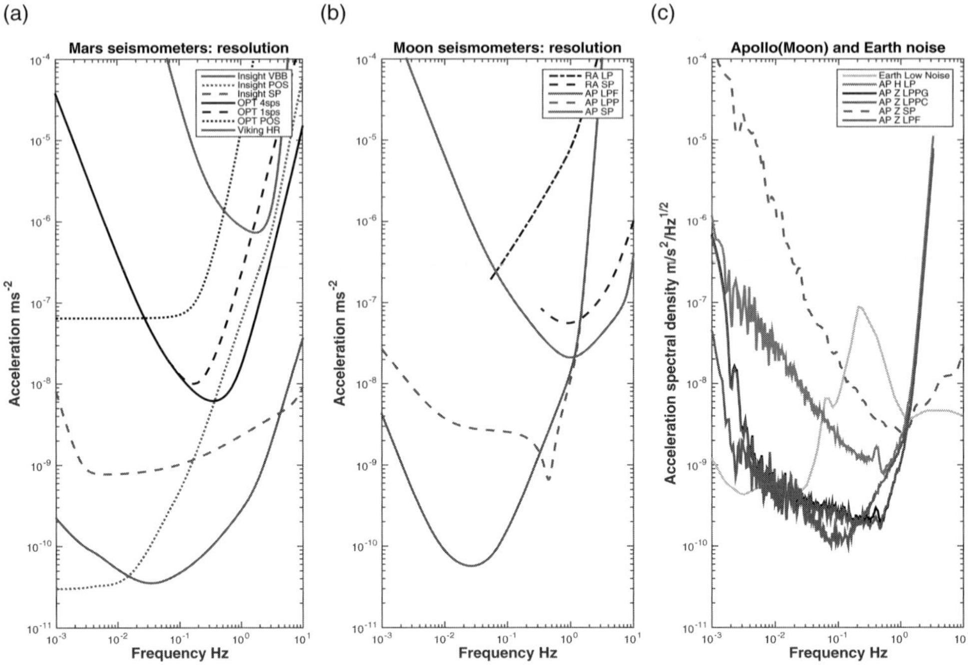

Figure 3.4 Response curves of the past (a) Mars and (b) Moon seismometers. (a) The resolution of the Viking in high data rate mode; the OPTIMISM seismometer in the long-period position mode (POS), in the velocity mode OPTVIT1 at 1 Hz sampling rate, and at 4 Hz (OPT VIT4). The figure also shows the expected resolution of the InSight seismometers, defined here as the RMS in $1/6$ of a decade (equal roughly to the product of $0.62 \times \text{ASD} \times \sqrt{f}$ where ASD is the amplitude spectral density in m s^{-2} $\text{Hz}^{-1/2}$ and f the frequency). The first two are the velocity flat and acceleration flat outputs of the VBB sensors (VBB and POS) while the third is the velocity flat output of the short-period (SP) instrument. VBB's components are sensitive to acceleration oblique at about 30° with respect to horizontal, while the SP are either vertical or horizontal. At 1 Hz, the expected resolution of the InSight VBB sensor is expected to be 10^3 times better than Viking. (b) The Moon case, with the long (RA LP) and short (RA SP) period analog outputs of the Ranger seismometer, the flat (AP LPF) and peaked (AP LPP) mode of the Apollo LP seismometer, and the short-period seismometer (AP SP). The transfer functions shown are the generic ones. (c) Different sources and levels of seismic noise. The noise levels recorded on the Moon by the different channels of the Apollo seismometers are compared with the Earth-based low noise model of Peterson (1993). Noise levels are from Apollo 12 vertical long-period flat (AP Z LPF) mode and from Apollo 14 vertical short-period (AP Z SP) mode. These noise levels are likely an upper estimate of the Moon noise and may be related to the instruments. For the other components, note that the peaks near 0.45 Hz of the peaked vertical (AP Z LPPG) and horizontal (AP H LP) peaked records might be related to differences between the Apollo 12 transfer function at the time of the recording and the pre-flight generic transfer function. The noise corresponding to the curve AP Z LPPC is obtained by changing the feedback parameters in order to minimize the noise peak (about 10% for the parameter K2 and h, see the Apollo transfer function at http://dartsjaxajp/planet/seismology/apollo/The_Apollo_Seismometer_Responses pdf). A black and white version of this figure will appear in some formats. For the color version, please refer to the plate section.

Other Apollo seismic instrumentation

Also deployed by Apollo were geophones, used for the Active Seismic Experiment on Apollo 14, 16, and 17. They had a flat velocity output from 8 Hz to about 100 Hz, and were operated at a sampling rate of 117.8 Hz. Their resolutions at 10 Hz were comparable to those of the PSE SP (with a system signal to noise of about 40 db at 10 Hz for a peak-to-peak signal of 1 nm). In addition, the Apollo 17 gravimeter, initially dedicated to the search for gravitational waves, was able to function as a seismometer (Kawamura *et al.*, 2010).

3.2.3 Viking

In 1976 the two Viking landers initiated the first seismic observations of Mars (Figure 3.3c; Anderson *et al.*, 1977). Only the Viking 2 lander's seismometer worked however, as the three components of the Viking 1 seismometer failed to uncage. The three-axis seismometer was an SP instrument, with a natural period of 0.25 s, a total mass of 2.2 kg, a volume of 1.4 liters and a nominal power consumption of 3.5 W. No recentering was necessary with a tilt range of 23 degrees. The sensitivity of this SP seismometer was ten times better than the SP Apollo seismometer, for periods smaller than 1 s (Figure 3.4). The seismometers were mounted on the body of the lander, with coupling to the surface though the shock absorbers on the legs. A further limitation was the low telemetry data rate, considerably lower than Apollo's. Compression of data was thus necessary and most of the monitoring was done in event mode, where the envelope of the seismic signal and the number of zero crossings was returned every 1.01 s. No convincing event detection was performed during the 19 months of nearly continuous operation of the Viking Lander 2 seismometer and the data have been mainly used to constrain the wind on Mars.

3.2.4 Mars 96

Twenty years later, the Mars 96 mission was expected to re-open the seismic exploration of Mars. The two small surface stations were equipped with a 0.1–1 Hz vertical seismometer (OPTIMISM experiment, Lognonné *et al.*, 1998), integrated inside the small stations (Linkin *et al.*, 1998) for thermal insulation reasons. These two stations were complemented by two penetrators equipped with very short period (100 Hz) vertical seismometers, unfortunately without overlap with the OPTIMISM frequency band. Nevertheless, the Mars 96 mission would have deployed the second operating seismic network on a terrestrial body if it had not been sadly lost shortly after its launch.

The design of the OPTIMISM seismometer was constrained by the very low power and mass allocation of the surface-station payload, as well as the high shock level of the landing. Among all the constraints, the most severe for OPTIMISM was the low power allocated (50 mW), leading the instrument to use a dual transducer design: a low-power velocity transducer for the bandwidth from 0.5 to 10 s and a very low power displacement transducer for the control of the mass position, as well as for surface wave detection (5–50 s). The pendulum was stored in a half sphere made of titanium with automatic leveling (Figure 3.3d). Nevertheless, OPTIMISM was expected to perform better than Viking (Figure 3.4) with only a mass of 405 g and a volume of one liter.

3.3 Current and future missions

3.3.1 Current instrumentation

Starting in the early 1990s a new generation of seismometers was developed for Mars with international efforts to gain approval for Mars network missions, under the proposals Mesur, MarsNet, Intermarsnet, Impact, and Netlander, amongst others (see Lognonné, 2005, for project references). Additional development was done in Japan for the Lunar-A penetrator mission (Shiraishi *et al.*, 2008), cancelled in 2007. However, it took two decades to see these efforts bear fruit with the selection by NASA of InSight (Banerdt *et al.*, 2012) under the Discovery program.

A single lander will be launched in March 2016, with a landing in late September 2016. As its prime payload element, SEIS (Seismic Experiment for Interior Structure) will be equipped with a dual VBB/SP three-axis seismometer, together with a geodetic beacon (RISE), a heat flow experiment (HP3) and a set of environmental sensors (APSS), to support the seismic measurements. SEIS is lead by the French space agency CNES and IPGP, with contributions from UK, German, Swiss and US space agencies and labs.

With only one lander, InSight will need to detect surface waves completing a full orbit around the planet, requiring a low instrument noise level, $<10^{-9}$ m s^{-2} Hz$^{-1/2}$, in the bandwidth 0.01–0.05 Hz. The performance and installation quality of the InSight seismometer will therefore be critical parameters to ensure success with a sensitivity higher than previous Mars seismometers (see Figure 3.4 for comparison). SEIS also can expect a much better deployment with the InSight lander providing a robotic installation of SEIS from the deck of the lander down to the ground followed by the lowering of a wind and thermal shield (WTS). Figure 3.5c illustrates the level of thermal protection and shows the evacuated titanium sphere in which VBB sensors are located. In addition, a thermal blanket wrapped around both the VBB sphere and SPs and the WTS will provide additional shielding. All together, these three thermal barriers produce an effective time constant at low frequencies of between 2 and 5.5 hr. It is likely that SEIS will be close to the best deployment and sensitivity that can be achieved by a robotic installation on the surface of Mars.

Looking at the seismometers themselves, SEIS VBB (see Figure 3.6) is based around an inverted pendulum, with the gravity acting against the spring to lower the pendulum natural frequency to about 0.45 Hz for a proof mass of 190 g. It uses a highly sensitive capacitive transducer to drive a feedback designed to provide a response flat in ground velocity to within 10 dB from 50 to 0.5 s, with an expected noise below 10^{-9} m s^{-2} Hz$^{-1/2}$ between 0.01 and 2 Hz. The output is digitized at 24 bits, giving 10 dB headroom over the instrument noise floor at the same time as a saturation threshold larger than Viking.

SEIS SP (Figure 3.7) consists of a set of micromachined sensor heads, one vertical and two horizontal axes. The suspension and proof mass of each sensor are etched from

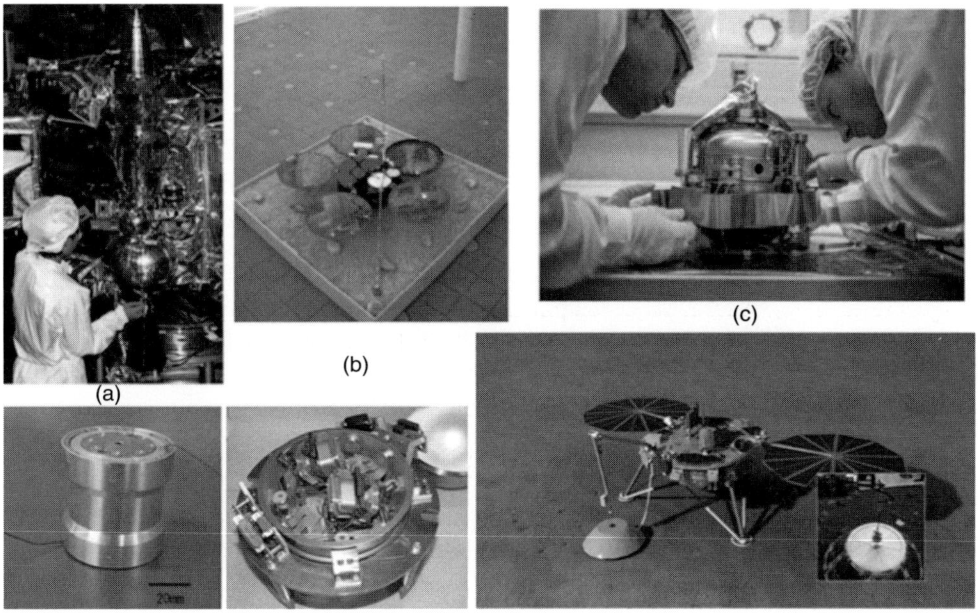

Figure 3.5 Past or current projects, again with the number of instruments expected. (a) Lunar-A (2 inst.); (b) Netlander (4 inst.), both cancelled and, respectively, aimed for the Moon and Mars. (c) The single station InSight SEIS system, which includes both the VBB and the SP axis, for launch in 2016. Note the differences in the deployment strategies. Lunar-A is based on penetrating the ground and enabling a buried configuration for the seismometers, while both Netlander and InSight are based on a surface deployment. The major difference between the latter two is the much superior control of the installation for InSight, which uses a robotic arm and covers the instrument with a windshield, while on Netlander the seismometer legs are deployed to the surface through holes in the lander base, without any choice in the placement. The seismometer was, however, expected to be mechanically decoupled from the lander after deployment and was protected from wind/temperature by the lander structure. (Credit: ISAS (a), CNES-DLR-IPGP-SODERN (b), CNES/J.Chetrit (top c), NASA (bottom c).)

single-crystal silicon wafers using deep reactive ion etching to produce a 7 Hz suspension and a 0.4 g proof mass. Both the period and mass are much smaller than conventional seismometers, though at the limit of micromachined devices. The motion of the proof mass is capacitively measured by the change in overlap between an array of electrodes on the proof mass and an opposed fixed array connected to the outer frame of the suspension. This geometry has several advantages for planetary operation. As a much larger motion of the proof mass can be accommodated compared to more conventional capacitive transducers, which sense a closing gap between their plates, the SP can be operated over a large tilt range of deployment. Combined with the very low cross-axis compliance of the micromachined suspension, this also allow SEIS SP to be fully tested on Earth under tilt conditions corresponding to Martian gravity. In addition, as sliding rather than squeeze-film gas damping occurs, a high quality factor of several hundred can be achieved without evacuation,

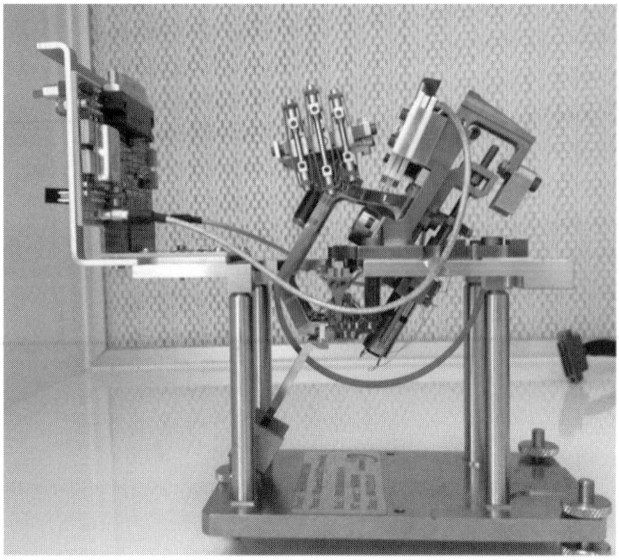

Figure 3.6 The InSight VBB sensor. Note the Earth counter mass, at the left, used on Earth to compensate the larger Earth gravity. This mass, removed for the Mars flight configuration, allows full tests on Earth. (Credit: IPGP-SODERN-CNES.)

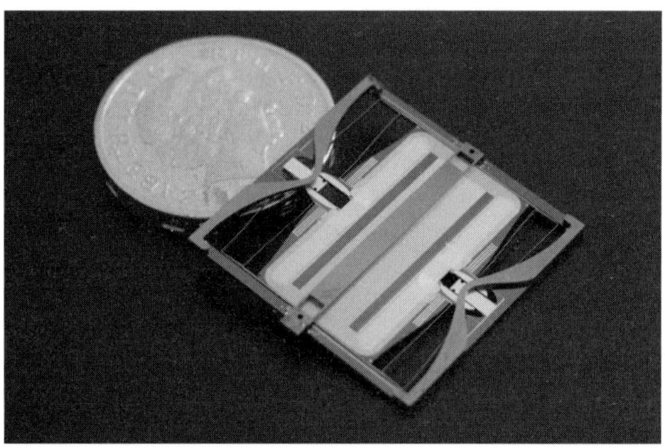

Figure 3.7 The micromachined sensor of the short-period seismometer of the InSight SEIS instrument. (Credit: Imperial College.)

giving a suspension Brownian noise of 4.5×10^{-9} m s^{-2} Hz$^{-1/2}$ at 300 K. SEIS SP's feedback produces a flat velocity output over a bandwidth from 40 Hz to below 0.1 Hz, producing an overlap with the VBB, with a target acceleration sensitivity of below 5×10^{-9} m s^{-2} Hz$^{-1/2}$.

3.3.2 Future seismic projects and instrumentation directions

The next decades' seismic projects will likely focus on return opportunities to the Moon and network opportunities on Mars alongside even more challenging deployments, such as Venus and Europa. While several seismic payloads have recently been studied for the Moon (JAXA/SELENE2, Tanaka *et al.*, 2013; NASA/Lunette, Neal *et al.*, 2011; ESA/FarSide, Mimoun *et al.*, 2012; ESA/Lunar Net, Smith *et al.*, 2009), Mars (ESA/INSPIRE, Chicarro, 2013) or Europa (Gowen *et al.*, 2011), only SELENE2 has gone through a full mission definition phase, though unfortunately not progressing further.

For these projects, the two major directions in terms of new instrumentation development are mainly mass/cost reduction for network missions and, for the Moon, developing the performance beyond Apollo required to completely elucidate the deep interior structure (Yamada *et al.*, 2013).

One mission approach that aims to greatly reduce costs below a conventional soft landing is the deployment of networks of seismometers using penetrators. These have the advantage of producing excellent coupling to the subsurface, while reducing environmental noise, but place severe constraints on instrumentation, particularly the suspension of a seismometer. A micromachined seismometer, with its small proof mass, can be armoured against high shocks. One approach, similar to the one used on the Ranger seismometer (Lehner *et al.*, 1962) is to encase the suspension in a sublimable wax that cleanly dissipates after deployment, and has been demonstrated for shocks up to 14 000 g (Hopf *et al.*, 2009). Another is to allow free motion of the proof mass but incorporate solder buffers to replace brittle with ductile contact at the extremes of motion. Such suspensions can survive drop tests to 5000 g (Delahunty and Pike, 2013).

While the determination of the level of the ambient seismic floor of Mars is awaiting the first data from the InSight seismometers, it appears that the Apollo seismometers may not have reached the lunar seismic floor (Lognonné *et al.*, 2009), with the ambient level possibly 100 times smaller than the Apollo resolution, perhaps 10^{-12}–10^{-11} m s^{-2} Hz$^{-1/2}$ from 0.1 to 1 Hz and even less at longer periods. Such a very low level would open up new perspectives in lunar seismology, including detection of gravitational waves, as envisaged for the Apollo 17 gravimeter, or even impacts of strange matter (e.g., Banerdt *et al.*, 2007). The corresponding demands on the instrumentation are high, with either superconducting quantum displacement sensing (Paik and Venkateswara, 2009) or possibly optical detection schemes (e.g., Zumberge *et al.*, 2010) offering the required sensitivity, with the latter giving the advantage of mass reduction through operation without feedback and fine leveling.

Landing on planets will however always significantly increase the cost of future projects, and remote sensing seismometry on telluric planets (see Section 14.4 for remote sensing seismology on giant planets) might be important for Venus, where surface seismic operation is extremely challenging, as well as airless bodies. For the latter case, the lack of atmospheric/ionospheric noise in the signal path between orbiting satellites and surface beacons or reflectors can give a resolution at the micrometer level, which would correspond to a ground acceleration of 2×10^{-10} m s^{-2} at 500 s, comparable to a surface-deployed seismometer. This might enable measurements of normal modes, surface waves, and even large quakes from space, and has been proposed for Europe using radar ranging (Panning *et al.*, 2006). In contrast, for Venus the thick atmosphere and resultant large coupling between the planet's interior and atmosphere (Garcia *et al.*, 2005; Lognonné and Johnson, 2007) offers the possibility of detecting the surface waves of quakes through induced vertical ionosphere oscillations, in ways similar to those now done routinely on Earth with Doppler sounding (e.g., Artru *et al.*, 2004), but with attenuation short-period limitation due to the carbon dioxide atmosphere (Garcia *et al.*, 2005).

Part II

Data and physical parameters

4

Asteroseismology: From the telescope to the modeler's desk

STEVEN D. KAWALER

In this chapter, I discuss the first steps in asteroseismic analysis – following data through the analysis to the products that are what stellar modeling tries to explain. For asteroseismology, the time-varying signal is generally either the total brightness of the target (i.e., time-series photometry) or the (disk-integrated) radial velocity as measured via Doppler shifts of spectral lines. In either case, we measure a differential quantity ($\Delta Y/Y$) at regular intervals. More precisely, this quantity is integrated over a finite but short time (shorter than any expected signal from the star), producing an average value over the time interval of the integration. This produces the time series of the signal which I refer to as $y_n(t_n)$ not in an instantaneous sense, but in a smoothed way.

Our time series represents an averaged, discretely sampled quantity as a function of time. From that time series, our goal is to determine the harmonic content of the signal – over what time scale(s) does the signal vary? Are there coherent (in phase and/or amplitude) sinusoidal variations? Are there preferred time scales for "noise" processes? Are there periodicities whose phase varies slowly over the entire interval? Are there "temporary" periodic signals? Answering these questions requires more and more sophisticated analysis techniques. In all cases, though, the sampling of the signal is critical in limiting the information content that we can derive.

There are many tools and techniques available for analysis of time-series data. In this chapter I will concentrate on Fourier analysis (i.e., power spectrum analysis), and also describe least-squares fits to amplitudes, phases, and frequencies. An equivalent analysis is via the Lomb–Scargle periodogram (Scargle, 1982; Appourchaux, 2014). Of course, more sophisticated tools are available but these have provided the dominant "one-two" punch for most asteroseismic studies. Various specific applications of asteroseismology employ subsequent analysis of the power spectra or frequency lists to derive oscillation parameters, but I leave those to subsequent chapters.

I begin with a *very* brief tutorial on producing a power spectrum using the Fourier transform, and then describe how the power spectrum behaves for various time-series properties of relevance. Asteroseismic data can produce clean power spectra but the nature

---

Extraterrestrial Seismology, eds. Vincent C. H. Tong and Rafael A. García. Published by Cambridge University Press.

of the pulsators, and that of the observational constraints, usually produces complications in the derived power spectrum that need to be disentangled before proceeding with further analysis and modeling. I discuss these various aspects and preview how they will influence the subjects coming later in this volume. I outline extraction of pulsation frequencies via least squares, and finish with an example of how phase information provides additional possibilities for probing stars.

4.1 Fourier transform/power spectrum

Harmonic analysis of a time series can reveal, quickly and clearly, the oscillation frequencies present in a data set. Hence the Fourier transform (hereafter FT) is a fundamental tool. With evenly spaced data, an extra bonus is the rapidity with which the FT can be computed.

In essence, one cross-correlates the time series with a sequence of sinusoids of fixed frequency, with the output being the amplitude of such a sinusoid that is the most highly correlated with the original data. Such a correlation is of the form

$$F(\nu) = \text{const.} \sum_{n=1}^{N} y_n(t_n) \times e^{-i2\pi \nu t_n}, \tag{4.1}$$

where $F(\nu)$ is the (complex) FT of the discrete time series $y_n(t_n)$. If the data themselves are periodic, such a correlation coefficient will be large. The (real) amplitude of the FT represents the amplitude of the oscillation about the mean value. By convention, we usually discuss the "power" $P(\nu)$, which is the sum of the square of the real and imaginary parts of $F(\nu)$. A peak in power $P(\nu)$ (or in amplitude $A(\nu) = \sqrt{P(\nu)}$) represents a possible true periodicity. Embedded in the FT is the phase at the given frequency; the ratio of the real and imaginary parts of $F(\nu)$ is the tangent of the phase.

The range of independent frequencies computed can cover from the sampling frequency (the so-called Nyquist frequency) down to the inverse of the observing timespan. The function $F(\nu)$ is usually evaluated at discrete frequencies $\nu_k = k/T$, where T is the time spanned by the data and k ranges from 1 to $N/2$. This is done in part for convenience but also because, formally, the data sampling and extent imply that only a finite number of frequencies are independent. The formal frequency resolution of a data set of length T is $1/T$, and the lowest practical frequency is approximately equal to the length of the observation – though in cases where the signal-to-noise is high, and the assumption of sinusoidal variation is sufficient, longer periods may be inferred.

Computation of the FT for actual data can be resource intensive, especially for the very long data sets coming from asteroseismic campaigns and space observations. Fortunately, an extremely efficient and fast technique for computing FTs was developed in the early days of computer analysis – the FFT, or fast Fourier transform, algorithm (Cooley and Tukey, 1965). The FFT requires data that are equally spaced in time and, formally at least, N must be an integer power of 2.

In practice, data are more likely to be unevenly sampled, as is frequently the case in multiple-night ground-based data, or long-duration space-based observations where the barycentric correction changes significantly over time. In these cases, the FFT cannot be directly implemented, and computation becomes a significant task. However, computational tricks can limit the damage here – in particular the extirpolation schemes outlined in *Numerical Recipes* and elaborated upon by Press and Rybicki (1989) – see also Section 13.8 of Press *et al.* (1992).

4.2 The ideal time series

While never actually realized, the ideal time-series data are of sufficient time span to adequately resolve signals of all frequencies of interest, from the shortest period variations to the longest time scales for changes in amplitude, frequency, and phase. Individual integrations need to be short enough to allow resolution of the fastest phenomena in the system, but long enough that the individual measurements have acceptable signal-to-noise (S/N). Ideally the integration time should match the sampling time, though some dead/readout time is often imposed by instrumentation.

To achieve such an ideal time series is clearly impossible (but we can always hope). Ground-based data are necessarily "gappy," with gaps caused by day/night cycling, clouds, and seasonal visibility. But let's not worry about all that now. What are the properties of an unbroken, finite time series, sampled uniformly?

Uniform sampling in time can be harder than one thinks. The frame of reference that matters is that of the target stars. Over a long enough time interval, for cosmic objects, the time must be corrected for the changing position of the observatory (the Earth itself for ground-based observers, or the position of the spacecraft taking the data). This is accomplished by computing the barycentric correction – that is, the time of arrival of the data at the barycenter of the Solar System. Given that the observatory is moving, the barycentric time intervals change over time.

Continuous observations are also ideal. If there are gaps in the data, then we lose information on what our target is doing, and effectively "zero out" the data during the gap. Analysis tools, facing these gaps, attempt to model the data as having no amplitude at those times by including additional periodicities at the appropriate phase to null the harmonic components present when data are available. More on this "aliasing" below.

4.2.1 Coherent pulsations: "Classical" pulsators

Certain classes of pulsating stars (most of the so-called classical pulsators such as Cepheids, RR Lyrae stars, δ Scuti stars, and pulsating white dwarfs) have a very stable energy source supporting the oscillation, with the stellar material responding in a predictable and deterministic way. These stars usually exhibit coherent pulsations (constant amplitude, phase, and frequency) producing peaks in the FT that reveal the frequency (to a resolution of about $1/T$) and amplitude. Phase information is also available from the FT itself in the ratio of the imaginary part to the real part (see above).

For coherent pulsations, the precision with which we can determine the oscillation parameters increases with time. In photon-counting environments (the Poisson regime),

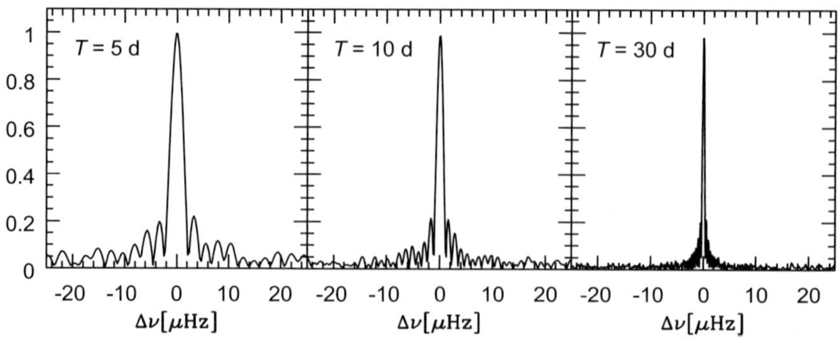

Figure 4.1 Fourier amplitude spectrum $A(\nu)$ of a single sinusoid with unit pulsation amplitude. From left to right, the time span of the time series increases from 5 days to 30 days.

the *amplitude* precision scales with T^2, and frequency precision increases as T. The latter is apparent from the frequency resolution – the width of a peak in the FT scales as $1/T$. Figure 4.1 shows how the peak in the amplitude spectrum stays at the proper frequency and amplitude, but narrows as the time span T of the data increases (from left to right in Figure 4.1).

4.2.2 Periodicities subject to phase and amplitude fluctuation: Solar-like oscillations

Unlike the classic pulsators, stars that show oscillations are driven by turbulence in their convection zones (known as solar-like pulsators) with stable oscillation periods, but the amplitudes and phases change on time scales of days. These pulsations also produce signals at the underlying frequency, but decay in the amplitude and are re-excited at random phases. This produces a signature in the FT of the form of a Lorentzian envelope (see Figure 4.2). When the observations span a time longer than the coherence time of the pulsations, the "envelope of power" is a measure of the coherence time. In the case of a stochastically excited oscillation, the phase (and amplitude) are effectively reset at random times. A transform of such a time series "sees" this as a beating phenomenon, resulting in multiple peaks and frequencies separated by the time scale for excitation/damping/re-excitation. Thus this mean reset time, corresponding to the coherence time, sets the width of the power bump. As might be expected, determination of precise oscillation frequencies is more challenging than for coherent pulsators – see Chapter 5 in this volume for further details.

For coherent pulsators, the detectability of a mode increases with time. In simple cases where photon-counting statistics dominate the noise, the noise level (as measured in terms of $\Delta Y/Y$) decreases with the square root of time – the more photons, the lower the noise. Since a coherent pulsation produces a peak of fixed amplitude, a longer observation (more photons) reduces the noise at all frequencies. Thus the signal-to-noise ratio increases as $T^{-1/2}$.

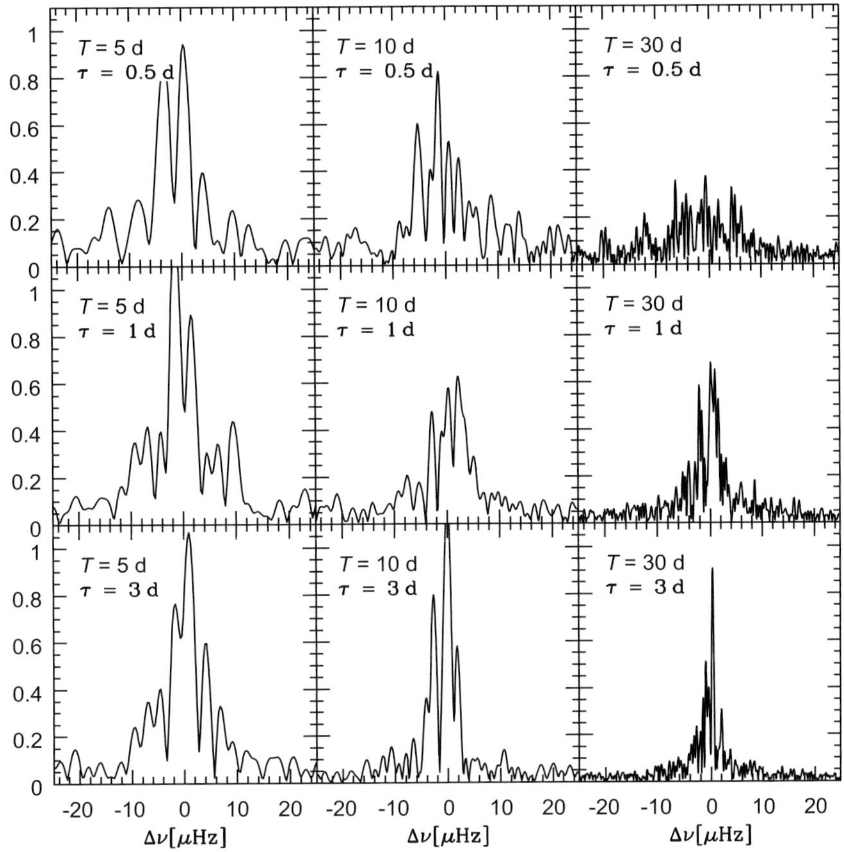

Figure 4.2 Amplitude spectrum of a single sinusoid with unit pulsation amplitude, but subject to damping and stochastic excitation. From left to right, the time span of the time series increases from 5 days to 30 days. From top to bottom, the damping time scale increases from 0.5 days to 3 days. The shorter damping times produce a wider spread in the signal area, with peaks falling within a Lorentzian envelope. Compare with Figure 4.1.

4.3 Observed time series: Gaps and aliasing

Real time-series asteroseismic data have gaps, uneven sampling, and systematic changes in instrument sensitivity that all produce artifacts in a harmonic analysis. Some of those artifacts can mimic stellar signals, and so are particularly obnoxious.

Periodic gaps in the data produce "aliasing" as they mimic beating between modes spaced at the gap frequency. During times when the target is not observed, the formal amplitude of the signal is zero (of course, the star is merrily pulsating along while we are not looking). Therefore, an FT of a singly periodic data set that has gaps displays multiple peaks, separated in frequency such that the periodicities destructively interfere with the real periodicity on the proper time scale to reproduce the gaps. This phenomenon is shown in

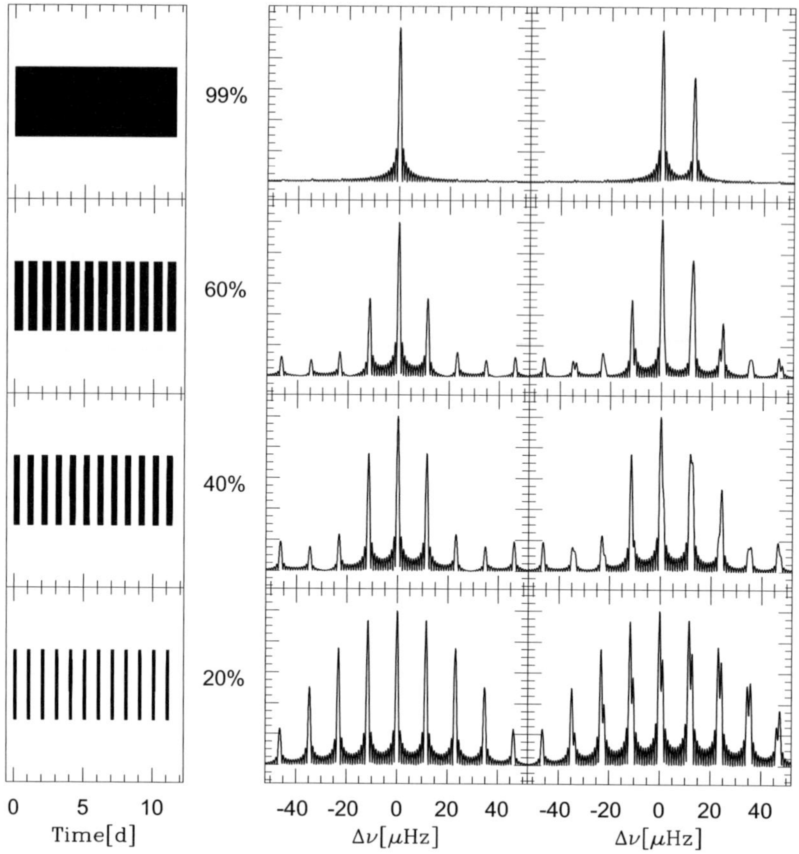

Figure 4.3 Impact of daily gaps on the amplitude spectrum of a signal. Sampling intervals are shown in the left column. The center column shows the amplitude spectrum of a single periodicity with the given duty cycle. The right column shows the amplitude spectrum of a signal with a second periodicity 12.5 μHz higher than the first, with an amplitude of 70%.

Figure 4.3. The top row of this figure shows the amplitude spectrum of a fixed amplitude sinusoid (with a period of 1000 s) in the center panel, adding a second sinusoid with a period of 1012.5 s in the third panel. The length of this simulation is 10^6 s, or about 11.6 days.

The amplitude of the side lobes depends on the duty cycle of the observation (that is, the ratio of the duration of observation to the duration of the gap). Figure 4.3 shows the growth of these sidelobes with worsening duty cycle. For a time series with regular gaps, these alias peaks (side lobes) decrease in amplitude as one moves away from the central peak. Clearly, if there are multiple true periodicities, side lobes of each might overlap, adding to the confusion.

A useful guide to interpreting these spectra is to sample a pure sinusoid of unit amplitude at the same time as the observations, and examine an FT of that time series. Thus the center column in Figure 4.3 represents this "spectral window" for the sampling shown in the first column. This spectral window is analogous to a one-dimensional point spread function for a single coherent periodicity.

Regular gaps such as those shown in Figure 4.3 are nearly unavoidable for ground–based observations because of the daily rotation of the Earth. One can try to overcome gaps in the time coverage by using a global network of observatories distributed in longitude (a favorite is the Whole Earth Telescope, Nather *et al.*, 1990). Such campaigns have had significant successes (Winget *et al.*, 1991), but are difficult to organize and resource (both human and instrumentally) intensive. Even with good longitude coverage, weather issues and inhomogeneities in instrumentation across a network can introduce some periodic modulation of the noise that leads to artifacts in the time-series analysis.

In principle, if one knows the spectral window precisely, one could try to deconvolve the signal from the aliased power spectrum in the frequency domain. In practice, though, this is a perilous path, as noise can conspire to thwart unique identification of aliases and real peaks. A much better way involves fitting, and then removing, established periodicities in the time (Zhao *et al.*, 2012c) domain, and redoing the analysis for the remaining periodicities that appear in the amplitude spectrum. A method to do this is discussed next.

4.4 From power spectra to frequencies, amplitudes, and phases

A set of periodicities (frequencies and amplitudes, along with phases perhaps), extracted from the time-series data, is what the modelers need to try to match with stellar models. Perhaps the most essential determination is the set of pulsation frequencies – they provide the primary constraint on stellar models. The amplitudes are of limited value to asteroseismic analysis, at least for classical pulsators. The phase can be very useful in coherent pulsators in determining the secular stability of the star and also to constrain the existence of binary companions (see below) which produce periodic variations in the position of the star that can be detected by varying light-travel time.

The FT is a good first step in determining the frequencies, amplitudes, and phases. However, we also require a quantitative estimate of the uncertainty in each of these parameters, and Fourier analysis can only provide an approximation for these uncertainties. Understanding the error budget, as well as the covariance between successive periodicities and for the parameters of each periodicity, requires other approaches. Least-squares fitting of the time series is mathematically equivalent to FT or Lomb–Scargle periodogram analysis, so such a fit provides consistent error estimates when the pulsations are coherent.

4.4.1 Least-squares fits to frequency, amplitude, and phase

For coherent oscillations, least-squares fitting to the individual periodicities is a powerful technique. For a star pulsating with a set of M independent periodicities, the differential light curve can be expressed in the following form:

$$y_n(t_n) = \sum_{j=1}^{M} a_j \cos(2\pi \nu_j t_n + \phi_j), \tag{4.2}$$

where the amplitude is a_j, frequency ν_j, and phase ϕ_j for each periodicity. The index j labels each periodicity – thus there are $3M$ free parameters for a signal with M periodicities. A word about phase is in order here. Clearly, this will depend on how one defines the start time of the observations. An alternative is to determine the time of first maximum after the start of the observation, $T_{\max,j}$ and replace the argument of the cosine in Equation 4.2 with $2\pi \nu_j(t_n - T_{\max,j})$.

Least-squares fits to the observed differential time series can provide estimates of these parameters; initial guesses at the values of the frequencies, amplitudes, and phases of each periodicity can be provided by the Fourier transform of the data. For isolated peaks on the power spectrum, this can be straightforward. However, as demonstrated in the left panel of Figure 4.3, this can be a challenge in gapped data with closely spaced periodicities.

Under these circumstances, a step-by-step, iterative process called pre-whitening is often successful. With some luck, the highest peak in the amplitude spectrum will represent a true periodicity in the star. One can fit that one peak via least squares, and remove the identified periodicity from the original data (in the time domain). The residual light curve will have all of the signal of the original, but with the dominant periodicity (and all of its aliases) filtered out. Repeating this iteratively until no peaks remain above the noise level in the amplitude spectrum of the residuals allows one to develop a list of periodicities that (hopefully) avoids diurnal aliases. One tool to facilitate this process is Period04 (Lenz and Breger, 2005).

Knowing when to stop pulling periodicities out of data is an important skill – mistakenly including noise peaks in a frequency list can send modelers on wild goose chases and cause them to question the credibility of observers. Least-squares analysis does provide a useful guide in that it can establish the mean noise level present in an FT. Experience has shown that for a typical asteroseismic data set, peaks that lie more than 4σ above the mean noise level in amplitude are highly unlikely to occur by chance. More precisely, as discussed by Breger *et al.* (1993) and Breger (1993), the noise in the range of the frequency is estimated in the absence of other periodicities, and a threshold is established at four times that mean value. A (somewhat self-serving) example of this process of successive pre-whitening to develop a frequency list can be found in Kawaler *et al.* (2010).

4.4.2 Solar-like oscillators: Estimating power, lifetimes

For solar-like pulsators that have very short coherence time, FT peaks are broadened as a result of the finite mode lifetimes and changing pulsation amplitudes. Clearly, the least-squares analysis that is effective for coherent pulsators is not available for stochastic pulsators, and the pre-whitening process described above is not appropriate. Other techniques to fit the envelope of power in the FT are needed. These do use the FT (or Lomb–Scargle periodogram) to determine the Fourier coefficients, allowing estimation of total power in the modes, mode lifetime, and other parameters.

In the frequency domain, with the expectation that areas of enhanced power represent stochastic modes, one can work in the frequency domain directly to determine the needed parameters. Fitting a Lorentzian profile to the envelope of extra power requires knowing the central frequency, maximum amplitude, and the width. The width is a measure of the coherence/damping time, and can be considered a free parameter. Optimizing these parameters in fitting the amplitude spectrum is discussed, for example, by Handberg and Campante (2011), Appourchaux (2014), and also Chapter 5 on solar-like oscillations in this volume.

4.5 Variations on a theme

4.5.1 Fine structure

As will be shown in other chapters, many of the signatures of non-radial pulsation in stars are manifest in the FT as "fine structure" in the distribution of peaks. Processes that depart from spherical symmetry and lift a degeneracy in mode frequencies and, in a way exactly analogous to Zeeman splitting of atomic energy levels, change a single peak into multiple peaks representing differing values of the quantum number m.

In the case of rotational splitting, the frequency difference between modes is of the same order as the rotation frequency of the star. As per the discussion above, to resolve these now separated modes, the duration of the observation must exceed the rotation period for those modes to be fully resolved. Put another way, beating between the multiplet components will generate a light curve with a modulation in amplitude at the splitting frequency. So to fully resolve the beating pattern, the observations must sample significantly more than a full beat cycle.

In ground-based data the time series will be periodically interrupted, so that distinguishing beating between closely spaced periodicities and diurnal aliasing can be significantly challenging for typical stellar rotation periods (from 1 day to 1 month or so). By retaining phase information, one can often distinguish between aliasing and rotational modulation.

4.5.2 Orbital modulation of the phase

With phase information, additional information about the mode physics is available, though the theoretical framework for exploiting the relative phases of modes is not clear at this time. While not strictly an asteroseismology application, phase information can be useful in probing for companions to the pulsating star if the phase of the oscillation itself is stable. Recall that phase can be expressed as a time of first maximum $T_{\max}$ for a given mode. If the star itself is accelerating with respect to the observer, then the change in light-travel time between the star and observer will result in a change in the derived $T_{\max}$, using the best period over that interval. Plotting T_{max} as a function of time can then provide a measure of accelerated motion of the star. As should be clear, if the star is moving at a constant velocity, the result would be a Doppler shift of the observed frequency with $\Delta\nu/\nu = v_{rad}/c$ which, for normal radial velocities, would be a tiny number that would not be distinguishable from a slightly modified pulsation period. If accelerating, though, then the frequency shift changes with time, which would generate a phase shift with time if the pulsation frequency is considered constant.

More conventionally, this kind of phase analysis can be done by employing a so-called $(O - C)$ diagram. This diagram is a plot of the (O)bserved T_{max} for a given interval of time compared to the (C)alculated value based on a firm ephemeris. If the star is not accelerating with respect to us, then $(O - C)$ will remain constant (at zero) as long as the period is correctly determined. If there is any acceleration in the motion of the star, however, then $(O - C)$ will show curvature with time. If the acceleration is caused by a body orbiting the star, then the $(O - C)$ changes will be cyclical with a period equal to the orbital period, and an amplitude given by the (projected) light-travel time across the orbit. The $(O - C)$ excursions for a star in a circular orbit with a companion are given by

$$(O - C) = 0.5\ \mathrm{s}\left(\frac{a}{1\ \mathrm{a.u.}}\right)\left(\frac{M_*}{M_\odot}\right)^{-1}\left(\frac{M_2}{M_{\mathrm{Jup}}}\right). \tag{4.3}$$

In our Solar System, Jupiter induces a 2.5 second signal over 12 years. More massive companions, closer to the host star, can produce a signal that is easily distinguished. This is analogous to the Røemer effect, which has been demonstrated in shorter-period binary star systems with an invisible component (see, for example, Telting *et al.*, 2012, in particular their Section 5). Similar analysis revealed a Jupiter-mass companion with a period of 3.2 years orbiting the pulsating sdB star HS 2201 (V391 Peg) (Silvotti *et al.*, 2007).

4.5.3 Secular changes from stellar evolution

A pulsation period can be be measured with exquisite accuracy. Clearly, a measurement of the period is not a measurement of the time between one maximum and the next. A short-period pulsator can undergo millions of cycles in just a few years, enabling determination of the period at much better than the parts-per-million level. As long as the number of cycles in a given time interval is known, the leverage of cycle count on the measurement provides this high precision.

Using the initial time of maximum in a time series following the procedure above, one can compute the successive times of maxima. The precision of the period projects to accuracy in the prediction of subsequent times of maximum. The uncertainty in the period is $\sigma_\Pi = \sigma_\nu/\nu^2$. The accuracy in the computed C is $\sigma_\Pi T/\Pi$ where T is the duration of the time series. The value of $(O - C)$ at some time T, the "local" time of maximum compared to that expected from the (fixed) pulsation period Π, should remain zero, within the uncertainties, if the period estimate is accurate. If $|(O - C)|$ is found to be significantly larger than $\sigma_\Pi T/\Pi$, the culprit could be an error in the assumed period, or changes in the period itself. This can be expanded around the estimated period as shown in Equation 4.4 (note that extra terms could represent orbital effects described above).

$$(O - C) = \Delta\Pi \, \frac{T}{\Pi} + \frac{1}{2}\frac{1}{\Pi}\frac{d\Pi}{dt}\, T^2 + \cdots \tag{4.4}$$

An error in the period ($\Delta\Pi$) produces an advancing error in the estimated time of a given maximum in the light curve, and so would show up as a linear trend in the $(O - C)$ diagram. Removing the linear trend can provide a more accurate estimate of the period. Care must be taken in that if the $(O - C)$ trend produces values in excess of the pulsation period, there is a wraparound effect – another form of aliasing resulting from, in essence, a miscount in the number of cycles that have elapsed over the course of a data set of length T. In practice, extending an $(O - C)$ analysis in time needs to be done with care so that the accumulated uncertainty in $T_{\max}$ is smaller than the pulsation period.

Pulsation periods measure the global properties of the stars. Since stars evolve with time through nuclear transformation in the core and gradual leakage of energy to space, these global properties (radius, density, mean temperature, etc.) also change. The time scale for stellar evolution, therefore, drives slow secular changes in the observing period on similar time scales. If the period is changing with time, that would produce *curvature* in the $(O - C)$ diagram, with a constant rate of period change producing a quadratic. A steadily increasing period causes an upward-pointing parabolic trend – the increasing cycle length results in maxima occurring at increasingly later times compared to the prediction based on a constant period. The departure of $(O - C)$ from zero increases "as time^2 goes by." Here too one must be careful, if the secular period change is relatively fast, to limit the expected change in $T_{\max}$ to be less than the pulsation period across gaps in observation.

As an example, the cooling pre-white dwarf star PG 1159-035 (GW Vir) has been observed for over 30 years, and was observed intensively from 1983 through 2002. This star evolves relatively quickly, with an e-folding time for luminosity or radius of about 10^6 years. Noting that the logarithmic time derivative of the period ($d \ln \Pi/dt = 1/\Pi \; d\Pi/dt$) is the inverse of the period change time scale, we can set this equal to the evolutionary time scale. For a pulsation period of 500 seconds, typical for a GW Vir star, $(O - C)$ is therefore $20\,\mathrm{s} \times T^2$ where T is measured in years. Therefore, measurement of a secular change in the period of a GW Vir star can be accomplished in just a few years if the precision of measuring $T_{\max}$ can be better than a few seconds over the course of a year.

The analysis by Costa and Kepler (2008) demonstrates that the prototype GW Vir indeed shows a secular period change that is approximately 10 times faster than expected. This is not entirely surprising, since the evolutionary period change represents a lower limit to the expected period change rate – any other process could only produce a larger period derivative.

An example where the evolutionary time scale and the time scale for period change are similar is the cooler DA white dwarf G117-B15A, for which over 30 years of data exist. Kepler *et al.* (2000) measure a change in the $(O - C)$ of 10 seconds over that time interval, and a period change time scale of 10^9 years. This is entirely consistent with models of cooling white dwarf stars.

4.6 Summary

In this chapter, I have briefly described the process of going from an observed time series to a list of parameters that feed into, and challenge, stellar models. The conditions under which the time series data are obtained can impact the derived parameters, so understanding how the properties of the data can mask signals, introduce spurious signals, or otherwise deceive is an important component in the asteroseismic process.

5

Global helioseismology and asteroseismology of solar-type stars

RACHEL HOWE AND WILLIAM J. CHAPLIN

5.1 Introduction

For almost all scientific uses of global helioseismic and solar-type asteroseismic data, from the lowest degree "Sun as a star" observations up to the 4096×4096 pixel images from HMI (Helioseismic and Magnetic Imager), it is necessary first to estimate the frequencies and other parameters of the modes. In global helioseismology this is typically achieved through non-linear maximum-likelihood estimation of small sections of the spectrum. In Section 5.2 we will discuss some important aspects of mode parameter estimation in global helioseismology, with particular reference to the algorithms used for data from GONG (Global Oscillation Network Group: Hill *et al.*, 1998) and MDI (Michelson Doppler Imager: Schou, 1992), on which most of the important results in resolved global helioseismology in the past couple of decades have been based. In Section 5.3 we turn to the special requirements for asteroseismology of Sun-like stars.

5.2 Mode parameter estimation in global helioseismology

Because of the stochastic nature of the excitation of the modes, the acoustic spectrum has the statistics of χ^2 with two degrees of freedom rather than Gaussian statistics, and so it is not appropriate simply to fit the model to the data by minimizing χ^2. Instead, the negative log-likelihood function is used, as given by Anderson *et al.* (1990):

$$S \equiv -\ln(L) = N \ln M + \frac{1}{M} \sum_{i}^{N} O_i, \tag{5.1}$$

where M is the model, O is the observed spectrum, and N is the number of data points (frequency bins) in the fitting window.

Formal uncertainties on the parameters are estimated using the inverse of the Hessian, or cross-covariance, matrix.

Extraterrestrial Seismology, eds. Vincent C. H. Tong and Rafael A. García. Published by Cambridge University Press.

5.2.1 Data preparation

Before the resonant peaks in the oscillation spectrum are fitted – a process colloquially referred to in the field as "peak bagging" – some preparation of the data may be needed. For a time series, this may include calibration (to convert a Doppler frequency shift to a velocity in meters per second), spike rejection, detrending to remove long-term drifts that would introduce low-frequency noise into the spectrum, interpolation to a uniform time cadence, zero-filling of missing data and filling of short gaps by interpolation. Time series may also be smoothed using a simple or a weighted running mean, but this does introduce some systematic errors, particularly in the line widths and in the separation of close peaks. As an alternative to smoothing, the GONG pipeline uses "multi-tapered" power spectra (Komm *et al.*, 1999).

The selection of the length of time series for analysis will depend on a number of factors. Longer series mean higher frequency resolution and better signal-to-noise (S/N) but risk missing short-term temporal variations; with shorter series, some modes or rotational components may be missed because they were not strongly excited in that interval. Historically, the GONG data have been analyzed in 108-day periods with start dates at 36-day intervals, while MDI and HMI have typically used non-overlapping 72-day periods. Analysis of the low-degree "Sun-as-a-star" data has been performed on a variety of different time series lengths, including sets lasting well over a decade that are crucial for the detection of very low-frequency modes and for elucidating fine detail of the oscillations spectrum (such as the frequency splittings due to rotation).

Finally, it should be noted that while many fitting algorithms operate on the power spectrum (in linear or logarithmic form), some use the full complex Fourier spectrum, making use of the phase as well as power information in the time series.

5.2.2 The model

In the most basic form, the model for a section of the acoustic spectrum will typically consist of one or more Lorentzian peaks whose frequencies ν_0, widths Γ (full width at half maximum, or FWHM, which is inversely related to the lifetime of the mode), and amplitudes A are to be determined. In interpreting the results of such fits, it may be important to distinguish between parameterizations that use the maximum height of the peak and those that use the area under the mode. The model will also include some parameterization of the background. For fits targeted on one or two peaks, the background model may be flat or linear. An asymmetry term (discussed in Section 5.2.5 below) may also be included.

The Sun's rotation breaks the symmetry between modes of the same degree l and different azimuthal order m, giving rise to so-called "rotational splitting" and turning peaks into multiplets. For low-degree spectra the rotational splitting may be included in the model; this is of the order of 0.8 μHz for $l = 1, m = \pm 1$ and is not fully resolved, apart

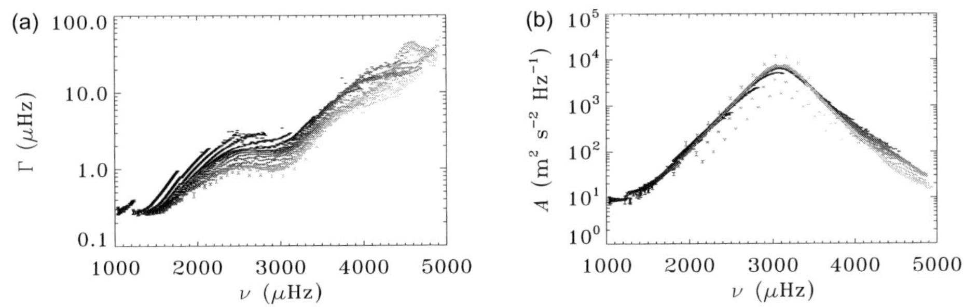

Figure 5.1 Width (a) and amplitude (b) for medium-degree mode fitting to GONG spectra, averaged over 17 years of data. Darker shades represent lower radial orders n. A black and white version of this figure will appear in some formats. For the color version please refer to the plate section.

from the lowest-frequency peaks, so it is very challenging to fit all the parameters of the split components individually. For medium- and high-degree spectra, the rotational splitting is handled by fitting a spectrum for each l and m; this is complicated by spherical harmonic "leakage" as discussed in Section 5.2.4 below, and the model will need to account for the leakage in some form. In the approach of Schou (1992), as used for the MDI medium-degree data, the whole of an l, n multiplet is fitted at once in the complex Fourier domain, with the frequency variation as a function of m treated as a series of orthogonal polynomials up to 6th, 18th or 36th order. In the GONG pipeline, by contrast, each spectrum is fitted independently and the polynomial expansion (up to about 14th order) is applied as a final step by fitting to the frequencies.

Finally, it may be necessary to include in the model the effects of daily modulation of the time series, such as occurs with single-site ground-based observations or in network observations when the duty cycle is low. This modulation manifests in the spectrum as "sidelobes" separated from the main peak by 1/day, or 11.57 μHz. The "window function" can be assessed by taking the Fourier transform of a pure sine wave that is set to zero where the data are missing; the observed spectrum is a convolution of the true spectrum with the window function. In the case of low-degree spectra, where the 1/day interval is close to the spacing of $l = 0, 2$ and $l = 1, 3$ pairs, the sidelobes cause serious difficulties with analysis (see, e.g., Chaplin *et al.*, 1997b). This can be addressed either by adding extra peaks to the model at the sidelobe positions, or (at much greater computational expense) by convolving the model with the window-function spectrum at each step of the optimization. Another common effect of a low duty cycle is that the line widths will be overestimated; this can be corrected in an empirical fashion, as was done for example by Komm *et al.* (2000).

Figure 5.1 shows the medium-degree width and amplitude estimates from the GONG pipeline, averaged over 17 years of observations; the amplitude peaks at about 3 mHz (or 5 minutes) while the width exhibits a "plateau" at a similar frequency but generally increases with frequency. Similar results have been shown for example by Libbrecht (1988).

5.2.3 Automation and rejection criteria

The transition from small, sporadic data sets to long series of systematic observations, particularly in medium-degree observations, required advances in the way in which mode parameters are extracted from the spectra. While it was practicable (although inefficient) to fit a low-degree spectrum "by hand," larger data sets require more automated procedures, and in particular an automated way of rejecting bad fits.

Hill *et al.* (1998) give some details of the approach used by the GONG team, including the automated criteria used to reject bad fits. Because the parameter space for a typical fit has many dimensions, the non-linear fit can appear to converge on a local rather than a global minimum of the quantity being minimized, so some further criteria need to be applied. For example, a fit is considered bad if the optimization routine reports that it failed to converge, but also if the mode frequency differs from the initial guess by more than 5 μHz or half the first-guess FWHM, the FWHM is more than twice or less than half the first-guess value, or the signal-to-noise is less than unity. In the tables provided by the project, bad fits and their error codes are listed, allowing the user discretion in which fits to accept. Even after these criteria have been applied, there are still fits from unresolved modes with severely underestimated errors. These multiplets can be excluded because they will fail at the polynomial-fitting stage, or simply by rejecting fits with $\delta\nu \leq 2\Gamma$.

5.2.4 Leakage

One of the pitfalls in mode parameter estimation, particularly in resolved-Sun helioseismology, is that the spherical harmonic time series, obtained by projecting a spherical harmonic mask on to the image of the solar disk, are not completely "pure," because we see only a fraction of the solar surface. Therefore, a time series for a given degree l and azimuthal order m will contain substantial contributions from modes with l' and m'. When these "leaked" modes are close in temporal frequency to the "target" modes, the fits can be contaminated, leading to systematic shifts or to fits that simply fail to converge, if the leakage is not properly accounted for in the model. Schou and Brown (1994) gave a detailed account of how to calculate the so-called "leakage matrix" that specifies how strongly each peak is present in the neighboring spectra; this is done by spatially convolving the spherical patterns of the target and leaked mode, including any apodization of the image and effects of centre-to-limb variation. The most important results to bear in mind are that (at least to first order) only leaks with $\delta l + \delta m$ even will appear, and that the strength of the leaks falls off with increasing δl and δm. Also, the "leakage" of a mode into its own spectrum determines the sensitivity of the observations to different spherical harmonics, which is lowest for the $m = 0$ mode and increases with $|m|/l$.

The detection of modes with $l > 0$ in Sun-as-a-star observations is a "benign" example of leakage. In general, leaks are problematic when the leaked and target peaks are not resolved. For an unconstrained fit to a stochastic spectrum, the two peaks must be separated by at least twice their FWHM for their frequencies to be successfully determined

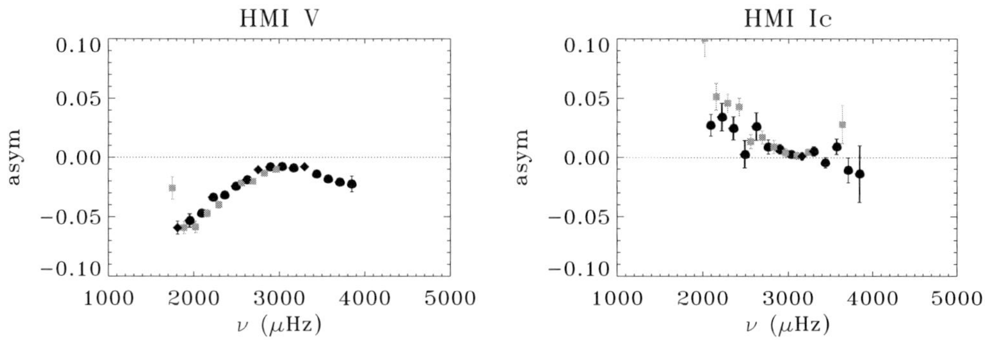

Figure 5.2 Dimensionless asymmetry parameter from fits to Doppler velocity (left) and continuum intensity (right) for Sun-as-a-star data from HMI. The dashed horizontal line indicates the zero level.

(Howe and Thompson, 1998). There are three main classes of leak affecting peak fitting. The l-leaks with $\delta n = 0$, $\delta l = \pm 1$, $\delta m = \pm 1$ particularly affect the higher-degree modes where the linewidth becomes comparable to the frequency separation between modes of adjacent l. These leaks make it impossible for a fitting algorithm that fits all peaks independently and neglects the leakage matrix, such as the one used by the GONG pipeline, to fit modes beyond about $l = 150$. The m-leaks, between spectra with $\delta n = 0$, $\delta l = 0$, $\delta m = \pm 2$, affect all of the modes and are not resolved across most of the p-mode spectrum, but they are most problematic where they are asymmetric, as happens when $|m|$ is close to l. As demonstrated by Howe and Hill (1998), this has a strong systematic effect on the estimates of the rotational splitting for low-degree modes. Finally, n-leaks occur where modes of different n but similar l coincide in frequency. Again, this occurs throughout the p-mode spectrum, but it has its worst effects for higher-frequency modes around $l = 20$.

5.2.5 Asymmetric peak fitting

The asymmetric nature of the peak profiles in helioseismic spectra was first observed by Duvall *et al.* (1993a), who explained it in terms of interference between inward- and outward-traveling waves in the atmosphere. A fuller explanation of the phenomenon, based on a study of MDI data, was proposed by Nigam *et al.* (1998); the observed pattern of asymmetry, with negative values for velocity and positive for intensity observables (see Figure 5.2), is believed to arise because of interference between the modes and the part of the noise background that arises from the solar granulation exciting the modes, and so is correlated with the modes. This contribution is dominant in the intensity spectrum and less important for velocity. Thus, the measurement of the asymmetry can give insight into the source of excitation of the modes. It also has a practical implication for the estimation of the other mode parameters, including the frequency. Fitting asymmetric peak profiles with a symmetric Lorentzian function will give rise to systematic errors, particularly where the mode amplitudes are weaker with respect to the granulation background and the asymmetry is therefore large. A physically accurate model of the profile should give the same value of the frequency for both velocity and intensity measurements, as the underlying oscillation cannot depend on the way in which it is observed. Nigam and Kosovichev (1998) proposed a parameterization of the asymmetric peak that is widely used, although not universally accepted:

$$P(x) = A\frac{(1 + Bx)^2 + B^2}{(1 + x^2)} + B_l, \tag{5.2}$$

where A is the amplitude, B is the asymmetry parameter, B_l is the background parameter, and $x = 2(\nu - \nu_0)/\Gamma$, Γ being the FWHM and ν_0 the central frequency. It should be noted that this approximation loses validity far from the central frequency, which can cause cross-talk between the asymmetry and background estimates. This can be mitigated by neglecting the quadratic term in B, as has been more commonly adopted, e.g., in fitting of Sun-as-a-star data (see Section 5.2.6 below).

Rabello-Soares *et al.* (1999) examined the effect of neglecting the peak asymmetry on global inversions for solar structure and concluded that this would not have significantly affected the results of prior inversions (Basu and Antia, 2000). In practice, asymmetric fitting has not been used in the standard analysis of medium-degree GONG and MDI data to date, but it has been used to good effect in the fitting of low-degree modes from Sun-as-a-star data, for example from the Birmingham Solar Oscillation Network (BiSON) and from the Global Oscillations at Low Frequencies (GOLF) and Variability of solar IRradiance and Gravity Oscillations (VIRGO) instruments onboard the Solar and Heliospheric Observatory (SOHO).

5.2.6 Fitting Sun-as-a-star data

Sun-as-a-star observations, like asteroseismic observations, are sensitive to the low-degree modes. It had been common practice to peak-bag Sun-as-a-star data a pair of modes at a time, i.e., by fitting together pairs of adjacent $l = 2$ and 0, or $l = 3$ and 1 modes. However, a series of detailed "hare and hounds" exercises performed with artificial data by the solarFLAG collaboration (Jiménez-Reyes *et al.*, 2008) showed that power from adjacent modes bleeding into mode-pair fitting windows (in frequency) is sufficient to give rise to small amounts of bias in the estimated frequencies of the target modes (since the power from the adjacent modes does not appear in the mode-pair fitting model). An obvious way to solve this problem is to fit all modes simultaneously, a "global" approach now commonly adopted in the analysis of asteroseismic data (as discussed in Section 5.3 below). This is computationally expensive, and the exquisite and subtle features that are observable in the solar data can present problems for achieving stable, converged fits of global models that include all the observable detail. A computationally more efficient alternative to global fitting – "pseudo-global" peak-bagging – was developed by Fletcher *et al.* (2009). The process starts from a global first-guess model for the spectrum, as per the fully global peak-bagging; but then the parameters are optimized one order at a time, with parameters describing the non-target orders held fixed. This iterative procedure typically demands two to three passes through the entire spectrum before adequate convergence is achieved, and is significantly faster (and potentially more stable) than the fully global approach.

Further work, including other hare-and-hounds studies in solarFLAG (Chaplin, 2006), has shown the importance of accurately modeling the visibility ratios of components of the non-radial modes (a leakage-matrix calculation by another name) in order to ensure accurate fitting of the frequency splittings due to rotation (and magnetic fields).

5.2.7 Fitting high-degree data

At higher degrees, and at higher frequencies, the peaks and their associated leaks blend into unresolved ridges, which cannot be fitted by treating the component peaks independently. Because of asymmetry in both the peaks and the leakage matrix, it is not sufficient simply to take the centroid of the ridge. Some progress has recently been made by Korzennik *et al.* (2013) on the analysis of a 90-day full-resolution data set from MDI. References in the article detail the long history of attempts to fit high-degree global modes, extending back as far as 1990. The approach involved a meticulous calculation of the leakage matrix, taking into account instrumental distortions, a heavy smoothing of the power spectrum using a high-order multitaper, and the use of forward modeling to recover the mode parameters from the fits to the position and shape of the ridge. They recovered frequencies for multiplets up to $l = 1000$, but still found imperfections in the match between the calculated leakage matrix and the observations.

5.2.8 Multispectral fitting

Multispectral fitting – the use of velocity and intensity information together – has the potential to improve both the estimation of mode frequencies and the understanding of how waves propagate in the solar atmosphere. To make the best use of this information, it is necessary to use not only the spectra of the individual observables, but their relative phases and cross-coherence functions across the spectrum.

The phase difference $\delta\phi_{A,B}$ between two time series A and B can be written as

$$\delta\phi_{AB}(\nu) = \arg\langle P_{AB}(\nu)\rangle, \tag{5.3}$$

where P_{AB} is the cross spectrum, $P_A(\nu) \cdot P_B^*(\nu)$. The coherence function, analogous to a frequency-dependent correlation coefficient, can be written in the form

$$\mathrm{COH}_{AB}(\nu) = \frac{|\langle P_{AB}(\nu)\rangle|}{\sqrt{\langle |P_A(\nu)|^2\rangle\langle |P_B(\nu)|^2\rangle}}; \tag{5.4}$$

see, for example, Barban *et al.* (2004b) and references therein. The angle brackets formally denote the expectation value; in practice this can be approximated by a mean over many short spectra, a smoothing by a running mean in the frequency domain, or, for resolved-Sun observations, by an average over all azimuthal orders m for a given degree l.

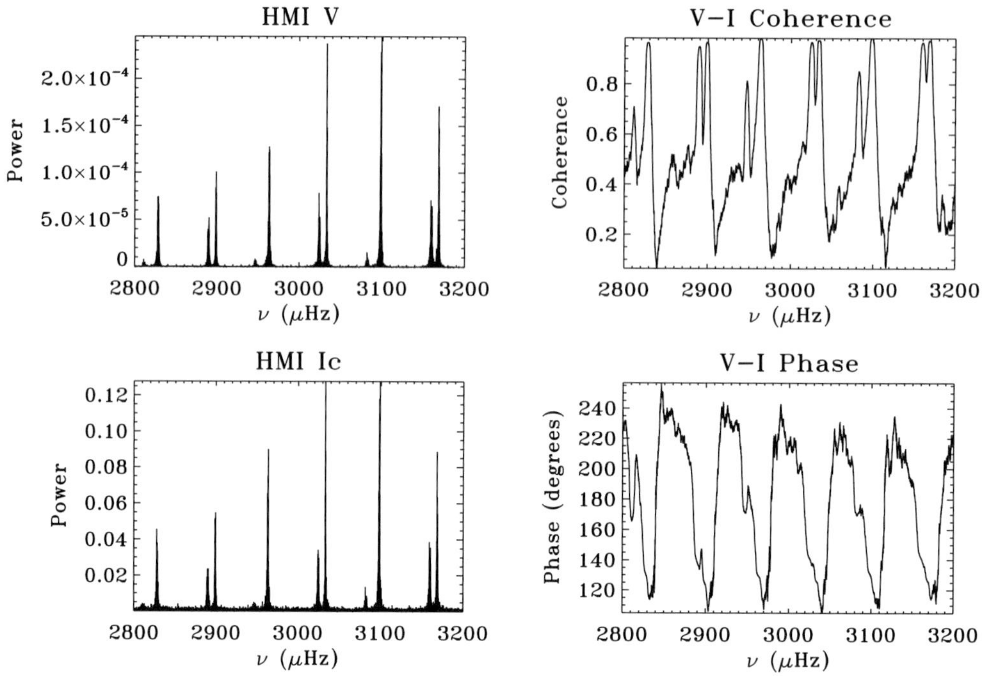

Figure 5.3 (Left) Sections of power spectra (arbitrary units) for HMI Doppler velocity (top) and continuum intensity (bottom). (Right) Coherence (top) and phase (bottom) for the two spectra.

For the case of Doppler velocity and continuum intensity, both phase and coherence vary across peaks in the five-minute spectrum, with a characteristic "saw-tooth" pattern in the phase. Examples of velocity and intensity spectra and their coherence and phase are shown in Figure 5.3 for HMI Doppler velocity and continuum intensity. A full multispectral fitting scheme needs to make use of the concept of correlated and uncorrelated noise backgrounds to model this behavior, as touched on above in Section 5.2.5. The use of such a scheme was demonstrated for GONG data by Barban *et al.* (2004b); however, until recently multi-observable data from the same instrument have only been available for relatively short time series. The advent of HMI, which produces both velocity and intensity data at all times, has made the routine use of multispectral fitting a more realistic possibility.

5.2.9 Future directions, and a note on practicalities

It may seem surprising that the analysis for the two major medium-degree helioseismic data sets from the last two decades, GONG and MDI, has depended on analysis codes that have barely been updated in that time and are far from reflecting the state of the art. Furthermore, these codes give slightly different results, as discussed for example by Schou *et al.* (2002). This situation becomes more understandable when we consider the need for homogeneous data sets for the study of long-term variations and the substantial computational expense involved in reprocessing large numbers of data sets, even once the modern improvements in fitting methods had been implemented in a sufficiently robust and automated form. An improved fitting method for all medium-degree data has been under development by Korzennik (2008), but is not yet a "production" code. Meanwhile, some improvements have been made to the code that was used for the MDI medium-degree fitting, and this is now being incorporated in more automated form into the HMI data analysis system (Larson and Schou, 2011). Other improvements to mode parameter estimation in general, such as the use of multispectral data, remain a matter of research.

5.3 Analysis of oscillation spectra of solar-type stars

Observational asteroseismology of solar-type stars has been revolutionized by data of unprecedented quality and scope, e.g., from the NASA *Kepler* and French-led Convection, Rotation and planetary Transits (CoRoT) satellites (e.g., see Chaplin and Miglio, 2013, and references therein), and from a new generation of spectrographs at large telescopes (e.g., see Bedding, 2014). Thanks to *Kepler*, exquisite quality oscillations data are now available on hundreds of solar-type stars (Chaplin *et al.*, 2011a). Over 100 stars have asteroseismic data spanning more than one year, with around 80 having more than three years of coverage. In the pre-*Kepler* era such high-quality, long-duration datasets were available on only one star, the Sun. The wealth of new data will take many years to fully mine and exploit, meaning that for now asteroseismology of solar-type stars remains an observationally driven field.

5.3.1 The oscillations spectra of main-sequence and solar-type stars

Low-mass, main-sequence stars show rich spectra of solar-like oscillations, acoustic (pressure, p) modes that are excited and damped intrinsically by convection in the outer parts of the star. The top left-hand panel of Figure 5.4 shows the frequency–power spectrum of the *Kepler* lightcurve of the G-type main-sequence star 16 Cyg A (HD 186408), the more massive component of the Sun-like binary system 16 Cyg. In addition to estimates of the parameters of the individual modes in the observed spectrum, data on global or average seismic parameters are important diagnostics of fundamental stellar properties, in particular when low S/N ratios prevent a more detailed analysis.

Just as for the Sun, the observed powers of the acoustic modes are modulated in frequency by an envelope that has a Gaussian-like appearance. The frequency at which the observed power is strongest, $\nu_{\rm max}$, carries information on the near-surface layers, and scales to good approximation as $g T_{\rm eff}^{-1/2}$ (Brown *et al.*, 1991; Kjeldsen and Bedding, 1995; Belkacem *et al.*, 2011), where g is the surface gravity and $T_{\rm eff}$ is the effective temperature of the star. The envelope tends to be flatter and broader in hotter F-type stars (e.g., Arentoft *et al.*, 2008; White *et al.*, 2012). Another important average seismic parameter is the average large-frequency separation, an average of the large separations, $\Delta\nu$, between consecutive

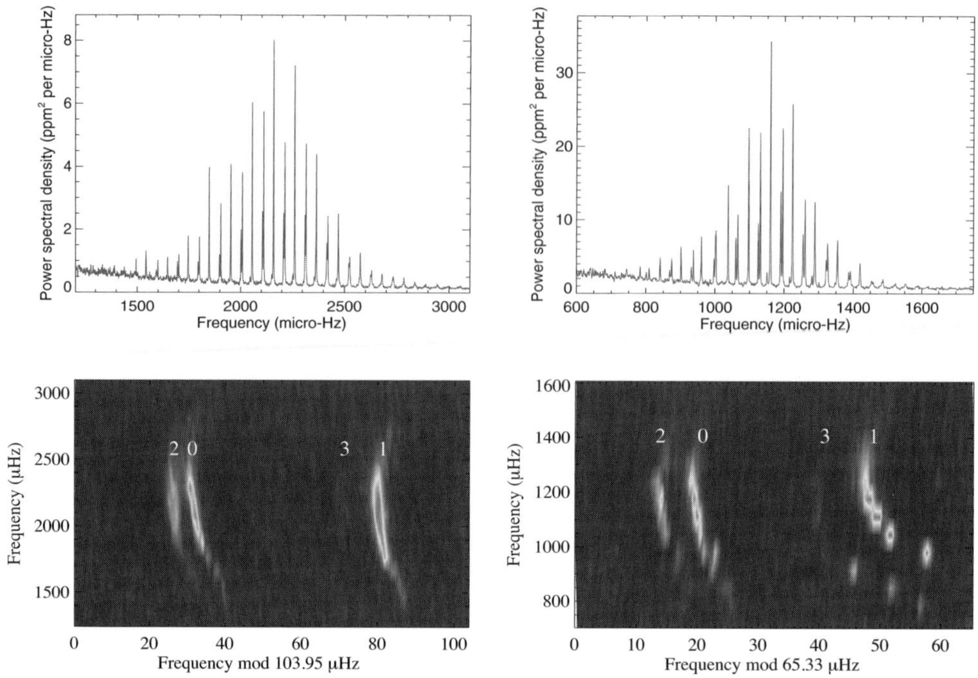

Figure 5.4 Frequency–power spectra (top) and échelle diagrams (bottom) for the G-type main-sequence star 16 Cyg A (HD 186408) (left-hand panels), and the G-type subgiant KIC 6442183 (HD 183159) (right-hand panels). Numbers on the échelle diagrams indicate where the putative ridge of each angular degree lies. A black and white version of this figure will appear in some formats. For the color version please refer to the plate section.

overtones n of the same spherical angular degree, l. The average large separation scales to very good approximation as $\rho^{1/2}$, $\rho \propto M/R^3$ being the mean density of a star having mass M and surface radius R (e.g., see Ulrich, 1986; Christensen-Dalsgaard, 1993).

The very regular frequency structure of the oscillation spectra of main-sequence stars (including the Sun) may also be shown in an échelle diagram, like the one in the lower left-hand panel of Figure 5.4. It was made by dividing the frequency–power spectrum of 16 Cyg A into frequency segments of length equal to the average $\Delta\nu$. The segments were then arranged vertically, in order of ascending frequency. The diagram shows clear "ridges," comprising overtones of each degree l, as well as the dominant large frequency separations (i.e., the spacings between ridges). The small frequency separations $\delta\nu_{02}$ between adjacent $l = 0$ and $l = 2$ modes are also apparent; as are the separations $\delta\nu_{13}$ between adjacent $l = 1$ and $l = 3$, albeit somewhat less conspicuously due to the weak observed amplitudes of the $l = 3$ modes. These small frequency separations depend on the gradient of the sound speed in the deep interiors of main-sequence stars and hence provide a sensitive diagnostic of the evolutionary state (for a given $\Delta\nu$ and chemical composition).

Once stars leave the main sequence and enter the subgiant phase – following the completion of core hydrogen burning – they begin to show detectable signatures of gravity (g) modes, where the effects of buoyancy provide the restoring force. The behavior depends on the buoyancy or Brunt–Väisälä frequency, the highest frequency at which g modes may be sustained. The buoyancy frequency in the deep stellar interior increases during the post-main-sequence phase, eventually extending into the frequency range of the high-order p modes. When the frequency of a g mode comes close to that of a non-radial p mode of the same degree l, the modes undergo an "avoided crossing" (Aizenman *et al.*, 1977). Interactions between the modes not only "bump" the observed frequencies but also change the intrinsic character of the modes, so that they take on mixed p- and g-mode characteristics (having g-mode-like behavior in the deep interior, and p-mode behavior in the envelope).

The frequency–power spectrum and échelle diagram of one such *Kepler* target, the G-type subgiant KIC 6442183 (HD 183159), are shown in the right-hand panels of Figure 5.4. Regarding the échelle plot, the $l = 0$ ridge is unaffected, because perturbations in buoyancy will not support radial modes. Some of the $l = 1$ overtones have been significantly shifted from the putative, undisturbed $l = 1$ ridge (including one mode that has been shifted close to the $l = 0$ ridge). The larger the shift in frequency from the undisturbed ridge, the stronger is the interaction between the p- and g-modes. The coupling is much weaker at $l = 2$. Whilst providing strong constraints on the internal structure, these departures from the near-regular structure shown by pure p modes pose additional challenges for analysis of the oscillations spectrum.

5.3.2 Estimation of oscillation parameters

The long datasets available from *Kepler* and CoRoT give the S/N and, crucially, the frequency resolution needed to extract accurate and precise estimates of the basic parameters of individual modes spanning several overtones about $\nu_{\max}$, e.g., frequencies, frequency splittings, amplitudes, and damping rates. Uninterrupted data offer huge advantages for the analysis, allowing clean extraction of frequency spacings that might otherwise coincide with diurnal artifacts given by ground-based observations with incomplete time coverage. Ground-based Doppler velocity observations do however provide one crucial advantage over the space-based photometric data: They are much less sensitive to signatures of convection at the surfaces of stars (so-called granulation), meaning it is possible to observe more of the weakest-amplitude modes, be they very low-frequency modes or modes of higher degrees l. The former are important because they are more lightly damped than their higher-frequency counterparts; the resonant peaks are hence narrow in frequency, making it possible to extract extremely accurate and exquisitely precise frequencies and rotational frequency splittings for making inference on the internal structure and dynamics. The latter are important because robust data on a wider range in l allow for a more sensitive discrimination of the internal structure.

Even when the quality of the observations is insufficient to allow estimation of individual mode parameters, it is often still possible to extract robust estimates of the global or average seismic parameters. Any analysis must seek first to establish the presence of significant power due to solar-like oscillations. Two approaches have been adopted (see Chaplin and Miglio, 2013, and references therein). In the first (and most common) approach, detection of signatures due to the near-regular spacing of the overtones (i.e., the large frequency separations) is sought. This may be readily accomplished via computation of the power spectrum of the power spectrum (formally equivalent to the autocorrelation of the time-domain data), which will yield a strong signal at $\sim\Delta\nu/2$, and multiples thereof; or approaches based on the use of the autocorrelation of the power spectrum, which yields a more complex structure but one from which it is much easier to extract estimates of the average small frequency separations. In the second approach, detection of a significant envelope of excess power, above the slowly varying power spectral density due to the stellar granulation and the "shot noise" (due to the finite number of photons detected), is sought. This demands accurate modeling of the background power.

These codes – many of which were tested extensively by the asteroFLAG collaboration – may be run in a blind, automated manner, scanning different ranges of the frequency–power spectrum to uncover the sought-for signatures. The application of statistical testing, and suitable a-priori information, may be used to establish the bona fides of any claimed detections. False-alarm and Bayesian odds-ratio tests are commonly applied (the latter involving the adoption of priors on the expected mode signal; see Appourchaux, 2014). Knowledge that the average large frequency separation $\Delta\nu$ shows a fairly tight power-law dependence on $\nu_{\rm max}$ is often used, meaning that if a candidate signal is extracted from a particular range in frequency, the estimated $\Delta\nu$ should be expected to lie within a certain range. Another example relates to the power search tests: Here, knowledge that the total observed power in the acoustic oscillation spectrum varies with $\nu_{\rm max}$ makes it possible to test whether the observed candidate mode power is significantly higher than expectation (suggesting an artifact). Clearly, any a-priori constraints should not be too strong, so as to avoid throwing away real, but unexpected, detections.

Peak-bagging may be attempted to extract estimates of individual mode parameters once the S/N and frequency resolution allows for sufficient discrimination of the overtone structure (e.g., see Appourchaux, 2014). The accumulated experience from Sun-as-a-star analysis has naturally fed into the development of peak-bagging for the asteroseismic analysis of solar-type stars. As per the helioseismic mode fitting, each mode component is usually modeled as a Lorentzian-like function. We note that application of asymmetric Lorentzian fitting functions is yet to be tested extensively on *Kepler* or CoRoT data. The ratio of the length of the time series T to the amplitude e-folding lifetime (i.e., the time for the amplitude to decay by the factor e) τ must satisfy $T/\tau \geq 10$ (or thereabouts) in order to resolve the mode profile. When $T/\tau \leq 2$, the observed profile tends to a sinc-squared function and a sine-wave fitting approach to extract the frequency and amplitude should be used. Typical peak widths $\Gamma = 1/(\pi\tau)$ in main-sequence stars are of the order of one to a few μHz. Observations spanning approximately one month are just adequate to resolve the peaks, although longer sets are usually needed to give robust (unbiased) estimates of the mode powers and linewidths. Datasets of a year or more are usually needed to extract robust estimates of frequency splittings from rotation or asymmetries of those splittings, due for example to the near-surface magnetic fields (total frequency splittings between adjacent non-radial components range typically from a fraction to several μHz in size).

A global approach to peak-bagging is usually applied to solar-type stars, wherein the parameters describing every observable mode component in the spectrum are fitted simultaneously. Stability of the fitting is of course demanded, given the large number of free parameters (which can be as high as 100), and hence regularization – e.g., demanding that the best-fitting mode powers be a smooth function of frequency – or simplifying constraints – e.g., fitting a single rotational splitting parameter to every non-radial mode – may be applied. The strength of those constraints will depend not only on the quality of the observed oscillations spectrum, but also the intrinsic characteristics of the star's seismic parameters. The slowly varying background due to the stellar granulation must also be fitted, and this is often done (and its best-fitting parameters fixed) prior to optimization of the oscillation parameters. Various functions have been used to describe the granulation, all variants on the form expected for a stochastic process having a characteristic time scale or memory. The computationally more efficient pseudo-global peak-bagging (see Section 5.2.6 above) is now also being applied to asteroseismic data, and can help to solve some of the above-mentioned stability problems.

The helioseismic heritage of the peak-bagging meant that maximum likelihood estimators were first utilized to obtain best-fitting parameters; however, Markov Chain Monte Carlo (MCMC) methods are now being used more extensively in the field (Gruberbauer *et al.*, 2009; Benomar *et al.*, 2009; Handberg and Campante, 2011). MCMC offers robust estimation of the posterior distributions of each parameter, which can be extremely important when those distributions are decidedly non-Gaussian (as may often be the case when the S/N for fitting is marginal). It is already possible to near fully automate the process of peak-bagging main-sequence stars (see Appourchaux *et al.*, 2012a, and references therein). However, more evolved solar-type stars, showing mixed modes – notably mixed modes with strong coupling and hence strong g-mode characteristics – present additional challenges. The regular frequency patterns are disrupted in the non-radial overtones, and this may shift some non-radial modes very close to the radial-mode ridge (as shown in Figure 5.4); without prior information, the fitting can very easily transpose the modes. In addition, mixed modes show smaller linewidths (less damping) than acoustic modes at the same frequency, with the disparity being a function of the strength of the p–g-mode coupling. This can create further complications, in particular if strong coupling means mixed modes are no longer resolved in frequency (which then leads to a reduction also of the maximum power spectral density, or height, of the peaks).

Coordinated efforts are now on-going to tackle the complicated problems presented by peak-bagging of red-giant stars, the ultimate goal being to automate the process so that the full potential of the oscillation spectra can be utilized on the large numbers of targets observed by *Kepler* and CoRoT. Looking ahead, there will be fresh challenges for analyzing data from the "second-light" *Kepler* Mission K2, and in the future data from the NASA TESS (Transiting Exoplanet Survey Satellite) and the ESA PLATO (PLAnetary Transits and Oscillations of stars) missions.

6

Measurements with local helioseismology

DOUGLAS C. BRAUN

6.1 Introduction

Local helioseismology encompasses remote observations, data analysis, and theoretical modeling of solar oscillations to infer the three-dimensional structure within localized regions of the solar interior. What defines a region as "local" is relative, however, since targets of interest have included sunspots and convective elements with spatial scales $\sim 10^{-2} R_{\odot}$ as well as large-scale plasma flows spanning much of a solar hemisphere. As a relatively new discipline first explored in the 1980s, local helioseismology has two main components: first, a research component to understand the interaction of solar oscillations (acoustic and surface gravity) with perturbations within the Sun and, second, the design and application of methods to infer the properties of the perturbations by modeling the measurements of those waves. Successful applications require a thorough understanding of the physics of the waves and their interaction with inhomogeneities inside the Sun. The research component is particularly critical. For example, the types of perturbations found in the Sun can include magnetic fields for which the wave interactions can be quite complicated. Currently, the types of structures most amenable to modeling using local helioseismic measurements consist of isotropic wave-speed perturbations and the three components of plasma flows. Assessing the subsurface magnetic field directly is a challenging, but largely unrealized, goal of the field. While the status of the field is evolving, the determination of plasma flows in the first few tens of Mm below the solar surface remains one of the primary practical applications.

We outline in this chapter the practical applications of, and resulting measurements made with, common local helioseismic methods. Broadly speaking, local helioseismology can be roughly divided into Fourier methods (which operate in the frequency–wavenumber domain) and cross-covariance based methods (which operate in the space–time domain). The former (Section 6.3) can be considered in many ways as extensions of the analysis of global oscillations (Chapter 5) to localized regions of the Sun. Another class of methods employs cross-covariance functions (Section 6.4) made between spatially separate signals

Extraterrestrial Seismology, eds. Vincent C. H. Tong and Rafael A. García. Published by Cambridge University Press.

obtained by sampling, spatial averaging, or other means. These measurements are sensitive to the temporal and spatial variations of the target wave-speed perturbations and flows. For each type of method we describe the *data products* obtained through its application to remote solar observations. These products provide modelers with the raw material for inferring subsurface flows and other physical perturbations.

6.2 Geometry, projections, and tracking

For many applications, local helioseismology is employed to probe solar regions spanning less than 200 Mm horizontally and extending less than a few tens of Mm in depth. In these smaller domains one can ignore the curvature of the Sun and assume that the background medium varies only with depth z from the surface. This focus on small regions also implies that one generally considers and observes solar oscillations with smaller wavelengths (or higher degree in a spherical harmonic representation) than what is utilized in global helioseismology. In this *plane parallel approximation*, these high-degree oscillations take the form of plane waves which have a horizontal propagation described by $\exp(i[\mathbf{k_h} \cdot \mathbf{r} + \omega t])$, where $\mathbf{r}$ is the horizontal position vector, $\mathbf{k_h}$ is a wavevector consisting of the two horizontal components (k_x, k_y) of the wavevector, t is time, and ω is the temporal frequency. As described in Chapter 5 the stratification of the sound speed with depth gives rise to vertical standing waves, which can be characterized by discrete values of the radial order (the number of wave nodes present in the vertical dimension). Sets of modes with the same radial order occupy parabolic-like *ridges* in the (k_h, ω) domain, where $k_h^2 = k_x^2 + k_y^2$. However, because of wave damping, which increases for waves of shorter wavelength, the horizontal wavenumbers k_h of the ensemble of oscillations form a continuous set, rather than take on discrete quantum values as for global modes.

Helioseismic operations and measurements are performed on remote observations of the surface wave field ψ. The highest quality and most frequently employed observations of the wave field used in local helioseismology consist of time series, spanning hours or days, of *Dopplergrams*. These are maps of the line-of-sight velocity of the solar surface obtained from telescopes either operating in space or as part of ground-based networks (see Chapter 2). Starting with observations that might span the entire observable hemisphere, a region is selected for analysis and tracked across the solar disk to remove most of the effects of solar rotation. This tracking is achieved by projecting a subset of the original image, using two-dimensional spatial interpolation, onto a local coordinate system centered on the region of interest. A Postel's, or azimuthal equidistant projection, is a common choice but is not exclusive in practice (e.g., Zaatri *et al.*, 2008). Typically, the tangent point of the Postel's projection (or other reference point) is placed at a point on the solar surface that moves with the solar rotation. Sometimes this tracking uses the Carrington rotation rate, for which the tangent point is at a fixed Carrington coordinate. But tracking rates that account for the latitudinal variation of the solar rotation may also be employed. Regardless of the rotation rate selected, it is to be understood that subsurface flows inferred from helioseismic analysis are only determined relative to this tracked rate. A three-dimensional

(3D) datacube results from the time series of projected images and forms the basis for many subsequent local helioseismic analyses.

6.3 Fourier-based methods

Spectral decomposition, through the application of Fourier transforms to the observed wave field, forms the basis of several local helioseismic methods. *Hankel analysis* and *ring-diagram analysis* use primarily transforms applied in polar and Cartesian coordinates, respectively. Both methods represent the earliest applications of local helioseismic procedures and were developed to study specific solar phenomena, consisting of the scattering properties of sunspots in the former (Braun *et al.*, 1987) and convective flows in the latter (Hill, 1988).

Because the wave sources on the Sun (believed to be caused by the turbulence of small-scale convection near the surface) are widely and randomly distributed, the propagation of the resulting waves is isotropic. Thus, for a given radial order (or ridge) the power for a given wavevector k_h is distributed uniformly in all directions. If a 3D Fourier transform (in two spatial dimensions and time) is performed of a tracked datacube, the amplitude (or power) spectrum shows a set of trumpet-like structures (see Figure 6.1). These concentric "trumpets" represent the extension into the 3D Fourier domain of the f- and p-mode ridges.

6.3.1 Ring-diagram analysis

Ring-diagram analysis (Hill, 1988) provides a good starting point for discussing Fourier-based methods of local helioseismology. At a constant temporal frequency ω a slice through the power spectra exhibits concentric "rings" of mode power. The measurement and analysis of these spectra, collectively known as ring-diagram analysis, seeks to determine and model the distortion of the rings due to flows and sound-speed perturbations beneath the local patch outlined by the tracked datacube.

The most common application of ring-diagram analysis has been the inference of sub-surface flows. A uniform flow $\mathbf{U}$ with horizontal components U_x and U_y relative to the tracking rate will produce a Doppler shift of the wave frequencies, at constant $\mathbf{k_h}$, by an amount $\delta\omega = \mathbf{k_h} \cdot \mathbf{U} = k_x U_x + k_y U_y$. Figure 6.1 shows an example of how the rings are distorted in (k_x, k_y) space due to a horizontal flow. A more general case is when the flow varies with depth. In this case, the frequency shift is proportional to a weighted average over depth of the horizontal component of flow. The weighting is described by a *kernel function* (see Chapter 9). A uniform change (δc) to the background sound speed beneath the area of interest will expand or contract the rings independently of direction, i.e., $\delta\omega = k_h \delta c$.

Data products

It is the goal of ring-diagram analysis to determine, through fitting a parameterized model to the power spectra, the values of the flow parameters U_x and U_y as well as the azimuthally

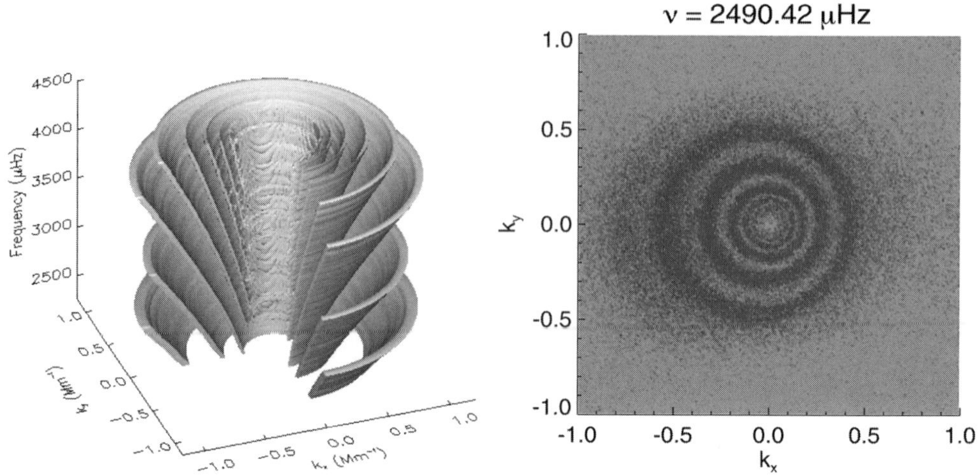

Figure 6.1 (Left) A nested trumpet-like model of how the power of solar oscillations is concentrated in Fourier space. The frequency axis points upward, and the two other axes are the horizontal components of the spatial wavenumber. (From Patrón *et al.*, 1997, and reproduced with the permission of the AAS.) (Right) A slice of the 3D power spectra of actual solar observations at constant frequency showing the distortion and displacement of the rings due to a horizontal flow, in this case, due to solar rotation. (Figure courtesy of M. Roth.)

averaged frequency ω_0. A variety of models and fitting procedures have been proposed. These include directly fitting the power in the rings at slices of constant frequency (Hill, 1988; Basu and Antia, 1999). Another approach is to fit the power in cylindrical slices defined by constant horizontal wavenumber k_h (e.g., Schou and Bogart, 1998). Other parameters in the models typically include mode amplitudes and a characterization of a background term, contributed by solar convective motions and noise. Typically, Lorentzian functions are used to characterize the resonant peaks of the p-mode ridge. Asymmetries in these peaks (see Chapter 5) can be included as well (Basu and Antia, 1999).

In principle, the offset of the mean frequency ω_0 from a theoretical value computed from a solar model can help infer sound-speed perturbations. In practice, the desire to minimize systematic effects makes it advantageous to consider relative frequency shifts between a target region and one or more *reference* regions whose power spectra have been subject to identical parameterized fits.

The selection of the size of the analysis region largely determines the spatial resolution. Specifically, the horizontal resolution is essentially the diameter (sometimes called the *tile* size) of a circular mask which is typically applied to the datacube (Birch *et al.*, 2007). The ability to resolve smaller scale flows improves with smaller tile sizes, but this happens at the cost of the ability to resolve individual p-mode ridges in the wavenumber domain, particularly at high k_h. Since the ability to infer deeper flows requires flow parameters of ridges with high wavenumber (or radial order), there is a general trade-off between horizontal resolution and depth sensitivity. Recently, progress has been made in modeling sets of tiles with multiple sizes to optimize both quantities (Featherstone *et al.*, 2011). Historically, most ring-diagram analyses have been performed with tiles spanning about 15 degrees or smaller and which employ the plane-parallel approximation. Applications (see Chapter 13) have included the probing of a wide variety of flows in the Sun, including differential rotation, meridional (north–south) circulation, convective flows, and flows associated with sunspot groups (also known as *active regions*).

6.3.2 Fourier–Hankel and Legendre decomposition

Early, and still widely studied, targets of local helioseismology include sunspots, which provide obvious localized imhomogeneities on the solar surface and have been observed visually for centuries. Arguably, the roots of local helioseismology can be found in observations and interpretations of waves inside sunspots and the accompanying theoretical analysis of wave propagation in magnetic fields (see, for example, Thomas and Weiss, 2012).

The conceptual basis for Fourier–Hankel methods (Braun *et al.*, 1987) of local helioseismology is the assessment of waves impinging upon, and scattering from, localized perturbations such as a sunspot. A cylindrical coordinate system is chosen, with the center located at the target of interest. In these (r, ϕ) coordinates, the wave equation has general solutions of the form

$$\psi(r, \phi, t) = e^{i(m\phi+\omega t)}[A_m(k_h, \omega)H_m^{(1)}(k_h r) + B_m(k_h, \omega)H_m^{(2)}(k_h r)], \tag{6.1}$$

where t is time, m is the azimuthal order, and ω is the temporal frequency. A_m and B_m are the amplitudes of wave modes that have a variation in r described by Hankel functions of the first and second kind and correspond to modes propagating towards and away from the origin, respectively. Applied to observations of the solar wave field, a straightforward modification of the Hankel transform (Olver *et al.*, 2010), which substitutes the functions $H_m^{(1,2)}$ for Bessel functions, is used to evaluate the amplitudes A_m and B_m over the (m, k_h, ω) domain. Figure 6.2 shows sample power spectra $|A_m|^2$ and $|B_m|^2$ for a set of observations centered on a sunspot. One of the first results of this application was the discovery of absorption of incoming waves by sunspots (Braun *et al.*, 1987).

Some considerations on selecting the annulus size for a given temporal duration, and other details, may be found elsewhere (Braun, 1995). Analyses over large annuli, for which the curvature of the Sun cannot be neglected, can be performed with basis functions consisting of Legendre functions of the first and second kind (Olver *et al.*, 2010). Obviously these methods are most effective for certain types of perturbations, typically those involving compact and isolated scatterers and which have a general azimuthal symmetry.

Data products

Absorption and emission estimates can be inferred from differences in amplitude between incoming and outgoing wave components. Phase shifts between A_m and B_m are additional parameters describing the scattering properties of the target. These shifts can be caused by refractive perturbations in the target. Observed phase shifts provided the basis for the first

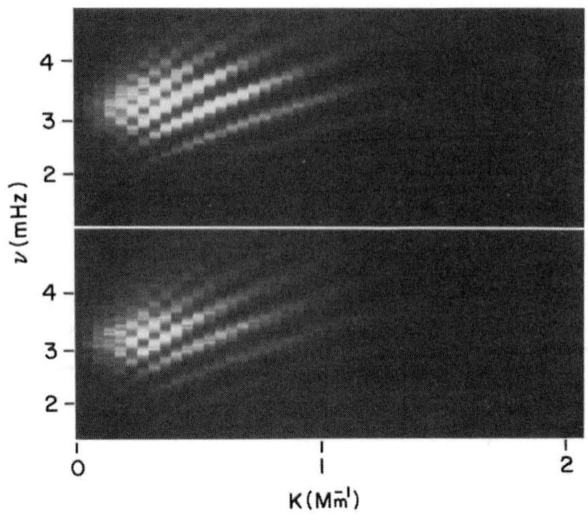

Figure 6.2 Power spectra for incoming (top panel) and outgoing (bottom panel) wave modes with respect to a target sunspot. The outgoing power is substantially reduced compared to the incoming power. (From Braun *et al.*, 1988, reproduced with the permission of the AAS.)

inferences of wave-speed perturbations below sunspots (Fan *et al.*, 1995). The apparent absorption seen in magnetic regions has also led to considerable theoretical study of mode conversion and simple models consistent with both absorption and phase shifts have been derived (e.g., Crouch *et al.*, 2005). It is worth mentioning that the amplitudes determined through these methods may be affected by a variety of physical mechanisms, including absorption, a reduction of emission, and local amplitude suppression (see, e.g., Chou *et al.*, 2009b).

Mode mixing can occur when incoming wave components are scattered into different azimuthal orders or horizontal wavenumbers. Evidence of this in sunspots has been observed (Braun, 1995). However, a practical problem persists in distinguishing true mode mixing from artefacts created by the limitations of the methods or measurements. Spectral leakage is one example of this, where a finite time series or spatial size produces a spread of wave amplitude from one discrete bin of frequency or wavenumber into another.

Frequency shifts can be measured for flows beneath the annular domain. These can be determined in a manner similar to how flow parameters are determined from ridge-shifts in ring-diagram analysis. Using Legendre functions appropriate for larger regions of the solar surface, this approach has been used to probe the depth variation of meridional circulation (see Chapter 13).

6.3.3 Cross-power-spectral and mode coupling methods

In a local volume that is stratified vertically but uniform in the horizontal directions and in time, the Fourier wavenumber components of the wave field are uncorrelated, or in other words, independent of each other. Horizontally varying perturbations within the volume

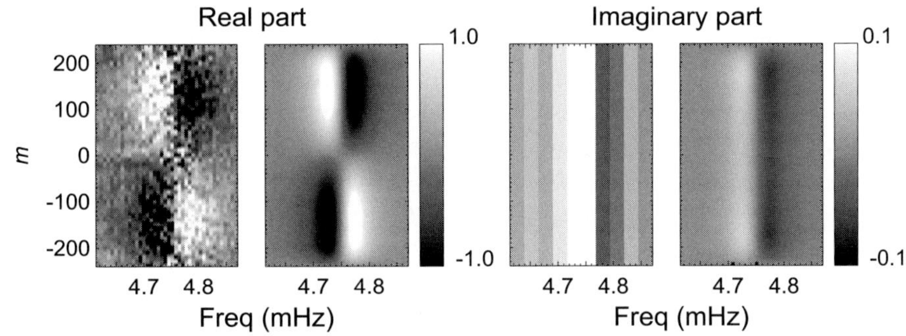

Figure 6.3 Observed and theoretical cross-power spectra for solar oscillations between spherical harmonic degree 200 and 202 and radial order $n = 8$ as a function of temporal frequency (horizontal axis) and azimuthal order m (vertical axis). The gray scales in the left and right pair of panels indicate the amplitudes of the real and imaginary part of the spectra respectively. The left (right) panel of each pair indicates the observed (theoretical model) of the cross-power spectra. The model assumes a solar-like differential rotation and meridional circulation flow. Note that the scale of the imaginary component is ten times less than the real part, and the observed imaginary component is averaged over vertical blocks in the frequency–azimuthal order domain. (Figure by the author from unpublished measurements provided by M. Woodard; see Woodard, 2009, for details of the analysis.)

introduce a coupling of wave components of the same frequency but different horizontal wavenumbers. This implies that, in principle, perturbations can be inferred from analyses of *cross-power spectra* (e.g., Woodard, 2002). Figure 6.3 shows an example of the cross-power spectra for solar oscillation modes correlated by the presence of solar rotation and meridional circulation.

Methods for measuring mode coupling and modeling the subsurface flows have been developed and applied to a variety of problems. These include the analysis of mode coupling in a local plane-parallel approximation to deduce small-scale flows (Woodard, 2002) as well as the analysis of global spherical-harmonic modes to study meridional circulation (e.g., Schad *et al.*, 2012). Analogous to the use of cross-power spectra to probe coupling of modes in the frequency–wavenumber domain, we discuss in the next section how local helioseismic methods using cross-covariances are used to assess correlations of the wave field directly in the space–time domain.

6.4 Cross-covariance methods

6.4.1 Basic considerations

Two local helioseismic methods we focus on here are based on the cross-covariance in time between signals $\psi(\mathbf{r}_1, t)$ and $\psi(\mathbf{r}_2, t)$ at positions $\mathbf{r}_1$ and $\mathbf{r}_2$:

$$C(\tau, \mathbf{r}_1, \mathbf{r}_2) = \int_0^T \psi(\mathbf{r}_1, t)\psi(\mathbf{r}_2, t + \tau)dt, \tag{6.2}$$

where T is the duration of the observed signals. Both time–distance helioseismology (Section 6.4.2) and helioseismic holography (Section 6.4.3), as well as the myriad flavors of these methodologies, can be defined in terms of which signals and which locations are used in the cross-covariance function. The generality that unites both of these methods is that, for a wide variety of local helioseismic applications, the cross-covariance function C is used to extract *travel-time anomalies* (Section 6.4.4) of acoustic or surface-gravity waves propagating between the two points. By *anomalies* we mean the difference between the wave travel times in the presence of subsurface perturbations to the solar interior and the travel times without those perturbations. These travel-time anomalies can be modeled, and in many cases *inverted* (Chapter 9) to yield the subsurface structure of the perturbations, which may include wave speed changes and plasma flows. The left panel of Figure 6.4 shows an example of an averaged cross-covariance function in the (τ, Δ) domain from an application of time–distance helioseismology to solar observations, where Δ is the distance between two points on the surface and the average is performed over many pairs of surface points.

It is helpful to understand how waves propagate between two points at the surface under the ray approximation where the wavelengths are smaller than the spatial scale of changes in the solar interior (Figure 6.4). From a starting point at the solar surface, downward propagating rays are refracted according to Snell's law as the sound speed increases with depth. At a lower *turning point* the rays are horizontal and the horizontal component of the phase speed (ω / k_h) is equal to the sound speed. At the solar surface, the returning upward propagating waves are again reflected when the wavelength becomes comparable to the vertical scale of the change in the density. Deeper lower turning points are achieved by waves with greater phase speeds. Waves that do not otherwise suffer significant damping can bounce several times from the surface or even around the entire Sun.

6.4.2 Time–distance helioseismology

Time–distance helioseismology primarily uses the cross-covariance function between observations of the wave field sampled between two points, or averaged over patches. Frequently, the signal-to-noise is enhanced for practical purposes by using many averages of point pairs and making the measurements over several hours or even days. The cross-covariance functions are the analog, in local helioseismology, of the seismogram in terrestrial seismology.

A common scheme employed in time–distance helioseismology is to combine cross-covariances between a point and the signal averaged over a concentric annulus (this is termed *center-annulus* geometry). In many cases, the annulus is broken up into four quadrants (Duvall *et al.*, 1997, see Figure 6.5). *Deep-focusing* methods may also be employed, which use cross-covariance functions from signals from the opposing quadrants. In all geometries, the positive and negative time lags τ in the cross-covariance functions distinguish the travel-time anomalies for waves propagating in opposite directions. Thus, the quadrant geometry

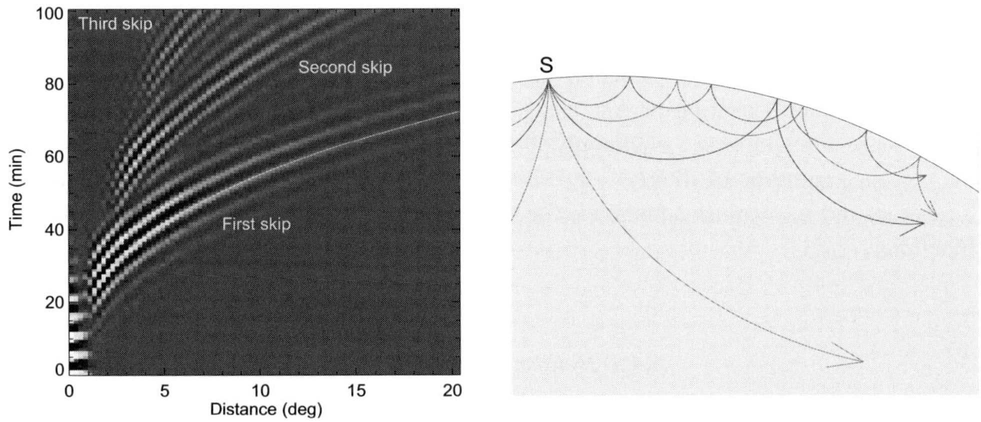

Figure 6.4 (Left) A time–distance cross-covariance function in time (horizontal axis) and between samples of the wavefield spaced different distances (horizontal axis) on the solar surface. The amplitude of the function, modulated by the five-minute solar oscillations, reaches maxima at times corresponding to the travel times of rays corresponding to different acoustic ray paths in the solar interior. (Figure by the author using unpublished cross-covariance measurements provided by T. Duvall, Jr.) (Right) Ray paths of sound waves propagating downward from a specific origin S near the surface, refracting back upward to the surface, and reflecting down again to advance a number of "skips" away. (Figure by the author.) A black and white version of this figure will appear in some formats. For the color version please refer to the plate section.

is ideal for assessing the two spatial components of directionally sensitive perturbations such as flows.

The cross-covariance functions in the (τ, Δ) domain made using time–distance helioseismology reveal ridge-like structures (see Figure 6.4 – these should not be confused with the p-mode ridges in the frequency–wavenumber domain). The lowest ridge (along the τ axis) corresponds to acoustic waves that reflect once from the lower turning point to arrive again at the surface a distance Δ away without additional surface reflections. At longer time lags τ, additional ridges represent the surface arrival of waves undergoing one or more skips to the surface.

Filtering

Filters are often applied to the datacube before cross-covariance methods are applied. *Phase-speed filters* are used to isolate wave modes with common ray paths through the solar interior, and to eliminate the noise caused by unwanted contributions of other waves or by solar convective motions. These are applied to the data in the 3D Fourier domain, and usually take the form of Gaussians in the variable ω/k_h which peak at the desired phase speed and have a constant width (Couvidat and Birch, 2006).

Filters that isolate surface gravity waves (f modes) are useful for probing imhomogeneities near the solar surface, where the energy of f modes is concentrated. *Ridge filters* can also be employed that isolate the set of p-modes of fixed radial order. Filters in temporal frequency are also in wide use in cross-covariance methods. Typically, it is desirable to remove low-frequency components of the observations which are dominated by solar convection. Filters covering narrow bands in frequency are also used in some instances. All of these filters are also commonly used in holography (Section 6.4.3). In principle, the presence and type of filter(s) used should be incorporated into the modeling and inference of the travel-time anomalies, for example, in the computation of the sensitivity functions.

6.4.3 Helioseismic holography

Helioseismic holography (also known as computational seismic holography and ambient acoustic imaging) is a set of procedures based on phase-coherent imaging of the solar interior. Conceptually, the idea is to start with the observed wave field at the solar surface and computationally regress it, either forward or backward in time, into a solar model to estimate the amplitudes of waves propagating into and out of a remote focus point. These amplitudes are called the *ingression* H_- and *egression* H_+ (Lindsey and Braun, 2000a) and are estimated at time t and position $\boldsymbol{\rho}$ by a convolution of the surface wave field ψ with appropriate Green functions (G_+, G_-):

$$H_\pm(\boldsymbol{\rho}, t) = \int_0^T dt' \int_P d^2\boldsymbol{\rho}' \, \psi(\boldsymbol{\rho}', t) \, G_\pm(\boldsymbol{\rho}', \boldsymbol{\rho}, t - t'), \tag{6.3}$$

where $\boldsymbol{\rho}$ is the three-dimensional position vector and $\boldsymbol{\rho}'$ defines the solar surface. The integral in time is performed over the duration T of the observations and the integral over the surface is made over an area P called the *pupil*, which is borrowed from optical terminology. The convolution expressed by Equation (6.3) is an analog, in local helioseismology, of the Kirchhoff integral theorem (e.g., Jackson and Fox, 1999). In general, Green functions are the basic solutions to particular differential equations and are subject to the type of equations and the boundary conditions of the system. For the wave equation appropriate for the solar interior, the helioseismic holography Green function $G_-(\boldsymbol{\rho}', \boldsymbol{\rho}, t - t')$ conceptually represents the response at the surface position $\boldsymbol{\rho}'$ and time t' due to an impulse located at position $\boldsymbol{\rho}$ and time t, while $G+$ is the time-reverse of this, i.e., $G_+(\boldsymbol{\rho}', \boldsymbol{\rho}, \tau) = G_-(\boldsymbol{\rho}', \boldsymbol{\rho}, -\tau)$. These functions regress the observed surface wave field either forwards (for G_-) or backwards (for G_+) in time and into the interior of the Sun.

One can consider a single-frequency component of the wave field ψ observed at the surface $\boldsymbol{\rho}'$ by taking the Fourier transform in time of Equation (6.3). If one also restricts the analysis to a local region and assumes a plane-parallel approximation, the convolution above simplifies to

$$H_\pm(\mathbf{r}, z, \omega) = \int_P d^2\mathbf{r}' \, \psi(\mathbf{r}', \omega) \, G_\pm(|\mathbf{r} - \mathbf{r}'|, z, \omega), \tag{6.4}$$

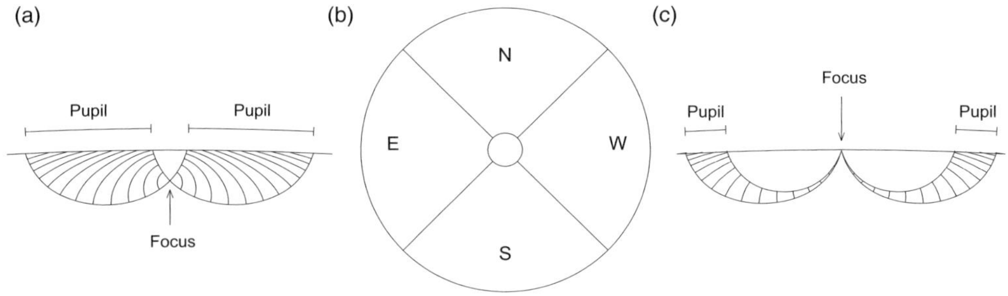

Figure 6.5 (a) The geometry employed with lateral vantage holography which selects waves that pass through a common focus point below the surface. Here the waves propagate through the focus at angles up to 45 degrees from the horizontal direction. Similar geometry is used for deep-focus time–distance methods, although frequently narrower annuli are used. (b) Looking from above, the annulus used in panel (a) is divided into quadrants extending north, south, east, and west on the solar surface to facilitate the assessment of the horizontal components of flows. (c) A side view of the waves selected with surface-focus methods of time–distance (also referred to as center-annulus geometry) and holography. Waves propagating from one side of the annulus bounce once at the surface focus point before reaching the other side. (Figure by the author.)

where z and $\mathbf{r}$ are the depth and horizontal position at which the amplitudes are to be assessed. Under these conditions it is noteworthy that the convolutions needed to compute $H_{\pm}$ using Equation (6.4) can be performed efficiently using fast Fourier transforms.

To study travel-time anomalies analogous to those determined by time–distance helioseismology, one considers cross-covariances between the ingression and egression, or between either of these and the wave field ψ. For example, if the ingression and egression are both regressed, from different pupils, to some subsurface *focus* at depth z and horizontal position $\mathbf{r}$, then the cross-covariance of the two amplitudes can be used to determine the travel-time anomalies of waves propagating between the two pupils through the focus. This is the basis for *lateral vantage* holography (Figure 6.5), and is highly analogous to deep-focus methods in time–distance helioseismology. Other common applications include *local control correlations*, which represent the cross-covariances between the surface-wave field ψ with either the ingression or egression assessed over a concentric annulus, which may be divided into quadrants. This is also termed *surface-focus* holography and is analogous to center-annulus time–distance methods.

Unfortunately, the limited information available from surface Dopplergram observations requires that helioseismologists use only approximate forms of the Green function. However, different types of Green functions have been derived numerically using models of the solar interior under a variety of approaches and approximations. One example of their construction may be found in Lindsey and Braun (2000a). Remarkably, when the point source is located at the surface the Green function closely resembles the time–distance cross-covariance function (e.g. Figure 6.4), which provides a conceptual linkage between holography and time–distance helioseismology.

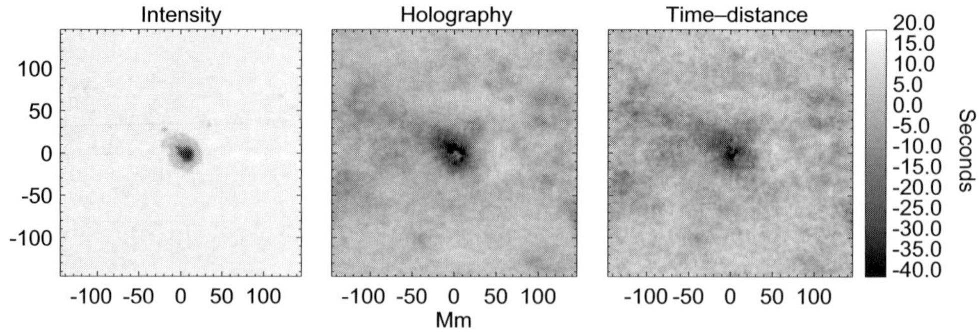

Figure 6.6 Sample maps of observed travel-time anomalies around a sunspot, employing different methods to extract the travel times. Both maps show a decrease in travel times within a sunspot. (Figure by the author from measurements by the author and T. Duvall Jr.)

6.4.4 Data products

For many purposes, the travel-time anomalies, extracted from the cross-covariance functions form the main data products of interest for further modeling and inferences of the solar subsurface. However, wave excitation and absorption can be studied with egression and ingression amplitudes. Deep-focus methods employing four quadrants, aligned with the cardinal directions on a region of the Sun (Figure 6.5), require the extraction of four travel-time anomalies (corresponding to propagation from east to west, west to east, north to south, and south to north). Surface-focused holography and center-annulus time–distance methods use eight measurements to determine the directional components of the travel-time anomalies. In addition, sums of cross-covariance functions over the four quadrants in the annulus can be used to assess the net inward and outward travel times. The difference of these (net outward minus inward) is sensitive to the horizontal divergence of flows and to vertical flow components. In general, the sets of measurements used for modeling flows consist of maps of the directional (east–west and north–south) travel-time differences and net (out–in) difference. The directionally averaged (or mean) travel-time anomaly is defined relative to mean travel times computed from a solar model or (more typically) measured empirically from solar observations. This quantity is sensitive to directionally invariant perturbations such as changes in the sound speed.

A main practical difference between the cross-covariance functions determined from each method is that, in time–distance helioseismology, one extracts for further analysis the portions of the two branches of the function centered around time lags $\pm\tau$ given by the time for wave propagation between the two measurement points in the quiet Sun; while for holography the time lag τ is near zero since the egression and ingression signals are computationally regressed towards each other in time by means of the Green functions. There are various methods to model or fit the cross-covariances to extract the travel-time anomalies and their uncertainties. Common methods include fitting cross-covariance to wavelet functions, such as the Gabor function, or to theoretical models.

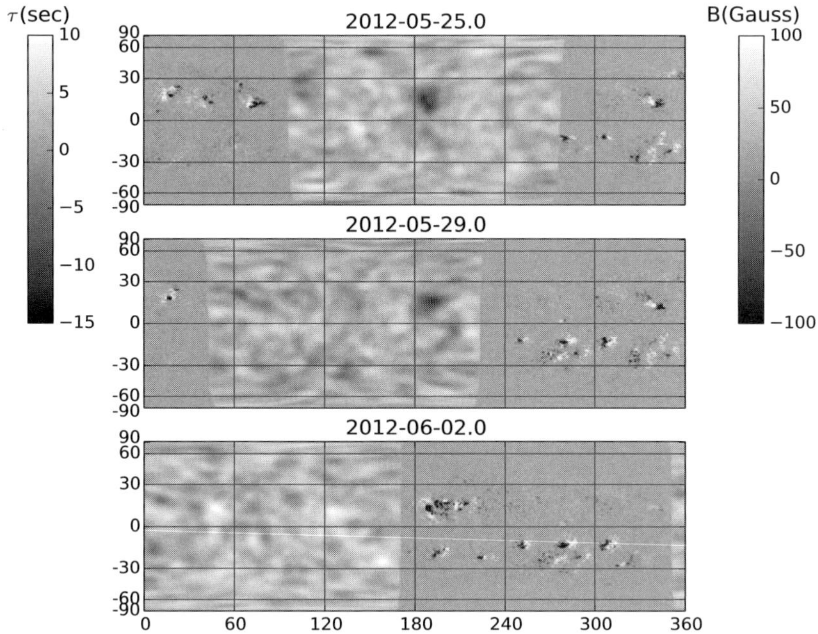

Figure 6.7 Composite farside-seismic (left color scale) and nearside-line-of-sight magnetic maps (right gray scale) show an active region (NOAA AR11498) passing the farside meridian (top panel) and rotating into view from Earth at Carrington coordinate (190W, 13N) in the near hemisphere (bottom) in May–June, 2013. (Figure provided by C. Lindsey.) A black and white version of this figure will appear in some formats. For the color version please refer to the plate section.

Couvidat *et al.* (2012a) provides a comparison of a number of popular methods. Figure 6.6 shows some examples of maps of travel-time anomalies around a sunspot as determined from different methods.

Travel-time anomalies from cross-covariance-based methods have been used to model wave-speed variations beneath active regions as well as flows on almost all spatial scales. An important application of multiple-skip cross-covariance analyses has been to map travel-time anomalies over much of the spherical solar surface, including the hemisphere on the opposite side to the Earth (Figure 6.7). This method makes use of the strong travel-time anomalies that are associated with magnetic regions, which can be seen even in global oscillation modes that propagate one or more times around the Sun within their mode lifetimes. Farside imaging was first achieved with holography-based methods (Lindsey and Braun, 2000b) and subsequently with time–distance helioseismology (Zhao, 2007).

6.5 Limitations and challenges

While instrumental noise and the effects of the turbulence in the terrestrial atmosphere can affect the measurements described above, for space-based observations the main source of random errors is typically solar *realization noise*, which results from the stochastic nature of the wave excitation process (e.g., Gizon and Birch, 2004). This noise provides the ultimate limits to the precision of both local and global helioseismology. While this noise can be substantially reduced by spatial and temporal averaging for some problems (e.g., the probing of long-lived and large-scale flows on the Sun), this is not practical for the study of smaller-scale flows or individual sunspots with short lifespans. Strategies that have been developed include the *ensemble averaging* of measurements that are employed to model the (averaged) properties of supergranules (Duvall and Birch, 2010) and small magnetic elements (Felipe *et al.*, 2012).

Data products for local helioseismology, much like those for global helioseismology, can be improved by long-duration observations free of the day–night duty cycle. Multiple vantage points away from the Earth–Sun line of sight, now routine for other types of solar observations, have not yet been achieved for helioseismic data and are clearly a desirable goal for future space missions. Recent theoretical developments are also facilitating the interpretation of local helioseismic data products and guiding improvements in analysis methods. For example, increasingly sophisticated magnetoconvective computations are making it possible to understand sources of systematic errors due to magnetic fields (e.g., Braun *et al.*, 2012) and convective motions (e.g., Baldner and Schou, 2012) in the Sun.

7

Planetary seismology: Early observational results

YOSIO NAKAMURA

7.1 Introduction

There have only been three occasions in the past where seismic data from the surface of an extraterrestrial body were successfully acquired, although several unsuccessful attempts have been made, as discussed in Chapter 3. The first was during and following the Apollo landing missions on the Moon in 1969 through 1977, the second was during the Viking mission on Mars in 1976 through 1978, and the third was the Soviet Venera missions to Venus in 1982. During the Apollo project, three seismic experiments were conducted at various landing sites: Passive Seismic Experiment (PSE) at Apollo landing sites 11, 12, 14, 15, and 16; Active Seismic Experiment (ASE) at landing sites 14 and 16; and Lunar Seismic Profiling Experiment (LSPE) at the Apollo 17 landing site. These landing sites are indicated on Figure 7.1. During the Viking project, seismometers were deployed at two landing sites, Viking 1 and Viking 2, but the seismometer at the Viking 1 site failed to uncage, and only at the Viking 2 landing site were records of ground motion obtained. The Venera 13 and 14 measurements lasted only briefly, 127 and 57 minutes, respectively.

In this chapter, we briefly describe the Apollo and Viking experiments and summarize key findings of the contemporary analyses of the acquired data. Included are some historical accounts of events experienced at the time. Since the turn of the century, with vastly more powerful computers now available and with developments of new analysis techniques, these decades-old data are attracting renewed interest, and new results are emerging. These new results are described in Chapter 15. The results of the Venera experiment will not be discussed here because information available to the western world is scarce. We understand that only a single event of unknown origin was counted (e.g., Ksanfomaliti *et al.*, 1982). The chapter will end with a brief account of remaining questions and suggested future directions in lunar and planetary seismology.

Extraterrestrial Seismology, eds. Vincent C. H. Tong and Rafael A. García. Published by Cambridge University Press.

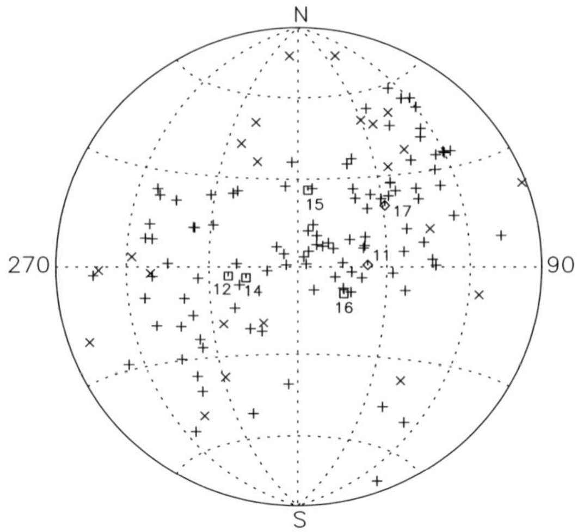

Figure 7.1 Locations of the Apollo landing sites (□ and ◇, numbered) and epicenters of deep moonquake nests (+) and shallow moonquakes (×) on the front side of the Moon, shown in Lambert azimuthal equal-area projection. The sites shown in squares constitute a four-station seismic network. (From Nakamura, 2010, reprinted with kind permission of the German Geophysical Society (DGG).)

7.2 Apollo Passive Seismic Experiment

As the only fully successful global seismic experiment on an extraterrestrial body to date providing a wealth of new information, the Apollo PSE is the key element of planetary seismology. The idea of emplacing a seismometer on the Moon was discussed in the late 1950s by a group of scientists and engineers at the Seismological Laboratory of the California Institute of Technology and Lamont-Doherty Geological Observatory of Columbia University (see Latham *et al.*, 1969a,b for a brief history). After the unsuccessful attempts with hard landers in the Ranger program and cancelled participation in the Surveyor mission, a seismometer was successfully deployed on the Moon as a component of the Early Apollo Surface Experiment Package (EASEP) when the first human being landed on the Moon with Apollo 11 in July, 1969. This was followed by successful deployments of PSE seismometers as a component of the Apollo Lunar Surface Experiment Packages (ALSEPs) at the landing sites of Apollo 12, 14, 15, and 16, between November 1969 and April 1972. The term "passive" is used here because the principal objective of the experiment was to passively observe naturally occurring seismic activities. It is to distinguish it from the "active" seismic experiment, which recorded seismic signals generated by explosive sources with an array of geophones.

7.2.1 PSE network

Although the seismometer at the Apollo 11 landing site became dysfunctional after 21 Earth days of successful daytime operations when the temperature became excessive as noon approached on the second lunar day, seismometers at the other four stations, which were protected with a thermal shield, operated continuously till they were turned off at the end of September 1977. These four formed a network of seismic stations near the center of the front face of the Moon (Figure 7.1) providing nearly eight years of continuous observation, with five and a half years of full network operation. In all, more than 600 gigabits of seismic data were sent back to Earth. In addition, the Lunar Surface Gravimeter (LSG) deployed at the Apollo 17 landing site in December 1972, designed to detect gravitational waves predicted by the theory of general relativity, functioned as a low-sensitivity short-period seismometer, thus providing a five-station seismic network while it was operational.

Each PSE station contained a set of three orthogonal long-period seismometers, two horizontal and a vertical, and a short-period vertical seismometer. The former was sensitive to ground displacement roughly from 0.07 to 1 Hz, while the latter was sensitive to ground velocity from 1 to 8 Hz.

7.2.2 Lunar seismic events

Despite the lack of recent surface faulting observable on the Moon, which suggested that the Moon has been tectonically inactive for a considerable amount of time, a study of thermal evolution of the Moon based on plausible models suggested that the Moon might still maintain strong seismic activity comparable to that of the Earth (MacDonald, 1960). These models indicated that the greatest rate of release of strain energy would be at depths of 100–700 km. On the other hand, there remained a possibility that the absence of topographic and structural features associated with earthquakes suggested that the mechanism of seismic energy release on the Moon may be quite different from that on Earth (Press *et al.*, 1960). So, how does the Moon's strain energy release differ from the Earth's? The seismic data acquired during the Apollo PSE gave us our first glimpse into answering this perplexing question.

Initially, we were hoping that the seismic data from the Apollo 11 landing site would give us some answers to this question by detecting natural seismic activity, if any, occurring in the Moon. However, this turned out to be not an easy task. Many seismic "events" were recorded, including some that were obviously generated by venting of gases in the lander, situated only 17 m from the seismometer. But there were several other events that looked different. None of them, however, looked anything like earthquake signals we are familiar with on Earth (Figure 7.2). It took another four months before this dilemma was partially resolved when another seismic station was deployed at the Apollo 12 landing site and the expended ascent stage of the lunar lander called the Lunar Module (LM) was dropped to the lunar surface to generate seismic waves.

Artificial impacts

Artificial impacts, if available, are a sure way of generating surface seismic sources of known location and origin time on a planetary body with which to derive its internal structure. The

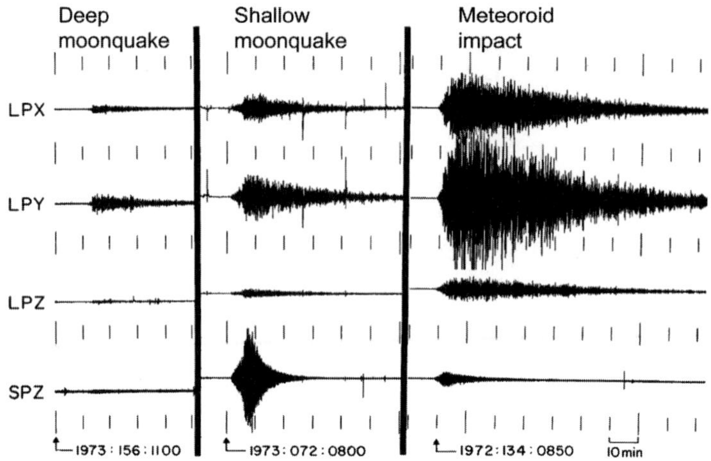

Figure 7.2 Representative lunar seismograms in compressed time scale at the Apollo 16 station. LPX, LPY, and LPZ stand for three orthogonal components of a long-period instrument with LPZ in the vertical direction, and SPZ stands for the short-period vertical component. (From Nakamura *et al.*, 1982.)

fact that their location and origin time are known gives them a clear advantage over natural seismic sources for deriving the seismic velocity structure of a planetary body, because there are four fewer unknown parameters for each impact.

During the Apollo project, nine such impacts were made with the final stage of the Saturn V rocket (S-IVB) and the ascent stage of the LM at distances ranging from 73 to 1750 km from a PSE station. They provided first-hand information on the crustal structure in the front central area of the Moon, as described later in Section 7.2.4.

The signal from the first LM impact as observed at the Apollo PSE station was a big surprise. It did not look anything like seismic signals generated by an explosion on Earth, but had a gradual beginning and a very long decay lasting over an hour. As described in Section 7.2.3, this unusual seismic waveform is caused by the extremely low attenuation and intense scattering of seismic waves inside the Moon, but we did not know this at the time. And when we looked back at the Apollo 11 seismograms recorded four months earlier, we realized that several similar events had indeed been recorded there. This led to the recognition of many meteoroid impacts on the Moon, which will be described next.

Meteoroid impacts

Meteoroid impacts are expected to be highly effective sources of seismic waves observable on an airless body such as the Moon, even though a very small fraction of the kinetic energy of an impactor is converted to seismic energy. The Apollo project did not defy this expectation by providing us more than 1700 such impact events recognizable on the long-period seismograms, representing impacts occurring far from a seismic station, during the eight-year operation of the seismic stations (Nakamura *et al.*, 1981). These impactors were estimated to have masses ranging between about 500 g and 50 kg (Duennebier *et al.*, 1975). Smaller impacts, which generate higher frequency signals observable only on short-period seismograms, were too numerous to be listed in a catalog.

Unless a meteoroid impact happens to be observed from Earth, usefulness of meteoroid impacts for deriving the internal seismic velocity of the Moon is more limited than that of artificial impacts because their impact locations and times are unknown. However, used in conjunction with data from other seismic sources, data from large meteoroid impacts that are observed at multiple stations will provide sufficient information to be useful.

Meteoroid impacts observed on the Moon also provide information on the distribution of small interplanetary objects crossing the orbit of the Earth–Moon system. By placing a seismic array on the Moon, we are essentially using the whole Moon as a large detector of such objects. A detailed examination of meteoroid impacts as detected by the Apollo seismic network revealed two distinct types of interplanetary objects: cometary and asteroidal (Oberst and Nakamura, 1991). The cometary impacts are mostly lighter ($<\sim 1$ kg) and less energetic, and many of them occur in clusters identifiable with known meteor showers, while the asteroidal impacts include many massive objects ($>\sim 1$ kg), mostly in prograde orbits and with a lesser degree of clustering. However, the asteroidal objects also include some prominent swarms, suggesting concentrated debris from relatively recent disintegration of asteroids in Earth-crossing orbits (Duennebier *et al.*, 1976). A more detailed discussion on impact seismology is given in Chapter 18.

Deep moonquakes

By far the most numerous seismic events observed on the Moon are deep moonquakes, occurring at depths about halfway to the center of the Moon. During the Apollo project, more than 1300 of them were positively identified, with many more events being suspected to be deep moonquakes. As described in Chapter 15, later re-analyses have confirmed this suspicion (e.g., Nakamura, 2003), bringing the number of positively identified deep moonquakes to exceed 7000, constituting an overwhelming majority of the seismic events detected, identified, and cataloged.

These deep moonquakes were entirely unexpected before the Apollo project. It was a pleasant surprise, shortly after the seismic station at the Apollo 14 landing site was established, to find seismic events detected by both Apollo 12 and 14 stations, representing natural events at great distances (Latham *et al.*, 1971). Often they had nearly identical waveforms, signifying that they originated from nearly the same location, and they occurred at monthly intervals, indicating that their occurrence was tidally influenced. It was an even greater surprise, after a third seismic station was established and thus the depths of these events could be determined, when we discovered that they were occurring at great depths, about halfway to the center of the Moon and deeper than any known deep-focus earthquakes on Earth.

As mentioned above, these deep moonquakes occur at monthly intervals and in groups having nearly identical waveforms. By the end of the Apollo project, about 100 such groups had been found through visual inspection of seismograms (Nakamura *et al.*, 1982), but a recent computer-based matching of waveforms extended the number of such groups to over 300 (Nakamura, 2003). The highly similar waveforms suggest that events in each group occur in close proximity to one another, in a "nest." Not all groups produce enough arrival times to determine the location of the nest, but 106 of them have provided a sufficient number of arrival times for their location to be determined (Nakamura, 2005). Their locations (Figure 7.1) are not evenly distributed over the Moon but are concentrated in certain areas forming a broad belt running in the WSW–ENE direction. It should be noted, however, that this belt is nothing like the narrow belt of seismic activity seen along a plate boundary on Earth and that deep moonquakes occur only in isolated nests in the broad belt. The maximum extent of a nest was estimated to be no more than a kilometer or so (Nakamura, 1978). Why they occur in this way is an important question that remains open.

The monthly periodicity with tidal correlation of deep moonquake occurrence was noted very early in the Apollo project (e.g., Ewing *et al.*, 1971; Lammlein *et al.*, 1974). For various nests, periodicities were clearly observed (Figure 7.3) corresponding to the anomalistic month (27.55455 days, relating to changes in the Earth–Moon distance), the draconic month (27.21222 days, relating to crossing of the Moon's orbit on the ecliptic plane), the interaction of the synodic month (29.53059 days, relating to the new-moon/full-moon cycle) with the anomalistic month resulting in a 206-day (~7 months) periodicity, and the interaction of anomalistic and draconic months resulting in a 2190-day (~6 years) periodicity. It is obvious that the occurrence of deep moonquakes is highly influenced by the tides raised by the Earth to a large extent and by those raised by the Sun to a lesser extent, but how exactly and why such influences are affecting the deep moonquake occurrences are other questions that are waiting to be answered.

Where, at what depths, and when deep moonquakes occur are distinctly different from how earthquakes occur on Earth. Undoubtedly these factors provide some clues as to the cause of these events. There are some distinct similarities in their occurrence to breaking of rocks and metals under cyclic loading, called fatigue failure, generating acoustic emissions in laboratory settings. There is a hypothesis that the presence of deep moonquakes may provide evidence for the presence of volatiles or partial melting at those depths (Frohlich and Nakamura, 2009). However, the true cause of these events has yet to be confirmed.

Shallow moonquakes

Another group of events distinctly different from impacts and deep moonquakes was observed. We called them shallow moonquakes because they occur at much shallower depths than deep moonquakes. Shallow moonquakes are rare. Only 28 of them were identified during the eight-year operation of the Apollo seismic stations. However, they include some of the most energetic seismic events ever observed on the Moon, with estimated equivalent body-wave magnitude of 5 or greater. [For various different definitions of magnitude, see e.g., Utsu, 2002.] They have significantly higher frequency content than most other events despite occurring at great distances from seismic stations as indicated by largely separated and relatively clear P- and S-wave arrivals, characteristics not found in any other

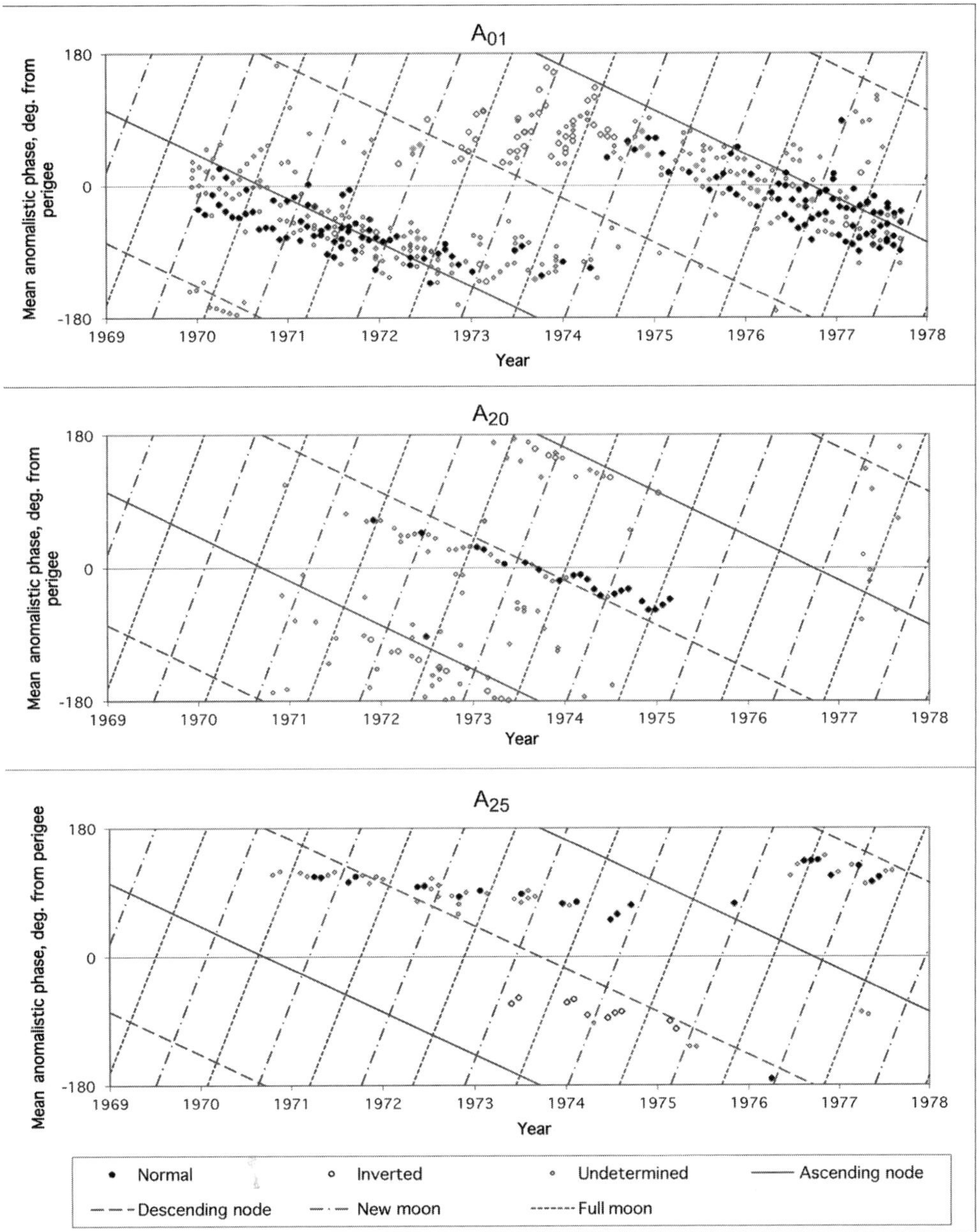

Figure 7.3 Occurrence time vs. anomalistic phase of deep moonquakes at nests A_1, A_{20} and A_{25}. An anomalistic phase of 0 means perigee crossing. Gently sloping lines give the draconic phase, with solid lines representing the Moon's northward crossing of the ecliptic, and the steeply sloping lines give the synodic phase, with dot-dash lines representing new moon. Note the specific ways deep moonquakes occur at each nest relative to the monthly tidal phases. (From Nakamura, 2010, reprinted with kind permission of the German Geophysical Society (DGG).)

types of events observed on the Moon (Figure 7.2). For this reason, they were originally called high-frequency teleseismic (HFT) events. The exceptionally high frequency content of these events may be indicative of extremely high stress retained in the upper mantle of the Moon. Oberst (1987) estimates the stress drop associated with the largest shallow moonquake to exceed 100 MPa. The high stress drop is consistent with a model of the Moon in which thermoelastic strain within a thick lithosphere is released by occasional shallow moonquakes as the Moon cools.

All of the 28 observed shallow moonquake events occurred outside the Apollo seismic network (Figure 7.1), and for this reason, their depths were not well determined except that they cannot be deeper than about 300 km and thus clearly shallower than deep moonquakes. Also, intense scattering in the seismic coda has prevented identification of surface-reflected phases, which would help to fix the source depth. An estimate based on their amplitude decay with distance put them in the upper mantle of the Moon (Nakamura *et al.*, 1979).

Having only a small number of events to guide our interpretation, past attempts to explain the cause of these enigmatic events have been inconclusive. Unlike deep moonquakes, no clear correlation of their occurrence with tides has been found. There appears to be a certain degree of correlation between their epicenters and large impact basins, but again this is highly questionable because of the paucity of data. There is a recent analysis result showing that most shallow moonquakes detected during the Apollo project occurred when the Moon was facing a certain direction in the celestial sphere, i.e., relative to stars. Frohlich and Nakamura (2006) hypothesize that high-energy objects from a fixed source outside the Solar System may be triggering or causing these events. Again, because of the paucity of data, this hypothesis does not go much beyond speculation. Thus, the real cause of shallow moonquakes must wait until we have more data. There is one possibility, however, that this could be achieved without waiting till we go back to the Moon. All of the 28 shallow moonquakes that were identified during the Apollo project were large enough to be recorded on both short- and long-period seismograms. Because of the unusually high frequency content of these events, many small shallow moonquakes are likely to have been recorded only on short-period seismograms, but they are difficult to identify visually because there simply are so many other events recorded there. A computer search with an appropriate algorithm may be successful in finding them in the near future.

Thermal moonquakes

A large number of very small seismic events were observed on short-period lunar seismograms every month starting about two Earth days after sunrise and the activity died down gradually as the lunar day progressed. Then, they repeated every month with similar waveforms. The diurnal changes of their activity suggest that they are caused by the large temperature changes occurring on the lunar surface. The lunar surface temperatures were observed to change from near 100 K to 290 K (Keihm and Langseth, 1973) from night to daytime. Similar waveforms suggest that they occur repeatedly at or near the same place every lunation. Many such sources within an estimated distance of about 2 km of each seismic station were identified (Duennebier and Sutton, 1974).

Judging from the way they occur, these events are caused either by fracturing or movement of rocks on the lunar surface or by slumping of lunar soil on slopes triggered by thermal stress. Either way, they may contribute to gradual degradation or erosion of the lunar surface, just as meteoroid impacts do, although it is unknown which of these is more effective.

As described later in Section 7.3.2, these small events serve as effective sources of ambient seismic noise with which to derive the lunar near-surface structure.

7.2.3 Seismic wave transmission in the Moon

As stated earlier, seismic signals that were recorded on the Moon did not look anything like those of earthquakes or explosions here on Earth. Signals start gradually with no clear beginning in most cases, slowly build up to maximum amplitude, and then decay very gradually, often taking hours to die down (see Figure 7.2). No distinct later arrivals, such as surface-reflected phases, are visually identifiable. The reason for this can be attributed to the extremely low attenuation of seismic waves in the lunar interior, especially at shallow depths, and to extensive scattering of seismic energy in the near-surface zone of the Moon. The low attenuation of seismic waves is attributable to the absence of water at least in the shallow part of the Moon, which, if present at grain boundaries, would dissipate seismic energy. The extensive scattering is likely to be the result of a highly pulverized, thick near-surface zone of the Moon, produced by numerous bombardments by meteoroids. Such a zone may extend to depths of tens of kilometers.

A simple way to describe transmission of seismic energy in such a medium is to treat the energy transmission as a diffusion process, governed by a diffusion equation, rather than propagation of waves, governed by a wave equation. With appropriate assignments of diffusion coefficient and attenuation factor, what we observed on the Moon can be approximated fairly well (e.g., Nakamura, 1977). Both impact signals at long distances and amplitude changes from a moving object, the Lunar Rover, at short distances (Nakamura, 1976) were well approximated by this formulation.

However, there is a clear drawback to such an approach. For example, the diffusion approach will not provide the high resolution often required in seismic investigations. Since diffusion of seismic energy is a statistical process while real wave propagation is a deterministic process, the diffusion approach cannot describe the detailed waveform of waves arriving at and observed at a seismic station. This may be detrimental to the analysis of seismic arrivals to determine the internal structure of the Moon. Perhaps some combination of these two approaches may be what we need. Further discussion of this problem and current efforts to better describe what we saw on the Moon are found in Chapter 10.

7.2.4 Internal seismic velocity structure

The Apollo project provided seismic signals from a number of sources, both artificial and natural, from which to derive the internal velocity distribution of the Moon. The artificial impacts, because of their known impact locations and times, were highly instrumental in providing the velocity profiles in the upper tens of kilometers of the Moon. From P-wave arrival times and the way the amplitude varies with distance, it became clear that the Moon has a crust of distinctly lower P-wave velocity than the mantle that lies below, proving that the Moon is a differentiated body. The data provided by large meteoroid impacts further strengthened this interpretation. Deep moonquakes, which occur at great depths, provided additional information about what lies between the lunar crust and the depths of deep moonquakes at around 1000 km. Some such velocity distributions determined earlier are included in Figure 15.8 in Chapter 15. One thing clear is that the seismic velocities in the lunar mantle above the level of deep moonquakes are nearly constant and even decrease somewhat with increasing depth. This suggests that this part of the lunar mantle is of relatively uniform composition. The decrease of shear-wave velocity with increasing depth in the upper mantle is likely to occur when the effect of increasing temperature exceeds the effect of increasing pressure. Recent updates of these velocity profiles through improved analyses are given in Chapter 15.

Mainly because the Apollo seismic network covered only the front center of the Moon, there were not many well-defined seismic rays crossing below the depths of deep moonquakes, and thus the seismic velocity structure below about 1000 km depth, that is the lower half of the Moon, was not well defined. Nevertheless, there were some data that gave certain clues as to what lies below. One such clue is the loss of shear-wave arrivals from deep moonquake and surface impact sources beyond certain distances. This strongly suggests that shear waves penetrating below the level of deep moonquake foci are attenuated. This may indicate that the lunar mantle beneath the level of deep moonquakes is either partially molten or retains enough volatiles to attenuate shear waves (Nakamura *et al.*, 1973). Another clue is a greatly delayed P-wave arrival from a meteoroid impact near the antipode. There was only one such case observed, but it suggests that there may be a molten core of at most 360 km radius at the center of the Moon (Nakamura *et al.*, 1974a). Figure 7.4 shows a schematic view of the lunar interior as known at the end of the Apollo project. The existence of the partially molten lower mantle and liquid core of the Moon was confirmed with recent re-analyses of the data. These developments are discussed in detail in Chapter 15.

7.2.5 Seismicity of the Moon

The reason why a large number of moonquake events were observed is not because the Moon is seismically as active as the Earth but because the Moon is lacking the atmosphere, ocean, and creatures that generate most of the ground noise on Earth, and thus the seismometers could be operated at a significantly higher sensitivity than possible on Earth. If the Apollo seismometers were operated at a similar sensitivity as most terrestrial seismometers, only a very few shallow moonquakes and large meteoroid impacts and none of the deep and thermal moonquakes would have been detected.

Seismic events equivalent to deep moonquakes and thermal moonquakes may occur on Earth, but even if they do, the high ground noise here prevents their detection. Most

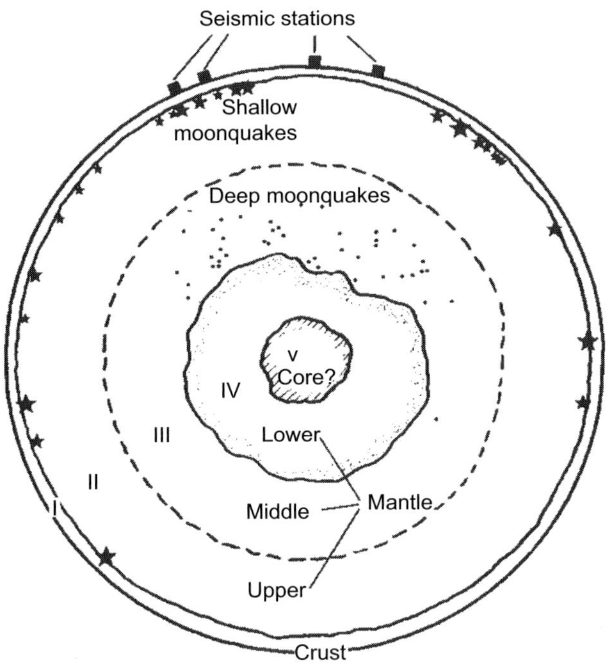

Figure 7.4 Schematic view of the lunar interior as known at the end of the Apollo project. (From Nakamura *et al.*, 1982.)

meteoroids falling on Earth disintegrate before reaching the ground and thus do not generate observable seismic signals. This leaves only shallow moonquakes for a possible basis of comparison with Earth's seismicity. However, keep in mind that most of the seismic energy on Earth is released along plate boundaries while the Moon's lithosphere is too thick to allow relative plate motion. If anything, shallow moonquakes may have some similarity to intraplate earthquakes.

The total amount of energy released by shallow moonquakes is estimated to be about 10^{12} J/year (Oberst, 1987). This is significantly less than the estimated energy released by earthquakes of about 10^{18} J/year (Kanamori, 1977). The Moon is a significantly smaller planetary body than the Earth, and thus has cooled faster, and the current tectonic activity is significantly reduced compared with the Earth. Nevertheless, the Moon is by no means tectonically inactive. The current rate of seismic energy release may provide a constraint on inferring how the Moon originated and evolved.

7.2.6 Free oscillation of the Moon: Normal modes

Prior to acquisition of seismic data, free oscillation of the Moon, with an estimated period of the ${}_0T_2$ mode at about 960 s, was seriously considered as a possible means of distinguishing various proposed internal structural models of the Moon (Derr, 1969). Thus, detection and identification of normal modes generated by large meteoroid impacts was one of the objectives of the PSE experiment. Unfortunately, this was not accomplished at a satisfactory level during the Apollo project. The main reason for this was that the PSE long-period seismometers turned out to be somewhat unstable, mainly because of the very large temperature fluctuations on the lunar surface, and no clear normal modes were observed following large meteoroid impacts. Even then, an attempt was made to extract normal modes among the noisy signals at the low-frequency part of the recorded data, but the result was inconclusive (Loudin, 1979).

A recent more sophisticated computer analysis, however, was able to extract some meaningful information on free oscillation of the Moon. This result is discussed in more detail in Chapter 15.

7.2.7 Roles that lunar seismology has played in advancing seismology in general

Observing a natural phenomenon in an environment significantly different from what we have here on Earth is in itself a highly rewarding exercise, especially in a discipline such as seismology, where a laboratory experiment to simulate an entire planetary body is virtually impossible, and the Apollo project provided the first opportunity for us to do this exercise. This happened when it was realized that most seismic signals recorded on the Moon were not easily explainable by using well-established wave propagation models governed by a wave equation. It became clear that recorded lunar seismic signals are explained more readily as a result of intense scattering of seismic waves, as described above. There had been some limited attempts to explain seismic coda here on Earth with single scattering, but the scattering observed on the Moon was much more intense than what can be explained by single scattering models, and different thinking was needed. This developed into a framework to describe very intense scattering of seismic energy governed mainly by a diffusion equation (Nakamura, 1977). Having lunar seismic signals with unusually prolonged codas, which in themselves contain many valuable pieces of information on what lies between the source and the receiver, helped us to realize the importance of looking at the seismic codas of earthquakes (Aki and Chouet, 1975; Dainty and Toksöz, 1981), and many developments that followed in earthquake seismology are still continuing today.

The long codas associated with lunar seismic signals were difficult to decipher for seismologists accustomed to dealing with impulsive arrival phases in earthquake signals. However, long codas do contain a wealth of information not found in impulsive arrival phases. The question is how to extract such information. Fortunately, the Apollo seismic data were acquired in pure digital forms, and thus were ready for computer processing. One approach that turned out to be highly successful was to cross-correlate the long codas to extract "hidden" information. By cross-correlating matching waveforms from two deep moonquakes occurring at two different times and observed at the same station, it was possible to determine their relative hypocenter locations very accurately, far more accurately than obtainable from reading the arrival times of impulsive phases (Nakamura, 1978). This was the beginning of what we now call seismic interferometry, now extensively used in earthquake seismology and exploration geophysics (e.g., Schuster, 2009).

These examples amply demonstrate that whenever we acquire data on an extraterrestrial body, which likely will present an environment significantly different from Earth, we learn something fundamentally new, and thus this develops into a new way to look at seismic phenomena from an entirely different perspective. This clearly is one important benefit of such an endeavor.

7.3 Apollo Active Seismic Experiment and Lunar Seismic Profiling Experiment

Two types of active seismic experiments, the Active Seismic Experiment (ASE) at the Apollo landing sites 14 and 16, and the Lunar Seismic Profiling Experiment (LSPE) at the Apollo 17 landing site, were conducted mainly to derive the shallow lunar structure from a few meters to at most a kilometer or so in depth. Thus, they complemented the results from the Passive Seismic Experiment described above, which mainly addressed much deeper structures and the seismicity of the Moon.

7.3.1 Active Seismic Experiment

Each of the ASEs conducted at the Apollo 14 and 16 landing sites consisted of a 91 m linear array of three evenly spaced geophones, a thumper device with small explosive initiators fired at 4.57 m intervals along the line of geophones, and a mortar package designed to launch grenades up to a distance of 1.5 km. The mortars, however, were not launched at the Apollo 14 site, and those at the Apollo 16 site were limited to a maximum distance of 450 m from the most distant geophone because it was feared that launching of mortars might damage the ALSEP instruments nearby.

The Apollo 14 site was in the Fra Mauro region in Mare Procellarum and the Apollo 16 site was in the southeastern highland region of the Moon. The results from these experiments showed that these sites were covered with an 8.5 m (Apollo 14) and 12.2 m (Apollo 16) thick layer of P-wave velocity 104 m/s and 114 m/s, respectively, suggesting porous or highly fractured and brecciated rock material representing the lunar regolith. This is underlain by a layer of P-wave velocity 299 m/s (Apollo 14) and 250 m/s (Apollo 16) and at least 1 km thick. This velocity is too low for a consolidated bedrock, suggesting highly fractured material (Kovach and Watkins, 1973a).

7.3.2 Lunar Seismic Profiling Experiment

The LSPE, conducted at the Apollo 17 site in Taurus-Littrow at the southeastern edge of Mare Serenitatis, consisted of a T-shaped array of four geophones, with one placed in the middle and maximum separation of ~100 m, and a set of eight explosive packages placed up to ~2.7 km from the array and detonated after the astronauts left the Moon. In addition, the impact of the ascent stage of the Lunar Module at a distance of ~9 km from the array was also recorded (Kovach *et al.*, 1973). The experiment was designed to investigate lunar near-surface structure somewhat deeper than those covered by the ASE.

In addition to the top layer with P-wave velocity 100 m/s, corresponding to the lunar regolith, layers of P-wave velocities 327 m/s, 495 m/s, 960 m/s, and 4700 m/s were found starting at depths of 4 m, 32 m, 390 m, and 1385 m, respectively (Cooper *et al.*, 1974). The top two layers are similar to those at the Apollo 14 and 16 landing sites; only the thickness of the regolith at the top is thinner than the other sites. The bottom layer of 4700 m/s was later found to be an artefact of incorrectly transcribed timing information, and the new analysis indicates the 960 m/s layer extends at least to a depth of 3 km or deeper (Nakamura, 2011).

The LSPE was also operated in its listening mode from time to time to listen to natural seismic events occurring in and on the Moon; because of its high-frequency response, it recorded mostly thermal moonquakes occurring near the station. Recently, with the advent of new analysis techniques in seismology, these events were used as ambient noise sources to evaluate the dispersion of Rayleigh waves and thus to infer the shallow structure under the station. This topic is discussed in more detail in Chapter 15.

7.4 Viking seismic experiment on Mars

During the Viking mission, conducted from 1976 to 1978, two seismometers, each containing three orthogonal short-period sensors, were deployed, one on Viking Lander 1 and another on Viking Lander 2. However, the seismometer on Lander 1 did not uncage, leaving only one operational seismic station at the landing site of Lander 2 in the Utopia Planitia region (Anderson *et al.*, 1977).

Besides the fact that only one seismic station was operative, denying the capability to positively identify naturally occurring teleseismic events, the experiment suffered from many deficiencies not experienced in normal seismic operations. To minimize the weight and complexity, the seismometer was not deployed on the ground but was mounted on the lander, thus increasing the intensity of noise originating from wind and lander activities. The amount of data allocated to the experiment was quite limited by the conflicting demands of other experiments of higher priority and also by the narrow bandwidth and time windows available for data transmission, forcing the use of extreme data compression with significant loss of information contained in normal seismic waveforms. In all, about 1 gigabit of seismic data was acquired.

Despite these severe constraints, the experiment did provide some information useful for future missions. Several earthquake-like events were observed during about one and a half Earth years of observation, excluding the periods when the wind noise was too high to detect most seismic events. Most of these events turned out to be associated with sudden wind gusts recorded by the meteorology experiment, but one occurred when the meteorology experiment was not operational and it had characteristics somewhat different from the others (Anderson *et al.*, 1977). This could be a marsquake, but its identity is highly uncertain. Goins and Lazarewicz (1979) compared this with what would be expected if the seismicity of Mars were about the same level as that of intraplate earthquakes. They concluded that detection of zero to one event is consistent with what might be expected, considering that the landing site is far from any areas, such as the Tharsis province, where high seismic activity is expected. It is now obvious that in future missions we will need seismometers of significantly higher sensitivity than the Viking seismometers if we hope to detect marsquakes useful for inferring internal structure of Mars in a reasonable observation period. Deriving the internal structure of Mars is one of the primary objectives of seismic observation on Mars. This is because theoretically derived internal structural models of Mars (e.g., Anderson, 1972; Johnston and Toksöz, 1977) are inherently non-unique.

Since the Viking seismometer was mounted on the lander, it responded to wind activities (Nakamura and Anderson, 1979). The Martian seismic data clearly recorded diurnal and seasonal variations in wind activity on Mars. There is a recent effort to use this to study Martian dust storms, as seismometers generally record activities such as this at significantly higher sample rates than did the instruments used in meteorology measurements (Lorenz and Nakamura, 2013). Unique old data such as the Viking seismology data often contain highly valuable information of which the original investigators were not aware.

7.5 Remaining questions in lunar and planetary seismology and future work

Even though the Apollo seismic experiments provided highly valuable data, they have many severe limitations in comparison with terrestrial seismic data. Major limitations are insufficient spatial extent of coverage, paucity of seismic stations, short duration of observations, narrow bandwidth, and large thermal noise. Because of these limitations, many questions remain, including: the petrologic composition of the lunar interior, which will require more precise knowledge of seismic velocity distribution; lateral structural variations, which would require a much larger areal extent of station coverage; size and composition of the core, which has been inferred but not fully confirmed; possible discontinuities within the lunar mantle, which may require more precise identification of arrival phases; and detailed structure of individual features, such as mascon basins, which would require regional and local networks.

Some of these questions can be addressed with further analyses of the existing data using new analysis techniques, as evidenced by recent works discussed in Chapter 15. One should be aware, however, that these data were acquired before any standard format for digital seismic data existed, and thus the original data are in pure binary strings. Furthermore, these old data are likely to contain many irregularities, such as bit errors, that need to be dealt with properly. Once these limitations are understood, one should not be surprised that they contain a wealth of hidden information.

Other questions, however, must wait till we have another chance to acquire additional seismic data from the Moon. How to address these questions with limited resources in future missions to the Moon will be a big challenge. It may require some unconventional, entirely new ways of acquiring seismic data.

The Viking seismic data from Mars were much more limited in all aspects. As with the Apollo data, further analyses of the existing data with new analysis techniques may lead to new discoveries, but answering most questions must wait till more seismic data are acquired. It is sincerely hoped that the InSight mission to Mars, described in Chapter 3, will give us long-awaited answers. The internal seismic structures of other planetary bodies are completely unknown beyond purely hypothetical modeling exercises at this time. As discussed above, past observations taught us that whenever we acquire data from an extraterrestrial body for the first time, we always find something entirely unpredicted. So, be prepared for the unexpected. The opportunity in extraterrestrial seismology is wide open.

Part III

Modeling approaches

8

Modeling approaches in asteroseismology

CONNY AERTS

As explained Chapters 4 and 5, even basic asteroseismic diagnostics, such as the frequency of maximum power combined with the large frequency separation for solar-like pulsators (e.g., Chaplin *et al.*, 2011a), or the period spacing for high-order gravity-mode pulsators (e.g., Degroote *et al.*, 2010), already lead to values of the global stellar parameters, such as the mass and radius, with a precision far better than that deduced from photometric color indices, spectral lines, or interferometric data. The combination of frequency and period spacings of the dipole mixed modes detected in evolved low-mass stars, allows us to distinguish between red giants with only hydrogen-shell burning or with core-helium burning in addition (Bedding *et al.*, 2011), a property that classical data cannot reveal.

The real benefit of asteroseismic versus classical data becomes even more obvious when modeling the individual oscillation frequencies $\nu_{n\ell m}$, since they are dependent on, and thus determined by, the details of the interior physics of the stars. Such seismic modeling of the solar oscillation frequencies led to a large step forward in the understanding of the structure of the Sun (Chapter 12), including its interior rotation profile (e.g., Kornilov *et al.*, 2012; Thompson *et al.*, 2003). Likewise, seismic modeling of the oscillation frequencies of other stars offers a unique opportunity to probe the internal physical processes of stars of various birth mass, initial chemical composition, and evolutionary stage. As an example, we shown in Figure 8.1 the observed frequency spectra from three months of *Kepler* photometry of the binary components and solar analogues 16 Cyg A and B. Finding the correspondence between the observed frequency peaks and those predicted by theory is not only an immensely powerful method to determine the basic fundamental properties of stars, such as their mass, radius, and age, but it also allows us to evaluate standard stellar structure models and improve their input physics. It is a lucky circumstance that stellar oscillations occur in very different types of stars in almost all phases of stellar life (e.g., Chapter 2 of Aerts *et al.*, 2010), offering the potential of a seismic calibration of the theory of stellar structure and evolution.

Extraterrestrial Seismology, eds. Vincent C. H. Tong and Rafael A. García. Published by Cambridge University Press.

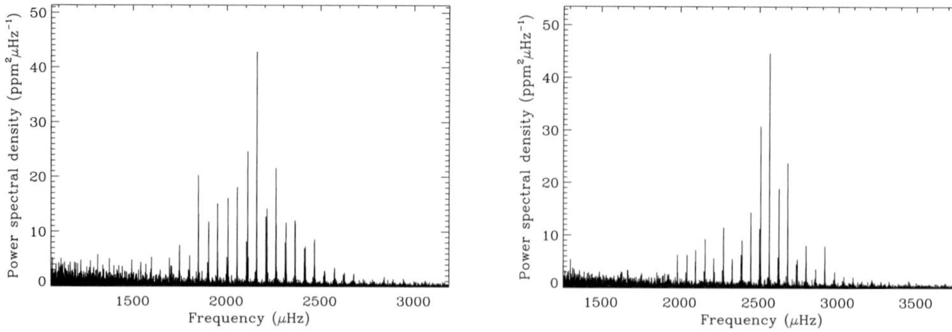

Figure 8.1 Observed pulsation frequency spectra of 16 Cyg A (left) and B (right) from *Kepler* photometric data. (Figure reproduced from Metcalfe *et al.*, 2012, with permission.)

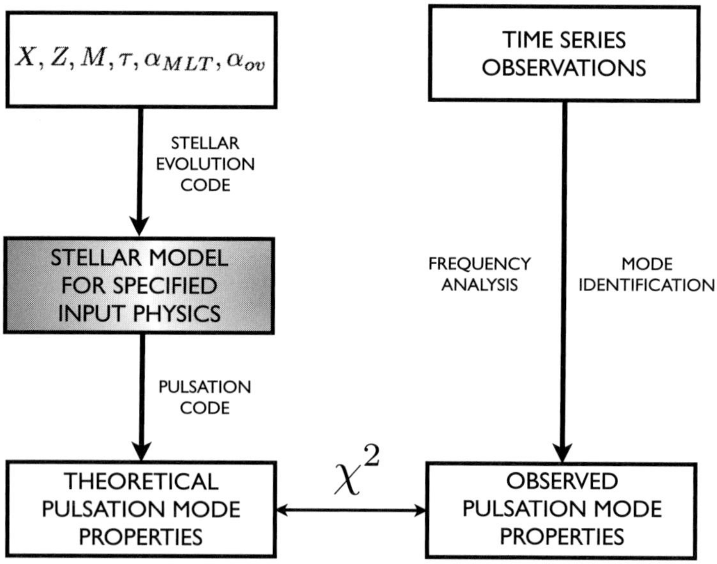

Figure 8.2 Context diagram showing schematically the adopted procedure when performing forward seismic modeling of a star.

8.1 The art of seismic modeling

In seismic modeling, the aim is to make a detailed comparison between the properties of observed and of theoretically predicted oscillation modes, up to the level of the precision implied by the data. This is schematically shown in a context diagram in Figure 8.2. In order to achieve that, one must thus rely on equilibrium models of stellar structure on the one hand, and on theoretical pulsation predictions on the other hand. In addition, observational seismic diagnostics must be derived from time series of data. Let us summarize the most important ingredients of the left side of Figure 8.2 while discussing a few case studies for stars of very different mass along the way.

8.1.1 Stellar models

Stellar models serve a great variety of research in astrophysics, encompassing not only stellar physics, but also exoplanetary, galactic, and cosmological physics. It is thus fair to state that stellar models are a fundamental basis for much of modern astrophysics. This is the reason why immense efforts have been put into the computation of such models, with various levels of sophistication on numerical issues and on the input physics, depending on whether the models should serve a general purpose or a particular concrete highly demanding application. Here, we give only a very brief summary of the various aspects that come into play in stellar model computations, referring the reader for details to the latest versions of textbooks on the subject and the numerous references therein, e.g., Maeder (2009); Hansen *et al.* (2004); Kippenhahn *et al.* (2013). Stellar model computations and comparisons with specific emphasis on asteroseismology have been discussed and described extensively in the context of the CoRoT mission in the compilation by Monteiro (2009).

Aside from the important numerical and implementation aspects of computer codes developed to compute stellar structure and evolution models, which we shall not discuss here, one needs to make assumptions on the input physics on which to rely. This concerns aspects of both *microphysics* and *macrophysics*. The general equations of hydrodynamics are used, such as the equation of continuity, the equations of motion, and the energy equation, with approximations valid for the conditions in stellar interiors. A lucky circumstance is that the photon mean-free path is short compared with the scale of the changes in the structure for the largest part of the star, which allows use of the diffusion approximation for the radiative energy transport. This is no longer valid in the stellar atmosphere, where a detailed treatment of radiative transfer is necessary. Therefore, the conditions at the outer layer where the diffusion approximation is still valid are usually matched to those of atmosphere models.

To treat the microphysics of stellar matter and compute the thermodynamic state of the gas appropriately, the initial composition in terms of hydrogen, helium, and metal mass fractions (X, Y, Z) has to be chosen, including the details of the metal mixture. Subsequently, an equation of state (EOS), which describes how the stellar density connects to the pressure and temperature for a given chemical composition, must be adopted. Given that density, pressure, and temperature may differ by many orders of magnitude inside a star, the EOS needs to cover various validity domains. This is usually ensured by adopting different formulations of the EOS and by smoothing the transitions between them (see, e.g., Paxton *et al.*, 2011, for a detailed discussion). Radiative and conductive opacities for the various temperature ranges are needed and, here as well, various opacity sources can be combined to cover the necessary ranges. Thermonuclear reaction rates for numerous reaction networks, as well as weak nuclear reactions and neutrino losses, must be considered to describe the nuclear energy generation and to compute the change in chemical composition as the star evolves, keeping in mind that thermal and concentration diffusion, gravitational settling, and radiative levitation might all come into play, depending on the kind of star and on its evolutionary stage.

While the microphysics relevant for stellar structure is fairly well known, except for the aspects of atomic diffusion just mentioned, this is far less so for various aspects of the macrophysics. Quite often, large simplifications are made, such as the omission of rotational motion, magnetic fields, and stellar winds. This is justified as long as the corresponding Coriolis and centrifugal, Lorentz, and dust- or line-driven radiative forces are small compared to the pressure and gravity forces. Fortunately, such simplified treatments work quite well for many of the stars and during large fractions of their life, but certainly not always.

Even in simplified models where the forces connected with rotation, magnetism, and wind are neglected, an additional major shortcoming is the simple treatment of convective motions. These occur in regions where the radiative temperature gradient exceeds the sum of the adiabatic temperature gradient and the gradient of the mean molecular weight. Convective motions and the transport they induce are usually treated by a highly simplified parameterized law. The one most often used is the so-called *mixing-length approximation* (Böhm-Vitense, 1958), in which the average scale of the convective eddies is described by a dimensionless free parameter $\alpha_{\rm MLT}$, which is expressed in units of the local pressure scale height denoted as H_p, although other treatments are also used (e.g., Canuto and Mazzitelli, 1991). Stellar matter is assumed to be fully mixed in convective zones, while partial mixing of chemical species may occur due to *semi-convection*. This term stands for the oscillatory motion of gas blobs in regions where the difference between the radiative and adiabatic temperature gradients is less than the gradient of the mean molecular weight.

From a physical point of view, it is necessary to take into account that convective blobs do not stop abruptly at the boundary of a convective region. Rather, their inertia causes them to "overshoot" into the adjacent stable area where radiative energy transport takes place. The amount of such overshooting is unknown. For this reason, it is customary to express it as $\alpha_{\rm ov}\, H_p$, where $\alpha_{\rm ov}$ is expected to be a small fraction of $\alpha_{\rm MLT}$. The unknown values of convective overshoot and semi-convection for stars with a convective core have a major effect on their evolution. That is why the determination of $\alpha_{\rm ov}$ is a key goal of asteroseismology.

Adopting all of the abovementioned physical ingredients as "true" leaves us with five free parameters in the simplest case of models for a non-rotating non-magnetic star without convective overshoot, without atomic diffusion, and without wind: the birth mass M, initial hydrogen fraction X and metallicity Z, the mixing length parameter $\alpha_{\rm MLT}$, and the age τ. The latter is often replaced by the central hydrogen fraction X_c for the core-hydrogen burning phase. Such equilibrium models are the starting point of seismic modeling. For stars that have a convective core at birth, one should add the core overshoot parameter $\alpha_{\rm ov}$ as a sixth parameter in view of the large effect of this phenomenon on the evolution and age of the star.

For an extensive compilation of the comparison between stellar structure and evolution codes for low-mass stars, including details on the adopted input physics and the numerical aspects, we refer to Monteiro (2009). We also point to the notable addition of the publicly available MESA code (Paxton *et al.*, 2011, 2013) since 2011.

8.1.2 Stellar pulsation computations

In order to compare the observed stellar pulsation properties with those predicted by models, the theoretical pulsation spectrum connected with each model must be computed. This is achieved by considering small perturbations to the equations of stellar structure, so that we can work in a linear regime, and compute the perturbed quantities. Depending on whether the dominant restoring force is the pressure force or buoyancy, we call the modes *pressure* or p modes and *gravity* or g modes. One can show that the perturbed equations constitute an eigenvalue problem whose eigenvalues are the pulsation frequencies (Aerts *et al.*, 2010, Chapter 3). The solution is a superposition of eigenmodes, some of which are excited to observable amplitude while others are not. Depending on the excitation mechanism, the modes have short or long lifetimes, but this quantity is usually not considered in seismic modeling. The quantity used for forward modeling is the set of frequencies connected with each of the detected pulsation modes. In the example of 16 Cyg A and B, it concerns 46 and 41 trustworthy frequencies, respectively (cf. Figure 8.1).

As extensively discussed in Aerts *et al.* (2010), to which we refer for details, each of the non-radial modes of a spherically symmetric equilibrium star can be described in terms of a spherical harmonic $Y_\ell^m(\theta, \phi)$, an amplitude $A_n(r)$, and a cyclic frequency $\nu_{\ell,m,n}$ whenever one chooses the distance r to the center, the co-latitude θ, and the longitude ϕ as the coordinates. Here, θ is measured from the axis of symmetry, which is usually chosen to be the rotation axis of the star except for pulsators with a strong magnetic field, in which case the magnetic axis is a better choice. The spherical harmonics are usually defined such that the integral of $|Y_\ell^m|^2$ over the unit sphere equals 1. The meaning of the *quantum numbers* is as follows: n is related to the number of radial nodes and is called the *overtone* or *radial order* of the mode; ℓ is the *degree* of the mode and specifies the number of surface nodes that are present; m is the *azimuthal order* of the mode, where $|m|$ specifies how many of the surface nodes are lines of longitude. Hence, the number of surface nodes that are lines of co-latitude is equal to $\ell - |m|$, where m ranges from $-\ell$ to $+\ell$, i.e., there are $2\ell + 1$ modes for each degree ℓ. For a non-rotating star, these $2\ell + 1$ *multiplet components* have equal frequency.

Rotation lifts this degeneracy and gives rise to multiplets due to rotational splitting of the pulsation frequencies. When considering the effect of the Coriolis force and assuming that the angular velocity only depends on the distance to the stellar centre and not on co-latitude, such splitting is equidistant, leaving the central frequency corresponding to the axisymmetric ($m = 0$) mode unaltered. Rotationally split multiplets are hence well recognizable for slowly rotating stars whose higher-order rotation terms can be ignored.

Depending on the aspect angle of the star, i.e., the angle between the axis of rotation and the line of sight of the observer, one detects all the multiplet components or only a few of them. Moreover, the various multiplet components may be excited to quite different intrinsic amplitudes depending on how the mode energy is distributed over the multiplet components. Thus one cannot simply assume that the difference in amplitude is only due to the geometrical projection on the line-of-sight, although such an assumption is quite often made for stochastically excited modes in solar-type stars. For this reason, and because the aspect angle is usually unknown, one uses the frequency values to perform seismic modeling, rather than the amplitudes of the modes.

Various pulsation codes have been developed and made available to the scientific community. In many of those codes, the pulsational computations are done in the adiabatic approximation and are restricted to axisymmetric modes ($m = 0$). The adiabatic approximation does not give any information on mode excitation. Some codes treat the fully non-adiabatic case and allow the prediction of whether the modes are excited or not with the given input physics, as well as delivering information on the detailed properties of the eigenfunctions in the outer atmosphere where observable quantities other than the frequency values are determined. Some codes also include higher order effects of the rotation in the pulsational computations in a perturbative approach, i.e., while still relying on spherically symmetric equilibrium models. This is a valid approach as long as the ratio of the rotational frequency to the pulsational frequencies remains modest (typically below 20%) and as long as the star can be considered as a sphere. Again, we refer to Monteiro (2009) for a compilation of pulsation codes, the comparison of their performance, and their basic assumptions on the mathematical, physical, and numerical treatment of pulsation theory.

One of the great challenges in seismic modeling is the labeling of each detected frequency ν with its appropriate wavenumbers (ℓ, m) from observations, i.e., the issue of *mode identification*. This is a necessary step before we can compare the observed frequencies with those predicted from the theory of stellar pulsations. Whenever complete quasi-equidistant multiplets occur, it is easy to identify (ℓ, m) and hence to find the frequency of the central axisymmetric mode component of $m = 0$, but this feature is not always available.

For low-degree solar-like oscillations in the asymptotic regime of high-order p modes, it is usually possible to identify ℓ from so-called *échelle diagrams*, introduced more than 30 years ago for the solar oscillations (e.g., Grec *et al.*, 1983). Such a diagram is obtained by cutting the frequency axis of the oscillation spectrum into pieces whose length equals the large frequency separation (for a definition, see Chapter 5) and stacking the pieces above each other, as shown in Figure 8.3 for 16 CygA and B and discussed in Metcalfe *et al.* (2012). For these particular stars, rotational splitting was not detected, given the slow rotation of the stars and the relatively short time base of the data (3 months).

Low-degree high-order g modes, on the other hand, give rise to almost constant period spacings and this also allows us to identify the degree of such modes. This was first put into practice for white dwarfs (Winget *et al.*, 1991) and is nowadays also within reach for massive core-hydrogen burning stars thanks to CoRoT (e.g., Degroote *et al.*, 2010).

Sophisticated methods to identify ℓ from high-precision multicolor observations and/or (ℓ, m) from high-resolution, high signal-to-noise spectroscopic time series have been

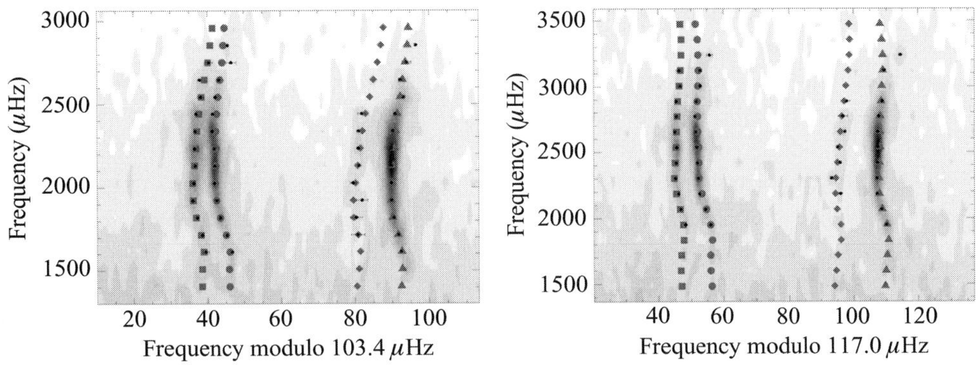

Figure 8.3 Échelle diagrams of 16 Cyg A (left) and B (right) from *Kepler* data. The observed frequencies are indicated as black points with error bars; the frequencies from the best fitting model for each of the two components are indicated with different symbols depending on the mode degree. (Figure reproduced from Metcalfe *et al.*, 2012 with permission.) A black and white version of this figure will appear in some formats. For the color version please refer to the plate section.

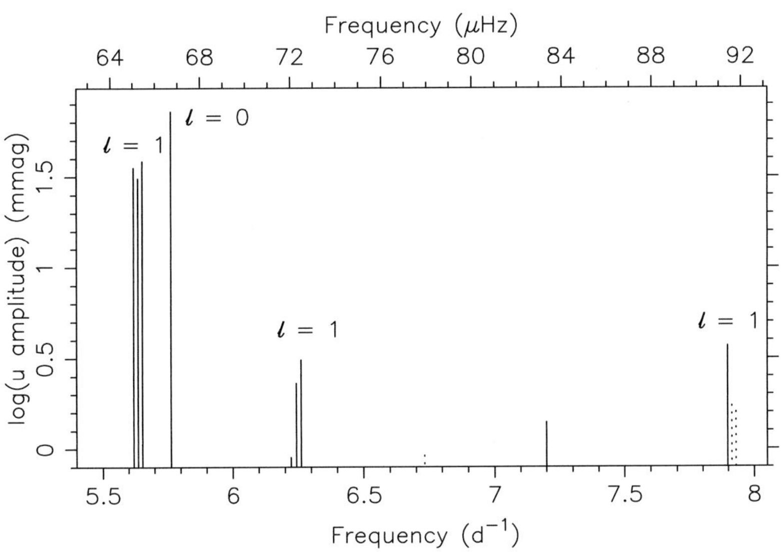

Figure 8.4 Mode identification of the p-mode frequencies detected for the B-type star ν Eri from combined photometric and spectroscopic time series. (Reproduced from Aerts *et al.*, 2010, with permission.)

developed for heat-driven pulsators with sparse frequency spectra and modes outside the asymptotic frequency regimes (see Chapter 6 in Aerts *et al.*, 2010). A nice example of such an application, including firm triplet detections, was achieved for the p-modes of the B-type star ν Eri (Figure 8.4).

Identification of the radial order n must necessarily come from comparison with models as it concerns a wavenumber connected with interior nodal surfaces. This is well achievable when ℓ is unambiguously labeled in a slowly rotating star and when good estimates of the effective temperature, gravity, and metallicity are available for the star.

We conclude that mode identification is well within reach for slowly rotating pulsators whose rotationally split multiplets are not merged and for which spherically symmetric equilibrium models combined with predictions of rotational splitting due to the Coriolis force are good approximations. In the rest of this chapter, we restrict the discussion to pulsators whose mode degrees ℓ have been unambiguously identified prior to seismic modeling.

8.2 Forward asteroseismic modeling

As a first step in the modeling of the star, the detected axisymmetric ($m = 0$) modes of the star are used to determine its global properties ($M, X, Z, \tau, \alpha_{\rm ov}$). This is done by computing equilibrium models, predicting their zonal ($m = 0$) oscillation properties from a pulsational analysis and comparing the predicted properties with the observed ones, taking into account the error of the latter. In order to achieve this, the wavenumbers (ℓ, m) of the oscillation modes must have been identified from observations, while the radial order n is usually sought during the minimization process. Depending on the type of star, the detected frequencies themselves are used, or else combinations thereof, such as the large or small frequency separations or period spacings, or even more suitable combinations.

For the modeling, "classical" observational diagnostics and their uncertainties, such as the effective temperature, gravity, metallicity, luminosity, distance, etc., are often used, provided that their errors are sufficiently small for them to be of practical use. In the case of eclipsing double-lined spectroscopic binaries, a model-independent mass of the components is available, while bright stars observed interferometrically may offer the advantage of having a model-independent radius determination (e.g., Cunha *et al.*, 2007, for an extensive review). These non-seismic observables can either be added to the pulsational properties used in the minimization procedure discussed below, or, if their errors are too large compared to those of the seismic observables due to unknown systematic uncertainties, be used to restrict the range of equilibrium models to consider by limiting to those that fulfill the classical observables.

As discussed extensively in Chapter 7 of Aerts *et al.* (2010), which includes numerous applications to various types of stars, one seeks to minimize a merit function for a given set of parameters $\mathcal{P} = \{P_i\}, i = 1, \ldots, N$ from a set of observables $\mathcal{O} = \{O_j\}, j = 1, \ldots, M$. It is assumed that the measured values $O_j^{\rm obs}$ are each accompanied by a standard error deduced from a proper statistical analysis that took into account correlations in the data as well as systematic uncertainties. These deduced errors σ_j are then assumed to be uncorrelated. Each observable O_j is assumed to be dependent on the parameters P_i: $O_j = O_j^{\rm theo}(P_i)$. A least-squares minimization is then performed:

$$\chi^2(P_1, \ldots, P_N) \equiv \sum_{j=1}^{M} \left[\frac{O_j^{\rm obs} - O_j^{\rm theo}(P_1, \ldots, P_N)}{\sigma_j} \right]^2 . \tag{8.1}$$

Finding the absolute minimum of this χ^2 is far from trivial. The problem is usually treated in two steps. First, one considers an extensive grid of stellar models, covering the appropriate ranges of the relevant parameters, with the help of the classical observables. With current CPU and data storage capacities, grids with millions of stellar models and their pulsational properties are feasible. Nevertheless, a second step is usually required, where non-linear optimization techniques such as genetic algorithms or neural networks are often considered, starting from the models that represent the few minima found in the grid and computing additional models and their oscillation properties in order to reduce χ^2 further and converge in this way to the absolute minimum. Some minimization codes skip the grid search and do a global search in a broad parameter range coupled at once to a genetic algorithm to find the global minimum.

8.2.1 Application to solar-like stars

In the case of solar-like oscillations in stars with a convective envelope, an additional step after frequency analysis and mode identification is usually necessary. Indeed, as is well known for the Sun, there is a global offset between the observed frequencies and those predicted by theoretical models, which is probably due to imperfect treatment of the physical processes in the outer stellar atmosphere, also known as *surface effects*. Kjeldsen *et al.* (2008a) adopted a pragmatic approach and deduced an empirical correction for the frequency offset, based on the solar case, which is used before the frequencies are matched according to the procedure defined in Equation (8.1).

The application of this method to 16 Cyg A and B, after correction of the surface effects, considered both the case of using only the detected and identified pulsation frequencies as additional observables $\mathcal{O} = \{O_j\}$, $j = 1, \ldots, M$ with $M = 46$ and 41 for components A and B, respectively, and the case where the effective temperature, gravity, metallicity, and luminosity based on the parallax were added as observables. The considered set of parameters to be estimated, $\mathcal{P} = \{P_i\}$, $i = 1, \ldots, 5$, consisted of the mass, age, initial metallicity and hydrogen fraction, and $\alpha_{\rm MLT}$, $\alpha_{\rm ov}$ not being of dominant importance for solar-like stars with a radiative core and representing *undershooting* from the convective envelope into the radiative zone. Both minimization procedures led to consistent results, which are listed in Table 2 of Metcalfe *et al.* (2012) and omitted here for brevity. In order to get a handle on the systematic uncertainties of the modeling due to various treatments of the input physics and numerics of the evolution and pulsation codes, six teams performed the modeling with independent codes and the results were used to derive the systematic uncertainty. This led to a final radius, mass, and age estimate with relative precisions better than 1%, 2%, and 5%, respectively. In such a successful case, the model obtained from the χ^2 minimization based on the axisymmetric mode frequencies forms the basis for further analysis, e.g., the study of the internal rotation of the star, as will be discussed in the next section.

The recent application of this same method to 22 somewhat fainter but still bright (to *Kepler* standards) solar-type stars has shown that the radii and masses can in general be determined with relative precisions of typically 1% to 2%, ignoring the systematic uncertainty due to the model dependency (Mathur *et al.*, 2012). Seismic ages are typically at the level of 10% relative uncertainty because $\alpha_{\rm ov}$ is of no importance for this stellar mass range. Such precisions are also within reach for the brighter *Kepler* exoplanet host stars (Fogtmann-Schulz *et al.*, 2014), but the fainter ones have lower signal-to-noise ratio data and typical uncertainties of 3% and 7% in radius and mass, respectively (Huber *et al.*, 2013a).

8.2.2 Application to slowly rotating massive stars

Let us now consider an entirely different mass regime and turn to massive OB-type stars, for which models are still highly uncertain. While relatively few such stars have been monitored by MOST, CoRoT, and *Kepler*, great progress has been made also from ground-based multisite campaigns. Seismic modeling of carefully chosen OB-type pulsators based on the identified axisymmetric modes worked in the sense that it delivered far more accurate model parameters than any other method, but led at the same time to the conclusion that changes in the input physics of the models are necessary. This situation has so far only been reached for pulsating OB stars in the core-hydrogen burning phase, for which $\mathcal{P} = \{P_i\}, i = 1, \ldots, 5$ consists of the mass, central hydrogen fraction (which is a good proxy for the age as the core is fully mixed), initial metallicity and hydrogen fraction, and $\alpha_{\rm ov}$, as in this case the precise value for $\alpha_{\rm MLT}$ does not have any impact on the models for stars with a convective core as long as reasonable values (typically between 1.5 and 2.5) are taken.

In the case of heat-driven pulsators, even a handful of well-identified frequencies, as in Figure 8.4, is sufficient to perform seismic modeling when it concerns low-order low-degree modes, because these probe the entire star. Aside from good global seismic parameters, shortcomings in the input physics were found in the sense that the lowest frequency g modes and the highest frequency p modes detected in all the carefully monitored OB pulsators are predicted to be stable. So either too low opacities of iron-group elements in the driving zone near 2×10^5 K occur and/or unknown physical processes causing extra mode excitation take place in such massive stars. A recent summary of the seismic estimates of $\alpha_{\rm ov}$ and of the rotational frequency for OB stars is available in Aerts (2013).

8.2.3 Application to low-mass evolved stars

It is probably fair to state that the largest progress in asteroseismology since CoRoT and *Kepler* became operational occurred in the area of evolved low-mass stars in the stage of hydrogen-shell burning (the so-called red-giant branch) and core-helium burning (in the red clump on the horizontal branch). This progress was made possible thanks to two major discoveries. The first one concerns the detection of stochastically excited non-radial modes whose identification of the mode degree ℓ is easily achieved from échelle diagrams, based on CoRoT data (De Ridder *et al.*, 2009). This was followed by the discovery of mixed dipole modes, spanning the whole range of gravity-dominated to pressure-dominated modes, from *Kepler* data (Beck *et al.*, 2011; Bedding *et al.*, 2011). The detailed analysis of red-giant pulsation spectra meanwhile allowed a direct application of the theoretical framework presented by Dupret *et al.* (2009) several years prior to the discovery of dipole mixed modes. For recent achievements in seismic modeling, and its input and implications for galactic structure studies, we refer to Chapter 11 and the numerous references therein, while we come back to the internal rotation of red giants in the next section.

8.3 The internal rotation profile of stars

As already mentioned, rotation modifies the frequencies of the star's pulsation modes. Rotation implies a splitting of the frequencies into $2\ell + 1$ multiplet components. In a slowly rotating star, the $m = 0$ modes are connected with standing waves whose frequencies are not affected by the rotation, while the $m \neq 0$ modes represent traveling waves. The shift in the frequencies of the modes in a multiplet due to rotation is different for modes with different degree and overtone. Moreover, the value of the rotational shift of the multiplet components is determined by the interplay between the interior rotation rate of the star and the propagation cavity of the modes (see, e.g., Aerts *et al.*, 2010, for a detailed description).

Whenever rotationally split multiplets are detected, as in Figure 8.4, we can go beyond a standard seismic analysis as discussed in Section 8.2, and deduce information on the internal rotation profile as a function of depth inside the star. This requires, in addition to identification of the mode degree ℓ, the identification of the azimuthal orders m of the rotationally split mode components.

Before the era of space asteroseismology, internal rotation properties were first claimed to be detected from rotational splitting of high-order g modes of pulsating white dwarfs from ground-based photometry (e.g., Winget *et al.*, 1991, 1994). An inversion of the frequencies to deduce $\Omega(r)$ as it was achieved for the Sun from helioseismology (Thompson *et al.*, 2003) turned out to be unfeasible (Kawaler *et al.*, 1999). Moreover, it was shown that non-linear resonant mode-coupling computations could also lead to the observed rotational splittings of pulsating white dwarfs, even for rigid rotation (Goupil *et al.*, 1998). This matter remains unsettled.

The first firm proof of non-rigid interior rotation in a core-hydrogen burning star other than the Sun was obtained for the massive B-type pulsator V836 Cen. It was based on one detected triplet and one incomplete quintuplet, after 21 years of single-site photometric monitoring, leading to immensely precise frequency values (Aerts *et al.*, 2003). Moreover, massive stars have a radiative outer envelope, hence modeling of the observed frequencies is straightforward and does not suffer from the unknown surface effects occurring for convective envelopes. The result obtained for V836 Cen was quickly followed by an analoguous interpretation for a similar star, ν Eri, based on the frequencies deduced from a multisite photometric and spectroscopic campaign, some of which are shown in Figure 8.4 (Pamyatnykh *et al.*, 2004). The core-to-envelope rotation $\Omega_{\rm core}/\,\Omega_{\rm env}$ of V836 Cen and ν Eri

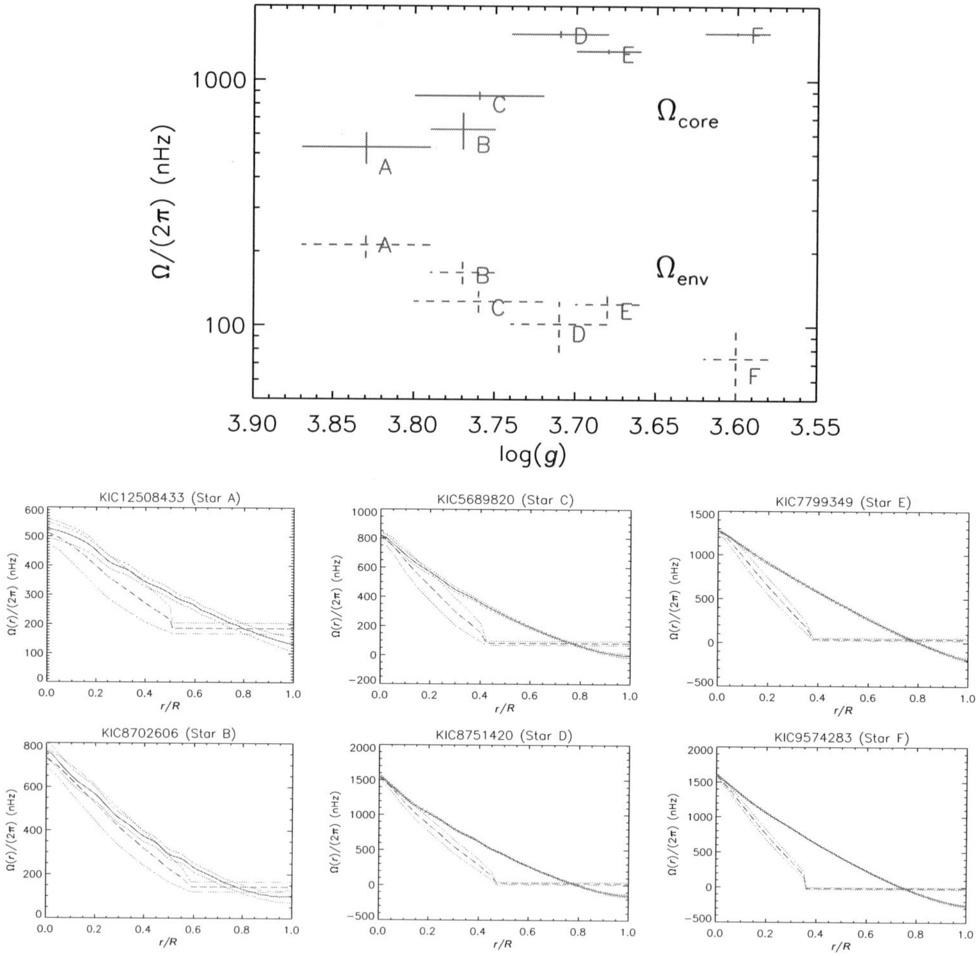

Figure 8.5 (Upper) Core and envelope rotation rates as a function of the surface gravity for six red giants, each specified by a character. The horizontal and vertical lines indicate 1σ error bars. (Middle and lower) Rotation profiles obtained relying on a smoothness condition on the rotation profile for the entire star (blue solid lines) or only in the radiative interior when the convective envelope is assumed to rotate rigidly (red long-dashed lines). The dotted lines indicate the 1σ error bars for both types of inversions. (Reproduced from Deheuvels *et al.*, 2014.) A black and white version of this figure will appear in some formats. For the color version please refer to the plate section.

amount to ~4 (Dupret *et al.*, 2004) and 5.5 (Dziembowski and Pamyatnykh, 2008), respectively. Rotational mixing in the near-core regions of massive stars is of crucial importance to improve their evolutionary models. Unfortunately, new insights into the internal rotation law of massive pulsators have not yet been achieved from CoRoT or *Kepler* data, because of lack of detections of rotational splitting for the few such observed stars in the field-of-view of these satellites. Several new case studies are underway, however, thanks to a complete variability classification of the entire *Kepler* database and the follow-up of study of the most interesting B-type pulsators by means of a Guest Observer program.

Two years of uninterrupted *Kepler* data allowed the derivation of $\Omega_{\rm core}/\Omega_{\rm env}$ for red giants (Beck *et al.*, 2012; Deheuvels *et al.*, 2012; Mosser *et al.*, 2012b). The ratio is roughly between 5 and 15, depending on the evolutionary stage of the star. This faster core than envelope rotation is typically an order of magnitude slower than theoretical predictions based on the assumption of local conservation of angular momentum without taking into account asteroseismic information (e.g., Eggenberger *et al.*, 2012a). The *Kepler* red-giant results for $\Omega_{\rm core}/\Omega_{\rm env}$ are all based on the detection of rotationally split mixed dipole ($\ell = 1$) modes. For such low-mass evolved stars, the level of mixed character of the dipole modes corresponds with the probing of different layers, the gravity-dominated modes being more sensitive to the rotation in the core region than the pressure-dominated modes.

In a pioneering study, Deheuvels *et al.* (2014) exploited the rotationally split dipole mixed modes of six young giants with a range of metallicities, including descendants of stars that had either a radiative or a convective core on the main sequence. The core and envelope rotation rates deduced from the seismic data, as well as the rotation profiles relying on two different conditions of smoothness, are depicted in Figure 8.5. The study resulted in estimates of the core rotation rates and upper limits for the envelope rotation. For this sample, the rotation contrast between the core and the envelope increases during the subgiant branch. For two of the giants shown in Figure 8.5, the discontinuous rotation profile with a deep discontinuity near the location of the hydrogen burning shell, which separates the layers that contract from those that expand, was found to reproduce the observed splittings better than a smooth rotation profile.

In conclusion, we have not yet reached the stage of mapping rotation laws inside stars in full detail. However, the exciting recent asteroseismic results pave the way to a much better understanding of interior rotation as well as the accompanying rotational mixing and angular momentum transport. Further exploitation of the full *Kepler* data sets is ongoing and will surely give rise to more as well as more accurate derivations of $\Omega(r)$ for stars with different masses and in various evolutionary stages in the near future.

8.4 Towards the improvement of stellar physics

The recent examples of seismic modeling discussed above and the additional ones in the recent literature and throughout this book barely relied on the full *Kepler* data sets. Moreover, the focus so far has been on the derivation of seismic estimates of basic stellar parameters, *assuming that the input physics of the models is correct*. While this way of working is certainly of major importance and led to essential insights, e.g., evolutionary stages, seismic ages, seismic distance estimates, limits on red-giant mass loss, etc., all of high relevance for various fields of research in astrophysics, *the improvement of stellar physics itself should be a major goal of asteroseismology* in the near future, just as helioseismology allowed us to drastically improve the physics of the solar model. This can only come from seismic modeling of carefully chosen individual pulsators through interpretation of their complete pulsational properties derived from the full *Kepler* lightcurves, *in combination with careful study of the impact of changes in various aspects of the input physics on the pulsations.* In other words, the gray central box in the left panel of Figure 8.2 must be tackled, now that CoRoT and *Kepler* have finished their nominal operations.

One particular aspect of the need for improved input physics did not come from the seismic modeling of individual stars but was derived for the class of subdwarf pulsators as a whole, where the *Kepler* data were complementary to ground-based data in showing that the instability strip for the g-mode pulsators among this class could not be explained by evolutionary models for such objects. This shortcoming was recently remedied by the full and consistent inclusion of thermal and concentration diffusion, as well as gravitational settling and radiative levitation for the chemical elements H, He, C, N, O, Ne, Mg, Fe, and Ni along the evolution (Hu *et al.*, 2011).

Similarly, the CoRoT and *Kepler* missions have shown that the majority of the heat-driven core-hydrogen burning pulsators of spectral types O to F are *hybrids*, i.e., pulsate in both p and g modes simultaneously. Current excitation computations for such stars predict far too few excited modes, which implies that the opacity is too low in the excitation layers. This is a clear illustration of shortcomings in the current input physics of the models. Even more challenging is the correct interpretation of the simultaneous occurrence of g-mode pulsations and of gravito-inertial modes in fast rotators, as recently discovered from CoRoT data of B-type stars (Neiner *et al.*, 2012; Pápics *et al.*, 2012).

Asteroseismology was very much observationally driven the past decade because the analysis teams were swamped with data of limited proprietory time, as is usually the case in space science. Now is a good time to change the research focus towards the improvement of the theory of stellar structure and evolution, starting from the observed asteroseismic properties revealed by thousands of stars.

9

Modeling approaches in helioseismology

S. P. RAJAGURU

9.1 Introduction

The theory of hydrodynamics or, where necessary, magnetohydrodynamics, provides an appropriate mathematical framework to study solar and stellar oscillations. Well-developed hydrodynamic descriptions of global oscillations of gravitationally bound gaseous spheres, aimed primarily at understanding stars such as Cepheid, delta-Scuti, and RR Lyrae variables, resulted from studies over half a century following that of pioneers such as Pekeris, Rosseland, and Chandrasekhar (see Ledoux and Walraven, 1958). The discovery of global non-radial oscillations on the solar surface (Leighton *et al.*, 1962) led eventually, though not immediately, to a successful utilization of such theoretical machinery for helioseismology. It was also realized early on (Goldreich and Keeley, 1977) that the excitation of small amplitude non-radial oscillations on the Sun is due to a mechanism different from that responsible for the typically large amplitude oscillations exhibited by the above variable stars (known as the classical variables), even though both these types of oscillations are acoustic in nature. Observed solar oscillations are now recognized as mainly due to acoustic waves generated by turbulent convection near the surface layers of the outer convection zone, intrinsically damped and stable, hence limited to small amplitudes and linear evolution (Goldreich *et al.*, 1994). These properties define the solar-like oscillations. There are also internal gravity waves, which, with buoyancy supplied by perturbations in gravity as the restoring force, propagate only in the sub-adiabatic (i.e., temperature gradients smaller than that required for convection) radiative interior beneath the convection zone, where they are evanescent. The surface gravity wave lives on the solar surface propagating horizontally.

9.2 Global helioseismology

Helioseismology consists of modeling the interior propagating acoustic and gravity waves, and the surface gravity wave, as small-amplitude linearly evolving perturbations to a background model of the Sun, and determining observable quantities to accuracies better than

Extraterrestrial Seismology, eds. Vincent C. H. Tong and Rafael A. García. Published by Cambridge University Press.

they are measured with. This enables updating the model iteratively for the best match with observations, thereby determining the internal structure of the Sun (Christensen-Dalsgaard, 2002). For global helioseismology, the main observables are oscillation frequencies. However, as we shall discuss later, modern methods, largely collected under "local helioseismology" aim at modeling a wider class of observables, such as travel times, wave phases, and power or amplitudes, and include numerical simulations of 3D wave propagation in a spherical domain. The basic equations are the hydrodynamic equations of conservation of mass, momentum, and energy (see, e.g., Unno *et al.*, 1989):

$$\frac{\partial \rho}{\partial t} + \nabla.(\rho \mathbf{v}) = 0\,, \tag{9.1}$$

$$\rho\left(\frac{\partial}{\partial t} + \mathbf{v}.\nabla\right)\mathbf{v} = -\nabla p - \rho\nabla\Phi + \rho\mathbf{f}\,, \tag{9.2}$$

$$\rho T\left(\frac{\partial}{\partial t} + \mathbf{v}.\nabla\right)S = \rho\epsilon_{\mathrm{N}} - \nabla.\mathbf{F}, \tag{9.3}$$

where $\mathbf{v}$ is the fluid velocity, p is gas pressure, ρ is gas density, T is temperature, and Φ is the gravitational potential. Forces denoted as f can include the electromagnetic, external forces, and, for a rotating star, Coriolis and centrifugal forces. In the energy equation, S is entropy, ϵ_{n} is the nuclear energy generation rate, and $\mathbf{F}$ is the sum of radiative and convective fluxes, $\mathbf{F}_{\mathrm{R}} + \mathbf{F}_{\mathrm{C}}$. We have neglected viscous forces and viscous energy dissipation in the momentum and energy equations, respectively, as they are generally small in the solar interior. Supplementary equations are the Poisson's equation,

$$\nabla^2\Phi = 4\pi G\rho, \tag{9.4}$$

an equation of state $p(\rho, T)$, and equations giving energy generation rates $\epsilon_{\mathrm{n}}(\rho, T)$ and tabulated opacities (or coefficients of absorption of electromagnetic radiation) $\kappa(\rho, T)$ for a given chemical composition.

9.2.1 Standard solar models

Standard solar or stellar models (SSMs) are solutions of equations determining stellar structure, and its quasi-static evolution (Christensen-Dalsgaard *et al.*, 1996). The stellar structure equations are the hydrostatic limits of Equations (9.1)–(9.3) satisfying spherical symmetry, and the supplementary equations described above. Solutions are obtained under certain physically justified assumptions on the initial chemical composition, nuclear energy generation, and on energy transport mechanisms (due to radiation, convection, and neutrinos), and with well-observed constraints on the current solar luminosity and radius.

9.2.2 Oscillation equations

Oscillation or wave equations are derived by perturbing and linearizing (i.e., neglecting all terms of order higher than one in perturbation variables) Equations (9.1)–(9.4). Global helioseismology relies mainly on accurate determination of oscillation frequencies to match the observations. Here, assumptions of adiabatic evolution of small amplitude oscillatory perturbations in a self-gravitating, spherically symmetric non-rotating star without magnetic field or any other external forces, and negligible Reynolds stresses, fare well. Under such assumptions, linearized equations for Eulerian perturbations (see Unno *et al.*, 1989), denoted with a prime $'$, are

$$\rho' + \nabla.(\rho\xi) = 0, \tag{9.5}$$

$$\rho\frac{\partial^2\xi}{\partial t^2} = -\nabla p' - \rho\nabla\Phi' - \rho'\nabla\Phi, \tag{9.6}$$

$$\frac{\rho'}{\rho} + \frac{1}{\rho}\xi.\nabla\rho = \frac{1}{\Gamma_1}\left(\frac{p'}{p} + \frac{1}{p}\xi.\nabla p\right), \tag{9.7}$$

$$\nabla^2\Phi' = 4\pi G\rho', \tag{9.8}$$

where ξ is the vector displacement of a fluid element (with $\mathbf{v}' = d\xi/dt$), and $\Gamma_1 = \left(\frac{d\ln p}{d\ln\rho}\right)_{ad}$ is the adiabatic index.

The above differential equations, because of the assumed spherical symmetry of the background structure, have the property that the coefficients appearing in them are all functions solely of the radial coordinate r. Further, the differential operations over the angular coordinates can be separated out using spherical harmonics. Seeking periodic solutions in time with $e^{i\omega t} = e^{i2\pi\nu t}$, where ν is the cyclic frequency, vector displacement ξ obeys the decomposition

$$\xi(r,\theta,\phi,t) = \left[\xi_{\mathrm{r}}(r), \xi_{\mathrm{h}}(r)\frac{\partial}{\partial\theta}, \xi_{\mathrm{h}}(r)\frac{\partial}{\sin\theta\partial\phi}\right]Y_{\mathrm{l}}^{\mathrm{m}}(\theta,\phi)e^{\mathrm{i}\omega\mathrm{t}}, \tag{9.9}$$

where the subscripts r and h denote the radial and horizontal components, respectively. The scalar variables obey the decomposition $p'(r,\theta,\phi,t) = p'(r)Y_{\mathrm{l}}^{\mathrm{m}}(\theta,\phi)e^{\mathrm{i}\omega\mathrm{t}}$. The spherical harmonic function $Y_{\mathrm{l}}^{\mathrm{m}}$ for degree l and order m is defined by

$$Y_{\mathrm{l}}^{\mathrm{m}}(\theta,\phi) = (-1)^m\frac{2l+1}{4\pi}\frac{(l-m)!}{(l+m)!}P_{\mathrm{l}}^{\mathrm{m}}(\cos\theta)e^{\mathrm{im}\phi}, \tag{9.10}$$

where $P_{\mathrm{l}}^{\mathrm{m}}$ is the Legendre function. The oscillation equations reduce to

$$\frac{1}{\rho}\frac{dp'}{dr} + \frac{g}{\rho c^2}p' + (N^2-\omega^2)\xi_{\mathrm{r}} + \frac{d\Phi'}{dr} = 0, \tag{9.11}$$

$$\frac{1}{r^2}\frac{d}{dr}(r^2\xi_{\mathrm{r}}) + \frac{1}{\Gamma_1}\frac{d\ln p}{dr}\xi_{\mathrm{r}} + \left(1-\frac{S_{\mathrm{l}}^2}{\omega^2}\right)\frac{p'}{\rho c^2} - \frac{l(l+1)}{\omega^2 r^2}\Phi' = 0, \tag{9.12}$$

$$\frac{1}{r^2}\frac{d}{dr}\left(r^2\frac{d\Phi'}{dr}\right) - \frac{l(l+1)}{r^2}\Phi' - 4\pi G\rho\left(\frac{p'}{\rho c^2} + \frac{N^2}{g}\xi_{\mathrm{r}}\right) = 0, \tag{9.13}$$

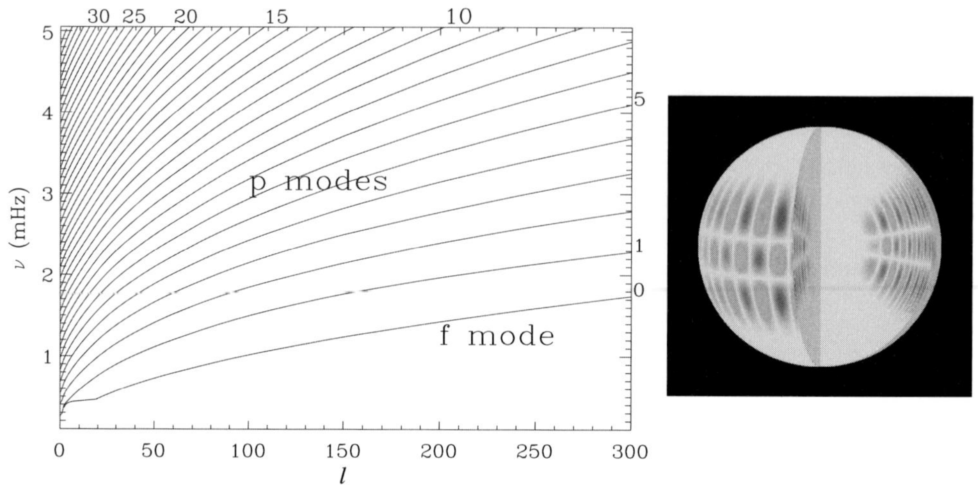

Figure 9.1 Theoretical spectrum of normal modes (p modes and f mode) calculated for a standard solar model (left, model S of Christensen-Dalsgaard *et al.*, 1996), and a normal mode eigenfunction (radial eigenfunction multiplied with the spherical harmonic function) for a p mode of angular degree $l = 20$, azimuthal order $m = 18$, and radial order $n = 14$ (right, Courtesy: A. G. Kosovichev). A black and white version of this figure will appear in some formats. For the color version please refer to the plate section.

where $g = GM/r^2$ is the acceleration due to gravity, $c = (\Gamma_1 p/\rho)^{1/2}$ is the sound speed. S_l and N are the Lamb and Brunt–Väisälä frequencies, respectively, and are given by

$$S_l^2 = \frac{l(l+1)c^2}{r^2} \tag{9.14}$$

and

$$N^2 = g\left(\frac{1}{\Gamma_1}\frac{d\ln p}{dr} - \frac{d\ln\rho}{dr}\right). \tag{9.15}$$

The above fourth-order system (9.11)–(9.13) constitutes an eigenvalue problem for the oscillation modes, with ω^2 as eigenvalues. It is, in general, to be solved numerically, subject to boundary conditions at the center ($r = 0$) and at the surface ($p = 0$), where it is singular. Normally, asymptotic solutions requiring regularity at the boundaries $r = 0$ and R serve as boundary conditions. Solutions of Equations (9.11)–(9.13), with suitable boundary conditions, determine the oscillation frequencies $\omega_{\rm nl}$ and radial eigenfunctions $\xi_{\rm r}^{(\rm n,l)}(r)$ and $p'^{(\rm n,l)}(r)$. Figure 9.1 shows solutions for the cyclic frequencies $\nu_{\rm nl} = \omega_{\rm nl}/2\pi$, and $\xi_{\rm r}^{(\rm n,l)}(r)$ multiplied with spherical harmonics (Equation 9.10) for a chosen set of $\lambda = (l, m, n)$, determined for a standard solar model.

9.2.3 Basic properties of solar oscillations

Under the Cowling approximation, which neglects perturbations in Φ, Equations (9.11)–(9.13) simplify to a second-order system,

$$\frac{1}{\rho}\frac{dp'}{dr} + \frac{g}{\rho c^2}p' + (N^2 - \omega^2)\xi_{\rm r} = 0, \tag{9.16}$$

$$\frac{1}{r^2}\frac{d}{dr}(r^2\xi_{\rm r}) - \frac{g}{c^2}\xi_{\rm r} + \left(1 - \frac{S_l^2}{\omega^2}\right)\frac{p'}{\rho c^2} = 0. \tag{9.17}$$

Basic dispersive properties of waves can be studied analytically by solving the above equations in the JWKB limit, i.e., by assuming solutions of the form $\xi_{\rm r} \propto \rho^{-1/2}e^{ik_{\rm r}r}$ and $p' \propto \rho^{1/2}e^{ik_{\rm r}r}$, with $k_{\rm r}$ as the slowly varying (in r) radial wavenumber. This analysis yields the following dispersion relation as the condition required for non-zero solutions,

$$k_{\rm r}^2 = \frac{\omega^2 - \omega_{\rm c}^2}{c^2} + \frac{S_l^2}{c^2\omega^2}\left(N^2 - \omega^2\right), \tag{9.18}$$

where

$$\omega_{\rm c} = \frac{c}{2H} \tag{9.19}$$

is the acoustic cut-off frequency and $H = (\frac{d\ln\rho}{dr})^{-1}$ is the density scale-height. Here, S_l and N appear as critical frequencies, which along with the sound speed c determine the modal frequencies ω as well as delineate the propagation regions against the solar structure. These critical frequencies calculated for a standard solar model are shown in Figure 9.2, which shows the regions of mode propagation. For $\omega^2 > S_l^2, N^2$ or $\omega^2 < S_l^2, N^2$, $k_{\rm r}^2 > 0$, i.e., the radial wavenumber $k_{\rm r}$ is real, and the waves can propagate in radial directions. When $N^2 > \omega^2 > S_l^2$ or $S_l^2 > \omega^2 > N^2$, $k_{\rm r}^2 < 0$, i.e., the radial wavenumber $k_{\rm r}$ is purely imaginary, then the waves become evanescent (i.e., exponentially decay with distance). Thus S_l and N define the boundaries or wave turning points that delineate the propagation regions (Figure 9.2). In terms of a horizontal wavenumber, defined as $k_{\rm h} = l(l+1)/r^2 = L^2/c^2$, the dispersion relation (9.18) can be rewritten as

$$k_{\rm r}^2 = \frac{\omega^2 - \omega_{\rm c}^2}{c^2} + k_{\rm h}^2\left(\frac{N^2}{\omega^2} - 1\right). \tag{9.20}$$

The high-frequency limit $\omega^2 \gg N^2$ corresponds to acoustic waves or p modes, and the low-frequency limit $\omega^2 \ll S_l^2$ to gravity waves or g modes. If r_1 and r_2 are the radii of the

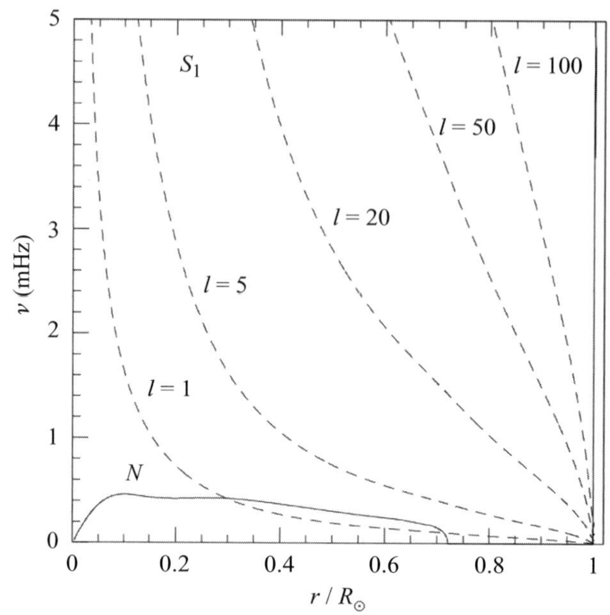

Figure 9.2 Brunt–Väisälä frequency N (solid curve) and the Lamb frequency S_l (dashed curves) against depth (in fractional solar radius) calculated for a standard solar model. Gravity waves propagate within the region enclosed by positive values of N and acoustic waves are trapped between the surface layers (as demarked by the variation of acoustic cut-off ω_c, which is nearly the same as N (solid curve) near the surface) and the locations of S_l, which are shown for a few values of l. (Credit: H. M. Antia.)

inner and outer turning points where $k_r = 0$, then the resonant condition determining the frequencies of normal modes is the quantization rule,

$$\int_{r_1}^{r_2} k_r dr = \pi(n + \alpha), \tag{9.21}$$

where n is the radial order (integer) and α is a phase shift suffered by the mode at the boundaries. The above rule in combination with two limits of the above dispersion relation can be used to derive the frequencies of solar p and g modes. The resonant conditions for the p modes can further be exploited to reduce the dispersion relation to one involving ratios, $F(L/\omega) = (n + \alpha)/\omega$, which is called the Duvall law (Duvall, 1982). Observed frequencies obey this relationship to a level that one can perform an asymptotic inversion for sound speed (Christensen-Dalsgaard, 2002).

9.3 Helioseismic inversion modeling

Solving oscillation equations, for realistic SSMs, is in general complicated. The true potential of helioseismology emerges when it is posed as a general inverse problem involving integral relations between the observables and internal properties. Such inverse analyses make it relatively easier, for practical as well as computational reasons, to probe realistic situations involving effects of stationary velocity (e.g., rotation) and magnetic fields on the solar interior structure. This approach exploits the variational principle, which was first shown by Chandrasekhar (1964) to be valid for oscillation frequencies determined as eigenvalues of adiabatic non-radial stellar oscillation equations (9.11)–(9.13). In operator representation (Lynden-Bell and Ostriker, 1967), the oscillation eigenvalue problem is written as

$$\omega^2 \xi = \mathcal{L}\xi, \tag{9.22}$$

where the linear operator $\mathcal{L}$ for adiabatic non-radial oscillations can be written as

$$\mathcal{L}\xi \equiv \frac{1}{\rho^2}(\nabla p)\nabla.(\rho\xi) - \frac{1}{\rho}\nabla(\xi.\nabla p) - \frac{1}{\rho}\nabla(c^2\rho\nabla.\xi) + \nabla\left\{G\int_{\mathrm{V}} \frac{\nabla_{\mathbf{x}'}.[\rho(\mathbf{x})\xi(\mathbf{x})]d^3\mathbf{x}}{|\mathbf{x}-r|}\right\}. \tag{9.23}$$

Since $\mathcal{L}$ represents a conservative non-dissipative oscillator system, it has Hermitian properties, and the variational principle leads to the determination of oscillation frequencies,

$$\omega^2 = \frac{\int_{\mathrm{V}} \xi^*.\mathcal{L}\xi\rho\, dV}{\int_{\mathrm{V}} \rho\xi^*.\xi\, dV}. \tag{9.24}$$

From the above relation, it is clear that contributions to first-order changes in eigenfrequencies are at least of order two in eigenfunction changes. In this formalism, we can determine changes to the oscillation frequencies, $\delta\omega_\lambda$, $(\lambda = n, l, m)$, that result from changes to the internal properties or to the physics governing the oscillations (e.g., rotation, magnetic fields, or non-adiabatic effects) expressible as $\delta\mathcal{L}$ that perturbes $\mathcal{L}$. To first order, the frequency changes are then given as

$$\left(\frac{\delta\omega}{\omega}\right)_\lambda = \frac{1}{2\omega_\lambda I}\int_{\mathrm{V}} \xi^*.\delta\mathcal{L}\xi\rho\, dV', \tag{9.25}$$

where

$$I = \int_{\mathrm{V}} \rho\xi^*.\xi\, dV \tag{9.26}$$

is the mode mass or mode inertia related to the mode energy $E = I\omega|\xi_{\mathrm{r}}(R)|^2$.

Adding background velocity, $\mathbf{v}$, and magnetic fields, $\mathbf{B}$, in the basic equations, perturbation and linearization yields (Gough and Thompson, 1990)

$$(\mathcal{L} - \omega^2)\xi + \omega\mathcal{M}\xi + \mathcal{N}\xi + \mathcal{B}\xi = 0, \tag{9.27}$$

where the operator $\mathcal{L}$ is as given earlier in Equation (9.23), $\mathcal{M}$ models the first-order effect due to the Coriolis force,

$$\mathcal{M}\xi = 2i\omega\mathbf{v}.\nabla\xi, \tag{9.28}$$

and

$$\mathcal{N}\xi = (\mathbf{v}.\nabla)^2\xi - 2\Omega_0 \times [(\xi.\nabla)\mathbf{v} - (\mathbf{v}.\nabla)\xi] - (\xi.\nabla)(\mathbf{v}.\nabla)\mathbf{v} \tag{9.29}$$

contains the second-order effects due to centrifugal forces as well as the distortion from spherical symmetry and first-order perturbations to the eigenfunctions, and

$$\mathcal{B}\xi = (4\pi\rho)^{-1}\left[[\rho^{-1}(\xi.\nabla)\rho + \nabla.\xi]\mathbf{B} \times (\nabla \times \mathbf{B}) - [(\nabla \times \mathbf{B}') \times \mathbf{B} + (\nabla \times \mathbf{B}) \times \mathbf{B}']\right] \tag{9.30}$$

includes second-order contributions from the magnetic field.

9.3.1 Sensitivity kernels and inversion methods

The system of integral equations (9.25) leads to a general inverse problem that can be written as

$$d_i = \sum_\alpha \int_V K_\alpha^i(\mathbf{r})\delta q_\alpha(\mathbf{r})dV + \epsilon_i, \tag{9.31}$$

where d_i is a set of observables or helioseismic measurements carrying contributions due to first-order changes in the solar models, δq_α, with errors ϵ_i, and $K_\alpha^i(\mathbf{r})$ are kernel functions determined by the eigenfunction solutions of the forward problem. In general, this system of equations is ill-posed, i.e., does not have unique solutions: solutions for δq_α are insensitive to arbitrary oscillatory functions that are added to them. A general approach is to apply smoothness constraints on the solutions, known as "regularization procedures." The two well-known methods of regularization in helioseismology are optimally localized averages (OLA) (Backus–Gilbert method), and regularized least squares (RLS) (Tikhonov method).

The OLA method chooses a set of coefficients $c_i(r_o)$ such that a resolution or averaging kernel determined as a linear combination, $\mathcal{K}(\mathbf{r}, \mathbf{r}_o) = \sum_i c_i(r_o)K_\alpha^i$, is highly localized about r_o while at the same time keeping the variance or errors, $\sum_i c_i(r_o)\epsilon_i$, small. A computationally more efficient version of OLA, named subtractive optimally localized averges (SOLA), was developed by Pijpers and Thompson (1992). In the RLS method, on the other hand, the solutions δq_α are parametrized by piece-wise linear functions or cubic splines along with a regularization operator $\mathcal{R}$ that acts on the unknowns through a parameter λ, and a least-square minimization of misfit to a set of observations d_i is carried out:

$$\chi^2 = \sum_i \frac{1}{\sigma_i^2}\left[d_i - \sum_\alpha \int_V K_\alpha^i(\mathbf{r})\delta q_\alpha(\mathbf{r})\right]^2 + \lambda\mathcal{R}\{\delta q_\alpha\}, \tag{9.32}$$

where σ_i are the noise estimates for the observables d_i. Both of these inversion techniques have been developed intensively over the last three decades for helioseismology, and have been tested extensively (see Chapter 12 by Basu and Thompson). These same methods, with calculations of 3D sensitivity kernels, are used in local helioseismology applications; we discuss them briefly in Section 9.4.2.

In terms of changes to speed of sound and density, the variational principle yields

$$\frac{\delta\omega_{n,l}}{\omega_{n,l}} = \int_0^R K^{n,l}_{c^2,\rho}(r)\frac{\delta c^2}{c^2}dr + \int_0^R K^{n,l}_{\rho,c^2}(r)\frac{\delta\rho}{\rho}(r)dr, \tag{9.33}$$

where $K^{n,l}_{c^2,\rho}$ and $K^{n,l}_{\rho,c^2}$ are the sensitivity kernels for sound speed and density, respectively, for a mode (n, l). In practice, application of the above equation requires inclusion of uncertainties due to unmodeled physics in the near-surface layers (due to, e.g., convection, magnetic fields, turbulence, etc.), which can be grouped under a third "surface term" in the right-hand side (Dziembowski *et al.*, 1990; Antia and Basu, 1994). Such seismic structure inversion analyses permit constraining material properties that go as important inputs via the equation of state, chemical composition, opacities, nuclear reaction rates, and neutrino fluxes (see Christensen-Dalsgaard, 2002, and references therein). An important recent development in this regard is the inconsistency between models incorporating newly determined heavy element abundances (of oxygen and other heavy elements, using 3D atmospheric models) that are significantly reduced (Asplund *et al.*, 2005) and the older SSMs that are seismically better matched (Antia and Basu, 2005). A resolution to this problem has not yet been achieved despite numerous attempts. Latest results with a detailed discussion of applications of inverse analyses, including that for solar internal rotation, are discussed in Chapter 12.

9.4 Local helioseismology

Probing localized, non-axisymmetric 3D structural changes and flows beneath the solar surface is possible through analyses of the wave-field observed over limited regions of interest on the solar surface. Such analyses fall under local helioseismology, and most often deal with a wider set of observables derived directly from the space-time domain, rather than the wavenumber–frequency domain employed in global analyses discussed above. Development of this field can be traced back to the suggestion by Thomas *et al.* (1982) that interactions between sunspots and helioseismic p modes could be used to probe the subsurface structure of sunspots. The analyses that followed this suggestion focused mainly on changes in the local frequency–wavenumber spectrum (ν–k) and in the modal power distribution. These studies led to the discovery of "absorption" of p modes by sunspots (Braun *et al.*, 1987): about 50% of the flux of acoustic wave energy impinging on a sunspot is not observed to return to the quiet Sun. Development of several local helioseismic techniques, namely ring diagram analysis (Hill, 1988), helioseismic holography (Lindsey and Braun, 1990) and time–distance helioseismology (Duvall *et al.*, 1993b), has since brought in new ways of probing the near-surface structure and dynamics of the Sun.

9.4.1 Forward and inverse modeling in local helioseismology

Forward and inverse modeling in local helioseismology follows the same principles as in global helioseismology described earlier except that the wave equations (Equations (9.5)–(9.8) or (9.22)–(9.23)) need to include explicitly source terms, which make them inhomogeneous. This is due to the fact that local helioseismology methods, except the local 3D power spectral (ring diagram) analysis, derive observables by directly following temporal correlations that a wave or wavepacket establishes between spatially separated points. Here, it is useful to consider the excitation or sources that give rise to waves, and often a given point from where a wave is followed is assumed to be a source for that wave. Since wave excitations are stochastic, the wavefield itself remains stochastic over space and time, and any given realization of wave correlation between two points is dominated by noise, and a sufficiently long temporal and spatial averaging is required to add up a sufficient number of realizations of the wave process that we intend to measure. In operator notation, the wave equation (9.22) is made inhomogeneous by adding a source term $\mathcal{S}$:

$$\mathcal{L}\xi - \omega^2\xi = \mathcal{S}. \tag{9.34}$$

The source term $\mathcal{S}$, in principle, can include wave excitation as well as damping, and a physical model for such processes often includes turbulent pressure (Reynolds stresses) and entropy fluctuations. Spatio-temporal covariance of the source function is an important ingredient in the determination of observables, e.g., wave cross-covariances and oscillation power spectra.

Solutions to Equation (9.34) can be written in terms of Green functions $\mathcal{G}$:

$$\xi(\mathbf{r}, t) = \int_V \mathcal{G}(\mathbf{r}, t; \mathbf{r}', t')\mathcal{S}(\mathbf{r}', t')dt'dV'. \tag{9.35}$$

The Green functions are the impulse responses of the background solar model that solves

$$(\mathcal{L} - \omega^2)\mathcal{G}(\mathbf{r}, t; \mathbf{r}', t') = \hat{\mathbf{e}}(\mathbf{r}')\delta_{\mathrm{D}}(\mathbf{r} - \mathbf{r}')\delta_{\mathrm{D}}(t - t'), \tag{9.36}$$

where δ_{D} is the Dirac delta function, $\hat{\mathbf{e}}$ is the unit vector. The Green functions appearing in the above equations are, in general, tensors $\mathcal{G}^i_j(\mathbf{r}, t; \mathbf{r}', t')$ representing vector displacements in the j-th direction at $(\mathbf{r}, t)$ resulting from a unit source acting in the i-th direction at $(\mathbf{r}', t')$. Solutions to Equation (9.36) are, in general, obtained numerically either via normal mode eigenfunctions satisfying Equation (9.22) or directly.

First-order perturbation analysis of Equation (9.34) can be done as in Section 9.3, substituting $\mathcal{L} = \mathcal{L}^0 + \delta\mathcal{L}$, $\xi = \xi^0 + \delta\xi$, and $\mathcal{S} = \mathcal{S}^0 + \delta\mathcal{S}$ and linearizing in perturbations. Here, the zero-order equilibrium solution, denoted with superscript ^{0}, is given by Equation (9.35). In general, Green functions and full solutions for observables, which most often are line-of-sight projections of velocities $\mathbf{v} = \partial\xi/\partial t$ acted on by instrumental transfer functions, can be written down by representing the transformations from ξ to the observable ϕ in operator form, and by defining a Green function for the observable ϕ (a detailed discussion can be found in Gizon and Birch, 2005).

Through explicit calculations of $\mathcal{L}^0$ and $\delta\mathcal{L}$, first-order changes of a set of observables, e.g., travel times (time–distance helioseismology), ring-fit parameters (ring diagram analysis), or changes in ingression–egression correlations (helioseismic holography), δd_i, can be written in terms of first-order changes in the solar models, δq_α. The resulting integral equation is the same as Equation (9.31) discussed in Section 9.3.1. Including an appropriate physical model for the source function $\mathcal{S}(\mathbf{r}, t)$ and first-order perturbations in it due to perturbations in the solar model, the kernel functions $K^i_\alpha(\mathbf{r})$ can be determined in terms of first-order perturbations in the solar model. This level of approximation is called the Born approximation, which thus includes first-order changes as well as the zero-order wavefield, and is written as

$$(\mathcal{L}^0 - \omega^2)\delta\xi = -(\delta\mathcal{L} - \delta\omega^2)\xi^0 + \delta\mathcal{S}, \tag{9.37}$$

which can then be written as solutions for $\delta\xi$ in terms of Green function solutions for ξ^0 given by Equations (9.35) and (9.36). Such solutions can be used to calculate first-order changes in cross-covariances, travel times, 3D power spectra, or for measurements made in helioseismic holography.

Similar to the variational principle treatment that determines first-order changes to global-mode eigenfrequencies due to background flows and magnetic field (Equation 9.27), such effects can also be modeled for local helioseismology observables via the Born approximation Equation (9.37).

9.4.2 Methods of local helioseismology

As mentioned earlier, development of local helioseismic techniques sprang from efforts at modeling the magnetic field–p mode interactions to understand the mechanisms responsible for the absorption of p modes by sunspots. One of the early analysis techniques is the Fourier–Hankel spectral method that decomposes the wavefield into those that enter and leave a region of interest, often chosen to enclose a circular area (Braun *et al.*, 1987). Helioseismic holography (Lindsey and Braun, 2000a) relates the amplitude and phases of the acoustic wavefield at the surface to scattering or absorption processes encountered by the waves at the interior target location. We will not discuss further the detailed modeling involved in these two methods due to space limitations, but describe that for the ring diagram and time–distance helioseismology methods below.

Ring diagram analysis

Measuring and modeling oscillation frequencies from observations over local areas, which are small enough to be treated in plane-parallel cartesian geometry, through 3D (k_x, k_y, ω) oscillation power spectra constitutes the ring diagram analysis. Hill (1988) developed this technique and showed that local 3D dispersion relations, $\omega = \omega(k_x, k_y)$, can be used to study local flows and thermal structure. Constant ω cuts in the 3D power spectra display concentric ring structures that correspond to the normal-mode ridges of global 2D power spectra, with the outermost ring corresponding to the f mode ($n = 0$) and the inner ones to the increasing radial order (n) p modes.

A horizontal flow field with a suitable depth average denoted as $\mathbf{U} = (U_x, U_y)$ would Doppler shift the frequencies of waves, which would satisfy

$$\omega(k) = \omega_0(k) + \mathbf{k}.\mathbf{U} \equiv \omega_0 + (U_x k_x + U_y k_y), \tag{9.38}$$

where $\omega_0(k)$ describes central frequencies that depend only on the magnitude of $\mathbf{k}$. The power spectra are fitted by a Lorentzian profile in ω at each $\mathbf{k}$,

$$P(\mathbf{k}, \omega) = \frac{A}{(\omega - \omega_0 + U_x k_x + U_y k_y)^2 + \Gamma^2} + \frac{b_0}{k^3}, \tag{9.39}$$

where A, ω_0, Γ, and b_0 are respectively the amplitude, central frequency, spectral line width, and a background noise parameter. The main fit parameters that are inverted in the ring diagram analysis are the flow parameters U_{x}^{nl} and U_{y}^{nl}, and $\delta\omega_{nl}$ for the difference between the fitted modal frequency and that from a model. The depth average flows $\mathbf{U}$ derived from U_{x}^{nl} and U_{y}^{nl} are related to the actual flows $\mathbf{u} = (u_x, u_y)$ via the integral relation

$$\mathbf{U} = \frac{\int \mathbf{u}(z)\rho\xi.\xi dz}{\int \rho\xi.\xi dz}. \tag{9.40}$$

The eigenfunctions ξ calculated using normal-mode solutions can then be used to derive appropriate flow kernels for inverting the above equation.

Apart from studying flows, inversions for sound speed and density can be done using Equation (9.33) (Basu *et al.*, 2004), as that in global mode analysis, for frequency shifts $\delta\omega_{nl}$ measured using ring diagrams (Rajaguru *et al.*, 2001). Here, the frequency shifts are relative to a quiet region or a region devoid of large-scale magnetic field. Both OLA and RLS methods have been applied to ring diagram inversions for structure and flows (Basu *et al.*, 1999b; Haber *et al.*, 2004).

Time–distance helioseismology

Time–distance helioseismology (Duvall *et al.*, 1993b), using temporal cross-correlations of the oscillation signals from separated points on the solar surface, measures and inverts travel times of acoustic waves (or wave-packets) to image changes in wave speed and flow velocities in the interior of the Sun, similar to what is done in seismology of Earth. The cross-correlation function is a suitably normalized cross-covariance of oscillation signals at two points $\mathbf{r}_1$ and $\mathbf{r}_2$ defined as

$$C(\Delta, \tau) = \int_0^T \Psi(\mathbf{r}_1, t)\Psi(\mathbf{r}_2, t + \tau)dt, \tag{9.41}$$

where $\Psi(\mathbf{r}_1, t)$ and $\Psi(\mathbf{r}_2, t + \tau)$ are, in general, filtered and suitably averaged (over space centered at $\mathbf{r}_1, \mathbf{r}_2$, respectively) oscillation signals (projected line of sight velocities $\phi(\mathbf{r}, t)$) at times t and $t + \tau$, respectively, Δ is the horizontal distance between $\mathbf{r}_1$ and $\mathbf{r}_2$, τ is the time lag and and T is the total time-length of observation. The cross-covariances $C(\Delta, \tau)$ have wave-packet-like structure, which is understood to be due to constructive interference

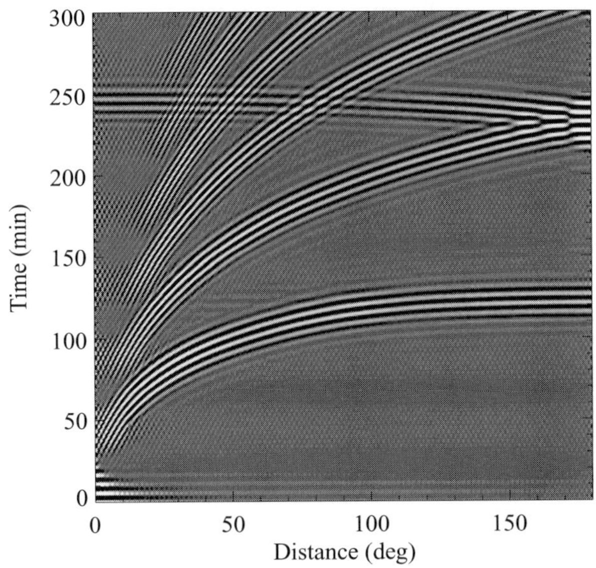

Figure 9.3 Theoretical time–distance diagram for p modes calculated for a spherically symmetric solar model (model S of Christensen-Dalsgaard *et al.* (1996)). Travel distances on the solar surface are on the x-axis, and the travel times are on the y-axis. See text for more details. (Courtesy: A. G. Kosovichev.)

of a group of acoustic waves with approximately the same horizontal phase–velocity (Bogdan, 1997), and represents a solar seismogram (it is often called a time–distance diagram). A theoretical time–distance diagram computed for a spherically symmetric solar model is shown in Figure 9.3. The lowest set of ridges corresponds to acoustic waves propagating between two points without additional reflection from the solar surface, the next set corresponds to waves arriving after one reflection from the surface, and the set of ridges at greater time delays result from waves arriving after multiple bounces.

The first step in time–distance helioseismology is the estimation of travel times by fitting the observational cross-covariance function computed using Equation (9.41). Normally, travel times are measured from the first-bounce ridges (of $C(\Delta, \tau)$); the central peak of this ridge corresponds to lag τ, which is estimated as the travel time. In reality, because of the stochastic nature of solar oscillation signals and finite wavelength effects, travel times measured from C are hard to interpret and model accurately, and the manner of fitting and estimating travel times also affect the inferences and hence have to be explicitly taken care of in further analyses. Applying the theory of normal modes to decompose the oscillation signal into the constituent resonant modes, Kosovichev and Duvall (1997) derived an approximate analytic evaluation of Equation (9.41):

$$C(\Delta, \tau) = A \exp\left[-\frac{\delta\omega^2}{4}(\tau - \tau_g)^2\right] \cos[\omega_0(\tau - \tau_p)], \tag{9.42}$$

where A is the cross-covariance amplitude, $\tau_g = \Delta/v_g$ is the group travel time, $\tau_p = \Delta/v_{ph}$ is the phase travel time, and ω_0 and $\delta\omega$ are the central frequency and width, respectively. Here, v_g and v_{ph} are the group and phase velocity of the wave-packet. An alternate fitting approach, based on one used in geoseismology, with a new definition of travel time, that employs a model cross-covariance fitting and identifying travel time with a single parameter that minimizes the mis-fit, was introduced by Gizon and Birch (2002).

Once the travel times are estimated, the forward calculations embodied in Equations (9.35) and (9.36) can be used to compute $\Psi(\mathbf{r}, t)$, with appropriate transformations to relate ξ with the observable ϕ (the line-of-sight projection of velocity $v = \partial\xi/\partial t$) and finally to Ψ. According to the Claerbout conjecture (Rickett and Claerbout, 2000), known in geoseismology literature, the cross-covariance function (Equation 9.41) represents the impulse response of the wave-propagating medium: waves received at $\mathbf{r}_2$ contributing to C are from a point source at $\mathbf{r}_1$, and hence C is just the Green function described by Equation (9.36). However, Gizon and Birch (2002) have shown that, for solar situations, explicit inclusion of source function $\mathcal{S}$ as in Equation (9.35) models additional physics involving wave-source scatterings and hence contributions to C. Changes in travel times due to changes in internal structure and flows are then related through kernels that are calculated based on Equations (9.36) and (9.37),

$$\delta\tau(\mathbf{r}_1, \mathbf{r}_2) = \sum_\alpha \int_V K_\alpha(\mathbf{r}_1, \mathbf{r}_2; \mathbf{r})\delta q_\alpha(\mathbf{r}) dV, \tag{9.43}$$

where the kernel functions $K_\alpha(\mathbf{r})$ represent all forms of perturbations (sound speed, density, flow magnetic field, etc.).

Computation of sensitivity kernels and inversions

Ray and wave theory calculations form two major ways of calculating travel-time sensitivity kernels $K_\alpha(\mathbf{r})$. In the ray approximation, the travel times are sensitive only to the perturbations along the ray paths and are determined by the equation of motion:

$$\frac{\partial \mathbf{r}}{\partial t} = \frac{\partial \omega}{\partial \mathbf{k}}, \tag{9.44}$$

and the phase travel times obey Fermat's principle:

$$\delta\tau = \frac{1}{\omega}\int_\Gamma \delta\mathbf{k}\, dr, \tag{9.45}$$

where Γ represents the path connecting two points. Substitution of the dispersion relation of the type in Equation (9.20) constitutes the ray-approximation treatment for travel times. One can separate the effects of flows and structural perturbations by measuring travel times of waves traveling in opposite directions along the same ray path. The differences, τ_{diff}, and the mean, τ_{mean}, of these travel times can be related to flows and wave-speed changes,

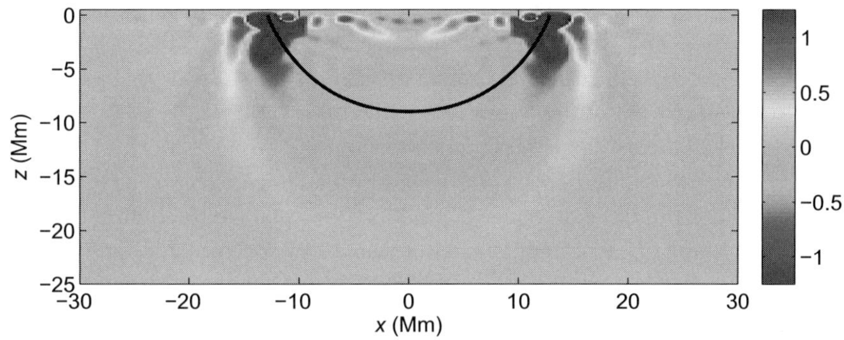

Figure 9.4 A depth slice, in the plane of the ray path, of the 3D Born kernel for sound-speed perturbations on travel times, from the results in Birch *et al.* (2004), for a surface travel distance of 25 Mm. This example shows that the sensitivity is small along the ray path, shown as a black solid curve, and negative in a shell surrounding it, resembling the traditional banana-doughnut shape. (Courtesy: A. C. Birch.) A black and white version of this figure will appear in some formats. For the color version please refer to the plate section.

respectively:

$$\delta\tau_{\rm diff} \approx -2\int_\Gamma \frac{\mathbf{n}U}{c^2}ds, \tag{9.46}$$

$$\delta\tau_{\rm mean} \approx -\int_\Gamma \frac{\delta w}{c}Sds, \tag{9.47}$$

where $\mathbf{n}$ is the unit vector tangent to the ray, and $S = \mathbf{k}/\omega$ is the phase slowness, and δw is the local wave-speed perturbation.

Computation of wave theory or finite-wavelength kernels in time–distance helioseismology has involved different approximations for treating the wave effects: Born, Rytov, and Fresnel-zone approximations, with each method further treating the wave excitations in single- or multiple-source approximations. The Rytov approximation estimates first-order corrections to the phase of the wavefield instead of to the wavefield itself (Jensen and Pijpers, 2003), while the Fresnel zone approximation parameterizes the kernels in terms of ray theory travel times and corrections to them (Jensen *et al.*, 2001). Comparisons of ray theory and Fresnel-zone kernel-based inversions were done by Couvidat *et al.* (2004).

In the Born approximation, solutions to Equation (9.37) can be obtained in terms of Green functions as explained in Section 9.4.1. Early calculations of Born kernels for travel times due to sound speed changes were done by Birch and Kosovichev (2000). A crucial difference between ray and Born kernels is that the latter have zero values along the ray path, with maximum values in a shell around the ray path – this feature is known in geoseismology and is referred to as the banana-doughnut kernel. An example of the Born kernel for sound speed changes, based on the results in Birch *et al.* (2004), is shown in Figure 9.4; here, the waves travel a surface distance $\Delta = 25$ Mm, and the corresponding ray path is shown as a black solid curve. Born kernels for flows in time–distance helioseismology have been computed by Gizon *et al.* (2000) and Jackiewicz *et al.* (2008).

In the RLS method, the travel-time inverse problem is the same as that given by Equation (9.32), where the set of observations d_i are the travel-time changes $\delta\tau_i$. RLS has been the most widely used inversion method in time–distance helioseismology. There have been extensive tests and validation of RLS via generation and inversion of artificial synthetic data as well as realistic 3D numerical simulations of solar convection and waves (see Kosovichev, 2011, and references therein). In recent years, performance and validation of the SOLA method for vector flow inversions from travel-time measurements have been carried out (see Jackiewicz *et al.*, 2012b, and references therein).

9.4.3 Local helioseismology of sunspots: Surface magnetic effects and related issues

Applications of time–distance helioseismology appeared to be a promising avenue with its 3D tomographic images of flow and sound speed structures beneath sunspots (Kosovichev *et al.*, 2000). These early results show an increased sound speed region extending from about 4 Mm down to about 18 Mm with a maximum change of about 1–2%, while the near-surface layers in the 1–3 Mm depth range show a decrease in sound speed of similar magnitude. The flow pattern consists of a shallow (1.5–3.0 Mm) converging flow that feeds a strong downflow beneath the sunspot (Duvall *et al.*, 1996; Zhao *et al.*, 2001) down to depths of about 5 Mm. Though these results have features indicative of the cluster model (Parker, 1979), new developments and improvements on several different fronts in local helioseismology have served to emphasize the inadequacy of such analyses (Gizon *et al.*, 2010). Recent new ways of travel-time measurement and inversion show that the moat outflows around sunspots extend much deeper (up to about 4–5 Mm) (Gizon *et al.*, 2010). Interesting, but not yet fully understood, revelations on the seismic disguises of the dominantly near-surface interactions between the magnetic field and acoustic waves came from analyses based on phase-sensitive holography (Lindsey and Braun, 2005a): upcoming acoustic waves in active regions undergo strong surface phase perturbations resulting in impairment of their coherence, similar to the blurring of images seen through a commercial showerglass. Corrections for such effects (Lindsey and Braun, 2005b) show that the sub-surface acoustic anomalies disappear below a depth of $\sim$5 Mm, in contrast to the time–distance helioseismic inferences (Duvall *et al.*, 1996; Kosovichev *et al.*, 2000).

The above developments have brought to the fore the dominant direct interactions between acoustic waves and magnetic fields, which leave too large a signal in measurements to be treated with the conventional methods of seismic inversions that model effects such as thermal perturbations (Schunker *et al.*, 2013). Furthermore, inadequate identification of the helioseismic observables due to complexities in the observation procedure itself, e.g., altered radiative transfer physics affecting spectral line formation in strong magnetic fields (Rajaguru *et al.*, 2010), makes the current helioseismic analyses questionable.

Closely related to the above realization of inadequacies in local helioseismology of strong magnetic field regions are the extensive studies of interactions between acoustic-gravity waves and structured magnetic field in the photosphere and above (Spruit and Bogdan, 1992; Cally and Bogdan, 1993; Cally *et al.*, 2003). While these studies successfully model acoustic wave absorption, mode conversion and transmission of wave energy by sunspot-like magnetic fields, in the form of different types of magneto-acoustic wave modes, quantifying their effects on the helioseismic inferences and accounting for them has proved difficult for lack of a consistent framework within the helioseismic modeling.

9.4.4 Numerical simulations of helioseismic wave propagation

The above progress in our understanding of complexities involved in performing local helioseismology of magnetic fields on the Sun has, in recent years, driven efforts to perform full 3D numerical simulations of linear wave equations (Parchevsky and Kosovichev, 2007; Schunker *et al.*, 2011) as well as fully non-linear convection and magneto-convection simulations with naturaly excited p-mode oscillations (Nordlund *et al.*, 2009). Three-dimensional simulations of wave propagation in a spherical domain have been used to study the effects of large-scale interior flows and structural perturbations on observables such as wave travel times (Hanasoge *et al.*, 2007; Hartlep *et al.*, 2013). These analyses help in calibrating the performance of helioseismic analyses and thus validate their use on observational data. Such numerical calculations essentially solve the linearized wave propagation equations (9.5)–(9.8) with suitable sources for oscillations added. On the other hand, the full non-linear equations (9.1)–(9.4) with MHD parts added are being solved to simulate sunspot-like magnetic configurations (Rempel *et al.*, 2009) with naturally excited p-mode oscillations, and the simulation data are used to check the performance of helioseismic analyses (Braun *et al.*, 2012).

The most rewarding way forward would be the use of the realistic 3D simulations mentioned above in developing appropriate physical models of the large perturbations that the acoustic waves suffer due to strong surface magnetic fields so that they are accounted for in local helioseismic analyses (Gizon *et al.*, 2010). This then paves the way for correct inferences on the subsurface structure of magnetic and flow fields.

10

Modeling approaches in planetary seismology

RENEE C. WEBER, MARTIN KNAPMEYER, MARK PANNING, AND NICHOLAS SCHMERR

10.1 Introduction

Of the many geophysical means that can be used to probe a planet's interior, seismology remains the most direct. Given that the seismic data gathered on the Moon over 40 years ago revolutionized our understanding of the Moon and are still being used today to produce new insight into the state of the lunar interior, it is no wonder that many future missions, both real and conceptual, plan to take seismometers to other planets.

To best facilitate the return of high-quality data from these instruments, as well as to further our understanding of the dynamic processes that modify a planet's interior, various modeling approaches are used to quantify parameters such as the amount and distribution of seismicity, tidal deformation, and seismic structure on and of the terrestrial planets. In addition, recent advances in wavefield modeling have permitted a renewed look at seismic energy transmission and the effects of attenuation and scattering, as well as the presence and effect of a core, on recorded seismograms. In this chapter, we will review these approaches.

10.2 Site selection for future planetary seismology missions

The ability of a seismic network to accurately locate an event improves as the number of seismometers increases. On Earth we take for granted that any given event will be relatively well located, due to the comparative ease of installation of seismometers. For planetary applications, we cannot count on a large distribution of stations. Various factors including cost, difficulty of installation, instrumentation longevity, and data transmission severely limit the number of instruments that have been or will be deployed on other planetary bodies. In this section, we will review various methods that can be employed to help determine the best landing sites for future planetary seismology missions, in order to maximize their scientific return. We focus here on the Moon and Mars, although many of these methods are adaptable to other planetary bodies.

Extraterrestrial Seismology, eds. Vincent C. H. Tong and Rafael A. García. Published by Cambridge University Press.

Key issues that feed into the site selection of future seismometer deployments include: (1) Theoretically establishing the seismicity and noise environment of a planetary body, including tectonic, volcanic, artificial, and natural impact sources; (2) Positioning of instruments relative to regions of expected seismic sources to maximize the detectability of important seismic waves that traverse the interior, including body, surface, and normal modes; and (3) Selection of landing sites and surface materials that maximize the longevity of a deployed instrument and minimize the seismic noise environment.

10.2.1 The Moon

The installation of seismometers on the Moon's surface during the Apollo era provided a wealth of information that transformed our understanding of lunar formation and evolution. Seismic events detected by the nearside network were used to constrain the structure of the Moon's crust and mantle down to a depth of about 1000 km. However, the lack of seismic ray paths penetrating the deepest Moon prohibited definitive identification of the Moon's core.

The presence of an attenuating region in the deepest interior, generally interpreted as a core, has been inferred from the paucity of farside events (Nakamura, 2005), as well as other indirect geophysical measurements (Wieczorek *et al.*, 2006). In addition, more recent works have made progress in the recognition of core-reflected phases in the stacked Apollo data (Weber *et al.*, 2011; Garcia *et al.*, 2011). Such phases typically arrive in the coda of the main P and S arrivals, hampering their identification on individual seismograms.

Because the presence of a lunar core has important implications for the origin and evolution of the Moon, future missions should focus on refining our knowledge of the deepest interior. Weber *et al.* (2010a) devised a technique to map ideal landing sites for the detection of core-interacting seismic phases from the known distribution of lunar seismic sources. The work focused on reflected (PcP, ScS) and converted (PKP) phases, but is easily adaptable to any seismic phase.

The method takes into account the predicted ray density and arrival amplitudes (Figure 10.1), as well as the level of seismicity from the known distribution of deep moonquakes. An example for the PKP arrival is illustrated in Figure 10.2a, using the velocity model of Nakamura (1983) modified to possess a core with radius 340 km and P-wave velocity of 5.0 km/s. At large epicentral distances, PKP is predicted as a first arrival, and hence should be easily identifiable on future seismograms.

10.3 Mars

The Viking landers of the late 1970s both carried a short-period three-component seismometer. On the Viking I lander, the uncaging mechanism failed to unlock the instrument. Viking II landed in Utopia Planitia and returned about 19 months of data (Lazarewicz *et al.*, 1981); unfortunately, because the seismometer was not well coupled to the ground,

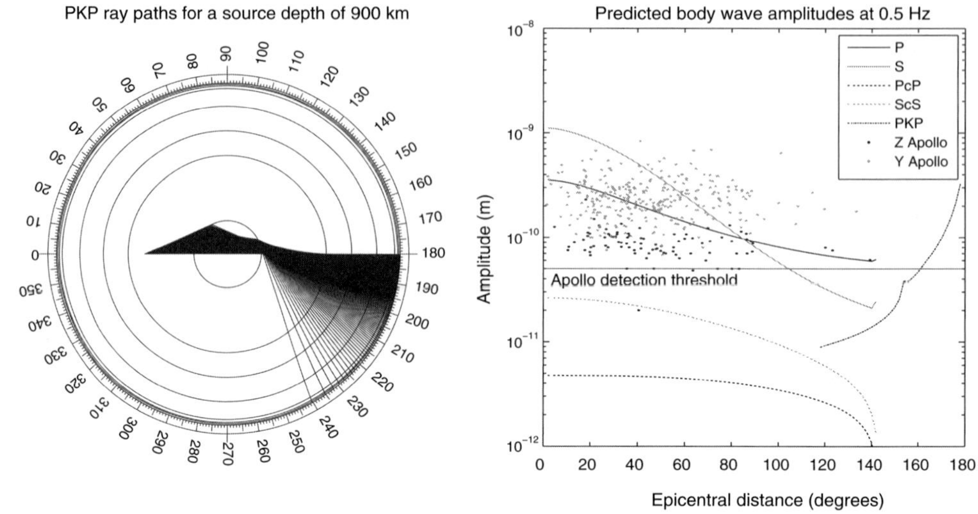

Figure 10.1 (Left) Cross-section of the Moon showing the PKP ray fan for a source depth of 900 km. (Right) Predicted arrival amplitudes for a number of seismic phases, including core-interacting phases. The dots show actual Apollo moonquake amplitudes. Note that core-reflected phases fall below the Apollo detection threshold. At large epicentral distances PKP is theoretically detectable by the Apollo seismometers, but such source–receiver geometries were lacking given the limited nearside extent of the array.

it recorded mostly (if not exclusively) wind events (Anderson *et al.*, 1977). Since Viking, several studies have aimed to estimate the amount of seismicity that can be expected on Mars, in order to optimize future seismic experiments.

In general, three questions need to be answered to model a planet's seismic activity: (1) How much deformation occurs in the lithosphere? (2) How does it translate into seismic events? and (3) Where do these occur?

Phillips (1991) analyzed several possible sources of deformation of the martian lithosphere. As no Earth-like plate tectonics occur today on Mars, secular cooling of the planetary interior is the most important source of deformation, with an estimated annual seismic moment release of 4.8×10^{18} Nm, whereas tidal stresses can be neglected. Analyses of the accumulated slip visible on the martian surface today (Golombek *et al.*, 1992; Golombek, 2002) yield moment release estimates of 10^{17} Nm and 10^{19} Nm per year.

Knapmeyer *et al.* (2006) investigated not only the total amount of moment release, but also its distribution among individual events and the distribution of these events over the planet. Their model is thus not only able to link the seismicity to geodynamic parameters but also able to produce synthetic event catalogs for the evaluation of seismic experiment concepts.

The model consists of three main parts. The annual moment release due to cooling of the lithosphere is first estimated. Then the moment–frequency relation (Gutenberg–Richter distribution) is modeled as a truncated Pareto distribution, giving the number N of events

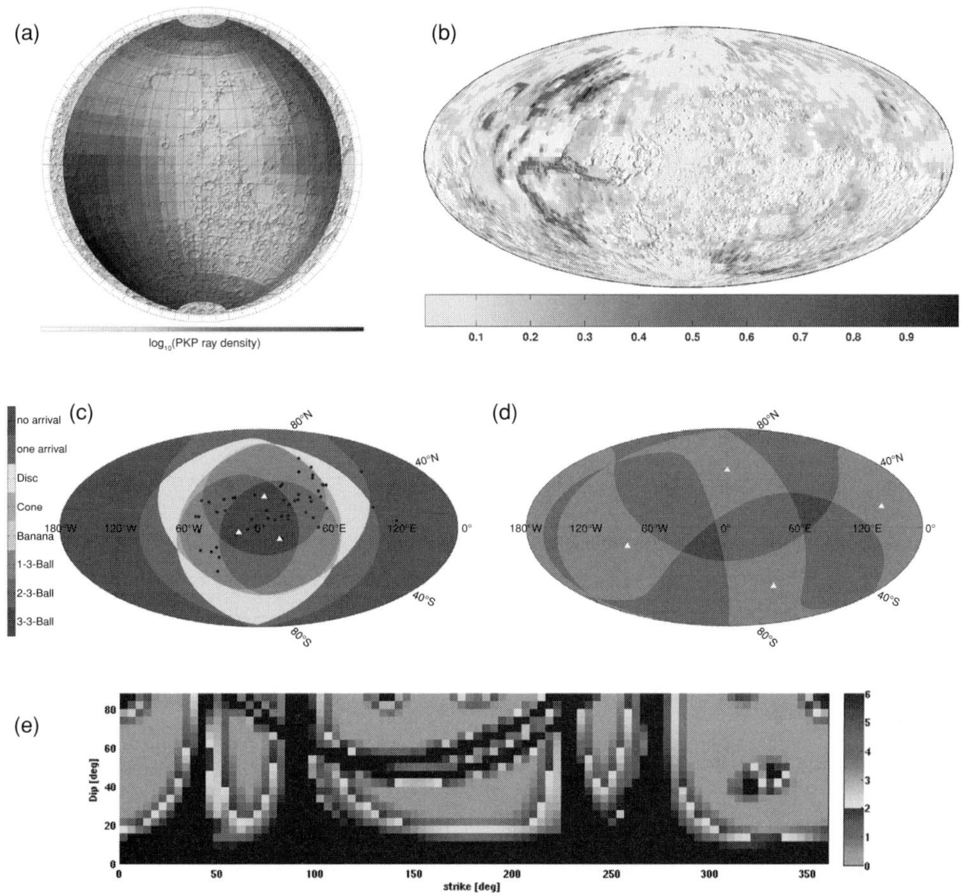

Figure 10.2 (a) Nearside map of PKP ray density from the known distribution of deep moonquake clusters as a function of landing site coordinates, taking predicted cluster occurrence activity and arrival amplitudes into account, shown in logarithmic scale. For this particular phase, landing sites near the limb are favored, particularly in the southwestern quadrant of the Moon, where the likelihood of detecting PKP from the northeastern farside events is greatest. (b) Map showing the probability that a $M_w \leq 4.8$ quake is assigned to 1 km^2 of surface, plotted on a $2.5^\circ \times 2.5^\circ$ grid (Knapmeyer *et al.*, 2006). (c) Location capability of an Apollo-like network. For comparison, real deep moonquake clusters with ball-shaped location clouds are shown (black dots). Triangles are station locations (Hempel *et al.*, 2012). (d) Location capability of a network with four equally distributed, modern seismometers (Hempel *et al.*, 2012). (e) Least-squares misfit between observed and calculated amplitudes on a regular grid of fault orientations averaged over all useable events from the A6 source region. The accepted regions (green) are selected by highlighting fault orientations that predict amplitudes within two standard deviations of the measured values. A black and white version of this figure will appear in some formats. For the color version please refer to the plate section.

Table 10.1 *Recurrence times for marsquakes of different magnitudes, for five seismicity scenarios. The first row gives the annual cumulative seismic moment release for the strong, medium, and weak models. Moment magnitudes are limited by the assumption of a largest possible event. Units day (d) and year (a) refer to the respective durations on Earth.*

Mag. $M_w \geq$	Moment $M_0 \geq$	str./few	str./many	med.	weak/many	weak/few
	$M_{\text{cum}} =$	4.78×10^{18} Nm		5.99×10^{17} Nm	3.42×10^{16} Nm	
1	4×10^{10} Nm	38 min	72 s	48.5 min	2.5 h	3.9 d
2	1.3×10^{12} Nm	5.5 h	10.5 min	7 h	21.9 h	33.5 d
3	4×10^{13} Nm	1.9 d	1.5 h	2.5 d	7.9 d	290 d
4	1.3×10^{15} Nm	17.1 d	13.1 h	21.8 d	68.5 d	6.7 a
5	4×10^{16} Nm	148 d	–	189 d	–	50 a
6	1.3×10^{18} Nm	3.6 a	–	4.5 a	–	516 a
7	4×10^{19} Nm	30.3 a	–	–	–	~4500 a
7.6	3.2×10^{20} Nm	115 a	–	–	–	~17 000 a

that exceed a seismic moment M_0. Finally, the seismic moment of an individual event is linked to the map length of the fault on which it occurred, by following the simplifying assumption that the stress drop is essentially constant for all marsquakes (e.g., Kanamori and Anderson, 1975a). Knapmeyer *et al.* (2006) then mapped visible faults on the martian surface to distribute the individual event sizes according to the chosen distribution on faults that are long enough to produce them, i.e., small events may occur on almost any fault while large ones are restricted to the longest faults (Figure 10.2b).

Five scenarios were named to refer to the annual moment release (strong or weak) and the number of events (many or few) resulting from the Gutenberg–Richter distribution, with an average model that evenly distributed both factors. Translated into end-member magnitudes, the model suggests that the shortest fault ($L = 4$ km) can produce events with magnitudes up to $M_0 = 4.7$, while on the longest one ($L = 1445$ km), a $M_0 = 9.6$ event is possible. In other words, the whole annual budget of the weak/few scenario may be released in a single event on each of the mapped faults, while there is no severe restriction for the largest possible event from the size of tectonic features. The recurrence times for quakes of different sizes, according to the five model scenarios, are given in Table 10.1.

This seismicity model and the epicenter distributions derived from the mapped faults rely on a number of assumptions that may or may not be fulfilled. The amount of lithospheric contraction considered is based on a numerical simulation of the cooling of the lithosphere alone. Cooling of the entire mantle, the core, as well as a possible freezing of an inner core may modify the radius changes. The model for the distribution of events on the martian surface assumes that only the visible faults exist, and especially the possibility of deep sources is neglected. The latter is based on the assumption that deep quakes such as those that occur on subduction zones on Earth cannot occur in the absence of plate tectonics, and that the state of the martian mantle makes very deep sources similar to the lunar deep quakes impossible. The model by construction assumes that a strongest possible quake exists. The magnitude of this can of course be set to very high values. However, one must not confuse the annual seismic moment budget with the strongest possible event: the largest events observed on Earth release a seismic moment several times higher than the average annual release. Finally, the possibility of seismicity linked to volcanic activity is neglected: volcanism is a mechanism to cool the planetary interior, so active volcanism would change the cooling rates, and related events would occur near volcanic centers. These two effects were ignored because the current rate of martian volcanism appears to be too low to be of significance for a seismic experiment in the near future.

In addition to tectonic seismic events, impacts can also be used to study the interior of planets, and are expected to be relatively common on airless or thin atmospheric bodies throughout the Solar System. For the Moon, over 1700 impacts large enough to be detected at far distances were recorded by the Apollo seismic array, translating to approximately one impact/day. Recent high-resolution imaging of fresh craters on the surface estimates a similar cratering rate will be present on Mars (Daubar *et al.*, 2013).

10.4 Sparse network or single-station seismology approaches

Future planetary seismology missions are unlikely to have large numbers of instruments, as discussed previously. Therefore, many standard processing techniques used in terrestrial seismology will be ineffective for planetary applications. Here we discuss specialized techniques that can be used to analyze data from sparse networks or single-station installations.

10.4.1 Event location by P-S differential time and back-azimuth

Rough source locations (event distance and back-azimuth) can be constrained with a sparse network using the differential time between the primary P- and S-wave arrivals, combined with polarization information. If the data come from high-quality, three-component instruments with accurately oriented horizontal components, location can be constrained using only two stations. Future seismic missions to the Moon can take advantage of this technique using established seismicity, namely, the deep moonquake clusters detected by Apollo, which are presumably still active today. A similar analysis might be appropriate for any body that has a tidally modulated seismic signal, such as the outer satellites.

Planetary seismic networks will record seismograms from active sources on N stations, and the P and S arrival times $t(P)$ and $t(S)$ from each station will result in $N \times 2$ unique observations. For a given event, we expect that with modern instrumentation we can measure $t(S) - t(P)$ with an accuracy of five seconds, back-azimuth with an accuracy of ten degrees, and the difference between the P arrivals at different stations with an accuracy of one second.

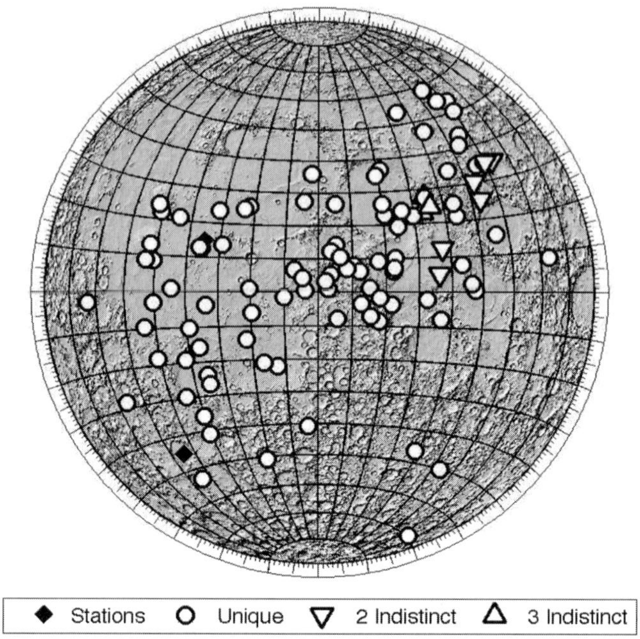

Figure 10.3 Number of neighbors with which a given deep moonquake cluster can potentially be confused. A large number of clusters can be uniquely identified using information from only two recording stations (diamonds). This number increases as the number of seismic stations increases.

Using this information for the deep events recorded by Apollo, we can map the clusters that can be uniquely identified using N desired landing sites. An example is shown in Figure 10.3 for $N = 2$.

Most of the clusters can be uniquely identified with the $S - P$ time difference and back-azimuth. However, the remaining clusters can possibly be identified by including consideration of the temporal distribution of their occurrence times relative to the tidal cycle. All clusters could be identified uniquely with four or five stations in reasonable network geometry, simply by location. It is important to note that the number of unique identifications strongly depends on the positions of the stations relative to the clusters. Future landing sites should be chosen carefully to take advantage of known seismicity.

10.4.2 Event location via timing of surface wave orbits

With a sufficiently large event where multiple orbit surface waves are recorded, epicentral distance and origin time can be determined at a single station without any requirement for an a-priori velocity model. In the traditional surface wave numbering scheme, orbits are numbered such that the Rayleigh wave propagating along the minor arc is numbered $R1$, the wave propagating along the major arc is $R2$, and the wave that propagates along the minor arc with a complete great-circle orbit is numbered $R3$. If we filter data containing multiple orbit surface waves using a series of narrow band filters, and then measure the times of the $R1$, $R2$, and $R3$ peaks on the waveform envelopes, we can determine the great-circle averaged group velocity dispersion, epicentral distance, and event origin time using simple geometry.

For terrestrial data, we need events of moment magnitude ~6 in order to reliably observe the weakest arrival, $R3$, but this approach can be extended to smaller events for smaller planetary bodies, since the amplitude decrease caused by geometrical spreading is less. Mars, for example, has a radius approximately half that of the Earth, and therefore $R3$ would be expected to be about an order of magnitude larger in amplitude simply due to geometric effects. A planet without an ocean would also likely have lower background seismic noise than the Earth (where the microseismic background noise is generated via ocean processes), and so multiple orbit surface waves will likely be measurable for magnitude 5 events on Mars with a high-quality broadband seismometer, and possibly to lower magnitudes. This epicentral distance observation can be made independently in multiple frequency bands, which allows for some estimates of the error.

Provided the surface waves are adequately recorded on the horizontal components, we can also determine back-azimuth using the elliptical particle motion of the Rayleigh wave by correlating rotated horizontal components with the Hilbert transform of the vertical component (Panning *et al.*, 2015). Using such an approach with terrestrial data between magnitude 6.0 and 6.5, epicentral distance was constrained within a degree, and origin time within 30 seconds for most events, despite some biases due to un-modeled 3D structure (Figure 10.4). While such location errors would be large for the purposes of state-of-the art terrestrial seismology, initial tests show that they are adequate for performing velocity inversions using small numbers of P and S body wave picks, and resolve mantle velocity structure within 5%. Finally, the group velocity dispersion measurements included in this approach can then also be used to constrain shallow mantle structure using traditional surface wave dispersion modeling. This shallow structure then can allow better location of smaller events using S and P differential times as discussed previously.

10.4.3 Deep structure via normal mode analysis

For situations in which there is only a single deployed seismometer, another effective technique for studying deep structure is to use normal mode frequencies, which do not require knowledge of the source location. The periods of free oscillations of a planet are dependent on the spherically averaged structure. In particular the state of the core affects the fundamental modes of free oscillation, with the gravest spheroidal mode (${}_0S_2$) being the most sensitive.

Spectral peaks that are sensitive to mantle structure (5–20 mHz) can be detected at a noise threshold level of 10^{-9} m s^{-2} Hz$^{-1/2}$, well within the capabilities of modern broadband seismometers being developed for planetary missions (Lognonné *et al.*, 1996). A single

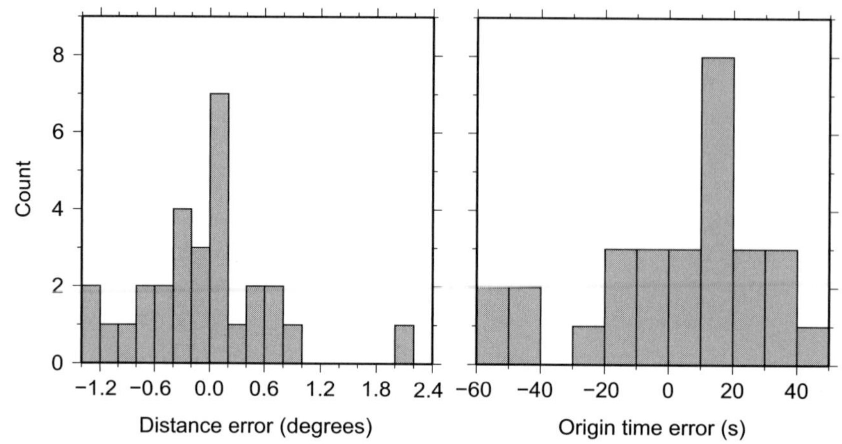

Figure 10.4 Histograms of epicentral distance errors (left) and origin time errors (right) for 29 events between magnitude 6.0 and 6.5 recorded at BFO in Germany, located by picking $R1 - R3$ arrival times. Positive errors represent distances that are too large, or origin times that are too late, respectively (Panning *et al.*, 2015).

large quake can excite a planet's normal modes, or multiple quakes with an equivalent cumulative seismic moment can be stacked. Single-seismogram analysis requires a quake with seismic moment $\geq 10^{18}$ Nm, equivalent to a body wave magnitude $m_b \sim 6$.

Khan and Mosegaard (2001) attempted to investigate lunar normal modes using Apollo seismic data recorded from large surface impacts, in order to constrain seismic velocities and densities in the deep lunar interior. Their observations were limited to the period during which the instruments were operated in "flat" mode, i.e., response extended to low frequencies; only five events during this time were large enough. Sufficiently large events on Mars are similarly predicted to be rare, with less than one event per year likely (Golombek *et al.*, 1992; Knapmeyer *et al.*, 2006); future seismic missions must rely on longevity to detect them.

10.4.4 Location capability for future lunar seismic networks

The traditional method for quake location converges towards the best-fit quake origin by minimizing the squared misfit between observed and theoretical travel times. This method assumes that arrival time readings are Gaussian distributed. This assumption is problematic for data from sparse networks; a lack of redundancy means arrival time readings with large errors are more likely to be included in the solution.

The LOCSMITH event location scheme (Knapmeyer, 2008) was developed for sparse seismic networks on planets other than Earth. Rather than solving for a best-fit location, it divides the solution set into falsified and non-falsified candidate locations using an adaptive grid search, and accounts for arrival time uncertainty using windows around the true arrival time. The iterative algorithm starts with an icosahedral grid over all depths of the planet, refining the grid size on each iteration.

Hempel *et al.* (2012) applied the LOCSMITH algorithm to the Apollo seismic data. For a candidate set of grid points and detecting stations, theoretical travel times were calculated via ray-tracing using several different known lunar seismic velocity models. The travel times were then compared with observed arrival time windows for both compressional wave (P) and shear wave (S) phases. The only assumption was that the chosen time window certainly contains the searched arrival, which is adequate for moonquakes given that the first arrivals are often emergent rather than impulsive.

If the gridded theoretical arrival times were found to lie within the arrival time windows for all stations and all phases, the grid point density around the corresponding grid point was refined for the next iteration. If not, the grid point was considered falsified. The solution sets of this algorithm are therefore clouds composed of all non-falsified grid points. The cloud size and shape depends on the number of observed arrivals. The best located clusters have ball-shaped clouds that are similar to traditional quake location error ellipses. Less well-located clusters have elongated clouds that resemble cones or bananas. The worst located clusters can only be constrained to disks that slice through the entire Moon.

This correlation between cloud shape and location uncertainty can be used to map the detection capability of future lunar seismic networks, assuming the known deep moonquake clusters are still active (and we have no reason not to expect them to be). Figures 10.2a and 10.2d show the global location capability of an Apollo-like network, using a uniform distribution of 10000 synthetic events. For a relatively small-aperture nearside array consisting of Apollo-era instruments (Figure 10.2c), only events occurring in the vicinity of the network will be well located (ball-shaped uncertainty clouds). For a globally distributed network consisting of modern state-of-the-art instruments (10 times more sensitive than Apollo), there is a dramatic improvement in the location capability (Figure 10.2d).

Rudimentary locations with near-surface epicenters can be used to search for visual evidence of ground disturbance such as landslides, fresh craters, fresh fault scarps, and displaced boulders, when orbital imagery is available. This can provide an independent constraint on source location.

10.5 Exploring the relationship between lunar seismicity and tidal forcing

As discussed in Section 7.2.2, deep moonquakes are known to occur periodically, with monthly (tidal) periodicity. To analyze the relationship between lunar seismicity and tidal forcing, several previous studies created models of the tidal stress field within the Moon (Toksoz *et al.*, 1977; Lammlein, 1977; Cheng and Toksoz, 1978; Goulty, 1979; Minshull and Goulty, 1988; Weber *et al.*, 2009). Solid-body tides are caused by the gravitational influence of the Earth and the Sun, resulting in the deformation of the Moon such that moonquakes are caused or triggered in the deep interior.

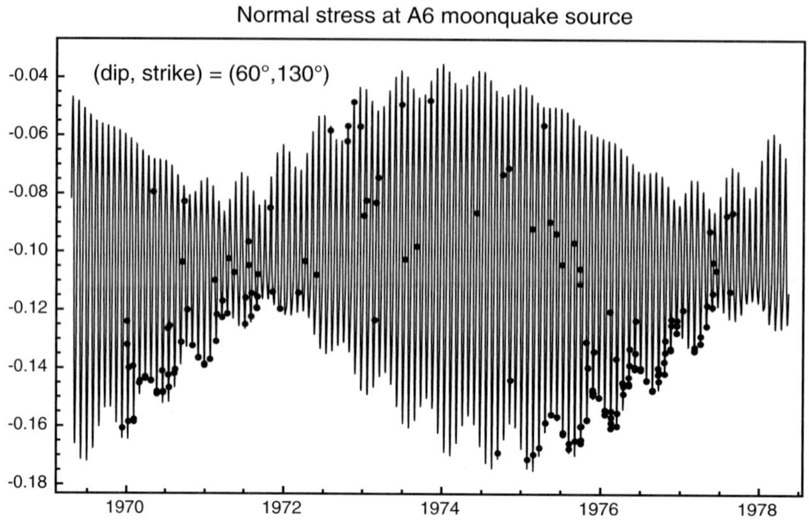

Figure 10.5 Normal tidal stress calculated at the A6 source location, on a fault with a dip of 60° and a strike of 130°, over the course of the Apollo experiment. Stress is calculated at one-day intervals (line) and at A6 moonquake times (dots). Vertical axis unit is bars.

In studies of tidal forcing, a key constraint is the focal mechanism: the fault parameters describing the type of failure moonquakes represent. Knowledge of the failure plane allows us to resolve the tidal stress into its shear and normal components and evaluate the state of stress at the observed moonquake occurrence times. The normal stress resolved onto an arbitrary fault plane calculated at the A6 deep moonquake source is shown in Figure 10.5. A6 moonquakes appear to occur at or near local minima in the normal stress, suggesting a relationship between moonquake occurrence and unloading on the fault.

In terrestrial seismology, the Coulomb stress criterion predicts failure when a linear combination of the shear and normal stress on a fault reaches some critical value. Weber *et al.* (2009) attempted to apply a similar criterion to deep moonquakes by searching for a fault orientation such that moonquakes occurred when a linear combination of shear and normal stress on that fault reached a constant threshold. The results of their attempt for the A1 source are shown in Figure 10.6. Fault orientations were searched in 10° increments of dip (δ) and strike (α). The linear combination C of normal and shear stress was computed on each fault plane such that the variance of the ratio between C_m computed at moonquake times and C_a computed on 1-day intervals was minimized. The absolute minimum of this variance ratio over all grid spaces was selected as the preferred fault plane.

This method worked better for some clusters than others, with a deep minimum around the best-fit solution. For others, the definition of the best-fit plane was more ambiguous, with a relatively shallow minimum, or several nearby minima in the strike-dip grid space. Since deep moonquake failure planes have no surface expression, there is no way to visually verify whether the best-fit plane is truly representative of the physical failure mechanism.

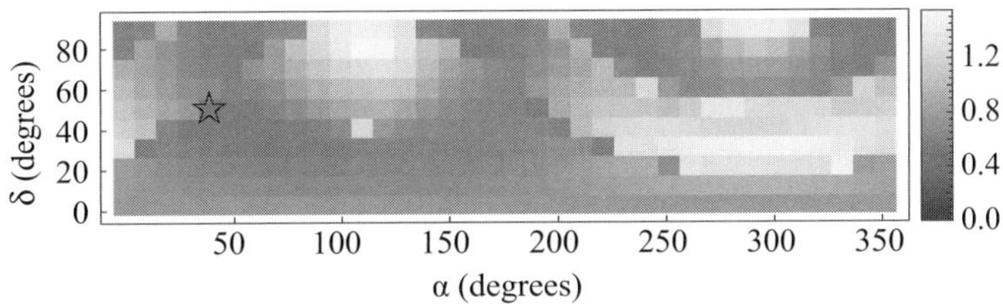

Figure 10.6 Variance ratio of the linear combination of shear and normal stress C_m computed at A1 moonquake times to C_a computed at 1-day time steps, computed on a grid of all possible fault orientations. The best-fit fault plane occurs at the minimum variance ratio, and is indicated here with a star. (Figure adapted from Weber *et al.*, 2009.)

In addition, the quality of the lunar seismic data combined with the limited source/receiver geometries of the Apollo network prohibit the determination of deep moonquake fault parameters using first-motion polarities, as is typically done in terrestrial seismology.

An alternative method for focal mechanism recovery examined the extent to which shear (S) and compression (P) wave amplitude ratios can constrain moonquake fault geometry by determining whether, for a given cluster, there exists a focal mechanism that can produce a radiation pattern consistent with the amplitudes measured by the Apollo instruments (Weber and Knapmeyer, 2012). Amplitudes were read in the ray coordinate frame, directly from seismograms for which the P and S arrivals were clearly identifiable on all long-period channels of the four Apollo stations. An empirical station correction accounted for site effects at the four Apollo stations, as well as the differences between P- and S-wave attenuation in the lunar interior.

Instead of focusing on the best fitting solution only (as in Weber *et al.*, 2009), the inverse problem was formulated using a falsification criterion: All source orientations that do not reproduce the observed S/P amplitude ratios within an error margin derived from the uncertainty of the amplitude readings were rejected. All others were accepted as possible solutions. Similarly to the Weber *et al.* (2009) method, the inversion was carried out using a grid search with predefined step size, encompassing all possible combinations of fault strike and dip, but also included slip (the direction of fault motion between strike and dip).

If we assume that all events from a given cluster share the same source orientation, combining their inversions reduces ambiguities in the results. Using only amplitude information, the fault plane parameter space for a given cluster was reduced on average by half (Figure 10.2e). We note that this assumption may not be true for all clusters.

10.6 Seismic wavefield modeling

Lunar seismograms differ greatly from terrestrial seismograms. The low attenuation of seismic energy combined with the highly scattering environment in the shallow subsurface results in recordings that are emergent rather than impulsive, with long ringing coda that obscure all but the primary seismic phases. For this reason, it is useful to attempt to better understand the transmission of seismic energy within the Moon through wavefield modeling. In this section we will discuss different methods for synthetic seismogram generation, which are useful both for comparison with existing seismograms (the Moon) or for the prediction of seismic signals that might be recorded on other planets and their satellites (Mars, Europa).

10.6.1 Normal mode seismology

Complete wavefields can be predicted efficiently using summation of normal modes calculated with a-priori models of planetary structure. Such an approach is not well suited for high-frequency data, as the computations become more expensive and unstable for high frequencies and it is not trivial to include the effects of scattering and 3D structure, but they are extremely useful when predicting long-period responses. Such calculations can be used in advance of particular planetary missions in order to explore likely candidate seismic signals for discriminating between models of planetary interiors.

For Europa, a moon of Jupiter, Cammarano *et al.* (2006) produced a range of possible thermodynamically consistent interior models by varying assumptions of initial chemistry, importance of tidal heating, and thickness of the ice shell overlying the likely global liquid water ocean (e.g., Kivelson *et al.*, 2000). Panning *et al.* (2006) used these to create normal mode catalogs complete up to periods of 10 seconds. By analyzing synthetic seismograms from mode summation, it was determined that long-period signals from internal events analogous to deep lunar events would be unlikely to be recorded at the surface, but fracturing events in the ice shell would likely produce very strong signals with characteristics strongly dependent on the thickness of the ice shell. In the frequency band between 1 and 100 mHz, the fundamental mode Rayleigh wave undergoes a transition from a nearly non-dispersive wave at high frequencies to a strongly dispersed flexural wave at lower frequencies. This flexural wave, also observed in terrestrial sea ice at higher frequencies (Press and Ewing, 1951), has a characteristic dispersion with phase velocity proportional to the square root of frequency, and group velocity equal to double the phase velocity. The transition between the non-dispersive Rayleigh wave and the flexural mode leads to a distinctive peak in group velocity at a frequency that depends on the thickness of the ice shell. This peak means that identification of a surface wave-train, even in the absence of event location, should be able to determine ice shell thickness simply by determining the frequency of the first-arriving Rayleigh energy.

10.6.2 Reflectivity and full wavefield methods

Seismic energy transmission in the Moon is governed not only by the very low attenuation of waves in the surface zone, but also the intense scattering and reverberation of waves in the low-velocity lunar megaregolith and regolith. These lowered seismic velocities are similar

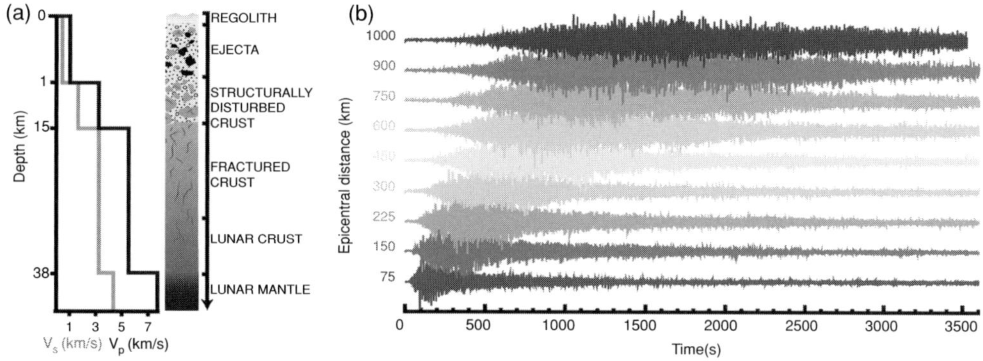

Figure 10.7 Seismic structure of the lunar crust and megaregolith. (a) The crustal velocity model after Weber *et al.* (2011), showing the geology of the subsurface. The megaregolith extends from the surface to just above the structurally disturbed crust. (b) 1D wave propagation for a model with a 1 km thick layer of low-velocity megaregolith showing the development of surface waves and reverberations trapped in the layer.

to those observed near impact craters on Earth, and are related to a high amount of pore space and fracture density lowering the rigidity and compressibility of impact brecciated materials (Figure 10.7a). Here we discuss methods to reproduce the long duration and pronounced coda observed in lunar seismic signals, and implications for seismic structure of planets/objects with surface megaregolith layers.

High-frequency (>1 Hz) synthetic wave propagation techniques are essential for properly replicating long duration codas that relate to scattering and reverberation of seismic energy within the megaregolith, which normal mode methods do not capture. Reverberations and trapping of seismic energy in a low-velocity wave channel can be achieved with a variety of 1D wave propagation codes, but scattering requires a much more computationally expensive 3D wavefield simulation, which has only recently become available. Below are two examples of how high-frequency wave propagation methods can be used to investigate the effects of low-velocity surface waveguides and scattering in planetary surface materials.

Propagation of waves within the low-velocity surface materials can be approximated using a variety of 1D seismic wave propagation codes. Typically these models allow specification of a layer-cake planet and are efficiently run on modern CPUs at frequencies of ~1 Hz. One example is GEMINI (Green function of the Earth by MINor Integration), made available through the SPICE (Seismic wave Propagation and Imaging in Complex media: a European network) project. It solves the equations of motion for seismic source(s) and convolves the result with the Green's coefficients calculated for layered 1D media. Source functions can be either empirically defined or from a standard set of basic source types. The code produces seismograms for anisotropic and anelastic media for any 1D layer-cake model structure; responses are calculated for a user-defined set of stations at dominant frequencies up to 10 Hz. From 1D simulations (Figure 10.7b) we can see that surface waves and multiple reverberations become trapped in the low velocities of the lunar

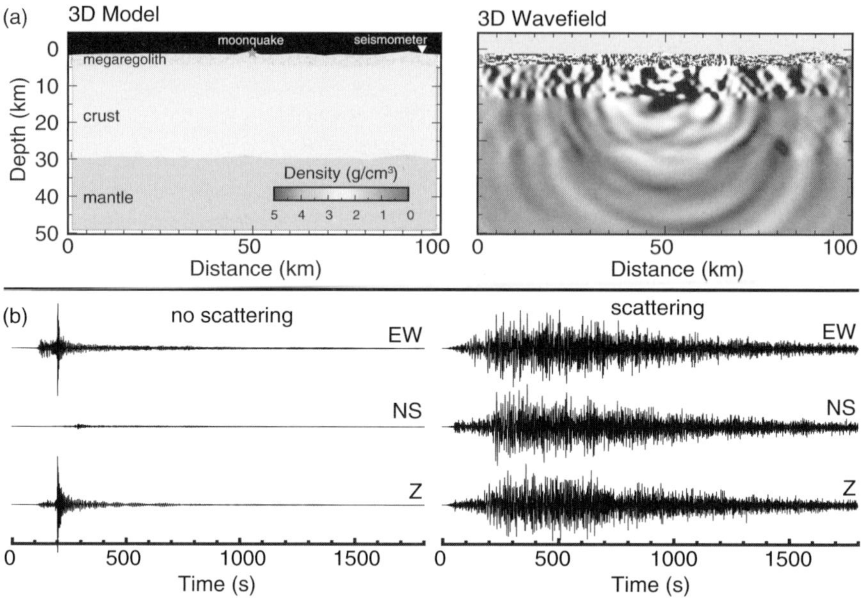

Figure 10.8 Seismic scattering in the lunar crust and megaregolith for an impact source. (a) Cross-section through a lunar model with 25%, 200 m scale-length, Von Karman random velocity heterogeneity distribution (Frankel and Clayton, 1986) within a 3 km thick megaregolith and time snapshot of vertical displacement at 40 s into a WPP simulation. Positive amplitudes are black, negative white. (b) Seismograms for a model with no scattering in the subsurface (left) and 25% heterogeneity (right) at 50 km from the source (location given in part (a).

megaregolith, serving as a waveguide for energy transmitted into this layer. This simulates the long duration of codas but does not fully replicate the scattering properties observed in the lunar megaregolith.

To properly address scattering, a high frequency, 3D wave propagation code, such as WPP (Wave Propagation Program) is required (Petersson and Sjogreen, 2010). WPP solves the equation of motion on a 3D finite difference, non-uniform cartesian mesh, and is designed to run on parallel multi-CPU clusters. The user may specify elastic structures in 3D, including topography, layering, lateral heterogeneity, attenuation, and multiple sources with varying source time functions. The code is primarily used for regional simulations of wave propagation in complex media, the elastic response of the medium is calculated for dominant frequencies up to 10 Hz or greater, though computational resources required grow for higher frequencies.

In Figure 10.8, we demonstrate the effect of scattering in the megaregolith using WPP to obtain lunar-like seismograms. By introducing a layer of scattering into the low-velocity megaregolith, several lunar-like wavefield features, including long duration codas, partitioning of energy on all three components of motion, and diffuse surface waves are reproduced. Additionally, body waves entering into the coda layer from below become scattered at the surface only. As computational resources grow, fully global simulations at high frequency are becoming available, allowing us to investigate the global effects of scattering at very long distances.

Scattering effects will vary from world to world, with airless bodies with well-developed megaregoliths (Moon, Mercury, asteroids) having the highest proportion of scattering within their crusts relative to bodies with atmospheres and intragranular fluids and ices (Mars, Earth, Venus, the outer satellites). WPP is appropriate for any planetary object with a solid surface and can be used to approximate the scattering from rocky and icy worlds. In addition to scattering and detailed crustal structure, synthetic codes can be used to study the expected amplitudes and arrival times of various seismic waves that sample the interior of a planet. This is very useful for planning mission geometries and developing single-station techniques for analyzing data returned in future missions. Numerical codes such as SPECFEM3D_Globe (Komatitsch and Tromp, 2002) are highly optimized and regularly used for calculations up to $\sim$1 Hz for Earth studies and can be run in spherical geometries, which is important for future modeling of long paths.

Part IV

Discoveries of physical structures and processes

11

Asteroseismology of red giant stars

RAFAEL A. GARCÍA AND DENNIS STELLO

11.1 Introduction

If appropriately excited, a star will oscillate like a giant spherical instrument. In most stars, including the Sun, surface convection provides the excitation mechanism (Goldreich and Keeley, 1977). With turbulent velocities reaching speeds comparable to the local sound speed near the surface of the star, the vigorous convective motions can excite standing acoustic waves. These are known as pressure or p modes because the restoring force arises from the pressure gradient. The broad frequency spectrum of this excitation mechanism gives rise to many oscillation modes, both radial and non-radial, excited simultaneously. These stochastically excited and intrinsically damped oscillations were first detected in the Sun (Leighton *et al.*, 1962), and hence are commonly known as solar-like oscillations.

Oscillation modes can be characterized by the number of nodes, n, in the radial direction, called the radial order, and a non-radial part described by spherical harmonics, each with an angular degree, l, which equals the number of nodal lines on the surface, and an azimuthal order, m, which is the number of those nodal lines crossing the equator of the star. Except for the Sun, we can generally not resolve these oscillations on the surfaces of cool stars and so the surface displacements of modes with high angular degree ($l \gtrsim 4$) cannot be observed because regions of opposite phase tend to cancel out.

When stars grow old and the supply of hydrogen fuel is exhausted in the core, their envelopes expand and cool: they become subgiants and eventually red giants. This transition is shown in the so-called HR diagram in Figure 11.1, which indicates the main stages of evolution of a one solar mass star from the "main sequence" where the Sun is currently located (between points 1 and 2) through to the red giant phase discussed in this chapter (beyond point 3). Like the Sun, red giants have convective outer envelopes, but the much longer convective time scale drives oscillation modes at much lower frequencies. The expansion and contraction of different parts of the stellar surface when a star oscillates gives rise to variations in temperature – and hence also luminosity – across the stellar

Extraterrestrial Seismology, eds. Vincent C. H. Tong and Rafael A. García. Published by Cambridge University Press.

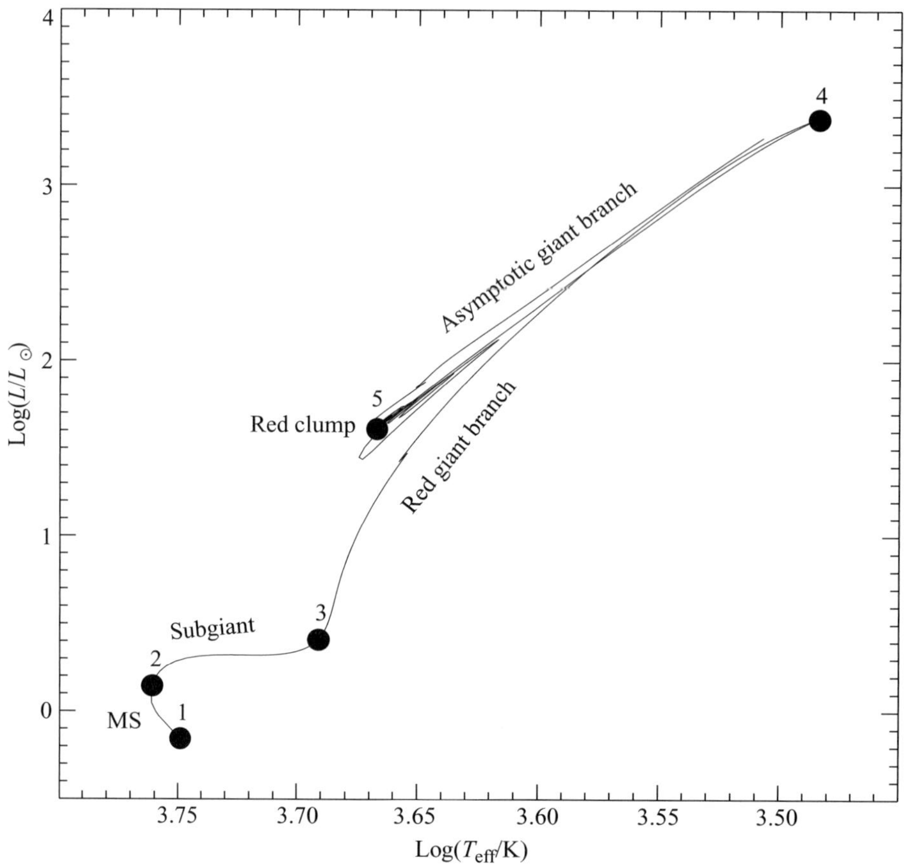

Figure 11.1 HR diagram of a stellar model for a one solar mass star of solar composition, showing luminosity in solar units versus effective (surface) temperature. The star starts its life burning hydrogen at the zero-age main sequence (point 1). Each following evolution phase is indicated as follows: 1–2: main sequence (hydrogen core burning); 2–3: subgiant (hydrogen shell burning); 3–4: red giant branch (hydrogen shell burning); 4: red giant branch tip (helium ignition); 5: red clump (helium core burning); 5 and up: asymptotic giant branch (helium shell burning).

surface. For the Sun, the surface moves at about walking speed and has an oscillation amplitude of the order of 100 m. The corresponding brightness variations are only of a few parts per million (ppm), equivalent to temperature fluctuations of about 0.1 K. In most red giants, those brightness variations have increased to several tens, even hundreds, of ppm, which is however still relatively small compared to the twinkling of stars we see at night caused by the Earth's atmosphere. From the ground, the most successful method for detecting the oscillations has therefore been using time-series spectroscopy to detect the surface velocity changes, through the Doppler shift of the light, as the stellar surface moves in and out. This method is not impacted by the twinkling of stars, but current limitations on the ultra-high-precision spectrographs required for these observations generally restrict investigations to one target at a time. Hence, to do large-scale investigations one needs to use photometry to measure the brightness of many stars all at the same time, which requires space-based observations.

Before asteroseismology entered the space age, a range of ground-based efforts were carried out using either spectroscopy or photometry. Measurements of solar-like oscillations had been pursued since the early 1990s (e.g., Gilliland *et al.*, 1993, and references therein), but it was nearly a decade later until we saw the first firm detection in a red giant (Frandsen *et al.*, 2002). Despite an increasing number of oscillating red giants detected from ground (e.g., Barban *et al.*, 2004a; De Ridder *et al.*, 2006) and early space missions (Buzasi *et al.*, 2000; Retter *et al.*, 2003; Barban *et al.*, 2007; Gilliland, 2008), at the time, it was still not clear whether these stars showed both radial and non-radial modes (Christensen-Dalsgaard, 2004), and hence how strong the mode damping was (Stello *et al.*, 2006). It was therefore not known how information-rich and useful red giants would be as seismic targets. In an ambitious attempt to detect oscillations in a large group of red giants in the open cluster M67 using 10 telescopes simultaneously for over one month, only marginal detections were achieved, which concluded almost two decades of ground-based photometric attempts (Stello *et al.*, 2007).

11.2 Red giants: The new frontier in asteroseismology

It was data from space missions that transformed the field and demonstrated that red giants show rich spectra of oscillation frequencies. The existence of non-radial modes (see Figure 11.2) were indicated in a few bright stars observed by the star tracker camera on the WIRE spacecraft (Stello *et al.*, 2008). But it was a sample of hundreds of stars observed by CoRoT that unambiguously demonstrated that red giants exhibit both radial and non-radial modes (De Ridder *et al.*, 2009). Shortly after this came the initial results from the *Kepler* mission, which provided more precise data for even larger numbers of stars (Gilliland *et al.*, 2010; Bedding *et al.*, 2010b), confirming they pulsate with both radial and non-radial modes, whose frequency patterns resembled that of the roughly equally spaced acoustic modes in the Sun, but at lower frequencies. However, in a small percentage of stars the dipole modes showed unexplained low amplitudes (Mosser *et al.*, 2012a; García *et al.*, 2014a). Over a period of only 2–3 years, the number of red giants with detected oscillations went from a handful (pre-CoRoT) into the thousands (Mosser *et al.*, 2010; Kallinger *et al.*, 2010a; Hekker *et al.*, 2011; Huber *et al.*, 2011; Miglio *et al.*, 2012b; Stello *et al.*, 2013). These long high-precision time series for large numbers of stars opened up a completely new way of exploring red giants using asteroseismology.

The large frequency separation computed from the equally spaced overtone modes provides information on the density of the star. Through scaling relations anchored at the solar values, this information combined with the frequency of maximum power, $\nu_{\max}$, and

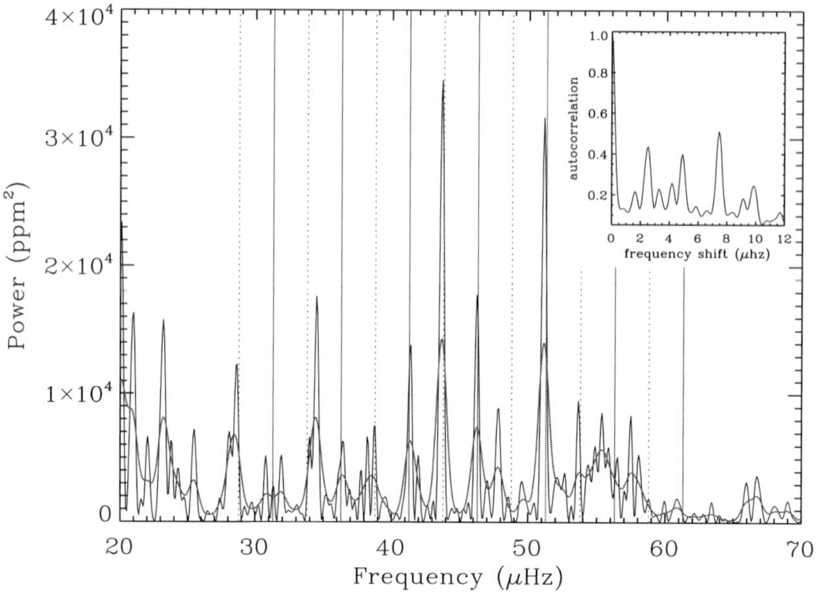

Figure 11.2 Power spectrum of the red giant βVol observed by the star tracker on the WIRE spacecraft (smoothed version in gray). Equally spaced vertical solid and dotted lines indicate radial and dipole modes, respectively. The inset shows the autocorrelation of the spectrum, showing peaks at multiples of half of the large frequency spacing: $\Delta\nu/2$.

the effective temperature allows estimation of mass and radius for stars pulsating from minutes to hundreds of days (see Figure 11.3). These relations have been extensively tested for both main-sequence and red giant stars, with additional constraints such as parallax measurements or individual frequencies (Stello *et al.*, 2008; Bedding, 2014; Miglio *et al.*, 2012a; Mathur *et al.*, 2012) and by using stellar models (Stello *et al.*, 2009b; White *et al.*, 2011; Miglio *et al.*, 2013a; Belkacem *et al.*, 2013). Recent independent methods have verified the scaling relations at the 4–5% level (Huber *et al.*, 2011; White *et al.*, 2013; Silva Aguirre *et al.*, 2012), although uncertainties remain about additional systematics for giants at certain stages of their evolution, which need further investigation (Miglio *et al.*, 2012a, 2013a). Masses and radii can therefore be inferred for a wide range of evolutionary stages from the main sequence through to the red giant branch, the clump, and even asymptotic giant branch stars (Corsaro *et al.*, 2012).

Among the most luminous red giants observed by *Kepler*, it was shown that they also pulsate in both radial and non-radial modes (Mosser *et al.*, 2013), concluding a decade-long debate about the possible presence of non-radial modes in these highly evolved stars. Stello *et al.* (2014) further demonstrated that the frequency pattern of these stars was markedly different from that of lesser evolved stars with dipole modes dominating the power frequency spectrum and being shifted relative to the radial modes.

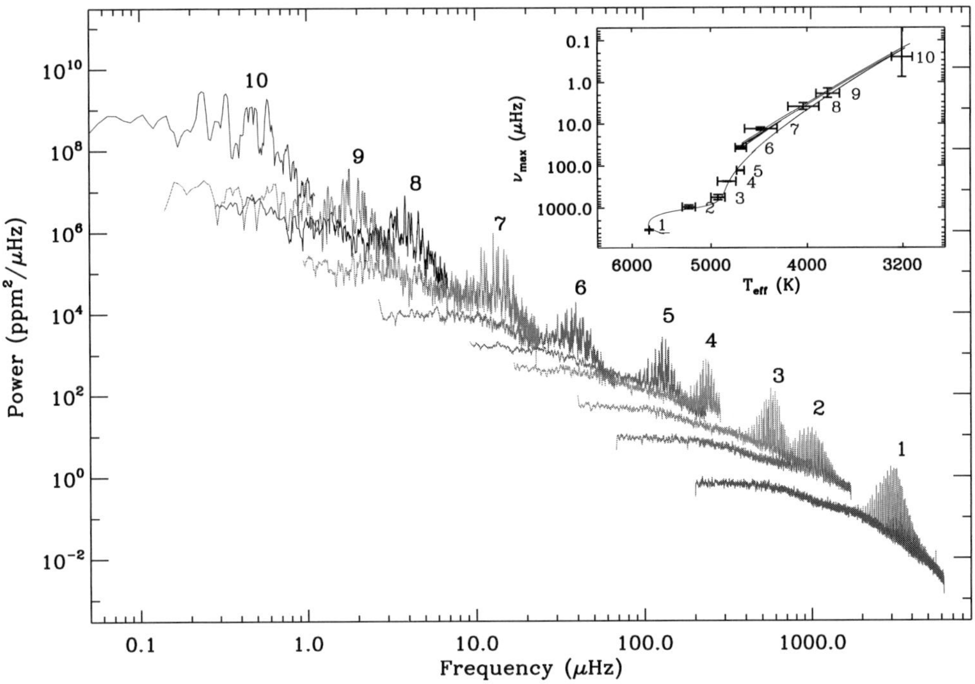

Figure 11.3 Sequence of stellar oscillation spectra starting from a main-sequence star, the Sun (1), to the tip of the red giant branch. Star 6 is a red clump star. Temperatures are from the *Kepler* Input Catalog (Brown *et al.*, 2011) and from Huber *et al.* (2014). The inset represents a seismic HR diagram. The solid line is the evolutionary track of a one-solar-mass stellar model computed using MESA (Paxton *et al.*, 2011, 2013) in which the position of each star is represented. The 1σ error bars in ν_{max} are multiplied by a factor of 10. A black and white version of this figure will appear in some formats. For the color version please refer to the plate section.

11.3 Mixed modes: A window into the inner radiative core of red giants

One of the key reasons red giants have been such a success story, reaching even beyond the field of asteroseismology in recent years, is the presence of so-called mixed modes, predicted by Dupret *et al.* (2009) and later observed by Bedding *et al.* (2010b). In the following, we will explain how these modes occur. The stellar interior of cool stars essentially consists of two cavities, an outer envelope where standing acoustic waves (p modes) predominantly reside, and a radiative core where standing gravity waves (g modes) propagate. For a star like the Sun, p-mode frequencies are much higher than those of the g modes – 3000 μHz or 5 minutes compared to 100 μHz or 2.8 hours, respectively. They therefore do not "sense" the presence of each other. When a star evolves to become a red giant the core contraction and the envelope expansion results in the p-mode frequencies becoming lower while the g-mode frequencies become higher, eventually resulting in overlapping p- and g-mode frequencies. This causes the two types of waves to couple. When a p mode couples with a g mode and forms a mixed mode its frequency is shifted – the mode is bumped from its

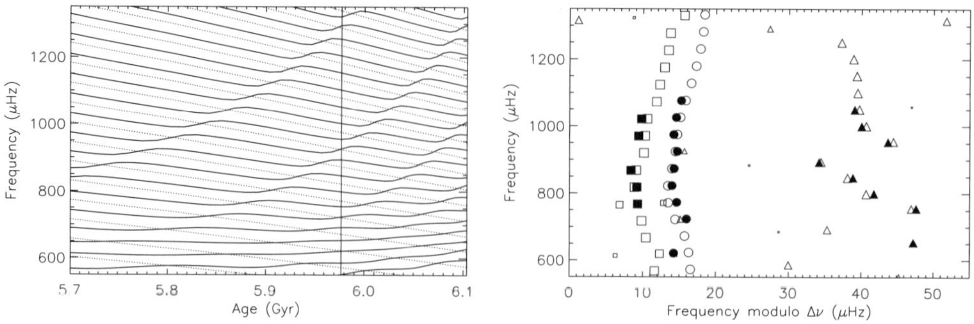

Figure 11.4 (Left) Temporal evolution of the $l = 0$ (dotted) and $l = 1$ (solid) oscillation frequencies based on KIC 11026764 (Metcalfe *et al.*, 2010a). The vertical solid line indicates the location of the model whose frequencies matches the observed. (Right) Échelle diagram using a frequency separation of $\Delta\nu = 50.5$ μHz. Open symbols are the frequencies from the best fitting model. Solid symbols are the observed frequencies from Metcalfe *et al.* (2010a). Circles are used for modes with $l = 0$, triangles for $l = 1$, and squares for $l = 2$ modes. (Figure courtesy of Dr. G. Doğan.)

original position – as the two modes undergo a so-called "avoided crossing." The regular frequency spacing between consecutive modes of the same degree and different order is broken. The first mixed modes occur during the subgiant phase where the associated mode bumping can be observed. Examples include stars such as η Boo (Kjeldsen *et al.*, 1995; Carrier *et al.*, 2005), β Hyi (Bedding *et al.*, 2007; Belloche *et al.*, 2011), the CoRoT subgiant HD 49385 (Deheuvels *et al.*, 2010), and many *Kepler* targets (e.g., Mathur *et al.*, 2011b, 2012; Campante *et al.*, 2011). An example, based on the *Kepler* target KIC 11026764 (Metcalfe *et al.*, 2010a), is shown in Figure 11.4. The left-hand panel shows the temporal evolution of radial and dipole modes of a series of models. Each instance where the dipole p mode (solid line) moves up in time, we have an avoided crossing with an underlying g mode whose presence can be traced by the series of mode bumpings of ever-increasing frequency. The model that best matches the observations is marked with a vertical line. The corresponding échelle diagram is shown in the right-hand panel, which is expected to show near-vertical ridges for modes of the same spherical degree l, meaning almost equidistant frequencies following the typical pattern of high-order low-degree acoustic modes described by the asymptotic relation (Tassoul, 1980). Bumped dipole modes are clearly seen around 900 μHz, revealing the presence of a g mode at that frequency.

When stars reach the red giant phase, the frequency spectrum of the g modes becomes denser than that of the p modes. Hence, for each p mode there are several g modes with the same angular degree and similar frequencies, each coupling to the p mode leading to several mixed modes with a p mode character in the envelope and a g mode character in the core (Osaki, 1975; Aizenman *et al.*, 1977; Dziembowski and Pamyatnykh, 1991; Beck *et al.*, 2011). Because of the coupling, mixed modes carry valuable information about the inner radiative interior directly to the surface, and due to their p-mode character, their amplitudes are high enough to be detected at the surface. This is in contrast to main-sequence stars like

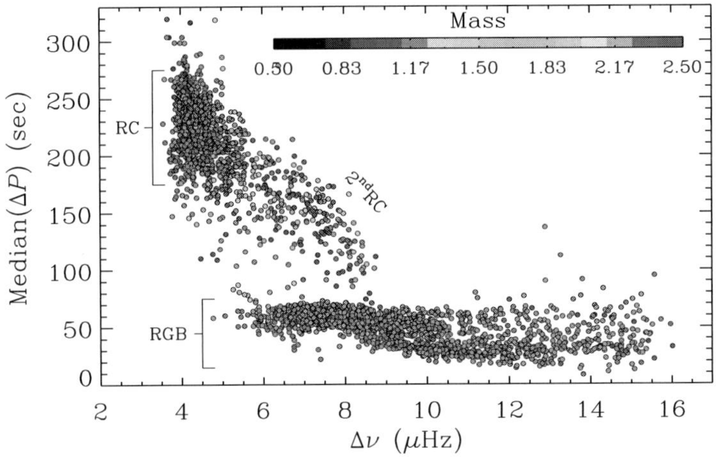

Figure 11.5 Median period spacing of mixed modes for each star analyzed by Stello *et al.* (2013) versus its large frequency separation. Red giant branch (RGB), red clump (RC), and secondary clump (2^{nd}RC) stars are indicated. The masses of the stars obtained from the scaling relations are color coded. A black and white version of this figure will appear in some formats. For the color version please refer to the plate section.

the Sun, where g modes reach the stellar surface with tiny amplitudes because these modes are evanescent in the convective envelope, making them difficult to detect (García *et al.*, 2007).

The closer in frequency a g mode is to a p mode, the more p-mode dominated the resulting mixed modes will be. A series of g modes coupling to the same p mode therefore generates mixed modes with different degree of p- and g-mode character, each probing different regions of the stellar interior, allowing a stratified "view" of the stellar interior.

11.4 Period spacings: A way to distinguish different stellar evolution phases

Classical observations of the surface properties of red giants do not allow us to disentangle stars ascending the red giant branch – only burning hydrogen in a shell – from stars in later evolution stages also burning helium in their cores, such as the red clump stars (Figure 11.1).

By measuring the period spacing of the dipole mixed modes, Bedding *et al.* (2011) realized that red giants fall into two well-defined groups, those with period spacings around 50 s (hydrogen-shell-burning stars) and those with period spacings ranging between 100 and 300 s (stars also burning helium in the core). This divide arises because g modes sense the difference in the core as it becomes convective and expands due to the onset of helium burning. Kallinger *et al.* (2012) demonstrated that the constant term of a linear fit to the radial modes versus their radial order also enabled them to characterize this divide.

This characterization can now be done automatically for very large samples of stars. In Figure 11.5 we show the period spacing, ΔP, against the large frequency separation of

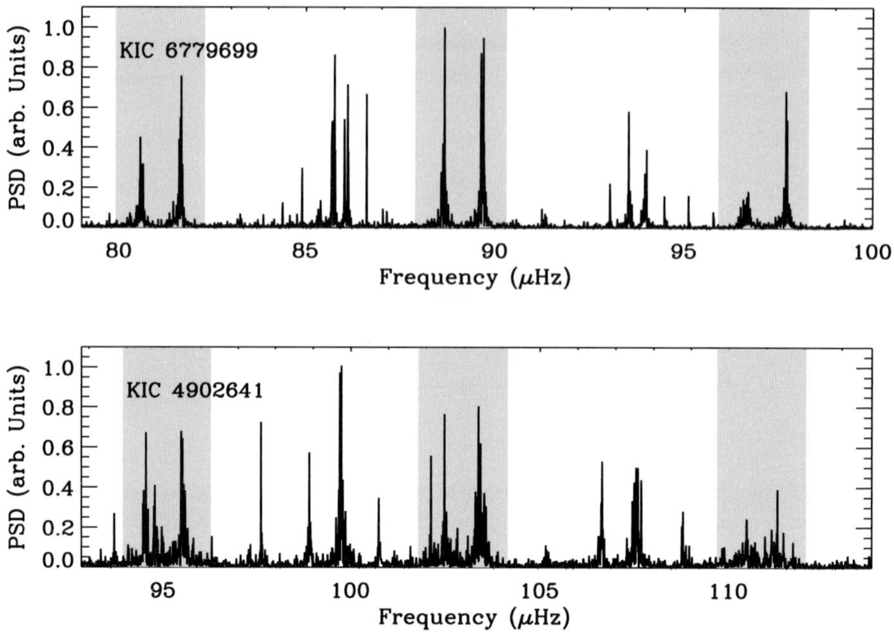

Figure 11.6 Normalized power spectral density (PSD) of two red giant stars, one on the red giant branch (top) and the other in the red clump (bottom). Shaded regions correspond to the $l = 0, 2$ modes, while the unshaded regions are dominated by dipole mixed modes.

several thousand red giant stars observed by *Kepler* (Stello *et al.*, 2013). Red giant branch stars have period spacings below ~80 s. Red clump stars are mostly concentrated above 180 s, being the less massive stars clustering on the left-hand side of the diagram (lower $\Delta\nu$), when they reach this evolutionary stage. The more massive helium-core-burning stars in the clump form the high-$\Delta\nu$ low-ΔP tail of the helium-core-burning stars (marked $2^{\rm nd}$RC).

In Figure 11.6 we show the power spectrum around the frequency of maximum power, $\nu_{\rm max}$, of two red giants observed over ~4 years by *Kepler*, one on the red giant branch and the other in the red clump (for more details see Kallinger *et al.*, 2012). The red clump star shows a relatively large spacing between consecutive dipole modes (large ΔP) while the red giant branch star has its dipole modes bunched closely together (small ΔP).

It was further noted by Montalbán *et al.* (2013) that ΔP to some degree follows the helium core mass for stars that have just ignited helium in their cores. This potentially gives a way to measure the amount of mixing that took place in these stars during their earlier evolution stages.

11.5 Differential rotation inside red giants

Mixed modes not only allow us to extract the structure of the inner core of red giants, they also probe the distribution of internal angular momentum during the evolution along the red

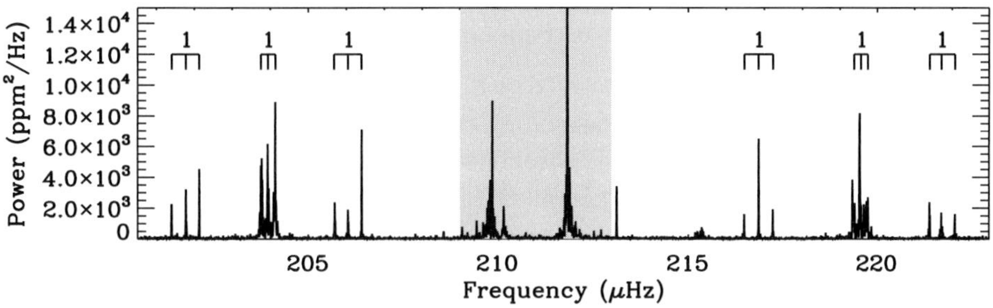

Figure 11.7 Power density spectrum of KIC 5356201 observed for nearly 4 years by *Kepler*. Rotationally split $l = 1$ mixed modes are marked. The shaded regions cover the $l = 0, 2$ modes.

giant branch. Information about the angular momentum distribution is inaccessible to direct observations, but it can be inferred from the effect of rotation on oscillation frequencies. An extensive overview on surface and internal rotation of stars can be found in Chapter 21. Here we will give only a brief explanation of the most important discoveries related to red giants.

Beck *et al.* (2012) detected non-rigid rotation in the interior of three *Kepler* red giants by exploiting the rotational frequency splitting of mixed modes. An example of the power spectrum of one of these stars, KIC 5356201, can be seen in Figure 11.7. Deheuvels *et al.* (2012) performed an inversion of the radial rotation profile in an early red giant, KIC 7341231, and recently Deheuvels *et al.* (2014) analyzed the rotation of six subgiants and red giants within a range of metallicities and masses. The latter study demonstrated that the stellar cores spun up during the subgiant phase, while the envelopes spun down. For two of the stars, the radial rotation profile showed a discontinuity located at the depth of the hydrogen-burning shell, which roughly corresponds to the boundary between the contracting and the expanding layers.

The ensemble analysis of around 300 red giants (see the next section for more details on ensemble studies), performed by Mosser *et al.* (2012b), showed the evolution of the rotation rate of the radiative cores from the subgiant phase to the red clump and the secondary clump. It demonstrated that the rotational splitting of dipole mixed modes is mostly sensitive to the core rotation (Goupil *et al.*, 2013). In the earlier stages of the red giant branch, the core rotation shows a slight decrease as the stars evolve. This seems to be in contradiction with core contraction when only local conservation of angular momentum is assumed (e.g., Eggenberger *et al.*, 2012a; Ceillier *et al.*, 2013; Marques *et al.*, 2013). Later, when stars reach the red clump, the cores have spun down even more. Part of this effect is related to the expansion of the helium-burning core (Iben, 1971; Sills and Pinsonneault, 2000). Nevertheless, this spin down of the core during the red giant branch and to the red clump seems to be in agreement with the slow rotation rate found in low-mass white dwarfs (Kawaler *et al.*, 1999), which are essentially the end product of red giant cores after the envelope has been stripped away during the late stages of red giants.

11.6 Ensemble asteroseismology and stellar populations

With more than 16 000 red giants observed by *Kepler* and about the same number observed by CoRoT, the space age of asteroseismology has opened up for the statistical analysis of large samples of stars usually called "ensemble asteroseismology." Helped by these large numbers, we can identify new and common features in asteroseismic diagrams revealing key properties of stellar evolution, while providing ideal tracers of Galactic stellar evolution history to relative large distances (Miglio *et al.*, 2009).

Using the scaling relations for $\Delta\nu$ and $\nu_{\max}$ (Kjeldsen and Bedding, 1995) on ensembles can provide valuable results on the mass and radius distributions of large cohorts of stars. Aided by stellar modeling, such ensemble analysis can also provide estimates of the age distributions. When combining radius with the apparent stellar brightness, we can further estimate distances out to several kiloparsecs (1 parsec $\sim$ 3 light years). In comparison, precise distances currently available from direct parallax measurements (van Leeuwen, 2007) reach only a few hundred parsecs. Hence, for red giants or any other solar-like oscillating star it is possible to write

$$\log d = 1 + 2.5 \log \frac{T_{\text{eff}}}{T_{\text{eff},\odot}} + \log \frac{\nu_{\max}}{\nu_{\max,\odot}} - 2 \log \frac{\Delta\nu}{\Delta\nu_{\odot}} + 0.2(m_{\text{bol}} - M_{\text{bol},\odot}), \qquad (11.1)$$

where the distance d is expressed in parsecs, m_{bol} is the apparent stellar bolometric magnitude, $M_{\text{bol},\odot}$ is the absolute solar bolometric magnitude, T_{eff} is the effective (surface) temperature, and $\odot$ denotes the solar values (Miglio *et al.*, 2013b). Therefore, it is possible to use red giants to map and date the Galactic disc in the regions probed by space-borne missions such as CoRoT and *Kepler*, and in the future TESS (Ricker *et al.*, 2010) and PLATO (Rauer *et al.*, 2013). The results from CoRoT and *Kepler* compared to models of synthetic populations of stars have already revealed significant differences in the stellar mass distributions. One interpretation of these results explains this as due to the vertical gradient in the distribution of stellar masses (hence ages) in the Galactic disk (Miglio *et al.*, 2013b), as well as on the star formation rate, and the initial mass function of stars (Miglio *et al.*, 2009; Chaplin *et al.*, 2011a).

Another important contributor to galactic evolution is the rate of mass loss when the stars evolve as red giants. There is currently no adequate physical theory on the mass loss occurring at late stages of the red giant branch that can properly explain observations. Ensemble asteroseismology can provide important information on this crucial phase of stellar evolution by comparing the mass distribution of red clump stars with stars on the red giant branch (see next section). Using CoRoT and *Kepler* observations, it has been shown that there is a population of low-mass red clump stars that is absent on the red giant branch (Mosser *et al.*, 2011). Assuming the scaling relations used to infer those masses are not introducing any systematic effects, this result could imply that these now low-mass red clump stars lost a significant fraction of their mass while ascending on the red giant branch. Therefore, after the helium flash, these stars appear in the red clump with a much lower mass than the one they had during the red giant branch.

11.7 Asteroseismology of open-cluster red giants

Open clusters have long served as universal calibrators in astronomy because stars in a cluster are considered a homogeneous sample. They are supposed to be formed from the same cloud of gas, which means they have roughly the same age, common initial composition, and are at the same distance. In stellar physics, these assumptions reduce the number of usual unknowns when matching stellar models to observations, which in combination with precise asteroseismic measurements promises more rigorous investigation of stellar models.

The scaling relations for $\Delta\nu$ and ν_{max} mentioned in Sections 11.2 and 11.6 allows us to assign membership of stars to a particular cluster. The relations depend on stellar mass, surface temperature, and luminosity, but only the last will be markedly different from star to star among red giants in a cluster. Hence, $\Delta\nu$ and ν_{max} are expected to form tight relations with apparent brightness, which for cluster members is a good proxy for luminosity, and outliers are therefore likely non-members due to them not sharing the cluster distance. Because oscillations only depend on the stellar properties, this technique is independent of the distance to the cluster and not very sensitive to the interstellar absorption and reddening (Stello *et al.*, 2011). Another important success of asteroseismology of open clusters was the determination of the distances to NGC 6791 and NGC 6819 based on the seismic studies of red giant branch stars using grid-based modeling (Basu *et al.*, 2011). Finally, Miglio *et al.* (2012a) estimated the integrated red giant branch mass loss in NGC 6791 by comparing the average masses of stars in the red clump and on the red giant branch.

12

A helioseismic view of the Sun's internal structure and dynamics

SARBANI BASU AND MICHAEL J. THOMPSON

12.1 Introduction

The frequencies of the global resonant modes of oscillation of the Sun depend on its structure and dynamics. Since we are able to observe and determine frequencies of many of the Sun's oscillation modes, the observed frequencies can be "inverted" to determine solar structure and dynamics. Inversions of solar oscillation frequencies have proved to be extremely successful in determining solar internal structure and dynamics as well as in testing solar models and inputs to solar models. We discuss here some of the major results that have been obtained through inversions. Recent reviews of inversions for solar structure and dynamics may be found in Christensen-Dalsgaard (2002), Basu and Antia (2008), and Howe (2009).

There are a number of sources, both ground based and space based, for helioseismic data. The Birmingham Solar Oscillation Network (BiSON; Chaplin *et al.*, 1996) is the predominant ground-based source of data for modes of low spherical-harmonic degree ℓ ($0 \leq \ell \leq 3$) while the Global Oscillation Network Group (GONG; Hill *et al.*, 1996) is the source of ground-based intermediate degree modes (ℓ up to about 150). Instruments such as Global Oscillations at Low Frequencies (GOLF; Gabriel *et al.*, 1997) and Variability of solar Irradiance and Gravity Oscillations (VIRGO; Lazrek *et al.*, 1997) on the Solar and Heliospheric Observatory provide space-based data on low-degree modes, while MDI was until recently the source of intermediate degree data. The Helioseismic and Magnetic Imager (HMI; Scherrer *et al.*, 2012) on board the Solar Dynamic Observatory (SDO) collects data on modes up to $\ell = 3000$, though the frequencies of all visible modes cannot be determined easily.

12.2 Solar structure

As described in Chapter 9, inversions to determine solar structure proceed through the linearization of the oscillation equations around a known solar model. What is obtained

Extraterrestrial Seismology, eds. Vincent C. H. Tong and Rafael A. García. Published by Cambridge University Press.

from the inversions is the relative differences in the structure – in particular between the sound speed, density, or adiabatic index Γ_1 profiles – of the Sun and the model. This allows us to test the models, and to localize where there are significant discrepancies between the models and the Sun. This also allows us to test the physics inputs of models; if a certain input, such as the equation of state (EOS), makes a model worse compared with the Sun, one needs to examine the input carefully. In this we are helped by the concept of the "standard solar model" (SSM). Although there is not a single universally accepted SSM, the term is generally understood to mean a spherically symmetric model of the Sun of one solar mass ($1\,M_\odot$), which has been evolved from zero age main sequence without mass loss or accretion of additional material. The model takes as assumed various microphysical inputs such as nuclear reaction rates, opacities, and the equation of state, and physical processes such as the gravitational settling of helium and heavy elements (commonly referred to as diffusion). The effect of convection on the 1D structure is generaly modeled in a highly simplified way using a prescription such as mixing length theory. The effects of convective overshooting may also be included in some fashion. Other macroscopic motions (rotation, circulations, etc.) are ignored, as are magnetic fields. SSMs are not fine tuned to match the helioseismic constraints, but there are typically two free parameters in the models, the initial helium abundance Y_0 and the mixing length parameter α, which are adjusted so that the model evolves to the correct radius ($1\,R_\odot$) and luminosity ($1\,L_\odot$) for the Sun at the solar age. In other words, there are two constraints ($R_\odot$ and $L_\odot$) to fix the two unknowns (Y_0 and α).

Figure 12.1 shows the relative differences in sound speed and density between a number of SSMs and the Sun. As can be seen, the sound speed of the models agrees to within fractions of a percent with that of the Sun, and density matches to within a small percentage. There are, however, notable differences. One feature is the sound-speed difference just below the convection zone (henceforth CZ). The decreased sound speed in the models compared with the Sun is believed to be a result of excessive settling of helium and heavy elements below the CZ (thus increasing the mean-molecular weight and thereby reducing the model's sound speed below the CZ base). The problem can be reduced if one introduces turbulent diffusion (see, e.g., Richard *et al.*, 1996). The presence of a layer of strong rotational shear co-incident with the CZ base (the tachocline: see below) means that turbulent diffusion is quite likely in these layers. However, none of the modern models include turbulent diffusion. Another region of concern is the core. As can be seen in Figure 12.1, the structure of the core of the models differs from that of the Sun. The difference is usually interpreted as being caused by the presence of extra mixing in the early life of the Sun that is not present in the models (Gough *et al.*, 1996).

The extraordinarily good match between the models and the Sun helped in solving the long-standing solar neutrino problem. For nearly 30 years, the detected flux of solar neutrinos was found to be about a third of that predicted by SSMs. This was for many years thought to be a result of deficiencies in the models. However, the advent of precise helioseismic data showed that the solar models agree with the Sun to an extraordinary degree and hence the problem could not lie with the models and their predictions. If

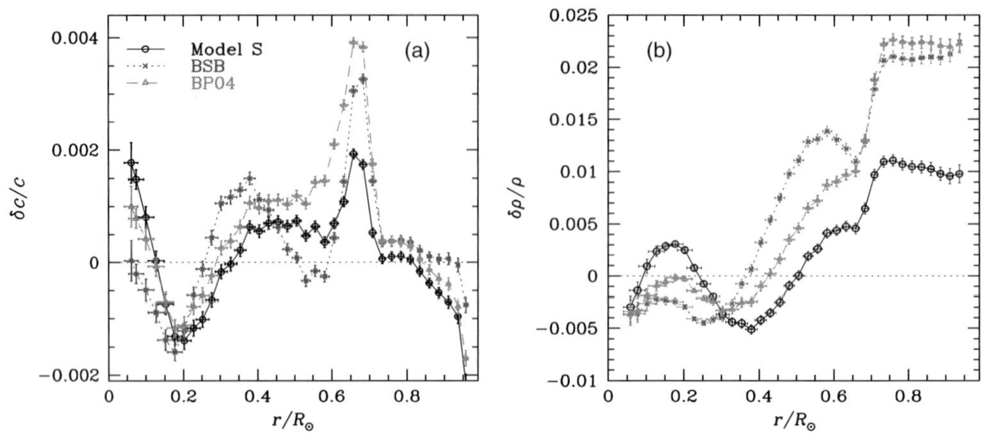

Figure 12.1 The difference in sound speed (a) and density (b) between the Sun and three solar models as determined by inversions of global helioseismic data. The inversion results are from Basu *et al.* (2009). Model S refers to the model from Christensen-Dalsgaard *et al.* (1996), BSB is model BSB(GS98) defined in Bahcall *et al.* (2006) and BP04 is from Bahcall *et al.* (2005b).

anything, models constructed by changing inputs so that the predicted solar neutrino flux matches the observed one have structures that agree less well with the Sun (see Bahcall *et al.*, 1998). We know now that the reason for the discrepancy lies not with solar models being incorrect but rather with the standard model of particle physics being deficient. The standard model of particle physics assumed that all neutrinos were massless and hence one neutrino flavor could not transform into another. In fact, as results from the Sudbury Neutrino Observatory (Ahmad *et al.*, 2002) showed, neutrinos do change flavor, and once neutrinos of all flavors are detected, the mismatch between the predicted and observed neutrino fluxes disappears.

Solar inversion results have been used to determine many properties of the solar interior such as r_b, the position of the base of the CZ, and $Y_{\rm CZ}$, the helium abundance of the solar CZ. Differences in r_b or $Y_{\rm CZ}$ between the Sun and models leave a signature in the sound-speed difference between the Sun and the models, and this difference can be calibrated to determine r_b and $Y_{\rm CZ}$ for the Sun.

In Figure 12.2a we show the sound-speed difference between a model with $r_b = 0.713\,R_\odot$ and similar models with different values of r_b. Note the sudden change in the sound-speed difference at the CZ base. This difference is caused by the change-over from the adiabatic temperature gradient in the CZ to the lower radiative temperature gradient below. Since the temperature gradient determines temperature, which in turn determines the sound speed, a mismatch of the depth at which this change occurs causes a sudden change in the sound speed. The magnitude of the change depends on the amount of mismatch and this can be calibrated to determine r_b for the Sun. Figure 12.2b shows the sound-speed difference between the Sun and solar models with different r_b. The figure immediately shows us that the solar value of r_b should be around $0.713\,R_\odot$. A careful

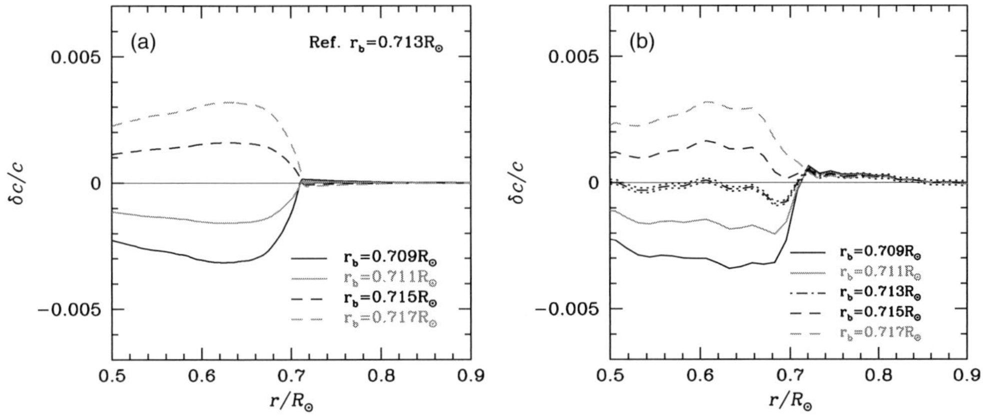

Figure 12.2 (a) The difference in sound speed between a model with CZ base location $r_b = 0.713\ R_\odot$ and models with different values of r_b. (b) The sound-speed difference between the Sun and models with different values of r_b. The solar sound-speed results are from Basu *et al.* (2009).

analysis of the sound speed, though using somewhat modified techniques, shows that the r_b for the Sun is $0.713 \pm 0.001\ R_\odot$ (Christensen-Dalsgaard *et al.*, 1991; Basu and Antia, 1997). The mismatch in CZ depth between models, or between models and the Sun, extends over a considerable region, and hence does not explain the rather localized sound-speed difference observed in Figure 12.1. As mentioned earlier, this localized difference is now believed to be a result of the fact that SSMs do not include rotational turbulent diffusion or rotation-induced mixing below the CZ base.

The CZ helium abundance is determined in a similar manner; the process of helium ionization depresses the adiabatic index Γ_1 which in turn causes a localized depression in sound speed (since $c^2 = \Gamma_1 P/\rho$). The magnitude of the depression increases with increase in helium. Thus there will be a difference in the sound speed between the Sun and models constructed with different amounts of helium. This difference can be calibrated to determine $Y_{\rm CZ}$ for the Sun. This leads to an estimate of $Y_{\rm CZ} = 0.248 \pm 0.003$ (Basu and Antia, 2004). Note that the helium abundance in the solar photosphere cannot be measured spectroscopically since the photospheric temperature is too low to produce helium lines. Helioseismology is the only means that we have today to reliably determine the helium abundance of the Sun.

Inversions of solar oscillation frequencies have also allowed us to test other aspects of the physics of the solar interior. One early result was that the gravitational settling of helium and heavy elements is essential to obtain a good solar model. In Figure 12.3 we show the relative sound-speed and density differences between the Sun and two solar models, one with diffusion and one without. It is clear that the model with diffusion is much better. The primary reason for the model without diffusion to be this discrepant is that it has a much shallower CZ than the Sun. Additionally, the model without diffusion has a much higher value of $Y_{\rm CZ}$ compared with the Sun. Such inversions have also been used as tests

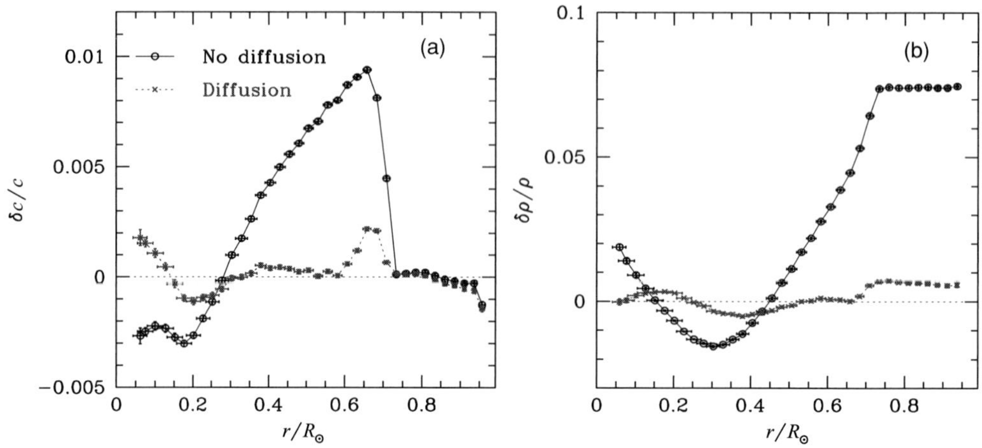

Figure 12.3 The relative sound-speed (a) and density (b) differences between the Sun and two SSMs with identical inputs, except that one SSM includes diffusion and the other does not. Models are from Basu *et al.* (2000).

of equation of state (see, e.g., Basu and Christensen-Dalsgaard, 1997), and even nuclear reaction rates (Antia and Chitre, 1999; Turck-Chièze *et al.*, 2001).

One of the more recent issues about solar structure is that of the solar heavy-element abundance, i.e, the metallicity Z/X at the surface. The solar heavy-element abundance affects its structure, principally through the sensitivity of the opacities to metallicity. The most commonly used heavy-element abundance was that of Grevesse and Sauval (1998, henceforth GS98) with $Z/X = 0.023$. The issue of solar abundances was rendered uncertain following updated analysis techniques that included the use of 3D model atmospheres, as a result of which the solar metallicity was revised downwards by Asplund *et al.* (2005, henceforth AGS05). With further improvement in analysis, the abundances were updated by Asplund *et al.* (2009, henceforth AGSS09) to $Z/X = 0.018$. These abundances were quickly adopted by some members of the stellar astrophysics community, since the 3D model atmospheres were considered to be superior to the 1D atmospheres used earlier. Additionally, the oxygen and carbon abundances derived from different lines were consistent with each other, and brought the solar abundance in line with some of the young stars in the solar neighborhood. However, an independent calculation using a different 3D model atmosphere found a higher metallicity, $Z/X = 0.0209$ (Caffau *et al.*, 2011, henceforth CAF), leaving the field even more confused.

The problem with the new lower abundances is that the standard solar models made with them do not match the Sun to the same degree as the higher-Z/X models shown in Figure 12.1. We illustrate this in Figure 12.4 where we show the sound-speed and density differences between the Sun and solar models constructed with different value of Z/X but otherwise identical inputs. As can be seen, the lowest-Z/X models do not agree with the Sun. The low-metallicity models do not satisfy solar constraints of the position of the

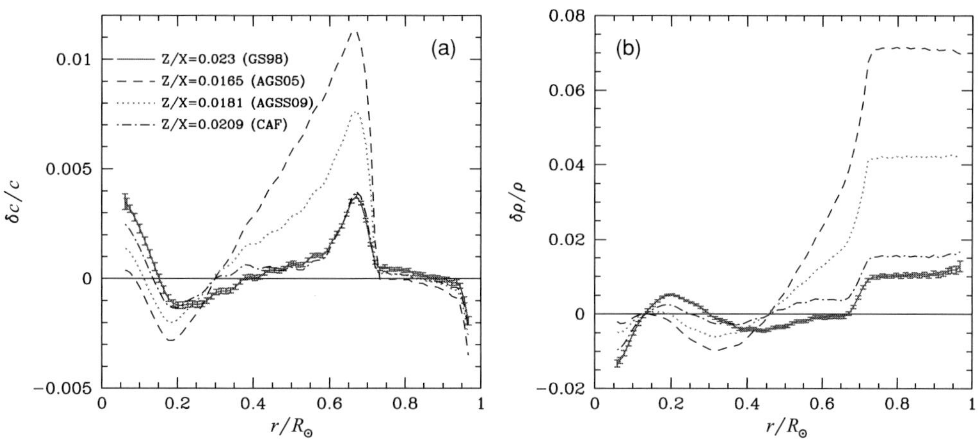

Figure 12.4 The relative sound-speed (a) and density (b) difference between the Sun and models constructed with different heavy-element abundances but otherwise identical inputs. The abundances used are GS98, AGS05, AGSS09, and CAF.

convection-zone base, the convection-zone helium abundance or even the initial helium abundance.

Thus far, we have talked only of the spherically symmetric structure of the Sun. Solar asphericity is small (after all, the difference between the polar and equatorial radius is only a few parts in 10^5) and difficult to detect. There is some evidence that there is an excess of sound speed at a latitude of 60° at a level of 10^{-4} (Antia *et al.*, 2001, 2003).

The structure of the Sun shows very little solar-cycle-related change. Current estimates suggest a relative change in the sound speed at the CZ base of $(3.61 \pm 1.04) \times 10^{-5}$ between the solar maximum at 2001 and minimum in 1996 (Baldner and Basu, 2008). The small change is most likely to be a result of the fact that the average magnetic field of the Sun, and changes thereof, are rather weak in terms of their contributions to the effective total pressure and wave speeds. We can actually find much larger changes if we look at differences between the structure of solar active regions and of a nearby quiet region. We cannot use global helioseismic techniques for this, and instead rely on local helioseismology, where we look at only a small region in the Sun. It should be noted that these techniques currently allow us to probe only the outermost layers of the solar interior. Inversions of frequencies obtained using the so-called "ring-diagram analysis" show that the sound speed in active regions is depressed in about the outer 7 Mm of the Sun, and then increases. We show an example in Figure 12.5. The amount of this decrease is a function of the strength of the magnetic regions, but it appears to saturate at high magnetic field strengths (e.g., Basu *et al.*, 2004; Baldner and Basu, 2008).

12.3 Large-scale dynamics

Determining the profile of rotation over much of the solar interior has been one of the triumphs of helioseismology. Rotation (and other departures from spherical symmetry, as

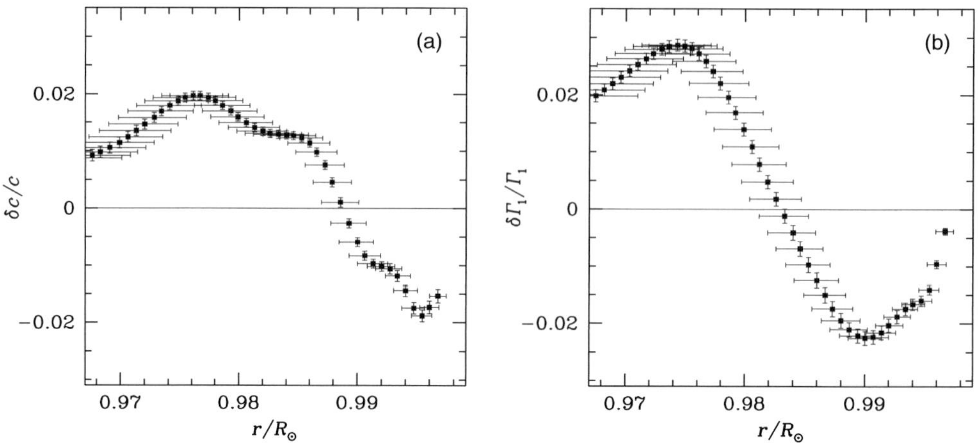

Figure 12.5 The inferred relative differences in sound speed (a) and Γ_1 (b) between active region AR9906 and a nearby quiet region. The inversion results are from Basu *et al.* (2004).

discussed above) causes frequency splitting, i.e., modes of the same radial order n and degree ℓ but differing in azimuthal order m have slightly different frequencies. Specifically, rotation causes the frequencies of modes with the same absolute value of m but with different sign – which correspond to waves propagating in opposite directions around the Sun – to be split by an amount that is equal to $2|m|$ times a mode-dependent spatial average of the rotation rate. These frequency splittings are inverted to infer the rotation rate, as a function of both radial and latitudinal location inside the Sun. Unlike for the Sun's spherically symmetric structure, prior to the advent of helioseismology we had no good theoretical models for the Sun's internal rotation. Yet the internal rotation is an important ingredient in the working of the solar dynamo, and such effects as rotationally induced instabilities and large-scale circulations may play an important role in transporting and mixing chemical elements in the solar interior over the Sun's lifetime.

It has long been known from e.g. the motions of sunspots on the solar disk that the visible surface of the Sun rotates differentially, with the equatorial regions having a rotation period of about 25 days and the high latitudes completing one rotation in 30 days or more. Before helioseismology began to reveal the true nature of the Sun's internal rotation in the 1980s, it was predicted from numerical models and from arguments based on the Taylor–Proudman theorem that the rotation would be constant on cylindrical surfaces throughout the convection zone, matching onto the observed surface rotation at the top of the convection zone. Because young stars tend to be fast rotators, and as they age lose angular momentum from their surface through the action of stellar winds, it was also predicted that the radiative interior of the Sun, and in particular its inner core, could rotate many times faster than its surface (see, e.g., Endal and Sofia, 1978; Pinsonneault *et al.*, 1989; Chaboyer *et al.*, 1995; Krishnamurthi *et al.*, 1997).

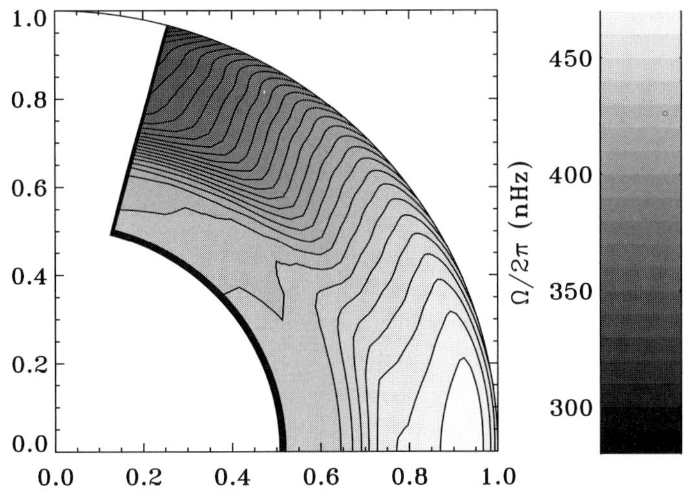

Figure 12.6 The mean rotation profile obtained with data collected by the GONG project from mid 1995 to mid 2004. The contours are at intervals of 5 nHz. The rotation rate in the deep interior has large uncertainties and hence has not been shown. The inversion results are from Howe *et al.* (2005). (Figure courtesy of R. Howe.)

The actual mean rotation profile of the inside of the Sun, as revealed by helioseismology, is shown in Figure 12.6. There is some hint of rotation on cylinders in the near-equatorial regions of the convective envelope, but for the most part the CZ rotation is more nearly constant on cones, i.e., at fixed latitude the rotation is almost independent of depth – there is a tendency for the rotation contours to lie at about 24° to the rotation axis (see, e.g., Howe *et al.*, 2005; Gilman and Howe, 2003). There are two prominent rotational shear layers, one at the bottom of the convection zone that is now called the tachocline, and a near-surface shear layer. The existence of the tachocline had been posited by Spiegel and Zahn (1992b), who called it the "tachycline." Beneath the CZ, the rotation of the radiative interior is consistent with rigid-body rotation, though the rotation in the inner 15% by radius is not really known because it is so poorly constrained by the observations of global acoustic modes. (If global gravity modes are unambiguously detected and their rotational frequency splittings measured, these would provide much stronger constraints on the rotation of the inner core.) There are some apparent spatial variations in the rotation profile of the radiative interior, but their statistical significance is weak.

The inferred rotation profile of the Sun shows some variations with time. Inverting frequency-splitting data taken at different epochs, and then subtracting from the inversion results some time-averaged rotation profile or some other reference profile, reveals some subtle but systematic variations in the convection zone. There are bands of faster and slower rotation that migrate from mid-latitudes towards the equator, in line with the well-known equatorward migration of the sunspot activity belts over the course of the solar cycle. These have been called "torsional oscillations." At the same time, there is a high-latitude branch that strengthens and migrates from mid-latitudes polewards. The physical cause of these migrating zonal flows is not completely understood, although both Lorentz forces and thermal effects have been evoked (see, e.g., Spruit, 2003; Covas *et al.*, 2005; Rempel, 2007). There are other temporal variations, in particular a reported episodic 1.3-year quasi-periodic variation in the rotation rate both above and beneath the base of the convection zone, but the statistical significance of that finding is disputed (Howe *et al.*, 2000; Antia and Basu, 2000).

As well as rotation, another class of large-scale flows that may play an important role in the solar magnetic activity cycle are meridional (i.e., north–south) flows. These have only a second-order effect on global-mode frequencies which is very hard to detect, but meridional flows as well as rotation can be measured at least in the upper part of the CZ using local helioseismic techniques such as ring-diagram analysis or time–distance helioseismology (see, e.g., Gizon and Birch, 2005, and references therein). In ring-diagram analysis, the effect of the flows can be thought of as introducing a Doppler shift to the frequencies of the waves. In the time–distance framework, the effect can be thought of as being a speeding up or slowing down of propagating waves depending on whether the waves are traveling with or against the background flow, which thus affects the measured travel times.

Haber and Hindman (2005), Basu and Antia (2010), Komm *et al.* (2013), and others have studied the subsurface meridional flow over multiple years, using the ring-diagram analysis technique and MDI data. It is thus now well established that in at least the upper 15 Mm of the convection zone (i.e., at least down to about 0.98 $R_\odot$) in both hemispheres there is a poleward meridional flow of 15–20 m/s. The flow extends from the equatorial region up to at least 50° latitude.

By mass conservation, corresponding to the poleward flow there must be a return equatorward flow at some depth. Where that occurs is of interest to the flux-transport dynamo model. Dikpati *et al.* (1999, 2006), for example, assume that the return flow is located in the deep convection zone, so that there is a single circulation cell spanning the whole convection zone in the depth direction. There have been a number of reports in the literature of the detection of the return flow at much shallower depths (e.g., Mitra-Kraev and Thompson, 2007; Hathaway, 2012). The latest and perhaps most robust such report is that of Zhao *et al.* (2013), who applied the time–distance technique to two years of SDO/HMI data. Zhao *et al.* find a poleward flow of 15 m/s extending from the surface down to about $0.91R_\odot$ and an equatorward flow of 10 m/s between 0.82 and $0.91R_\odot$, and evidence of a poleward flow again below $0.82R_\odot$. Thus, their analysis indicates a shallow meridional circulation cell beneath the surface, with a second meridional circulation cell below the first one.

12.4 Other stars

Determining the structure and dynamics for other stars is extremely challenging – very few modes are observed, and those that are observed are low-degree modes. Also, we have independent estimates of the solar mass and radius, but in the case of other stars, these two global properties have to be determined from the frequencies themselves. Any error in the assumed stellar mass and radius leads to systematic errors in the inversions (Basu, 2003). Although space missions such as CoRoT and *Kepler* have provided very precise stellar oscillation frequencies for $\ell = 0, 1, 2$ modes, none of the sets is yet large enough to perform a proper inversion. As a consequence, there are as yet no stellar inversion results. Data for stellar rotation are even more sparse since the radial ($\ell = 0$) modes do not split in response to rotation. However, that has not stopped researchers from trying. Deheuvels *et al.* (2012) used frequency splitting of an early red giant KIC 7341231 to deduce that the core of the star is rotating at least five times faster than the envelope. Results such as these will shed light on the evolution of angular momentum in stars.

13

Recent discoveries of structures and physical processes in local helioseismology

TAKASHI SEKII AND HIROMOTO SHIBAHASHI

13.1 Introduction

One of the greatest remaining mysteries surrounding the Sun is the mechanism of its magnetic activity cycle. Most prominently, the number of sunspots repeatedly increases and decreases with an approximate period of 11 years. This must be a result of interactions between various flows and magnetic fields in the Sun, broadly called a dynamo mechanism. Pinning down the mechanism of the solar dynamo would require concerted efforts between theory and observation. Therefore it is of paramount importance, as an observational approach in our attempt to understand the solar dynamo, that we measure these internal flows and, if possible, magnetic fields in the Sun.

Among the various flows, of particular interests are the two components of large-scale flows, i.e., differential rotation and meridional circulation. Convective flows, in particular supergranules, the flows associated with supergranulation, are also of great interest. The origin of supergranules is still debated, but they are normally considered as convective cells, typically with a spatial scale of 3×10^4 km and a lifetime ($\sim$ the turnover time) of about a day.

The Sun rotates with a period of about a month, but the equator rotates faster than the poles – the Sun rotates differentially. The solar differential rotation is directly responsible for the so-called ω effect, in which the toroidal component (the ϕ component in spherical coordinates) of the magnetic field is produced by shearing the poloidal component (the component in the r–θ plane). It also provides the Coriolis force, which is important in the less understood α effect, in which the poloidal component is produced from twisting the toroidal magnetic field. Meridional flows transport magnetic flux latitudinally by advection. Supergranulation is believed to be important in diffusive transport of the solar magnetic fields. For general reviews of large-scale flows in the Sun and the solar dynamo, readers are referred to Miesch (2005) and Charbonneau (2010), respectively.

Global helioseismology, primarily based on information contained in eigenfrequencies, has been successful in mapping the solar differential rotation in the low- to mid-latitude

Extraterrestrial Seismology, eds. Vincent C. H. Tong and Rafael A. García. Published by Cambridge University Press.

convection zone and in the outer part of the radiative interior. Global helioseismology however has its limitations, since the waves that form eigenmodes sample the solar interior following the symmetric pattern determined by eigenfunctions (or, more precisely, the sensitivity kernels for the particular quantity that is studied). Not only is it unsuited for measuring small, localized features such as sunspots and small-scale convective flows, it is not sensitive to certain global (in the sense that the spatial scales involved are large) but asymmetric structure. For example, to the first order of the differential rotation, only the component that is north–south symmetric can be measured, because eigenmodes sample the northern hemisphere and the southern hemisphere with equal weights. If one is interested in measuring the north–south asymmetry of the rotation, a local technique will have to be employed.

While the typical rotational flow speed at the surface is about 2 km/s, and the fluctuating surface convective flow speeds are of the order of 1 km/s, meridional flows are much slower, only up to around 20 m/s, and are therefore difficult to measure even at the surface, evidenced by early disparate results. In addition, unlike rotation, there is no net momentum flux for meridional circulation, as a result of which, once again to first order, meridional circulation does not affect eigenfrequencies. The global techniques of analyzing eigenfrequencies therefore would not help us in mapping meridional flows. The eigenfunctions, on the other hand, are distorted by the meridional flows and attempts to observe such distortions to measure the meridional flows have been made (one recent example is the work by Schad *et al.*, 2013). These are relatively recent developments, however, and the first measurement of the meridional flow in the solar interior came from local helioseismology.

Measuring supergranular flows including the subsurface velocity profile, not restricted to what is visible on the surface, is also an important target for local helioseismology as they cannot be directly measured by global helioseismology.

The formation, evolution, and decay of sunspots are another topic of importance on its own as well as in the context of understanding the solar dynamo, as they are the most visible direct products of the dynamo action in the Sun. They are often the sites of various activity phenomena such as solar flares. The structure and flows around sunspots also are good targets for local helioseismology because they cannot be studied by global helioseismology. Below, we start by reviewing the history of this exciting and controversial topic.

13.2 Structure and flows in and around sunspots

When a magnetic flux tube rises to the surface of the Sun, and part of the tube is above the photosphere, the cross-sections of the tube are observed as sunspots, regions with low temperature (∼4000 K) responsible for their dark appearance, and strong magnetic fields (∼kG) that are believed to be suppressing convective heat transport beneath them, leading to the low temperature. Thus the formation of sunspots is initiated by flux emergence, but details of the process are still not fully understood. Once sunspots form, there are questions of how they are maintained, and how, eventually, they decay; to answer these questions it is crucial to investigate the subsurface structure of sunspots and flows around them. The strong magnetic fields in the sunspots, however, produce magnetic pressure, which is comparable to gas pressure of the surrounding plasma. Adding sunspot magnetic fields to the unmagnetized structure therefore cannot be treated as a weak perturbation. This, as we will see below, is the main source of difficulty in our investigation.

The beginning of sunspot seismology can be traced back to the work by Braun *et al.* (1987), before the term "local helioseismology" was coined. They expanded waves around sunspots in Hankel functions and examined the amplitudes of inward travcling waves and outward traveling waves. They discovered that the outward components were of lower amplitudes, indicating that up to as much as 50% of the wave power is absorbed by sunspots. It was pointed out that the power decrease could be due to reduced excitation, as the local magnetic field would suppress turbulence, which is the source of the solar five-minute oscillation. So far there is no evidence that either of these two possibles processes really dominates the apparent absorption. Chou *et al.* (2009a) discovered that, for the sunspots they studied, the amplitude changes are largely due to absorption, but they could not determine whether the rest are coming from a reduced excitation or a local suppression of amplitudes (without changes in power). Study of wave propagation around sunspots is still continuing. Zhao and Chou (2013) used a different technique to study the waves scattered by sunspot magnetic fields. It was pointed out (Moradi *et al.*, 2010) that this kind of analysis can also be useful in studying flows around sunspots. These studies are in any case helpful in understanding interactions of acoustic waves and magnetic fields (see, e.g., Bogdan and Cally, 1995), and are eventually to be fused with more direct efforts in mapping the structure and flows around sunspots.

More direct investigation began when Duvall *et al.* (1996) reported their detection of downflows under sunspots using the time–distance technique. Then Kosovichev (1996) formulated a full inversion recipe based on the time–distance technique, which was used by Kosovichev and Duvall (1997) and then by Zhao *et al.* (2001).

The main points of the early pictures of the subsurface structure of sunspots, drawn by these authors, are as follows. The flow pattern beneath sunspots is generally converging near the surface, before it turns downwards (flow speeds involved are typically up to a few km/s). It was speculated that this pattern was holding the magnetic flux tubes together, as in Parker's cluster model of sunspots (Parker, 1979). How deep these downflows reach varies from around 5 Mm to a few tens of Mm.

Note that the converging flow near the surface would imply a large shear near the surface, as the well-known Evershed flow (for a recent review, see, e.g., Bellot Rubio, 2010) is a diverging flow with a typical speed of 5 km/s.

Wave speed, the speed of waves of acoustic nature but in a strongly magnetized plasma, is lower immediately beneath sunspots, but further down there is a region where it is increased. It was argued that even the strong magnetic fields in the sunspot cannot explain these variations without an increase or decrease in temperature. Thus, it is implied that beneath sunspots there is a layer cooler than the surrounding shallow layer (just as

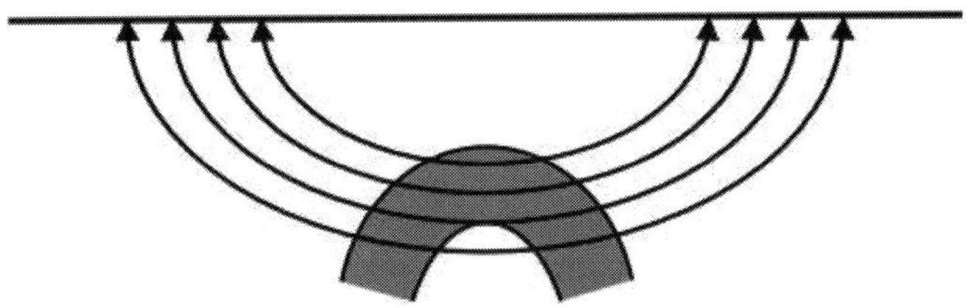

Figure 13.1 Illustrations of the typical observational setup for detection of magnetic flux tubes (grey area) before their emergence at the surface (horizontal line), based on the simple ray theory. Waves travel along various rays (curves with arrows on both ends, indicating that they can travel both ways) but those that travel deeper can be captured by observing oscillations further from the center of the eventual emergence site.

in the photosphere) and then a hotter layer deeper down. It was speculated that the latter layer results from convective heat from the interior blocked by a strong magnetic field (at least in terms of the ratio of magnetic pressure to gas pressure) closer to the surface.

These initial developments were very stimulating and encouraging, leading to more and more efforts. Later, however, a few issues were raised. The most obvious is that of the simple ray-theory treatment of wave propagation in magnetized media, in which neither the finite-wavelength effect nor the mode conversion was taken into account. To improve on the former, instead of the ray theory, a treatment based on the Born approximation, for example, was proposed (Birch *et al.*, 2001). They found that this would certainly be an improvement on the ray theory, but it is still not satisfactory for strong perturbation cases, which is of course understandable because the Born approximation is a weak-scattering approximation, designed for weak perturbations. This is not the only problem that the strong perturbation by magnetic fields causes. Probably most importantly, it may not permit approaches based on linearized inversions. The strong perturbation was also mentioned as a possible source of difficulty in the context of filtering methodology; a typical local-helioseismology analysis uses various filters in $k - \omega$ (horizontal-wavenumber–angular-frequency) space to isolate waves of specific properties (such as penetration depth, which is a function of ω/k), and if the perturbation is not small, to ensure that the same wave components are captured in the unperturbed and the perturbed states, these filters may have to be changed ("perturbed"); but this is not possible without carrying out the inversion first (in principle, one can adopt an iterative procedure). From a more practical perspective, it has been reported that various local-helioseismic methods applied to the same dataset led to inconsistent results (e.g., Gizon *et al.*, 2009; Moradi *et al.*, 2010). Braun *et al.* (2012) used a numerically modeled sunspot to test the procedure and concluded that the current treatment is insufficient.

There have also been attempts to detect magnetic fluxes before they emerge (Figure 13.1). Ilonidis *et al.* (2011) reported detection of travel-time anomalies, using their new technique, in regions where sunspots were to appear about a day later, when the flux tube was estimated to be about 65 Mm deep. The typical travel-time shift of 12–16 s, however, seems too large to be explained easily. According to Birch *et al.* (2010), the main source of the travel-time shift is in fact the horizontal flow, and the shift should be of the order of 1 s. Given that Kholikov (2013) was largely successful in reproducing the results, it seems that the detection is not due to an error, but the interpretation remains an issue.

From a different angle, Toriumi *et al.* (2013), expanding on a previous work by Hartlep *et al.* (2011), looked for the oscillation power deficit caused by subsurface magnetic flux, and found evidence that this is indeed detectable up to a few hours before the emergence. By using several filters selected for different depth sensitivity (corresponding to different rays in Figure 13.1), they also estimated the speed of the upward motion by noting when the power deficit began for various depths. In the depth range of 10–15 Mm the speed was several km/s, but as the center of the power deficit (tentatively assumed to be where the flux tube is) approached the surface, it slowed down eventually to around 0.5 km/s. This deceleration is consistent with the two-step emergence model (Toriumi and Yokoyama, 2010), in which a rising tube expands near the surface to be decelerated, before its eventual emergence due to the Parker instability. It is noted in numerical simulations that sometimes, depending on various parameters, a flux tube fails to emerge – therefore it is interesting to study the emergence rate of flux tubes that float to a certain depth. Once again, however, the interpretation is phenomenological so far, and whether the amount of detected power deficit is reasonable or not is still to be investigated. The power deficit could be due to waves reflected off the magnetic flux, or it could be due to waves carried off by flow associated with the rising flux tube.

Detecting emerging fluxes before their final appearance in the photosphere is useful for space weather prediction too. One question about both works (Ilonidis *et al.*, 2011; Toriumi *et al.*, 2013) is that, if the detected anomalies in either travel time or power are indeed due to the magnetic field in the flux tubes, why do we not see double peaks, as the cross-section of an Ω-shaped flux tube at particular depth would point to? The flow associated with the tube, on the other hand, seems to be fairly concentrated near the apex (Birch *et al.*, 2010). Another question, particularly about the power deficit result, is whether it is not simply a surface effect of precursor fields that appear on the surface before the actual emergence. This however is unlikely, because the power deficit begins first further from, rather than close to, the site of the emergence (see Figure 13.1).

It would be fair to say that probing sunspots and active regions using local helioseismology is proving to be less straightforward than was initially hoped. By comparing the difficulty with the success of global helioseismology, it is clear that the source of the difficulty is partly in complex physical processes of wave propagation in magnetized media, and partly in the lack of understanding of the sunspot structure on which we can build. In global helioseismology, wave propagation mostly in non-magnetized, spherically symmetric structures was considered. For structure inversions we had a good spherically symmetric model, and for rotation inversions the non-rotating model is good enough to justify a linearized approach, as the Sun is a slow rotator. For five-minute oscillations, the typical ratio of the oscillation period to the rotation period is $\sim 10^{-4}$.

If we are not to depart from a linearized treatment of the problem, what we eventually need is a good model around which linearization of the wave propagation problem is justified. However, unlike the case for the solar equilibrium structure, of which there is just one, sunspots are all different. We need, therefore, the means to construct a good enough sunspot model from non-seismic observations. This obviously requires much better understanding of sunspot structure in general, and the physical processes relevant. To reach that stage, we would need iterative approaches to numerical modeling of sunspots, local-helioseismic inversions (however rough they might be initially) based on the models, and interpreting the results so as to give feedback to the modeling. The last part may be the most difficult.

On the other hand, Hanasoge *et al.* (2011) proposed an approach for solving a non-linear inverse problem by carrying out successive linearized inversions, exploiting the so-called adjoint method, a mathematical method used broadly across many areas in mathematical physics, for efficient computation of sensitivity kernels.

13.3 Meridional circulation

Meridional circulation in astrophysics was first discussed in the context of the thermal equilibrium of rotating stars (see, e.g., Kippenhahn *et al.*, 2013). The Sun being a slow rotator, the importance of meridional flows in the Sun may not be so obvious in this original context, but not only do they transport angular momentum to play a prominent role in shaping the solar differential rotation, they also transport magnetic flux to be doubly important for the solar dynamo.

Giles *et al.* (1997) first reported helioseismic measurement of meridional flows, using the time–distance technique. A poleward flow of up to $\sim$20 m/s was measured, which was argued to be as deep as 4% of the solar radius. The flow was detected up to around the 60° latitude, and one might be tempted to assume that the flow is poleward all the way to a certain latitude. Then the counter flow must be equatorward, and this flow may be transporting the magnetic flux in the interior of the Sun towards the equator, which explains the equatorward migration of sunspots as a solar cycle proceeds.

It is known from numerical simulations that meridional flows vary over relatively short time scales. What matters for the solar dynamo, however, is the mean flow averaged over the time scale that is relevant for the 11-year solar cycle. Variation over the solar cycle is also of interest. Chou and Dai (2001) used the time–distance technique to find that the meridional flow speed increased in the declining phase of solar activity from 1994 to 1997, and that there was an additional component present in each hemisphere in the rising phase of 1997 to 2000. This component is a divergent flow centered around the 20° latitude, and it is of such a high amplitude that around 2000 the overall flow was equatorward in the low latitudes. This trend reaches down to a depth of 70 Mm or beyond. Chou and Dai argued that the additional component could be an outflow from active regions. Haber *et al.* (2002) used the method of ring-diagram analysis and discovered that, in addition to the general

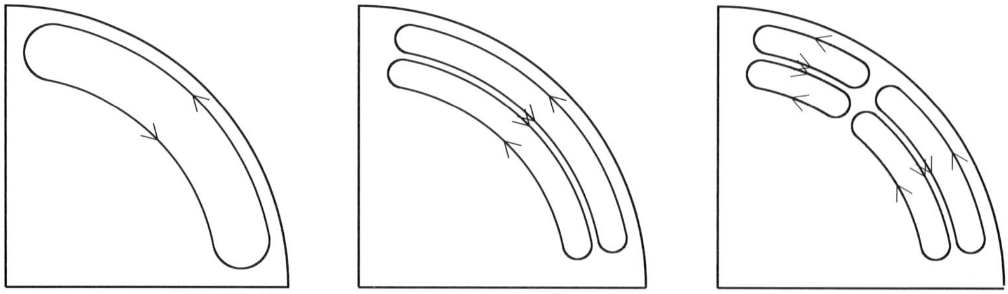

Figure 13.2 Illustrations of three possible configurations (among infinitely many others) of meridional circulation in the Sun. It could be a single-cell configuration (left), a radial double-cell configuration (middle) or a quadruple-cell configuration (right).

poleward flow, there is a submerged equatorward flow below ~5 Mm depth. Such a cell was found in datasets from 1998 to 2001, only in the northern hemisphere. Mitra-Kraev and Thompson (2007) also reported detection of an equatorward flow beneath ~40 Mm depth.

The results quoted above do not point to a clearly unified view, but in the steady state the surface poleward flow has to be accompanied by deeper equatorward flow in any case (Figure 13.2), and what was discovered could be part of the counterflow captured with a relatively low spatial resolution.

Recently, by analyzing data from the Helioseismic and Magnetic Imager (HMI) aboard the Solar Dynamics Observatory (SDO), which offers higher resolution in a broader portion of the solar disc, Zhao *et al.* (2013) also found evidence for a double-cell structure (Figure 13.2, middle). In many kinematic dynamo calculations (e.g., Dikpati and Gilman, 2006), a single-cell structure was assumed. Pipin and Kosovichev (2013) argue, however, that a double-cell meridional flow can be consistent with the observed behavior of the solar cycle. Moreover, there have also been dynamical computations that predict double-cell structures (e.g., Browning *et al.*, 2006). Therefore we should not be surprised by a double-cell structure, but we note that the SDO-HMI result depends on how one removes a systematic effect in the travel-time measurement (Zhao *et al.*, 2012b). Kholikov *et al.* (2014), however, applied the same method to data acquired from the Global Oscillation Network (GONG) project to obtain results consistent with the SDO-HMI effort, so the problem and the cure are not specific to the HMI data. There are indications that the method is working, but the origin of the systematic effect remains unknown, which is not completely satisfactory. Baldner and Schou (2012) pointed out that this may at least partly be explained by the upward–downward asymmetry in the convective flow field (this, in addition to helping remove the systematic effect, might lead to a method of studying certain properties of convection).

The "global" method by Schad *et al.* (2013), mentioned earlier, also reports a multi-cell structure, though in their case the total flow pattern is fairly complex with multiple cells in the latitudinal direction too (Figure 13.2, right, may be too simple).

13.4 Convective flows

Convection is the main mode of energy transport in the outer 30% of the Sun. The Reynolds number of the solar convection zone is thought to be of the order of 10^{14} (see, e.g., Miesch, 2005) and therefore the convection in the Sun is turbulent. Of convective motions, the smallest scale easily identified observationally is that of granulation ($\sim 10^3$ km, with a typical lifetime of $\sim$10 min), and the largest speculated is that of a so-called giant cell ($\sim 10^5$ km, with a typical lifetime of $\sim$1 month). The power spectrum across spatial scales is more or less continuous but the granular scale shows up as a broad peak. Another prominent peak is found around 3×10^4 km, corresponding to supergranulation. Supergranulation is thought to be important in diffusive transport of magnetic flux, in which magnetic fields are transported in a random-walk fashion.

Supergranules are normally considered to be convection cells, and the horizontal divergent flows at the surface "sweep" out the surface magnetic fields, resulting in field enhancement at the supergranular boundaries. Their origin, however, is still being debated. It used to be speculated that supergranules derive their typical size from the depth of the second ionization zone of helium, which is about 2×10^4 km deep. However, local helioseismic measurements of the interior velocity fields reveal that supergranulation is in fact a fairly shallow phenomenon (see, e.g., Kosovichev and Duvall, 1997; Sekii *et al.*, 2007), only a few $\times 10^3$ km deep.

Another mystery is that the supergranules are moving faster than the background medium; they rotate around the Sun faster than the surrounding plasma. Gizon *et al.* (2003) discussed the wave-like nature of supergranulation, which explains this superrotation by the waves predominantly being prograde.

The outflow component of supergranulation, typically $\sim$0.5 km/s, is relatively easily measured, but the weak vertical flow is difficult to measure. Duvall and Birch (2010) averaged over a thousand supergranules near the disk center, determining the cell location using horizontal velocity maps derived by an f-mode time–distance technique, to obtain upflow speed at the center of $\sim$10 m/s and down-flow speed at the boundary of $\sim$5 m/s. If this vertical speed persists through the entire cell, even for a cell only a few Mm deep the vertical crossing time would be a few days. If the lifetime of a supergranule, typically about a day, is to be comparable to its convective turnover time, this is obviously too slow. Recently, Duvall and Hanasoge (2013) followed this up by applying time–distance technique for waves that travel into or out of supergranulations, and found that the vertical velocity is much greater beneath the surface. They estimate that at the depth of 2300 km it is as fast as 240 m/s.

Regarding the spatial distribution of the supergranular cells, Lisle *et al.* (2004) reported a north–south alignment at low to mid latitudes, which they argued could be induced by the elusive giant cells. Nagashima *et al.* (2011) studied supergranules in the polar region of the Sun, exploring the high-resolution of the Hinode Solar Optical Telescope, and found that supergranules are also lined up in a curious manner, predominantly in north–south directions. Further investigations are awaited, but it is strongly suspected that the Hinode data are suffering from the same systematic effect mentioned in the previous section. This may not matter for this particular investigation but we need to address this issue. Using the same data, Nagashima (2010) reported on evidence that supergranules are smaller and deeper in high-latitude regions compared to low latitudes. Although the effect of rotation is not so large (the ratio of supergranular turnover time to the rotation period is about a few percent), this is in accordance with what we know about rotating fluids. This too, however, needs follow-up investigations.

As for the solar convective flows at other scales, granules are probably difficult, because their typical size is comparable to the acoustic wavelengths of five-minute waves, although this does not rule out seismic investigation based on high-degree f modes. On the larger-scale side, Hanasoge *et al.* (2010, 2012b) attempted to estimate convective velocity. They used the time–distance technique to search for the signature of convective flows. Unfortunately, they found that the result was dominated by noise, but they used this non-detection as means to constrain convective velocity as function of spherical-harmonic degree l ($l < 60$) at various depths in the outer 10% of the Sun. They concluded that the convective velocity is much weaker than numerical simulations predicted, by a factor of 20–100. For example, at $r/R_\odot = 0.92$ and for $l = 30$, numerical simulations predict $\sim$40 m/s flow speed, but their upper limit is as low as a few m/s. In other words, if the convective speed was faster than this limit, they should have detected a signature but they did not. This is a serious statement with many impacts on various aspects of the dynamics of the Sun, including the convective transport of energy and the solar dynamo.

13.5 Final remarks

In this brief chapter we overviewed what we have learned about certain aspects of the solar interior – sunspots, meridional circulation, and convective flows – from local helioseismology. It is apparent that there are many on-going investigations and unanswered questions in this field. In some cases, in spite of its promises, we are finding that the task involved is much tougher than it initially seemed. In some cases, although we should still be careful because we have not yet fully understood the situations, certain pictures are emerging, waiting for physical interpretations.

Needless to say, understanding physical processes requires both observational and theoretical approaches, and it is hoped that efforts on both sides continue, to bring us knowledge that is not accessible by any other means.

14

Seismology of giant planets

PATRICK GAULME, BENOÎT MOSSER, FRANÇOIS-XAVIER SCHMIDER, AND TRISTAN GUILLOT

14.1 Introduction

Seismology of giant planets has long been considered as both a potentially powerful tool for probing their interiors and a natural extension of helioseismology. Giant planets are mostly fluid and convective, which makes their seismology much closer to that of solar-like stars than that of terrestrial planets. For this, we refer the reader to the introductory chapters about helio- and asteroseismology for basic concepts and vocabulary. By being the biggest and closest, Jupiter has attracted most of the efforts in this domain. Theoretical studies started in the late 1970s and the first observational attempts were undertaken in the late 1980s. So far, the two major results are a clear detection of acoustic oscillations of Jupiter (Gaulme *et al.*, 2011), and the signature of Saturn f modes in the rings by the NASA Cassini spacecraft (Hedman and Nicholson, 2013).

This chapter first examines the theoretical motivation for developing seismology of giant planets, which mainly stands on an inaccurate knowledge of their interiors (Section 14.2). The next sections focus on two crucial points: why seismology can be done on giant planets (14.3), and what can it bring in terms of physics (14.4). We then present the observation techniques that have been used or envisioned to detect oscillations (14.5), and the main observational results (14.6).

14.2 Interior structure

Giant planets are planets massive enough to have retained the hydrogen and helium initially present in the circumstellar disk that led to the formation of the central star and its planets. The study of their composition is important in understanding both the mechanisms enabling their formation and the origins of planetary systems, in particular our own. Unfortunately, composition determination is complicated by the fact that their interiors are thought not to be homogeneous, so that spectroscopic determinations of atmospheric abundances are probably not representative of the planet as a whole. Instead, the determination of their composition and structure must rely on indirect measurements and interior models. Figure 14.1 summarizes the current models of interior structure of the four giants.

Extraterrestrial Seismology, eds. Vincent C. H. Tong and Rafael A. García. Published by Cambridge University Press.

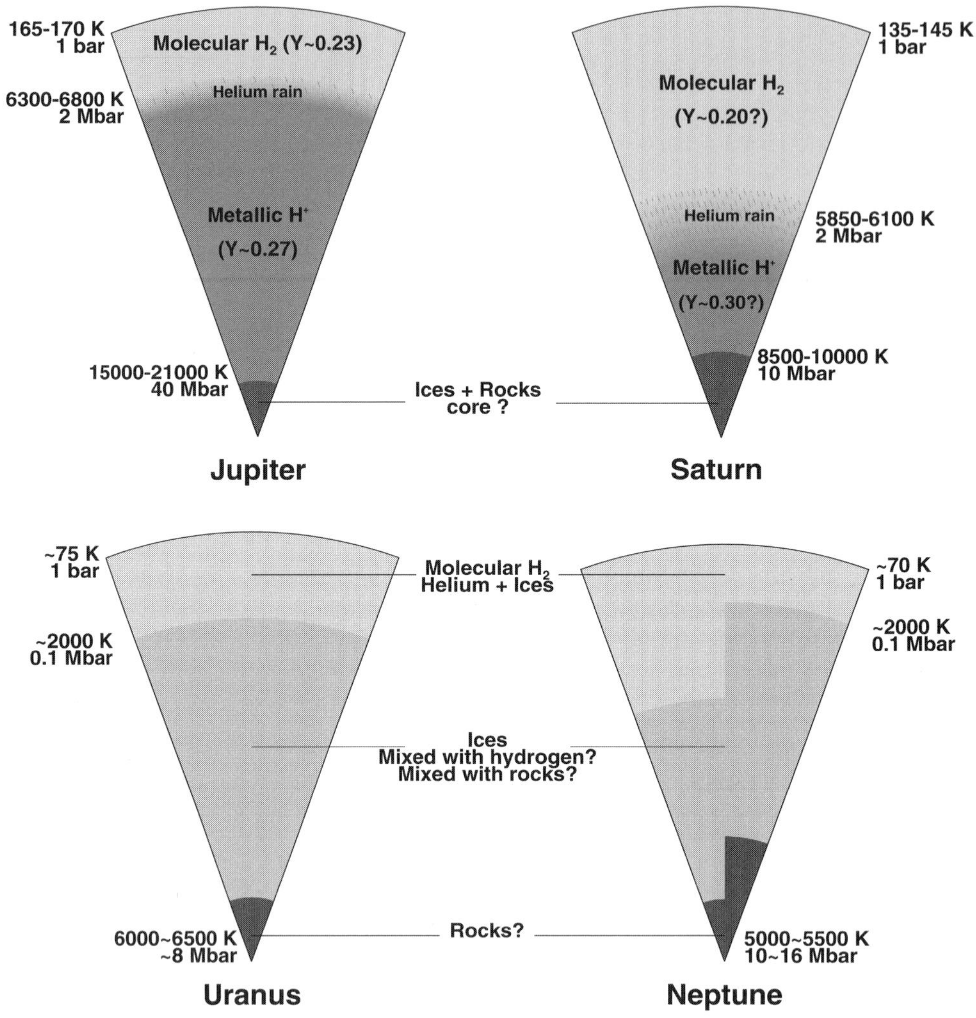

Figure 14.1 Schematic representation of the interiors of Jupiter, Saturn, Uranus, and Neptune. The range of temperatures for Jupiter and Saturn is for models neglecting the presence of the inhomogeneous region (adapted from Guillot, 1999). Helium mass mixing ratios Y are indicated. The size of the central rock and ice cores of Jupiter and Saturn is very uncertain (see text). Similarly, for Uranus and Neptune, considerable uncertainties exist (Nettelmann *et al.*, 2013a). Recent models of Neptune indicate a wider range of possible solutions, of which two representative ones are shown. A black and white version of this figure will appear in some formats. For the color version please refer to the plate section.

The determination of the fundamental parameters (mass, radius) is a prerequisite, prior to any other characterization, both for the giant planets in our Solar System and for giant exoplanets. The mean density thus provided is a first-order constraint on the mean composition of the planet. Fortunately, for the Solar System's giants, a second kind of measurement is available in the form of the departure of their gravity field from sphericity due to their spin. These departures, measured through the planets' gravitational moments, are linked to the interior density profile and therefore allow, through modeling, for global constraints on the structure and composition (e.g., Zharkov and Trubitsyn, 1978). The uncertainties remain high however. When separating hydrogen and helium from all the other, so-called "heavy elements," modern models predict Jupiter has between 10 and 42 $M_\oplus$ (e.g., Saumon and Guillot, 2004; Militzer *et al.*, 2008; Fortney and Nettelmann, 2010), Saturn between 16 and 30 $M_\oplus$ (Saumon and Guillot, 2004; Helled and Guillot, 2013; Nettelmann *et al.*, 2013b), Uranus and Neptune being mostly made of heavy elements, with only 1 to 4 $M_\oplus$ in hydrogen and helium (Podolak *et al.*, 2000, see also Nettelmann *et al.*, 2013a). Even these uncertain estimates may be changed if semi-convection is widespread in the envelopes of these planets (Leconte and Chabrier, 2012). Part of these heavy elements may be in the form of a central dense core of mass below about 15 $M_\oplus$ for Jupiter and 20 $M_\oplus$ for Saturn, with no core being an extreme possibility allowed in some cases.

In 2017, the Juno and the Cassini Solstice missions will yield much improved determinations of Jupiter's and Saturn's gravity fields, respectively. This should provide ways of better constraining the interiors of these planets. In the case of Juno, the combination of gravity, magnetic, and radiometric measurements (the latter providing the possibility to constrain the abundance of water in Jupiter's deep atmosphere) will be especially important in providing direct insight into the behavior and properties of Jupiter's deep interior. However, the amount of information carried by the gravitational moments is limited, for example only the mean density and second-order gravitational moment J_2 are sensitive to the presence of a central core, the higher order moments yielding constraints on layers that are progressively closer to the "surface" (Guillot, 2005). The constraints thus imposed will remain intrinsically model dependent.

Seismic studies of these planets should be a very powerful and complementary way of probing their interiors. As shown for the Sun or the Earth, a carefully determined oscillation spectrum provides a much richer wealth of information on the properties of the interior. It is also very sensitive to discontinuities on the density profiles, which is extremely valuable information for understanding the planetary structure. For example, in Jupiter and Saturn, discontinuities (or steep variations) of the density profiles may arise at the core/envelope interface, and also at the location where helium separates from hydrogen and forms droplets, and in double-diffusive interfaces. The discoveries of discontinuities and the determination of their depth would tell us directly which model can be ruled out and provide additional very strong constraints.

14.3 Why seismology is possible

The possibility of applying seismology to giant planets is conditional on the existence of a resonant cavity for acoustic waves, and the presence of a mechanism able to excite acoustic waves.

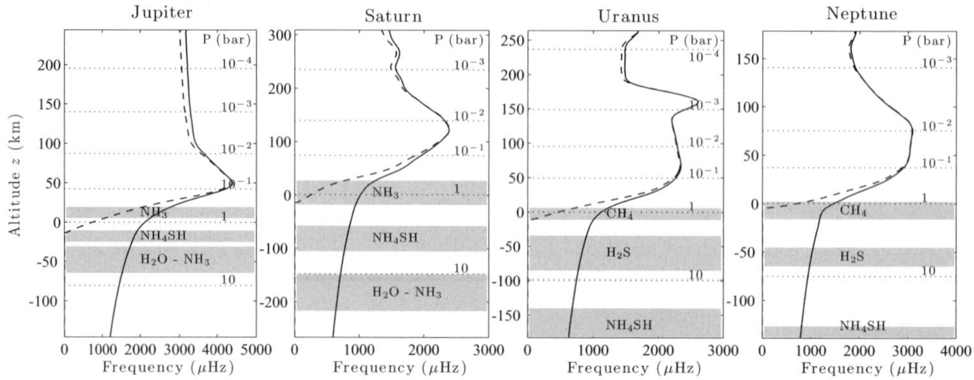

Figure 14.2 Acoustic cut-off ν_c (solid line) and Brunt–Väisälä $N/2\pi$ (dashed line) frequency profiles in the upper atmospheres of the four giant planets. Gray regions indicate cloud locations and their chemical compositions (e.g., Irwin, 2009). Above the 1-bar level ($z \equiv 0$ km), pressure–temperature profiles come from radio occultation of the Voyager spacecraft (Lindal, 1992, and references therein). Below $z = 0$, pressure–temperature profiles are adiabats, which imposes $N = 0$.

14.3.1 Existence of a resonant cavity

The existence of resonant modes of oscillations in a planet requires a trapping mechanism for the waves. As first demonstrated by Vorontsov *et al.* (1976), the atmospheres of the giant planets indeed reflect acoustic waves, but only if their frequency is below a cut-off value that varies from about 2500 to 4200 μHz. As shown in Figure 14.2, detailed calculations of the vertical propagation of acoustic waves by Mosser (1995) show that the level at which the waves are reflected is a strong function of their frequency. In all four giant planets, it turns out that most waves are reflected just above the main cloud deck. Besides, from reflectivity factors, Mosser (1995) put an upper limit on Jupiter's mode lifetime, which is strongly dependent on the mode frequency, from about 300 years at 1000 μHz to 10 days at 3000 μHz. Also, the Brunt–Väisälä frequency profile is similar to the acoustic cut-off above the tropopause, because both terms vary as g/H_p – the ratio of local gravity and pressure scale height, which means that part of the acoustic wave dissipation might transfer into gravity waves in the stratosphere.

14.3.2 Excitation mechanisms

Confidence that acoustic modes are excited in giant planets mainly relies on the large reservoirs of energy for Jupiter, Saturn, and Neptune, because the infrared excess luminosity respectively represents 67, 78, and 161% of the incident solar flux. This inner energy drives a significant convective flux. Uranus remains apart, with 6% infrared flux in excess, which indicates that the planet has almost reached thermal equilibrium.

As known for the Sun (e.g., Goldreich *et al.*, 1994), convection transfers part of the energy into acoustic waves. But contrary to the case of the Sun for which the forcing is inefficient, Bercovici and Schubert (1987) show that convective eddies with velocities of 10–80 m s^{-1}, and length scales on the order of a scale height (~20 km at 1 bar for Jupiter and Saturn) would have similar periods as those of the standing waves ($\tau_w = 4.5$–9 min) and would thus be well coupled. In that case, they argue that the power per unit volume supplied to the acoustic waves by the turbulent motions would be

$$W \sim \frac{\rho u^3}{h} M^3, \tag{14.1}$$

where ρ is the density, u and h the typical velocities and scale height of eddies respectively, and $M = u/c_s$ is their Mach number (c_s is the sound speed). (We neglected the M^5 term since $M < 1$ is expected). By equating the energy flow from turbulence, Wh, to that of the waves, $F_w = \rho c_s u_0^2$, they obtain the velocity amplitude of the waves:

$$u_0 \sim u^3/c_s^2. \tag{14.2}$$

By adopting $u \leq 80$ m s^{-1} and $c_s \sim 1$ km s^{-1}, Bercovici and Schubert (1987) obtain $u_0 \leq 0.5$ m s^{-1}.

It is not clear at present whether these estimates are realistic. The convective velocities predicted from mixing-length arguments are typically significantly smaller ($u \sim 1$ m s^{-1} at 1 bar – see Guillot *et al.*, 2004). However, values of the vertical velocities obtained in large water clouds typically reach tens of m s^{-1} (e.g., Hueso *et al.*, 2002) on the appropriate scale height. Zonal flows also are characterized by velocities over 100 m s^{-1}, but this implies much larger (horizontal) scales and it is not clear that they would efficiently couple to the waves.

14.3.3 Energy balance and mode lifetime

Another approach is to look at the energy balance implied by the presence of $u_0 \sim 0.5$ m s^{-1} seismic waves, as implied by the observations of Jupiter (Gaulme *et al.*, 2011, see Section 14.6.2). In this case, with the above value for c_s and $\rho \sim 10^{-4}$ g cm^{-3} in the atmosphere, the energy flux in the waves is $F_w \sim 25\,000$ erg cm^{-2} s^{-1}. This is to be compared to Jupiter's intrinsic and absorbed solar fluxes of 5440 and 8140 erg cm^{-2} s^{-1}, respectively (e.g., Guillot, 2005, and references therein), implying a maximum available flux $F_{tot} = 13\,600$ erg cm^{-2} s^{-1}. This implies that the energy in the waves must be concentrated spatially, or that there exists a coherent mechanism to excite the waves and that the damping time scale for these is long enough.

Let us assume that such a mechanism exists and transforms a small fraction ϵ of Jupiter's energy flux into acoustic waves. In this case the lifetime of the modes is obtained by calculating the time required to channel the energy flux ϵF_{tot} into waves of total energy per unit area $\tau_w F_w$:

$$\tau_{lifetime} \sim \tau_w \frac{F_w}{\epsilon F_{tot}}. \tag{14.3}$$

If we assume rather arbitrarily $\epsilon \sim 1\%$ (see, in a different context, Showman and Guillot 2002), then with the above assumptions $\tau_{\mathrm{lifetime}} \sim 2\,\mathrm{days}$. In this case, the lifetime of the modes is a measure of the damping of the waves, which must balance their excitation. Because it is difficult to imagine that the generation of acoustic waves would be more efficient, this implies that the modes should be long-lived (for at least days or more), which is compatible with the quality factor of Jupiter determined by Mosser (1995).

14.4 What seismology can reveal about interior structures

Given that main-sequence solar-like stars and giant planets are fluid and mostly convective, their oscillation spectra are expected to look similar. They are characterized by a series of peaks that are regularly spaced as a function of frequency, whose amplitudes roughly follow a Gaussian shape. The mean separation $\Delta\nu$ between consecutive mode overtones of a given degree is proportional to the square root of the planetary mean density, whereas the frequency at maximum amplitude ν_{max} is proportional to the acoustic cut-off frequency. It is therefore possible to estimate the "global seismic parameters" $\Delta\nu$ and ν_{max} from the asteroseismic scalings. However, two major differences between giant planets and solar-like stars that may affect their oscillation spectra are the possible density discontinuities (Section 14.2), and the fast rotation, which induces a non-negligible oblateness e (Vorontsov and Zharkov, 1981; Mosser, 1990). The oblateness turns the mean planetary radius into $R_{\mathrm{eq}}(1 - e/3)$ and thus introduces a relative change of $e/3$ in individual frequencies, $e/2$ on $\Delta\nu$, and $2e/3$ on ν_{max}. Table 14.1 presents the global seismic parameters of the four giant planets. The Sun and Jupiter have similar $\Delta\nu$ since their mean densities are equal, whereas ν_{max} largely differ as do their surface gravities and effective temperatures.

Two kinds of theoretical studies have been done to fully understand the power of seismology for giant planets. Forward modeling should identify relevant features in simulated oscillation spectra computed from planetary models. Inverse modeling should quantify how much models are constrained from seismic observations, by inverting simulated oscillation spectra. As regards forward modeling, the oscillation patterns of giant planets have been depicted in many papers (e.g., Vorontsov *et al.*, 1976, 1989; Provost *et al.*, 1993; Gudkova *et al.*, 1995; Jackiewicz *et al.*, 2012a). In particular, Provost *et al.* (1993) showed that a density jump at the core boundary of Jupiter would induce a modulation of the regular frequency pattern with amplitude up to $\Delta\nu/3$ and period of about 10 radial orders, from which it is possible to infer the core size. Also, the influences of the upper atmosphere, fast rotation, differential rotation, or chemical phase transitions on individual mode frequencies have been investigated. Overall, comparative works have shown that oscillation patterns based on different planetary models and performed with different seismic codes largely agree (e.g., Gudkova *et al.*, 1995; Jackiewicz *et al.*, 2012a).

The inverse problem has been investigated by fewer authors, and Jackiewicz *et al.* (2012a) led the most comprehensive work. From three models of Jupiter's internal structure typical of the current state of the art, they used linear forward modeling to compute the theoretical oscillation spectrum up to $l = 25$ and build a set of sensitivity kernels,

Table 14.1 *Comparison of physical and seismic parameters for the four giant planets. Asymptotic $\Delta\nu_{as}$ are obtained as the inverse of the acoustic travel time across the planet, computed from internal structure models (Saumon and Guillot, 2004; Helled and Guillot, 2013; Nettelmann* et al., *2013a, for Jupiter, Saturn, and Neptune respectively, and Nettelmann private communication for Uranus.). $\Delta\nu_{sl}$ and ν_{max} are extrapolated from solar values using asteroseismic scalings with correction of planetary oblateness e. For Jupiter, observed values from Gaulme* et al. *(2011) are shown for comparison.*

Planet	M (M_E)	R_{eq} (km)	e (%)	T_{eff} (K)	$\Delta\nu_{as}$ (μHz)	$\Delta\nu_{sl}$ (μHz)	$\Delta\nu_{obs}$ (μHz)	$\nu_{max,sl}$ (μHz)	$\nu_{max,obs}$ (μHz)	$\delta\nu_{rot}$ (μHz)
Sun	3×10^5	798 000	0	5777	138				3100	0.4
Jupiter	317.9	71 492	6.5	125	152–156	134	155	2030	1250	28
Saturn	95.1	60 268	10.2	95	111–115	99		1040		26
Uranus	14.53	25 559	2.4	57	~172	130		1030		17
Neptune	17.14	24 764	2.7	59	198–213	148		1270		19

i.e., linear combinations of mode eigenfrequencies and eigenfunctions designed to probe selected regions of the interior. Then, they applied inversion techniques developed for helioseismology to infer details on the expected discontinuities in the profiles of the jovian interior. They found that interior boundary layers with sound-speed jumps of a few percent could be detected with a sufficient number of observable modes. For deep transitions in the core boundary where fewer modes probe (low l), it is very likely that the locations of the large discontinuities could be accurately found, but the magnitudes of the jumps will be more difficult to ascertain because of stochastic noise. They stress that identifying modes up to $l = 25$ is important for fine-inverting Jupiter's oscillation spectrum, even for probing the core region. A similar conclusion is expected for the three other giants.

Beyond pressure oscillations, which have been considered so far because only acoustic waves propagate in convective envelopes, recent asteroseismic observations have shown the huge potential of mixed modes in stars with a radiative core, especially red giants (Bedding *et al.*, 2011). Such mixed modes result from the coupling of gravity waves in the core with pressure waves in the envelope. The possible presence of a dense and certainly radiative core in Jupiter and Saturn may give rise to such mixed modes: detecting them would provide even finer constraints on core size, nature and rotation, with respect to pure pressure oscillations.

14.5 Techniques to observe acoustic oscillations

Several approaches can be used for detecting acoustic modes: visible photometry for measuring reflected solar flux changes, infrared photometry for temperature fluctuations, and Doppler spectrometry for radial velocities.

14.5.1 Visible photometry

A natural approach is to search for variations of the solar light reflected by Jupiter, resulting from distortions of the planetary external radius R by acoustic modes. Mosser (1995) showed that a radial mode $l = 0$ of velocity v leads to a variation of the radius $\delta R = c_s v/g$, where c_s is the sound speed in the upper troposphere and g the gravitation. For Jupiter, a $1\ \mathrm{m\,s^{-1}}$ velocity implies $\delta R = 40$ m, with a corresponding variation of reflected luminosity, $(\delta\Phi/\Phi)_{\mathrm{refl}} = 2\delta R/R$ of the order of 1 ppm.

Gaulme and Mosser (2005) showed that the photometric fluctuations related to planetary radius variations are likely to be negligible with respect to albedo fluctuations of clouds that are caused by acoustic modes on their thermodynamical equilibrium. They demonstrated that acoustic waves drive a shift in the gas/condensate phase equilibrium that defines the cloud, which is damped by the kinetics of condensation. The resulting perturbation of the cloud equilibrium can be converted to albedo variations. For acoustic waves of $1\ \mathrm{m\ s^{-1}}$ amplitudes, photometric fluctuations reach several tens of ppm, which comfortably fits within the photometric performance of current space-based imagers.

Even though additional knowledge on cloud properties of the giant planets is required to assess this result, the photometric approach looks promising for a space mission. Indeed, photometric imaging instruments are much simpler and lighter than high-precision spectrometers and their feasibility and accuracy has been demonstrated by the remarkably successful asteroseismic observations of CoRoT and *Kepler* (Michel *et al.*, 2008; Borucki *et al.*, 2010). The first experiment will be realized by the extended *Kepler* mission K2, which is planned to observe Neptune in early 2015 for 40 days with minimal interruptions.

14.5.2 Thermal infrared photometry

In the infrared, the photometric detection of global oscillations is based on the approach used in helio- and asteroseismology in the visible domain: acoustic waves are temperature disturbances and lead to fluctuations of brightness temperature. The black-body emission of most stars peaks in the visible, while it peaks in the mid-infrared for the four giant planets. Thermal emission dominates the reflected solar spectrum at about 6.5 μm for Jupiter, 10.6 μm for Saturn, 18 μm for Uranus, and 15 μm for Neptune.

The relative Lagrangian temperature perturbation associated with a wave of velocity v is expressed as $\delta T/T = (1-\gamma)v/c_s$, where γ stands for the adiabatic index (e.g., Mosser 1995). For Jupiter, the conversion from wave velocity to temperature is about 0.07 K for $1\ \mathrm{m\,s^{-1}}$. The Lagrangian temperature variation induces a relative flux variation of the black body of

$$\left(\frac{\delta\Phi}{\Phi}\right)_{\mathrm{bb}} = \frac{hc}{\lambda k_{\mathrm{B}} T}\,\frac{\delta T}{T}. \tag{14.4}$$

Hence, we expect relative photometric variation of about 0.5% for $1\ \mathrm{m\ s^{-1}}$ at 10 μm, in the case of Jupiter.

Thermal infrared photometry looks the most favorable technique, because the flux fluctuations are driven by the ratio of wave velocity to the sound speed, instead of the speed of light for Doppler spectrometry (Section 14.5.3). However, the vertical extension of the atmospheric contribution function in the thermal infrared is likely larger than that of the visible (limited by clouds), which tends to average out the signature of p modes, whose vertical wavelengths are about 10 pressure scale-heights in the upper troposphere. In addition, helioseismology has demonstrated that photometric data of solar p modes are noisier by a factor 10 than Doppler data. Also, detectors are not as sensitive in the mid-infrared as in the visible, and ground-based performance is limited by Earth atmosphere emission.

Deming *et al.* (1989) observed Jupiter in the 8–13 μm bandwidth, using a 20-element linear array placed parallel to the planet's equator at about 20° in the northern hemisphere. Such an observational set-up totally filtered modes with degrees $l \leq 10$. Observations were performed with the 3-meter Infrared Telescope Facility (IRTF) in Hawaii, lasted 5 days, and did not lead to p-mode detection at the 0.07 K level.

Infrared photometry was then used in attempting to detect acoustic modes caused by the impact of comet Shoemaker–Levy 9 on Jupiter in 1994 (Lognonné *et al.*, 1994). Walter *et al.* (1996) reported imaging of the planet in a methane emission band at 7.8 μm with the IRTF, while Mosser *et al.* (1996) observed Jupiter with the mid-infrared camera of the ESO 3.6-meter telescope at La Silla. Both observations did not lead to any detection of impact-excited waves. Negative results were interpreted a posteriori. First, the wavelength of the waves might have been too short with regard to instrumental resolution, so that their signal was likely averaged out. Second, impacts occurred on the night side of Jupiter, so that waves were expected at the limb, where the photon noise is maximum. Third, observations occurred during daytime, which does not favor infrared observations. Since then, this technique has never been considered again either for ground-based observations or for space instrument projects.

14.5.3 Doppler spectrometry

The use of Doppler spectrometry for giant-planet seismology was inspired by its success in helioseismology. The basic principle relies on monitoring the position of a spectral line that probes an atmospheric level where the amplitude of acoustic modes is maximum. As for exoplanet search, the minimum spectral resolution to fully resolve spectral lines is about 70 000. So far, resonant cell and Fourier transform spectrometry have been considered and both allow for spatial resolution.

Atomic resonant cells are used in magneto-optical filters (MOF), whose use for velocity measurements was first studied by Cacciani and Fofi (1978). The incoming solar light goes through a cell containing an atomic gas that is present in the solar atmosphere and displays a strong absorption line. The solar atomic line is thermally broadened with respect to that of the atomic cell, which behaves as an extremely thin optical filter. Thus, a shift of the solar line produces a photometric variation of the transmitted light. The instrument sensitivity is the relative flux variation associated with a 1 m s^{-1} Doppler shift:

$$S = \frac{1}{c}\frac{\mathrm{d}\ln I}{\mathrm{d}\ln\lambda}, \tag{14.5}$$

where λ is the cell wavelength, c the speed of light, and I the solar spectrum.

The use of Fourier transform spectrometry (FTS) for Doppler measurements also derives from helioseismology (Scherrer *et al.*, 1995). The output signal (interferogram) is the Fourier transform of the incoming light spectrum. The presence of deep and evenly spaced absorption lines in the incoming spectrum generates interference fringes at large optical path difference (OPD), of which phase is a function of the Doppler shift. The output is expressed as $\Phi(\Delta) \propto 1 + \mathcal{C}\cos[2\pi\sigma_0\Delta(1 + v/c)]$, where $\mathcal{C}$ is the fringe contrast, σ_0 the wavenumber, and Δ the OPD. The instrument sensitivity is given by the phase shift of an interference fringe per a 1 m s^{-1} radial velocity:

$$S = \frac{2\pi\sigma_0\Delta}{c}. \tag{14.6}$$

In a MOF, all photons from a typical 0.01 nm photometric bandwidth contribute to the signal. In a FTS, the filter bandwidth is generally larger (1 to 10 nm), but only $\mathcal{C}/\sqrt{2}\,N$ contribute to the signal. When observations are limited by the photon noise only, the amplitude of a velocity signal that is detectable at the 1-σ level is $\delta v = (1/S)(1/\mathrm{SNR})$, i.e.

$$\delta v = \frac{1}{S\sqrt{N}} \text{ for a MOF, and } \delta v = \frac{\sqrt{2}}{\mathcal{C}}\frac{1}{S\sqrt{N}} \text{ for a FTS.} \tag{14.7}$$

Therefore, a high-performance seismometer must optimize the product of the sensitivity by the square root of the bandwidth.

Both types of instruments are sensitive to temperature changes. In a MOF, the reference wavelength is intrinsically stable, but not its transmission. In a FTS, the OPD is temperature dependent. Thermal effects can be minimized by working at a temperature close to an extremum of the sensitivity. The major drawback of a MOF is that its sensitivity is a function of the Doppler shift, making it not uniformly sensitive on a spinning object. The typical sensitivity range of a solar atomic line is ± 10 km s^{-1}. The rotational velocities of Jupiter and Saturn vary by 25.1 and 19.5 km s^{-1} along the equator, so that the shift of a reflected solar line is larger than its own width, making a MOF sensitive to a fraction of the planetary surface only. As we are about see, Doppler spectrometry is the only technique that has led to the detection of Jupiter's oscillations.

14.6 Observational results

14.6.1 Preliminary detections of Jupiter's oscillations

In 1987, Schmider *et al.* (1991) conducted full disc observations of Jupiter with a sodium cell MOF mounted on the 1.5-meter telescope of the Observatoire de Haute Provence (OHP), providing 7 days of observations with a 23% duty cycle. The data analysis exhibited a periodic signal of jovian origin, as the signature of the jovian rotation was detected, between 1000 and 2200 μHz. However, the absence of spatial resolution and direct amplitude calibration did not permit this signal to be clearly attributed to acoustic modes. Later, Cacciani *et al.* (2001) designed a double-cell MOF with imaging capability. They tested it on Jupiter, but its actual performance appeared significantly lower than expected, and no jovian signal could be detected.

In 1991 and 1996, observations were conducted with the Fourier transform spectrometer of the 3.6-meter Canada France Hawaii Telescope (CFHT) (Maillard and Michel, 1982), which was specially set at a fixed OPD, corresponding to the methane absorption band at 1.1 μm (Mosser *et al.*, 1993, 2000). Two circular regions of 10-arcsec diameter, symmetrically located around Jupiter's equator, were sequentially observed to get rid of drifts of the instrument and the spurious velocity signal due to jovian rotation. These observations confirmed the existence of a velocity signal of jovian origin, in the same frequency range as previous observations and with amplitude compatible with theoretical expectations.

In both MOF and FTS detections, a tentative comb-like structure with a 139 ± 3 μHz mean spacing was also identified. This was probably an artifact due to poor SNR. This frequency corresponds to the least-common multiple of terrestrial and jovian rotation frequencies, and is incompatible with theoretical estimates of $\Delta\nu$. The main lesson of these observations is the need for spatial resolution. First, it avoids the broadening of solar spectral lines by fast planetary rotation, which drastically alters the velocity sensitivity. Second, analyzing Doppler images by applying masks corresponding to each acoustic mode allows the detection of higher-degree modes. Third, the comparison of photometric images with Doppler maps permits removal of photometric contamination in the velocity signal. These arguments motivated the SYMPA project.

14.6.2 Detection of Jupiter's global modes with SYMPA

The SYMPA instrument is a Fourier-transform imaging spectrometer with fixed OPD, whose main optical device is a Mach–Zehnder interferometer (Schmider *et al.*, 2007). It produces four interferograms of the solar spectrum reflected by Jupiter, in a 5 nm spectral range centered around the Mg triplet at 517 nm. Each interferogram modulates the jovian photometric figure. Both photometric and interferometric maps are extracted and disentangled from a linear combination of the four images. Next, the velocity maps are decomposed into spherical harmonics to create a set of time series. The power spectrum of each time series is sought for oscillations.

Two SYMPA instruments were built and they were used simultaneously during two observation runs of Jupiter in 2004 and 2005 at San Pedro Martír and Teide observatories. The best observation sequence was acquired as part of the 10-day run performed at Teide in April 2005. The data were found to be of much higher quality than all previous observations: at the end of the run, with a 21.5% duty cycle, the mean noise level reached 12 cm s^{-1}, which was five times lower than previously achieved (Mosser *et al.*, 2000). However, analysis of

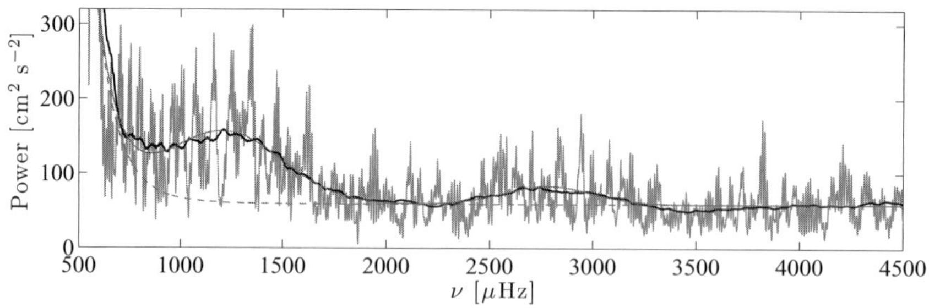

Figure 14.3 Power spectrum of the dataset obtained with the SYMPA instrument for Jupiter in 2005 at Teide Observatory (from Gaulme *et al.*, 2011). The oscillations of Jupiter are characterized both by the increase in power between 800 and 3500 μHz, and the regular spacing of the modes ($\Delta\nu = 155$ μHz). The decrease of the mode amplitude around 2000 μHz suggests that the propagation of acoustic modes in the upper atmosphere may be a function of frequency.

the instrument performance (Gaulme *et al.*, 2008) restrained the search of oscillation modes to low-degrees with antisymmetric pattern. Gaulme *et al.* (2011) focused on the time series with the best signal-to-noise ratio, which corresponded to modes of degree $l = 1$ with significant leakage from $l = 2, 3,$ and 4. As in previous Doppler measurements, the time series exhibited excess power between 800 and 2000 μHz and secondary excess between 2400 and 3400 μHz, but this time the power spectrum was modulated by a clear, high signal-to-noise comb-like structure (Figure 14.3). The frequency of maximum amplitude of the main excess power was measured at $\nu_{\max} = 1213 \pm 50$ μHz and the mean large spacing at $\Delta\nu = 155.3 \pm 2.2$ μHz. By assuming that modes are coherent along the observation run, the maximum mode amplitude is 49^{+8}_{-10} cm s^{-1}.

The large error bar in the peak frequencies (± 7 μHz) prevented any mode identification. However, the peak structure matches the signature of modes of consecutive degrees (e.g., $l = 1$ and $l = 2$). Thanks to the thermal control and the imaging capability of SYMPA, the authors could rule out that other effects caused the measured signal, such as temperature fluctuations in the interferometer, telescope pointing errors, albedo features on Jupiter, and solar p modes reflected by Jupiter's atmosphere. The measured $\Delta\nu$ value agrees with the asymptotic value (151–156 μHz), inferred from the sound speed integral from modern models of the interior of Jupiter (Saumon and Guillot, 2004). Better constraints will be provided only by the comparison of observed with theoretical individual mode frequencies.

14.6.3 Detection of Saturn f modes by Cassini

As proposed by Marley and Porco (1993), "the rings of Saturn may act as a seismograph, recording gravitational perturbations associated with acoustic oscillations modes of the planet." Twenty years later, observations of stellar occultations of stars by the rings made with Cassini showed that six density-wave structures detected in the C-ring were compatible with resonances due to Saturn f modes (Hedman and Nicholson, 2013). By comparing the phase in the density wave for the ingress and egress of the occulted star, the authors roughly identified the azimuthal number of the modes responsible for the observed ring structure. The frequency of the associated mode is easily derived from the position of the feature in the ring. However, a multiplet of three waves associated with a given azimuthal number was found, in contradiction to expectations.

Only a few f modes have low enough frequencies to excite resonances in the rings. Since f-mode frequencies are mainly dependent on surface gravity, they are sensitive to the structure of the most external layers of the planet, which includes differential rotation as noticed by Hedman and Nicholson (2013). They carry little information about the deep internal structure of the planet. However, Fuller *et al.* (2014) proposed that a coupling between surface f modes and spheroidal or toroidal modes of the solid core could cause the observed multiplets. Although they did not successfully reproduce the multiplet structure, they argued that a comprehensive description of a giant-planet oscillation spectrum should include couplings with solid body modes – as for terrestrial planets – as it might provide valuable information about the existence and the structure of the solid core.

Regarding the f-mode amplitudes, Marley and Porco (1993) expected that displacements less than 1 m could not produce any noticeable waves in the rings. From the Cassini observations, Fuller *et al.* (2014) estimated the f-mode amplitude to be 30 cm. Because their frequencies range between 50 and 100 μHz, the induced velocity variations are well below 1 mm s^{-1}, which is impossible to detect with Doppler measurements. Actually, such low-degree f modes cannot even be detected on the Sun.

These results also give a hint about the amplitude of Saturn's acoustic modes. Goldreich *et al.* (1994) observed the amplitude of solar acoustic modes to be proportional to ν^4, which would mean – if that applies to giant planets – that Saturn's acoustic modes are larger than 30 cm s^{-1} near ν_{max} (~1000 μHz, Table 14.1). This is coherent with what was measured on Jupiter and should be detectable with Doppler measurements.

14.7 Prospects

Seismology of giant planets is still in its infancy, and the situation reminds one of helioseismology in the late 1970s and asteroseismology in the 1990s. After the recent results on Jupiter and Saturn, there is a clear need to improve the observations and the theoretical models. Improvements in the observations could be done from the ground for Jupiter and Saturn, where oscillation modes with amplitude larger than 5 cm s^{-1} and degree up to $l = 25$ could be measured. However, such observations would require a network of 2–4-meter telescopes evenly placed around the world observing continuously for several weeks. Uranus and Neptune would require even larger telescopes. Detection attempts might be envisioned from the ground, but only space missions could achieve the full potential of the technique.

The studies in the framework of the ESA JUICE mission demonstrated the possibility of reaching the objectives with a dedicated instrument on board a spacecraft approaching the planet. Two instrumental concepts, a MOF and a Doppler imager (Soulat *et al.*, 2012), have reached a sufficient technology readiness level (TRL) and are available for future missions, even though no such payload was included in the mission. Several Solar-System mission projects contemplate seismic observations of Saturn, Uranus, or Neptune. Such missions could achieve continuous observations for global seismology over several weeks or even months, including during flybys of Jupiter or Saturn by a mission aiming at the ice giants. Observations from an orbiter or during a close fly-by would also allow for studying shallow acoustic waves with solar local-seismic techniques, which give access to the physical conditions just below the surface.

Alternative approaches may be envisioned. Impact seismology could be performed if an event similar to the SL9 comet happened. A strong impact (with an energy deposit of 10^{20} J) would produce detectable tsunami waves and could launch waves probing the upper envelope, and a very strong impact ($>10^{21}$ J) should excite P-waves that probe the entire planet. In 1994, the strongest impact remained unfortunately about 20 times too small to excite detectable waves, but today's new infrared detectors with enhanced sensitivity and smaller pixels would detect such an event.

In a broader context, Le Bihan and Burrows (2013) examined the possibility of giant exoplanet seismology. This is out of reach with any current observation facility. However, observations such as the monitoring of molecular bands on the day-side of the hot Jupiter τ-Bootis along its orbit (Brogi *et al.*, 2012) allow us to believe that some of the brightest hot Jupiters might be observable with sufficient SNR to detect oscillations, when extremely large telescopes equipped with high-resolution infrared spectrometers become available.

After success with the Earth, the Sun and many stars, seismology of giant planets is entering a new era, and will transform our understanding of giant-planet physics and our view of Solar System history.

15

Seismicity and interior structure of the Moon

MARTIN KNAPMEYER AND RENEE C. WEBER

The beginning of the seismological investigation of the Moon dates back to the beginning of space flight: A working group implemented by NASA in 1959 (Hall, 1977) suggested the development of a seismometer for a hard landing on the Moon. This resulted in Ranger missions 3 to 5, which all unfortunately failed for technical reasons (Hall, 1977). The first measurement of elastic properties of lunar soil was conducted by the Surveyor landers a few years later (Christensen *et al.*, 1968).

Besides these early attempts, seismological studies of the Moon divide into two phases: The first one saw the installation of a seismometer network on the Moon, starting with Apollo 11 on July 20, 1969 (*Apollo 11 Mission Report*, 1969), followed by the collection of continuous data until network shutdown on September 30, 1977 (Bates *et al.*, 1979), and, in parallel and ongoing until the early 1990s, the analysis of the data. The second phase began in the late 1990s, when cheap computer power allowed for massive data processing on desktop workstations and the application of new methods.

This chapter aims to give a sketch of the Moon as it results from these two phases. The following sections will first describe the different types of seismic events observed on the Moon, and then detail the structure of the main layers of the lunar interior, i.e., the crust, mantle, and core. A summary section finally gives an overview of the present-day concept of the interior structure of the Moon.

15.1 Seismic sources and seismicity

Both endogenous and exogenous sources create seismic waves on the Moon. It is thus common to speak of "events" rather than "quakes," unless the type of source has been identified. However, analysis of the spatial and chronological distribution of events, and of seismogram characteristics, leads to the identification of several classes of sources.

Astronaut activities were recorded during the extravehicular work of the Apollo crew. Footsteps on the ground, astronauts climbing the ladder of the Lunar Module, and

Extraterrestrial Seismology, eds. Vincent C. H. Tong and Rafael A. García. Published by Cambridge University Press.

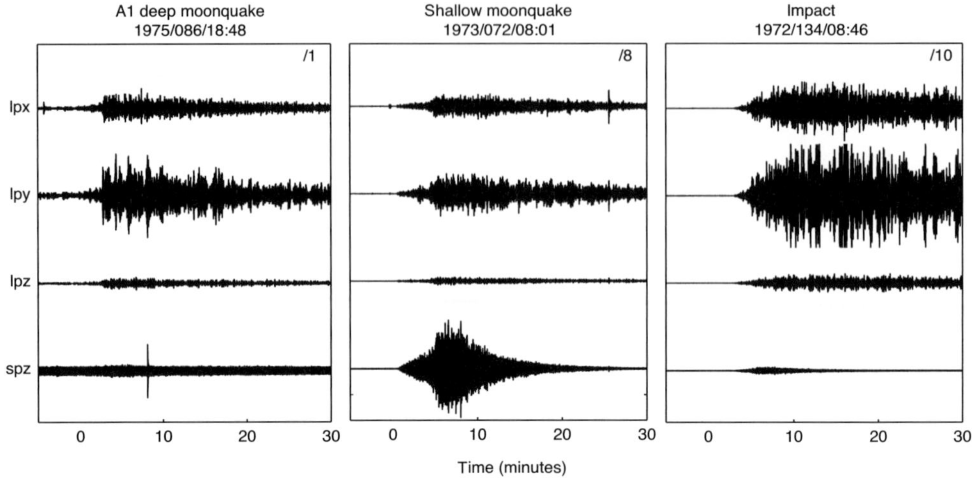

Figure 15.1 Comparison of seismograms from a Cluster A01 deep moonquake, a shallow moonquake, and a meteoroid impact. The four channels shown from top to bottom are the three long-period (two horizontal and one vertical) and the single short-period vertical channels of the Apollo 16 seismometer. Amplitudes were divided by factors given in the upper right corners of each plot. (Redrawn after Nakamura *et al.*, 1974a.)

vibrations from driving the Lunar Roving Vehicle (LRV) were all identified in seismograms (Latham *et al.*, 1969a,b; Nakamura, 1976).

Artificial impacts are impacts of abandoned spacecraft or spacecraft parts (such as booster rocket stages) that were targeted to the lunar surface on purpose as powerful sources at known times and places.

Meteoroid impacts are impacts of small particles and big boulders that orbited through the Solar System before colliding with the Moon.

Thermal moonquakes release the stress resulting from the harsh temperature changes during sunrise and sunset, when the lunar surface temperature in equatorial regions changes by several hundreds of kelvins in relatively short time. They are recorded on short-period (SP) seismometers and geophones only.

Shallow moonquakes were originally termed "high frequency teleseisms" or HFT-events due to their significant amplitudes on the SP channel (Figure 15.1). Their source region is between the surface and about 300 km depth, which is shallow compared to the deep moonquakes.

Deep moonquakes are very weak, occur at depths below 700 km, and come in clusters that exhibit repeatable activity patterns parallel to the Moon's tidal deformation.

Unclassified events are those that could not be assigned to any of the known source types. The main obstacle for classification is the low magnitude of many events: Even some deep moonquakes have amplitudes only slightly above the least significant bit of the digital registration.

Volcanic sources have not yet been observed on the Moon.

The *Long-Period Event Catalog*, first published in 1981 (Nakamura *et al.*, 1981), is currently (2014) maintained at the University of Texas in Austin. The first release listed a total of roughly 12 000 events, about 2800 of which were classified as deep moonquakes, while the majority of close to 7500 events (about one-third of the catalog) remained unclassified (numbers after Nakamura *et al.*, 1980). Event classification in the 1970s was based on visual inspection of paper printouts of the seismograms, since automatic real-time processing proved to be too slow to handle the incoming data (Nakamura *et al.*, 1980). Statistical cluster analysis based on waveform cross-correlation led to several revisions of the original event catalog (the latest revision at the time of writing is that of October 2008): The number of recognized events was increased to 13 058 and the fraction of unclassified events reduced to 25% at the same time. Also, the assignment of individual deep moonquakes to clusters and thus specific source regions could be revised, resulting in the definition of 319 deep moonquake clusters (Nakamura, 2003; Bulow *et al.*, 2005).

A comparison with terrestrial events (Table 15.1) shows that lunar seismicity is much weaker, both in terms of the largest observed event and the average annual seismic moment release. A detailed comparison of seismic energy release with heat flow data and tidal dissipation (Goins *et al.*, 1981a) shows that the seismic energy release of the Earth is about 10^9 times that of the Moon, yet the Earth produces many more quakes. The Earth, however, dissipates only 10^5 to 10^6 times the tidal energy dissipated by the Moon, and releases only ten times as much heat (Goins *et al.*, 1981a). The reason for this discrepancy in seismic efficiency appears to be that the Moon is a stagnant lid planet while the Earth is in the plate tectonic regime.

Another important difference between the two bodies is that all types of natural lunar seismicity exhibit some kind of periodicity that can be tied to orbital periods of the Moon, while the temporal distribution of earthquakes is random.

15.1.1 Artificial impacts

When left unattended, the likely trajectories of spent Saturn V upper stages (S-IVB stage) led to a 40% probability for impacting the Moon. Yates (1967) thus suggested furnishing the S-IVB with a tracking transmitter and introducing a midcourse maneuver to increase the impact probability to >90%, and also considered impacting the Lunar Module (LM) ascent stages. Yates estimated that an S-IVB impact could be detected by the Apollo seismometers to distances up to 500 km, and up to 200 km for the LM ascent stage. Four S-IVBs were indeed impacted at pre-planned sites under remote control from Earth, and a total of nine artificial impacts were recorded by the then operative seismic instruments, including the Apollo 17 geophones (Latham *et al.*, 1973). It is noteworthy that the S-IVB impacts occurred prior to the astronauts' landing and thus prior to the installation of that mission's seismometer, while the LM impacts occurred afterwards. The impact location of the Apollo 16 S-IVB was determined from seismic data a posteriori, as tracking contact was lost and the intended trajectory disturbed by a propulsion leakage (*Apollo 16 Mission*

Table 15.1 *Main lunar event classes and event counts according to the* Long-Period Event Catalog, *revision October 2008 (thermal moonquakes after Duennebier and Sutton, 1974).*

Type	Count	Strength	Reference
Artificial impacts	9	$3.1\ \text{GJ} \le E_K \le 47\ \text{GJ}$	Toksöz *et al.*, 1974
Meteoroid impacts	1744	E_K up to 820 GJ	Oberst, 1989
Shallow moonquakes	28	$4.7 \times 10^{12} \le M_0 \le 1.6 \times 10^{15}$ Nm $2.38 \le M_W \le 4.07$ $0.03\ \text{GJ} \le E_S \le 6900\ \text{GJ}$	Oberst, 1987
Deep moonquakes	7400	$M_0 < 5 \times 10^{13}$ Nm $M_W < 3.0$ $E_S < 0.001$ GJ	Goins *et al.*, 1981a
Thermal moonquakes	>1000	$E_S < 0.000001$ GJ	Duennebier and Sutton, 1974
Unclassified events	3322	?	
For comparison			
VW Golf (1300 kg) at 100 km/h (≈ 60 mph)		≈ 0.0005 GJ	
100 g chocolate		≈0.002 GJ	
Sumatra/Andaman earthquake December 26, 2004		$M_0 = 1.15 \times 10^{23}$ Nm $M_W = 9.3$ $E_S \approx 10^8$ GJ	Tsai *et al.*, 2005

Source strength is given in appropriate units: M_0, seismic moment; M_W, moment magnitude; E_S, seismic energy; E_K, impactor kinetic energy (1 GJ = 10^9 J).
References given are for source strengths.
For meteoroid impacts, the largest observed impact is reported here.

Report, 1972). Impact sites of the Apollo 13, 14, and 15 S-IVBs as well as a tentative site for the Apollo 14 LM impact were identified in photos taken with the Apollo 16 metric camera from orbit; craters with 41 and 39.5 m diameter were identified for the Apollo 13 and 14 S-IVBs, respectively (Whitaker, 1972). The Apollo 16 LM was also planned for impact, but after separation from the command module, attitude control was lost and it remained in lunar orbit with an expected orbit lifetime of one year (*Apollo 16 Mission Report*, 1972).

15.1.2 Meteoroid impacts

About 13% of all events listed in the event catalog, i.e., the second largest event group, are meteoroid impacts. Impactor size as well as the size of the produced crater, if any, can be estimated from the equivalent magnitude of the seismic event and empirical relations between magnitude and kinetic energy (see Chapter 18 of this volume). Impact velocity and angle, however, remain unknown.

Early estimations assuming an impact angle of 45° and a velocity of 16.9 km/s (average velocity of NEOs modified by lunar gravity; Oberst, 1989) result in masses for the largest impactors from 19 kg to about 5700 kg, creating craters of up to 50 m. A more recent evaluation yielded masses of up to 33 000 kg and resulting crater diameters of up to 73 m (Gudkova *et al.*, 2011).

Impactors with masses exceeding 10 kg can be detected globally by Apollo-class long-period seismometers (Dorman *et al.*, 1978), but because of emergent P and S arrivals, only a few impact locations could be determined (91 locations are given by Oberst, 1989). A set of 19 relocated impacts was used by Gagnepain-Beyneix *et al.* (2006).

Two main effects cause a deviation of the impact distribution from uniformity: first, the population of candidate impactors is not distributed isotropically, i.e., impactors tend to come in from the plane of the ecliptic (LeFeuvre and Wieczorek, 2008). Second, velocity addition for impacts on the leading hemisphere increases the efficiency of an impact, both in terms of crater production and seismic efficiency. Due to the synchronous rotation of the Moon, this efficiency increase always occurs on the same hemisphere (LeFeuvre and Wieczorek, 2008; Kawamura *et al.*, 2011). These two effects modify the cratering rate from 80% of the lunar average at the North Pole to 125% of the average at the center of the leading hemisphere. Gravitational focusing by the Earth results in a 0.1% modification (LeFeuvre and Wieczorek, 2011).

The distribution of observed impacts in time is also not uniform: About 20% of all impacts occur in only 3% of the time (Nakamura *et al.*, 1982). For small impacts with masses less than about 1 kg, this concentration increases to 31% of falls within 4% of the time. The largest falls predominantly occur from April to July (Dorman *et al.*, 1978), corresponding to the observation of meteor "showers" on Earth: the Aquarids, β Taurids, Perseids, Orionids, and Leonids were all identified by Oberst and Nakamura (1991). Others, such as the Quadrantids, Lyrids, and Ursids, were not detected in the Apollo data, which might be due to their radiants (the direction of their orbit relative to the Moon) and low impact angles, which reduce their seismic efficiency (Oberst and Nakamura, 1991). Additionally, some showers that cannot be observed from Earth because of radiants on the day-side have been found on the Moon. Some of the largest impacts observed on the Moon in June 1975 belong to one of these showers (Oberst and Nakamura, 1991).

15.1.3 Thermal moonquakes and surface erosion

The short-period vertical seismometers (SPZ) of Apollo 14 and 15 as well as the Apollo 17 geophone array recorded thousands of seismic events (the Apollo 12 SPZ malfunctioned right from the deployment; Vostreys, 1980) that were too weak to be detected by more than one station and too high in signal frequency (4–8 Hz, Criswell and Lindsay, 1974) to be recorded by the long-period sensors. These events were termed thermal moonquakes shortly after their discovery, since their characteristics fit well to stresses induced by insolation.

The activity typically starts about 48 h after sunrise and ceases after sunset (Duennebier, 1976). Seismograms recorded at a given station can be grouped into clusters according to waveform similarity. The Apollo 14 SPZ, for example, recorded 825 such events between deployment and early 1974 (Duennebier and Sutton, 1974), which group into 48 clusters (identified here by a number with prefix "TM" to distinguish them from deep Moonquake clusters). Since highly similar waveforms in a strongly scattering medium can occur only when neither the source time function nor the source position change, the clustering is strong evidence for a repeating source. The correlation between occurrence time and the time since sunrise allowed prediction of the activity of certain source clusters, which ultimately led to the discovery of previously undetected events. The events of Apollo 14 cluster TM330, for example, occur between 64 h and 67 h after sunrise, while cluster TM340 is active between 443 h and 456 h after sunrise (Duennebier and Sutton, 1974). Of five selected clusters, the activity of which was predicted for a time window in 1976, three were found to be active with essentially unchanged waveforms (Duennebier, 1976). Some clusters, e.g., Apollo 14 cluster TM300, show a slight waveform modification from one event to the next, and the pattern of this change repeats every lunation. This is interpreted as a movement of the source during the lunation. It has never been observed that the direction of motion of a source reverted (Duennebier and Sutton, 1974).

Although the Apollo LM produces its own thermal crackling, it can be excluded as a source: since signal rise time in a highly scattering medium is a function of range, a rise time longer than that of events clearly ascribed to known LM activity (e.g., astronaut activity) points to sources at greater distances and thus of a natural origin (Duennebier and Sutton, 1974). Source location was possible only at the Apollo 17 site, where the geophone array of the Lunar Seismic Profiling Experiment (LSPE, e.g., Cooper *et al.*, 1974) was used not only to record artificial sources but also to listen to the lunar background. Classical location schemes using arrival times failed due to the emergent nature of all signals, and although events from a given cluster and recorded at the same station are highly similar, there is no waveform similarity between recordings of a given event at different geophones of the LSPE: Seismograms are shaped by scattering, not by the source time function. The explosives used for the LSPE provided a means of calibrating the decay of the signal spectrum with distance, and thus the source location of about 200 events within about 400 m of the array could be determined with an accuracy of about 50 m (Duennebier, 1976). The largest craters, notably Rudolph crater with 50 m diameter, and Geophone Rock, a boulder of approximately 2 m height at the site of the central station of the array, are pronounced centers of activity (Duennebier, 1976), while areas of previous astronaut activity do not show ongoing seismic activity, thus excluding man-made mechanisms as a cause for the majority of the signals. Also, the events located at Geophone Rock are too small to be detected over distances of more than 100 m – rocks of this size are thus not the source of the large events originating hundreds of meters away.

Comparison with seismograms from the LSPE explosives also yielded estimations of the source size: Fault areas of a few square meters and slip of a few centimeters, corresponding to Earth magnitudes of -2 to -2.6 resulted. Magnitude–frequency statistics (b-values) showed that the main energy release is due to the *small* events (Cooper and Kovach, 1975), contrary to the energy distribution of tectonic earthquakes. Duennebier (1976) estimates that the annual energy release of thermal moonquakes is about 10^8 J and thus comparable to a single 500 g meteoroid impacting at 20 km/s. It is also possible to distinguish two populations of thermal moonquakes with different *b*-values and different rise times that point to the existence of two different source mechanisms.

Since the diurnal thermal wave penetrates the lunar regolith no further than about 80 cm (Langseth *et al.*, 1976), thermal stresses thus being confined to the uppermost part of the soil, any source mechanism proposed for the thermal moonquakes must be triggered in the shallow subsurface, or at the very surface.

It has been suggested (Criswell and Lindsay, 1974) that one mechanism behind the thermal moonquakes is the same as in booming dunes: from terrestrial deserts it is known that dunes sometimes produce sounds with audible frequencies, and a harmonic spectrum sometimes compared to a cello or violin. These sounds even occur in areas where no motion of the sand is visible and may last for several minutes (see review by Hunt and Vriend, 2010). The origin of these sounds is the subject of an ongoing debate. A narrow grain size distribution appears to be mandatory, but also the existence of a low-velocity waveguide near the surface of the dune which, by resonance, produces the harmonic spectrum. Both these prerequisites, however, have not been demonstrated for thermal moonquakes.

Nevertheless, slumping of regolith on tilted surfaces is considered as the main mechanism. Estimations of erosion rates differ throughout the literature (Duennebier and Sutton, 1974; Cooper and Kovach, 1975; Duennebier, 1976), but this mechanism clearly operates on the scale of millions of years.

15.1.4 Shallow moonquakes

One rare type of teleseismic event has an amplitude on the Apollo SPZ seismometer channel more than four times higher than any other event type (Nakamura *et al.*, 1974b), which is why this event class was initially termed high-frequency teleseism (or HFT for short). Further characteristics are P and S arrivals more impulsive than those of artificial impacts and many other events (Nakamura *et al.*, 1974b), which indicates that the HFT sources are situated in areas with less scattering than observed for surface foci. The *b*-values of their Gutenberg–Richter distributions differ from those of known impacts and deep moonquakes, separating the HFT from both these classes. Furthermore, there are no intermediate cases that might indicate a relationship to other event types (Nakamura *et al.*, 1974a). During the operation of the Apollo network, only 28 HFT events were observed (Nakamura *et al.*, 1979; Nakamura *et al.*, 1981).

As with all other event classes, the seismograms make source location challenging. There appears to be only one publication that gives locations for all 28 sources in a consistent way (Nakamura *et al.*, 1979), but subsets have been located by different authors, e.g., by Lammlein (1977, 17 hypocenters) and Gagnepain-Beyneix *et al.* (2006, 8 hypocenters).

A source depth significantly above that of the deep moonquakes was inferred from the observation of S-waves from one HFT that occurred at an epicentral distance of about 120° – ray geometry suggests that S-waves for sources deeper than 300 km would cross the zone of increased attenuation in the deeper interior, and thus the conclusion that this HFT was shallower than that depth (Nakamura *et al.*, 1974a). The published hypocentral depths (see Figure 15.3) comply with this (Lammlein, 1977; Gagnepain-Beyneix *et al.*, 2006).

The HFTs are the strongest natural events on the Moon besides impacts. Not only are their magnitudes (moment magnitude up to 4.1; Oberst, 1987) higher than those of the deep moonquakes, the associated stress drop is also sometimes exceptional. While the stress drop of earthquakes is usually about 10 MPa or lower (e.g., Kanamori, 1994, and references therein), some of the HFTs reach values far above 100 MPa (Goins *et al.*, 1981a; Oberst, 1987). These stress drops suggest that HFTs may be similar to intraplate quakes on Earth, which usually show higher stress drops than interplate quakes. Oberst (1987) also concludes that the high stress drop indicates a high stress within the source area of the HFT, which complies with an initially completely molten state of the Moon.

With only a small number of events at hand, any kind of statistical evaluation is challenging. A connection between the HFTs and the lunar tides cannot be established by means of a Fourier transform of the event sequence. Other analysis methods have been applied by Shirley (1985) and by Frohlich and Nakamura (2006). Both found that a certain correlation between HFT occurrence times and the orientation of the Moon in terms of the geocentric ecliptical longitude exists. According to Frohlich and Nakamura, the HFTs predominantly occur at epicenters that roughly point to the direction of the Virgo and Leo constellations. While Shirley argues that this might be a hint of a gravitational effect other than tidal, Frohlich and Nakamura prefer the hypothesis of impacts of extrasolar particles on the Moon.

These two hypotheses are among the more exotic ones for the origin of the shallow moonquakes. Others suggest the impact of meteorites on extraordinarily competent materials or unusually deeply penetrating meteorites (Nakamura *et al.*, 1974a), or tectonics near impact basin rims (Nakamura *et al.*, 1979). None of these hypotheses is entirely compatible with all the properties of the seismogram observations, and given the paucity of events and the uncertainty of their hypocenters, it must be admitted that their origin is not understood.

15.1.5 Deep moonquakes

Deep moonquakes are the most numerous event class observed on the Moon. They were found to originate in distinct, mostly near-side clusters (Figure 15.2), at depths between approximately 700 and 1200 km (Figures 15.3 and 15.10). Currently, 319 clusters are known, of which roughly a third have been located. Cluster location suffers from the effects of scattering in the lunar regolith, which produces a long seismogram coda that obscures all but the direct P and S arrivals. Because each cluster produced its own characteristic waveform, most studies took advantage of waveform similarity by stacking seismograms from a given source to improve the visibility of arrivals. In general, the clusters with the

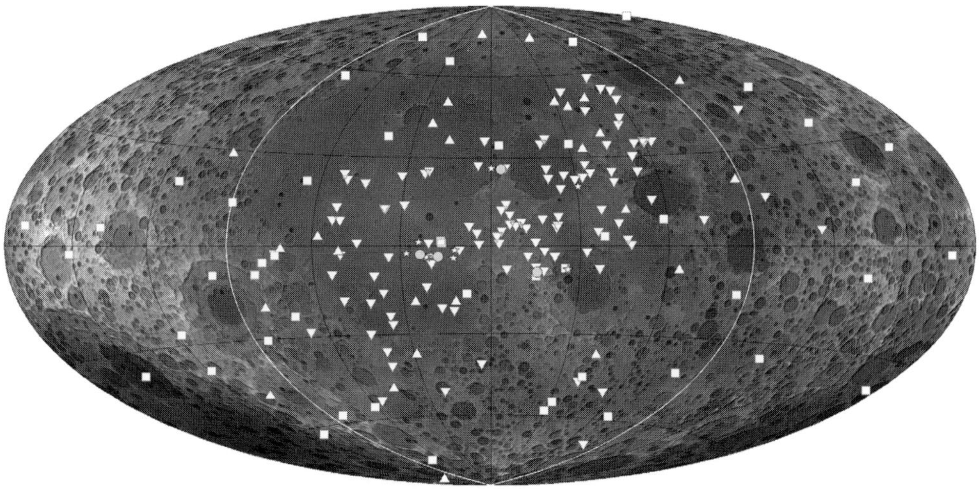

Figure 15.2 Epicenters of deep moonquakes (down-pointing triangles, Nakamura, 2005), shallow moonquakes (up-pointing triangles, Nakamura *et al.*, 1979), the 48 largest meteoroid impacts (squares, Oberst, 1989), and artificial impacts (pentagrams, Toksöz *et al.*, 1974). Circles denote the Apollo seismometer sites. The white meridians separate the near and far sides. Map in equal-area Hammer projection, background topography as in Figure 15.6. A black and white version of this figure will appear in some formats. For the color version please refer to the plate section.

largest numbers of events are the best located, while the more poorly populated clusters tend to have larger uncertainties in arrival time readings, which translates to larger location uncertainty (in some cases over 10 degrees in latitude and longitude).

The distribution of deep moonquake epicenters does not appear random; the well-located clusters have been found to correlate with the nearside distribution of mare basalts (Qin *et al.*, 2012). Only a few clusters have been proven to reside on the lunar farside. Given that the Apollo network was located exclusively on the near side, this argues strongly for a region of high attenuation in the deep interior that shadows seismic energy of farside origin, such as would be provided by a liquid core and/or a region of partial melt (Nakamura, 2005). Alternatively, if the active deep moonquake zone coincides with the mare basalt source region, this may argue for an aseismic farside, since mare basalts are largely constrained to the near side (Qin *et al.*, 2012).

Deep moonquakes are very weak, which translates into very small source radii. Most clusters are likely constrained to active regions on the kilometer scale (Nakamura, 1978). Attempts to construct fault plane solutions exist but suffer from limited coverage of the focal sphere and seismogram complexity (Toksöz *et al.*, 1977; Lammlein, 1977; Nakamura, 1978; Koyama and Nakamura, 1980; Araki, 2001; Weber *et al.*, 2009; and Chapter 10 of this volume).

Interestingly, deep moonquakes from a given cluster occur with monthly, 7-month, and 6-year periodicity, as dictated by the lunar orbit. This periodicity strongly suggests that deep moonquakes are triggered by variations in tidal stress, and this relationship has been

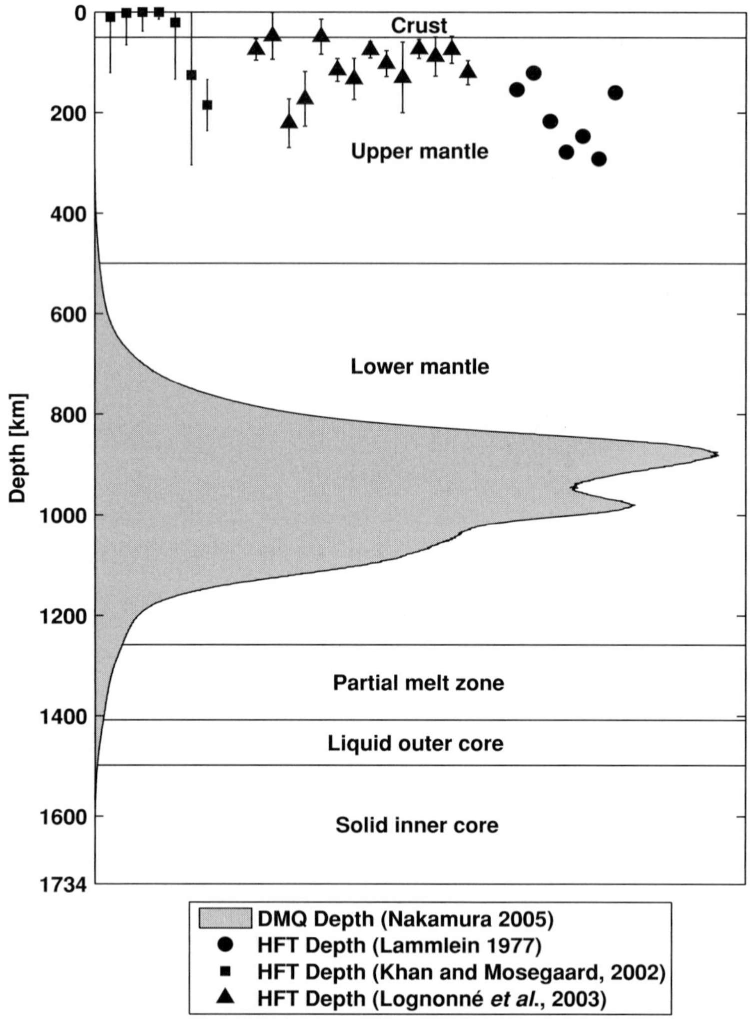

Figure 15.3 Depths of deep moonquake clusters and selected shallow moonquakes. The DMQ distribution was computed by summation of normal distributions based on mean depths and variances given by Nakamura (2005), weighted by the number of events attributed to each cluster in the 2008 revision of the *Long-Period Event Catalog*. Some of the HFT events were located by several authors, thus some events may be represented by more than one symbol here. The layering follows the discussion in the following section.

explored at length (see Chapter 10, and references therein). Using a special form of Fourier transform suitable for sequences of delta functions, it has been demonstrated that cluster activity correlates with tidal forcing at both the anomalistic and draconic periods (Bulow *et al.*, 2007). Some clusters prefer the anomalistic period, some the draconic period, some are active with both periods (Figure 15.4).

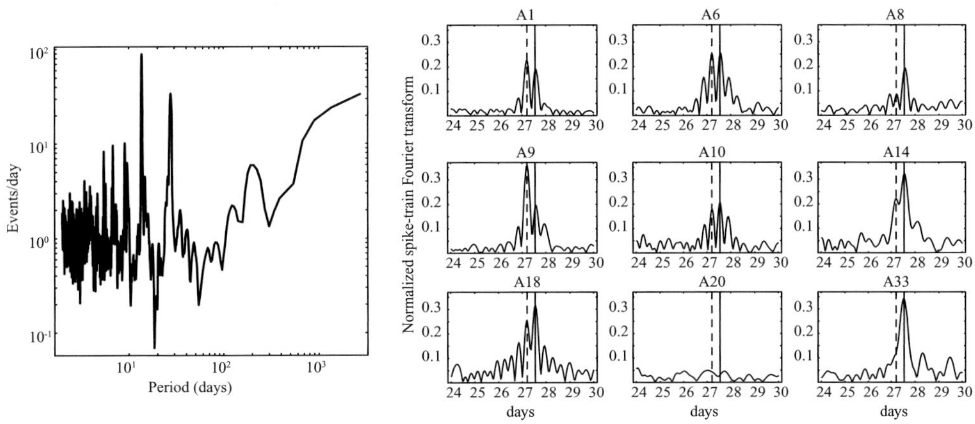

Figure 15.4 (Left) Regular spectrum of all deep moonquake occurrence times. (After Figure 2 in Bulow *et al.*, 2007.) (Right) Spike-train spectra of selected clusters. (After Figure 4 in Bulow *et al.*, 2007.)

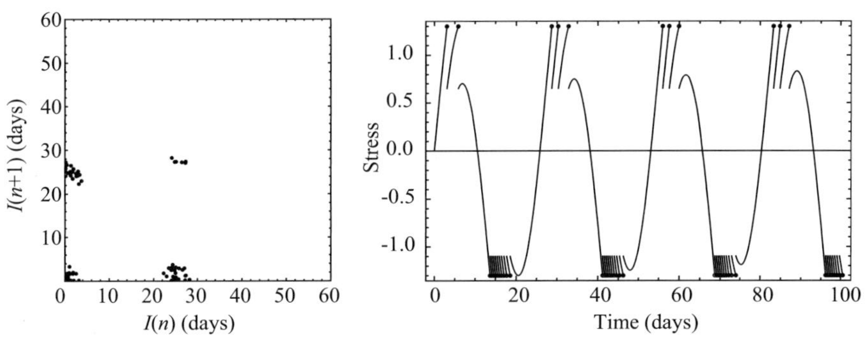

Figure 15.5 (Left) Delay–coordinate plot of deep moonquake times. (Right) Stress buildup and release model that explains observed temporal occurrence behavior. (After Figure 11 in Weber *et al.*, 2010b.)

Weber *et al.* (2010b) explored the temporal behavior of deep moonquake occurrence in more detail, through a delay-coordinate approach. When the interval between two events $I(n)$ is plotted against the successive interval $I(n+1)$, a pattern emerges (Figure 15.5, left) which can be statistically recreated using a periodic sequence of synthetic events of which only a certain percentage are detected. By using a driving stress that closely approximates the tidal stress observed within the Moon, and by assuming that moonquake occurrence relieves some of that stress, a physical model for failure can be achieved (Figure 15.5, right).

While many of the unique characteristics of deep moonquakes have been explored, we still have no definitive answer as to why they occur. Frohlich and Nakamura (2009) note that deep moonquakes share many similarities with intermediate-depth earthquakes, and suggest additional physical mechanisms. These include dehydration embrittlement, in which the motion of fluid phases or partial melts allows cracks to form and grow, cyclic stress and fatigue, and the presence of pre-existing zones of weakness. Clearly, more data are needed to determine which of these models is most likely. Future seismic missions to the Moon should focus on making significant improvements over Apollo, ideally by increasing the number and distribution of seismic stations.

15.2 Interior structure

The interior structure of the Moon is usually subdivided into crust, mantle, and core. Radio science measurements result in gravity maps and moment of inertia factors. Hence we know that crustal thickness on the near and far sides differs significantly, and that only a small (iron) core can exist. The following sections will focus on some selected topics of special interest to planetary seismology.

15.2.1 Crust

Regolith and scattering

The first experiments to determine the elastic properties of the lunar surface were carried out by the Surveyor landers: The deformation of the landing gear during touchdown, as recorded by strain gauges on the legs of the lander, allowed deduction of shear wave velocities of only 16 to 28 m/s and P-wave velocities of 30 to 70 m/s, with a bulk density of the soil between 1000 kg/m^3 and 1500 kg/m^3 at the very surface (Christensen *et al.*, 1968).

The Apollo missions were the first to return seismograms. In addition to the passive seismic experiments, active seismic experiments targeted at shallow crustal structure were conducted by the astronauts of the Apollo 14, 16, and 17 missions. The landing sites were Fra Mauro, Descartes, and Taurus-Littrow, respectively. As seen in the long-period data, waveforms from the active experiments show emergent arrivals, with a slowly decaying scattering coda with no obvious phases except P and S. Two mechanisms were debated to explain this behavior; they can be summarized as *surface scattering* and *volume scattering*.

Gold and Soter (1970), on the basis of the Apollo 12 LM impact, as well as Gangi and Yen (1979), on the basis of Apollo 14 and 16 active seismic data, suggested that the lunar surface is the only scatterer: A layer of self-compacting powder would show a smooth sixth-root increase of velocity with depth. Full compaction would be achieved in a depth of 6 to 9 km. This velocity profile would bend all body-wave rays back to the surface. A random surface topography would then reflect the rays back and forth and thus generate a coda. Gold and Soter suggest that this powder layer must have a thickness of a few kilometers to trap a sufficient part of the seismic energy. The surface scattering model is also supported by a study of Apollo short-period seismograph (SPZ) recordings of Lunar Rover activities at the Apollo 15 and 16 sites (Nakamura, 1976), which found that the size–frequency distribution of scatterers, as derived from the development of seismogram coda with time, strongly resembles the size–frequency distribution of impact craters in the surroundings of the landing site. This leads to the conclusion that the scattering in higher frequencies is due to both the structural and the topographic modification of the crust by impacts.

On the other hand, Dainty *et al.* (1974) conducted ultrasound experiments in 2D aluminum media with artificial scatterers distributed over the entire volume and hold that only volume scattering explains all the properties of the Apollo 12 LM impact seismograms. The scattering layer would reach a depth of at least 1 km, but a thickness of 20 or 25 km better fits the observations from the Apollo 12 LM impact, i.e., volume scattering requires thicker layers than surface scattering. The average distance between scatterers is inferred to be 2 km at 14 km depth and 5 km at 25 km depth. Volume scattering complies with the refraction-seismics interpretations of Kovach and Watkins (1973a,b), Watkins and Kovach (1973), and Cooper *et al.* (1974), who favor a layer of brecciated material or impact debris ($V_P \approx 200$–300 m/s) underneath a few meters of regolith ($V_P \approx 100$ m/s). Layers with typical solid-rock velocities were not found at any Apollo site (it must be noted here that the $V_P = 4700$ m/s layer described by Cooper *et al.* (1974) at the Apollo 17 site is affected by a clock error in the data recording; see Nakamura, 2011). Comparison with velocity measurements on Apollo rock samples and velocity models for deeper parts of the crust also shows that self-compaction of powder is not compatible with observed velocity changes with depth (Kovach and Watkins, 1973a).

The refraction analyses attribute the scattering to fractures within the breccia, and tend to produce constant-velocity layers separated by first-order discontinuities. Surface wave studies based on spectral (Horvath *et al.*, 1980) and correlation techniques (Larose *et al.*, 2005; Tanimoto *et al.*, 2008; Sens-Schönfelder and Larose, 2010) permit the assessment of the uppermost structure without the need to identify arrival times. They confirm the very low S-wave velocities of the regolith (the mentioned studies all find 40 to 100 m/s for the top 10 to 15 m), but surface waves cannot distinguish the smooth velocity profile necessary for surface scattering from the layered structure favored by the refraction analysis.

Despite the uncertainty about the cause of scattering, the result holds that the properties of the regolith layer vary little between Apollo landing sites. Velocity structures and waveform envelopes at all sites are remarkably similar and appear to be controlled by the regolith production process rather than by the host rock.

Crustal thickness

The thickness of the lunar crust is a matter of interest for the interpretation of gravity data, which, in contrast to seismic data, can be obtained from orbit both with very high spatial resolution and on a global scale. The localized mass concentrations (mascons for short) that were discovered after analysis of tracking data from Lunar Orbiter missions (Muller and Sjogren, 1968) are widely accepted to signify mantle uplift and thus crustal thinning underneath impact basins. Data from the Apollo 15 Laser Altimeter led to the determination

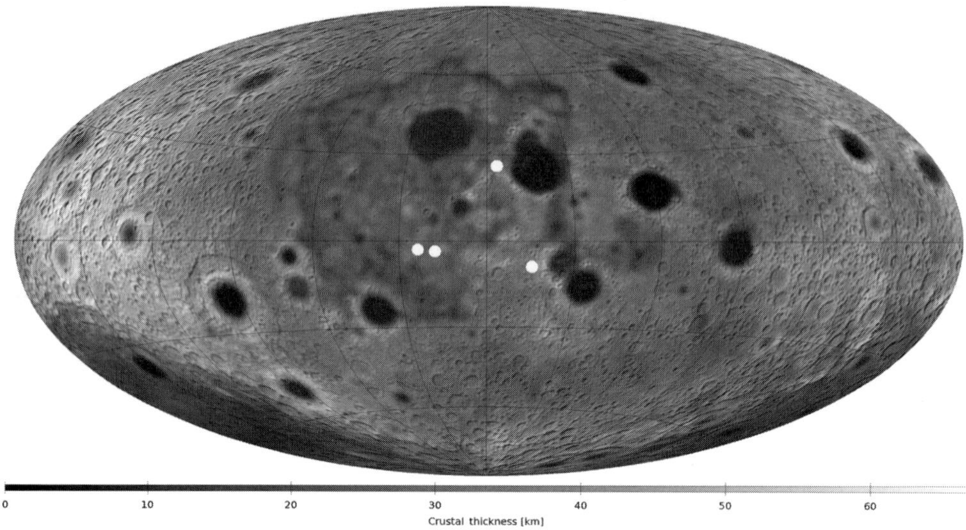

Figure 15.6 Gravity-based map of the lunar crustal thickness, based on the GRAIL gravity field (Wieczorek *et al.*, 2013). The model is adjusted such that crustal thickness underneath the Apollo 12 site is 30 km. White dots represent the Apollo seismometer sites. Gray scale represents crustal thickness in kilometers. Both crustal thickness and underlying LOLA topography model provided by GRAIL Crustal Thickness Archive at http://www.ipgp.fr/~wieczor/GRAILCrustalThicknessArchive/GRAILCrustalThicknessArchive.html

of the offset between the lunar center of mass and its center of figure (Roberson and Kaula, 1972): The latter is shifted away from Earth (a recent value for the shift is 1.9 km; Smith *et al.*, 2010), which means that a nearside–farside asymmetry in the internal density distribution of the Moon must exist.

The asymmetry can, to first order, be attributed to variations of the crustal thickness (Lingenfelter and Schubert, 1973), and this is where seismology enters the stage: For a determination of the crustal thickness on the farside, knowledge of its thickness on the near side is necessary as the two combined must result in the proper mass imbalance with respect to the center of figure.

The artificial impacts constitute a set of controlled sources, as their impact times, locations, and kinetic energies are well known from independent observations. Distances of a few tens to about 1100 km between impacts and seismometers were well suited to constrain velocities within the crust. The first structural elements could be deduced from the impact data after completion of the Apollo 15 mission (Latham *et al.*, 1972). With the additional data available after Apollo 16, Toksöz *et al.* (1972a) constructed a detailed model for the Oceanus Procellarum area (Apollo 12 and 14 sites) from P and S first arrivals and a few crustal multiples. The most prominent features were a velocity discontinuity at about 25 km depth, an almost constant velocity between 25 and 65 km depth, and a quick increase to about 9 km/s below 65 km depth (Figure 15.8, right). The latter velocity increase was interpreted as the bottom of the lunar crust, i.e., the lunar Moho. This model did not only explain the observed travel times, it was also backed up by laboratory measurements with solid rocks and rock fines from the Apollo sample collections, which indicated that seismic velocity in the upper crust (down to the 25 km discontinuity) should gradually increase due to crack closure under pressure, and remain constant in the lower crust if it consists of unfractured gabbroic, noritic, or anorthositic rock. A slightly smaller, but similar, crustal thickness of 58 ± 8 km for the same area was found by Nakamura (1983). Under the lunar highlands, as represented by the Apollo 16 site, Goins *et al.* (1981b) derived a Moho depth of 75 km from the conversion of P-waves to S-waves at the Moho, and multiples of these conversions. The average Moho underneath the entire Apollo network is at 60 km, as Nakamura *et al.* 1976 estimated from meteoroid impacts as well as shallow and deep moonquakes.

A first suggestion of a crustal thickness of only 45 km was given by Koyama and Nakamura (1979), but with little impact on the scientific discussion. But thin crusts became more popular with the application of massive numerical signal processing and inversion techniques in the more recent literature. Khan *et al.* (2000) and Khan and Mosegaard (2002) found, by means of a Markov Chain Monte Carlo (MCMC) inversion that returned 50 000 velocity models compatible with observed travel times, that structures with crustal thickness between 35 and 45 km are more likely to produce the observed data than structures with thickness between 50 and 70 km. This is usually interpreted in favor of a shallow Moho, although models with a thick crust – and especially those published in the 1970s and 1980s – reproduce the data equally well. From the data misfit point of view, the existing seismic data are insufficient to distinguish the two model classes. Nevertheless, as the data used by Khan *et al.* (2000) and Khan and Mosegaard (2002) were the same as used by Nakamura (1983), their model ensemble can be viewed as an account of the structural non-uniqueness inherent to this data set.

During the 1990s, the receiver function method, which isolates conversions between shear and compressional waves that occur at velocity discontinuities, became a standard tool to estimate Moho depth and the crustal Poisson ratio on Earth. The S-wave receiver function method, which has the advantage that the Sp phase arrives prior to S and its coda, has been applied by Vinnik *et al.* (2001) and Lognonné *et al.* (2003). As end-member interpretations, the arrival time of the Sp conversion can either be attributed to seismic velocities and density of the uppermost 1 km that are significantly higher (a factor 4 in the case of V_P, following Vinnik *et al.*), or by reducing the crustal thickness to 30 km underneath Apollo 12 (i.e., a factor ≈ 2 compared to Toksöz *et al.*, 1972a), and 40 km in the global average (following Lognonné *et al.*). A thickness of 30 km underneath the Apollo network is also favored by Gagnepain-Beyneix *et al.* (2006) as a result of a different parameterization and MCMC inversion. In an innovative attempt to get information from beyond the network radius, Chenet *et al.* (2006) evaluate travel time residuals to obtain estimates for the crustal thickness at the sites of both artificial and natural impacts (Figure 15.7). The approach maps all residuals to variations of crustal thickness and is thus likely to overestimate thickness variability, but nevertheless reduces the variance of the a-priori model and suggests shallow

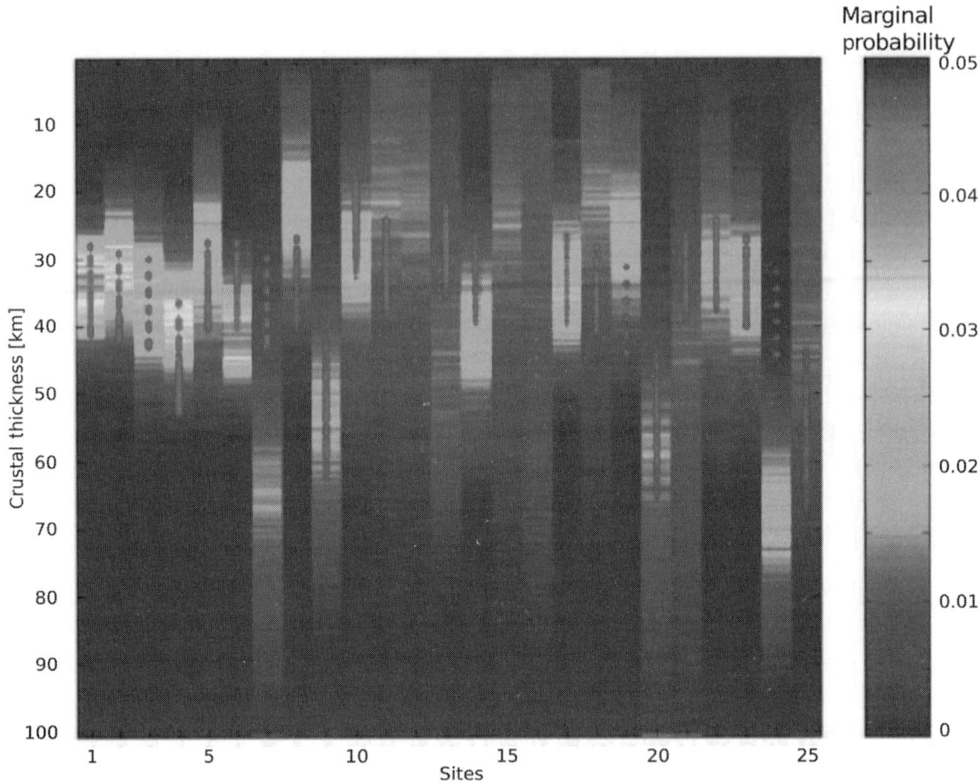

Figure 15.7 Posterior marginal probabilities for the sites investigated by Chenet *et al.* (2006). Sites 1–4 are Apollo sites 12, 14, 15, and 16, Sites 5 and 6 are the impacts of the Apollo 15 and 17 S-IVB stages, respectively. Magenta markers depict possible crustal thicknesses from gravity modeling for each site. A black and white version of this figure will appear in some formats. For the color version please refer to the plate section.

Mohos for most of the sites (38 ± 13 km averaged over 25 impact sites and the seismic stations). For each individual impact it was also demonstrated that the crustal thickness result is compatible with the Lunar Prospector gravity field variations.

Computer power alone is not a sufficient argument for the credibility of a model. Both the "old school" and the "modern times" approaches have their problems and merits. The main limitation of all interpretations is the data set itself. It has been noted several times in the literature (e.g., Nakamura, 2005) that arrival-time readings on lunar seismograms are very uncertain, with a median of several seconds, but up to 100 s in single cases. Moreover, correlation between arrival times at Apollo 12 and 14 stations is larger than between other station pairs because of the small distance between these stations, but they are usually treated as independent. Therefore, variations in the near-station properties of the crust will bias any structural inversion.

Some of the early studies of the mantle use the crustal model of Toksöz *et al.* (1972a) and its descendants, but vary the subcrustal mantle velocity. The sub-Moho velocity of the

Toksöz model, however, was chosen mainly to obtain a sufficient contrast to the lower crust velocity, which in turn trades off with crustal thickness. Thus a modification of the mantle requires a modification of the crust to follow the tradeoff curves.

These early studies are based on a close reading of the (then few) seismograms, comparison with lab data of the returned rocks, and intricate physical and geological reasoning, while recent studies are more focused on waveform processing.

The recent MCMC methods for the first time have the ability to search the parameter spaces much more exhaustively and are less prone to premature convergence, and can thus investigate the uncertainty about the inversion result that is inherent in the data. Sampling a multidimensional parameterization usually returns thousands or even millions of models that all reproduce the data within their error bars. But the interpretation of such a model ensemble has its own intricacies (see Khan and Mosegaard, 2002 for discussion). One pitfall of being guaranteed to find the globally best fitting model is overfitting, i.e., explaining the noise in the data by geological structure.

Travel times can be considered as weighted averages over propagation velocities, and as such they do not distinguish between gradients and discontinuous changes of velocity. Several types of observations may serve as the smoking gun for the detection of a discontinuity: the identification of secondary arrivals, head waves, travel time triplications, and converted waves. The early large crustal thickness relied to a large extent on the identification of secondary P-phases, while the modern shallower depths are tied closely to the S-receiver functions. The latter, as a waveform processing method, leaves less room for subjective interpretation than arrival-time picking. Since low propagation velocities of regolith are backed up by the Apollo active seismic experiments, laboratory measurements of returned rocks, and the interferometric studies cited in the previous section, the shallow-Moho interpretation of the receiver functions appears more convincing than the fast-regolith interpretation.

15.2.2 Mantle

Mantle discontinuities

The lunar mantle is often described as being subdivided into an upper and a lower portion, sometimes a mid-mantle is also mentioned. These structural levels, however, must not be confused with the Earth's layers of the same name: Pressure at the base of the lunar mantle is only about 4.5–5 GPa (e.g., Sohl *et al.*, 2009), which corresponds to a depth of about 150 km within the Earth. These pressures are insufficient to transform olivine, which is thought to be a major constituent of the lunar mantle, into the high-pressure mineral phases that make up the Earth's lower mantle.

The structural and geochemical model of the lunar mantle has undergone some revision in the recent literature, but to a lesser degree than the lunar crust. A feature of special interest is a seismic discontinuity about in the middle of the mantle (Figure 15.8, left), the existence of which has been predicted from models for the thermal history of the Moon (Toksöz *et al.*, 1972b): Anorthosite found in the lunar highland samples in large amounts is derived from a basaltic melt, and the deposition of an anorthositic crust of about 10 km

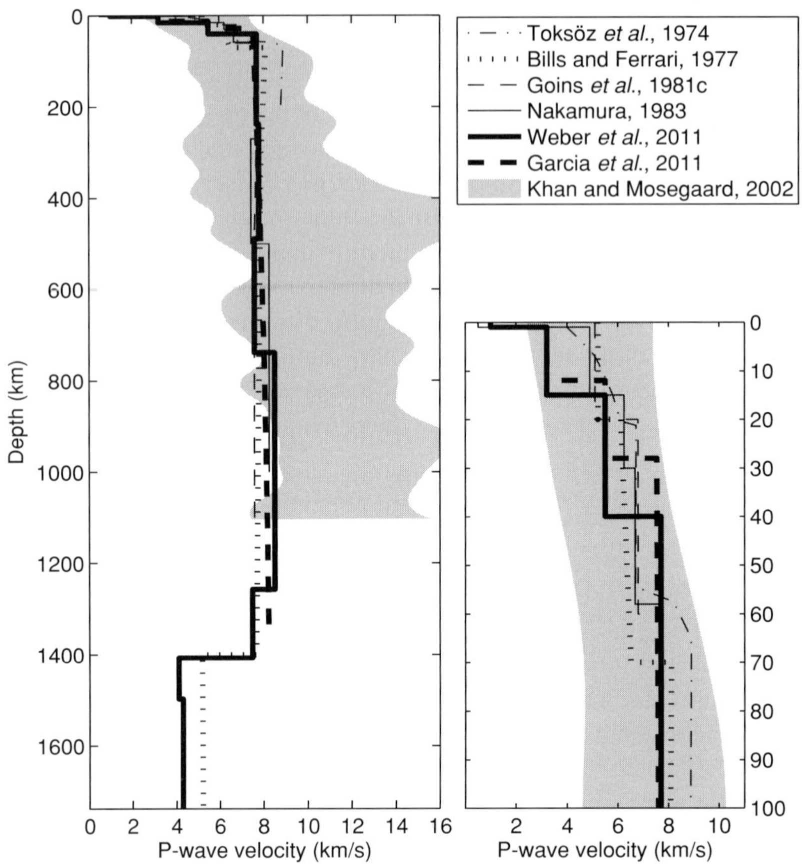

Figure 15.8 Comparison of P-wave velocity models. (Left) entire Moon; (Right) upper 100 km. The gray area in both diagrams is the 99% credibility interval of the Khan and Mosegaard (2002) model ensemble. The crust and mantle of the Weber *et al.* (2011) model is adopted from Lognonné *et al.* (2003).

requires the outer about 500 km of the Moon to be molten ("magma ocean"). The differentiation history of the interior thus suggests the existence of a seismic discontinuity at this depth.

Seismogram interpretation and velocity modeling resulted in several different depth values for a mid-mantle discontinuity, from 300 km (Nakamura *et al.*, 1974a, 1976), via 480 km (Goins *et al.*, 1981c), and 500 km (Goins *et al.*, 1977; Nakamura *et al.*, 1982) to 560 ± 15 km (Khan *et al.*, 2000) and, finally, 520–580 km (Khan and Mosegaard, 2001). The most recent models confirm the greater depths given in the early literature: Most models in the Khan and Mosegaard (2002) ensemble exhibit a sharp increase of both V_P and V_S at depths between 500 and 580 km. Gagnepain-Beyneix *et al.* (2006) arrive at a depth of 488 km. The VPREMOON model by Garcia *et al.* (2011, 2012) does not show any mantle discontinuity, but was aimed at a precise determination of the core radius.

The approach of Khan and Mosegaard (2001) is independent of that of the other studies in that it analyzes the spectrum of spheroidal free oscillations as resulting from vertical component seismograms. Both the resulting S-wave velocities and densities point to a discontinuity depth clearly below 500 km, and possibly a decrease of V_S below 1000 km depth, i.e., in a region that is almost inaccessible to body-wave studies due to the limitations of the source distribution.

Shear wave attenuation

The pronounced coda in lunar seismograms is an indication not only of the presence of scatterers but also of very low anelastic attenuation (i.e., high Q) in the outer portions of the Moon. The amplitude of P and S arrivals from large distances shows that the deepest parts of the lunar mantle, in contrast, are much more attenuating. The investigation of the deepest interior suffers from a lack of sources antipodal to the Apollo network, but a few impacts and deep moonquake cluster A33 allow delineation of the boundary of a low-Q region (Nakamura *et al.*, 1973; Goins *et al.*, 1977, 1981c). While P waves from the A33 source region are recorded at all four Apollo stations, S waves are observed only at the Apollo 15 and 16 sites, but not by the Apollo 12 and 14 seismometers. Similar non-detections were made for some meteoroid impacts. Ray tracing of the respective phases then shows that a highly attenuating zone exists below about 1100 km depth (Goins *et al.*, 1981c).

The increased attenuation has been attributed to partial melting of the lowermost lunar mantle in the mentioned studies. In that case, not only the Q factor, but also the seismic velocities decrease and the resulting velocity contrast to the overlying mantle may cause reflected phases. Such reflections have indeed been found by Weber *et al.* (2011), indicating a boundary depth of about 1250 km. Moreover, the decreasing V_S found by Khan and Mosegaard (2001) can also be attributed to the presence of partial melt below 1150 km depth.

15.2.3 Core

The size and existence of a small core has been a subject for debate since the Apollo missions. The lunar moment of inertia allows for a small iron core. Geodetic observations from both laser ranging and radio science suggest the existence of a liquid layer at great depth. Induced magnetism while passing through the Earth's magnetotail shows the existence of an electric conductor at similar depth (see Wieczorek *et al.*, 2006, and references therein).

Seismic evidence for a lunar core was published in 2011 (Weber *et al.* and Garcia *et al.*). Massive data processing and application of terrestrial array methods by Weber *et al.*, and careful source selection by Garcia *et al.* resulted in two independent observations of seismic

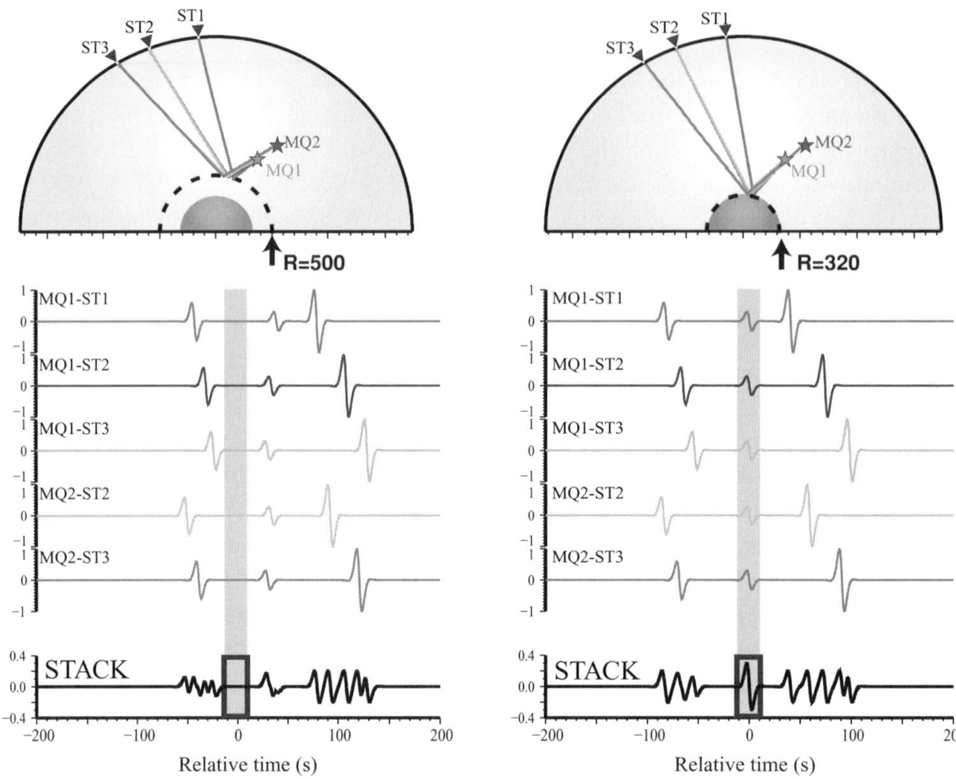

Figure 15.9 The array stacking technique illustrated for two hypothetical deep moonquakes (stars) recorded at three hypothetical seismic stations (triangles). Seismograms are time shifted in windows centered on predicted core reflection times. If energy is present from a layer boundary at a given depth, a coherent arrival will be observed in the stack (right panel). Stacking for an incorrect radius results in an incoherent trace (left panel).

phases that were reflected from the core–mantle boundary, and, in the Weber *et al.* study, also of phases reflected from an inner core boundary (Figure 15.9).

Small-amplitude core-interacting phases arrive after the primary P-wave, and hence are obscured by the P coda. Array processing methods enhance such subtle seismic arrivals by stacking seismograms that have been time shifted to predicted core arrival times. The technique as employed by Weber *et al.* (2011) is illustrated in Figure 15.9. Their iterative approach sought the best-fit radii and overlying seismic wave speeds of a multi-layered core consisting of a solid inner part, a liquid outer part, and a partial melt boundary layer. It is of special interest to note that from the lowermost discontinuity, identified with the boundary of the inner core, no SKiKS phases in SH polarization are observed: The existence of $SKiKS_{SV}$ phases (among others) demonstrates the existence of this discontinuity, while the non-existence of $SKiKS_{SH}$ shows that the medium above must be liquid, as SH energy does not convert into K-legs and is reflected as ScS completely.

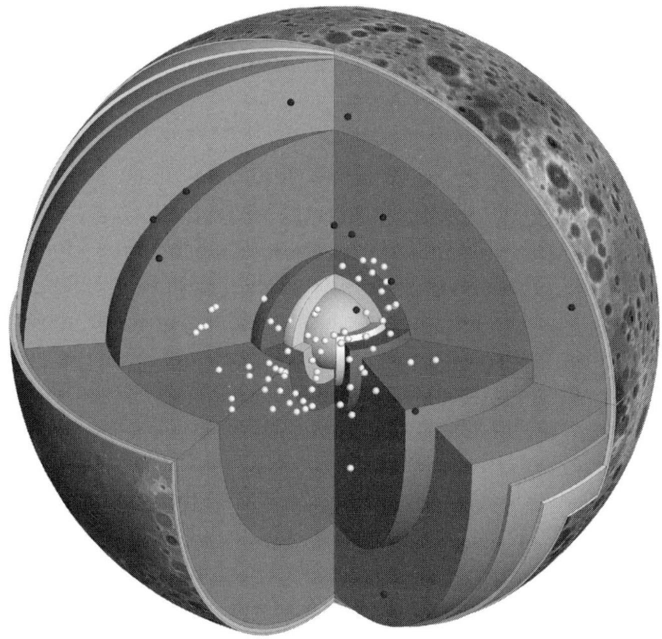

Figure 15.10 Schematic view of the lunar interior. The crust is subdivided into a regolith layer (not to scale), and an upper and a lower portion. The mantle consists of three layers: upper and lower mantle, and a low-Q zone at its base. The solid inner core is drawn as a full sphere. Dark and white dots represent shallow and deep moonquakes, respectively. Surface is from the KAGUYA digital terrain model, rendered at 4 degrees/pixel (Araki *et al.*, 2009). Layer boundaries as in Figure 15.3.

15.2.4 Structural summary

The seismic data available from the Moon so far are incomplete and challenging to interpret, but nevertheless give rise to the picture of a planetary body that is differentiated into three main layers, each of which consists of several sublayers (Figure 15.10). The radial extension of these layers is sometimes tentative and subject to further investigation, but the general structure is robust.

Seismologically, the lunar crust can clearly be subdivided into a thin surficial layer of strongly scattering material and a lower part where pressure smoothes out the scatterers. Its density differs between near and far sides, but 50 km appears to be a reasonable global average. Petrologically, there is evidence that the crust is subdivided into an anorthositic upper and a noritic lower crust.

Underneath the crust, follows a mantle that might be subdivided into an upper and a lower part at a discontinuity some 500 km below the surface. The global extent and significance of this discontinuity has yet to be demonstrated. The attenuation of shear waves suggests that the lowermost portion of the mantle is partially molten. Seismic velocities throughout the mantle still carry significant uncertainties.

The Moon possesses a small liquid core and a solid inner core. A precise determination of their respective radii requires a reliable mantle model. The liquidity of the outer core and the existence of the inner core are more robust results: Both are established by the observation of reflections that must have traveled the outer core as P waves, but only a liquid core returns pure SV polarizations from its interior.

The deepest interior of the Moon as deduced from seismic data challenges our understanding of its origin, evolution, and dynamic behavior in several respects. Thermal evolution modeling has not yet found a way to explain how partial melt could be preserved at great depth until the present. Karato (2013) pointed out that the increased attenuation of shear waves might well be due to an increased content of water – which should have been evaporated during the big impact that is at the heart of the accepted hypothesis for the origin of the Moon. The observation of a solid inner core was entirely unexpected, since modeling suggested that freezing out an inner core would result in a geodynamo, which is not observed in the Moon today. This observation thus challenges our understanding of planetary magnetic fields. These observations are directly connected to what we believe to be the origin and early evolution of our own planet.

Part V

Interdisciplinary research involving planetary and astrophysical sciences

16

Diskoseismology of accretion disks

SYLVAIN CHATY

16.1 Introduction

In this chapter we describe the recent field of diskoseismology: the study of extraterrestrial seismology in accretion disks. Accretion disks, formed by accumulation of matter falling toward a gravitating object, are ubiquitous in the universe, surrounding a wide variety of celestial objects.

They are a key ingredient in binary systems which host compact objects orbiting a so-called companion star, and attracting matter from it. In these objects, an accretion disk forms through accumulation of gas coming from the companion star, tidally attracted and passing through the inner L1 Lagrange point (see Figure 16.1). The compact object can be either a neutron star or a stellar-mass black hole (these systems are then called *compact X-ray binaries*, XRBs), or a white dwarf (systems called *cataclysmic variables*, CVs). Furthermore, they surround the *supermassive black holes* (SMBH) at the heart of active galaxies, we also detect them at cosmological distances in the extremely energetic *gamma-ray bursts* (GRBs), and finally, we find them around protostellar and *young stellar objects* (YSOs).

From the discovery of the first extra-solar X-ray source – Sco X-1, an XRB in the constellation of Scorpius – reported on December 1, 1962 by Giacconi *et al.* (1962), to the measurement of quasi-periodic oscillations and broad lines in accretion disks surrounding compact objects, our knowledge of accretion disks has greatly improved, ultimately leading to potential tests of general relativity, due to high gravitational forces naturally created by the presence of compact objects. Putting together these observations, simulations, and predictions, will eventually lead to unique results based on general relativity.

We will show in Section 16.2 that accretion disks are ubiquitous in the celestial objects of our universe, before reviewing in Section 16.3 the general characteristics of accretion disks. We will then introduce in Section 16.4 the field of diskoseismology, reporting the existing observations, numerical simulations, and theoretical predictions of the models, concerning the various celestial objects hosting accretion disks, before concluding in Section 16.5.

Extraterrestrial Seismology, eds. Vincent C. H. Tong and Rafael A. García. Published by Cambridge University Press.

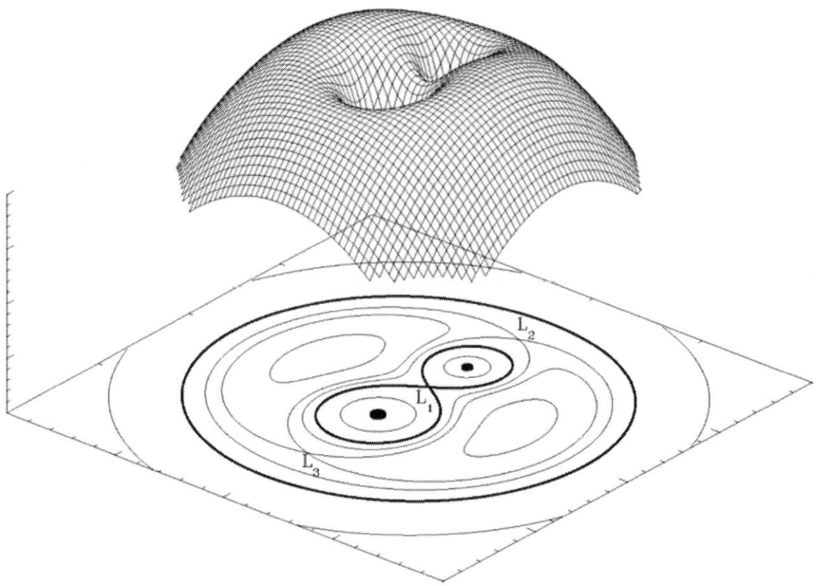

Figure 16.1 Sketch of Roche geometry (named after the French astronomer Edouard Roche): the Roche lobe is the region of space delimited by gravitational equipotentials around a star, where the orbiting material is gravitationally bound. In the case of a binary system, the material falls in through the inner Lagrangian point, accumulating to form an accretion disk.

16.2 Ubiquity of accretion disks in celestial objects

16.2.1 Compact X-ray binaries

Most of the stars in our Galaxy do not live in isolation: a recent census shows that more than 70% of the massive stars live in pairs (Sana *et al.*, 2012). In compact binary systems, the more massive (so-called "primary") component evolves faster, to become a compact object (either a neutron star or a black hole), while the less massive ("secondary" or "companion") reaches the giant state and exceeds its Roche lobe. Then, depending on the mass ratio and orbital parameters, these pairs might exhibit transfer of mass, with gas flowing from the companion star to the primary, potentially leading to emission of X-rays (see Figure 16.2, left). We will use the term compact X-ray binaries for all binary systems hosting the different types of compact objects: white dwarfs (in CV), neutron stars and black holes (in XRB).

The companion star is either an evolved low-mass star (typically $\leq 1\ M_\odot$, with a stellar spectral type later than B, usually K or M, a system called an LMXB for "low-mass X-ray binary"), or a luminous high-mass star (typically $\geq 10\ M_\odot$, of early spectral type OB, called an HMXB for "high-mass X-ray binary"). We refer the reader to two reviews on LMXBs and HMXBs, respectively: Chaty (2007, 2013). Three hundred compact X-ray binaries are known in our Galaxy: 187 LMXBs and 114 HMXBs (respectively 62% and 38% of the total number; Liu *et al.*, 2006, 2007), and the Galactic distribution of LMXBs

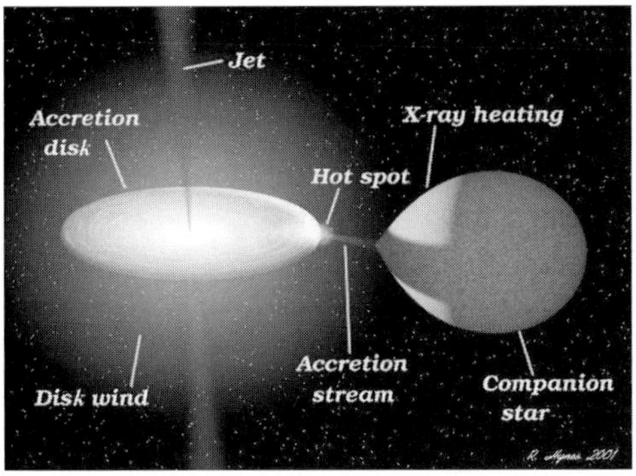

Figure 16.2 Artist's view of a compact X-ray binary (courtesy R. Hynes, left) and of a supermassive black hole (right). A black and white version of this figure will appear in some formats. For the color version please refer to the plate section.

and HMXBs is given in Grimm *et al.* (2002) and Coleiro and Chaty (2013), respectively. It is likely that this population is very common in our Galaxy – as in other galaxies – but that many have not been detected yet because of the episodic nature of their X-ray emission (they are not persistently accreting matter, and therefore not constantly emitting in X-rays). The formation and evolution of both types of compact X-ray binaries is described in Tauris and van den Heuvel (2006), and it is clear from their evolution (see for instance their Figures 16.12 and 16.15) that the accretion phase duration (i.e., the phase during which these binary systems are X-ray emitters) concerns a small fraction of the entire life of a compact X-ray binary.

Accretion of matter occurs differently for the two types of sources. In the case of HMXBs, accretion usually occurs through the wind of the active star, as shown in Chaty (2013) (except for some systems with an accretion disk, such as the black hole system Cygnus X-1). In the case of LMXBs, the small and low-mass companion star fills and overflows its Roche lobe, therefore accretion of matter always occurs through the formation of an accretion disk, surrounding compact objects of various masses: neutron stars of 1.4 $M_\odot$, and black holes between 3 and 14 $M_\odot$ (see, e.g., McClintock and Remillard, 2006).

16.2.2 Supermassive black holes

Supermassive black holes (SMBHs), located at the centers of probably all galaxies, have masses of between hundreds of thousands and billions of solar masses (see Figure 16.2 right). Our Galaxy, the Milky Way, hosts a SMBH of $\sim 4 \times 10^6\ M_\odot$, coincident with the Galactic center Sagittarius A*. The material constituting the accretion disk and surrounding the compact object, comes from the host galaxy itself, from molecular clouds, or even stars. It is believed that the SMBH formed at the same time as the host galaxy itself. We detect accretion–ejection phenomena occurring in these SMBHs, with jets extending for a few thousands of parsecs (1 parsec = 3.26 light years). It was in the 1960s that the first SMBHs

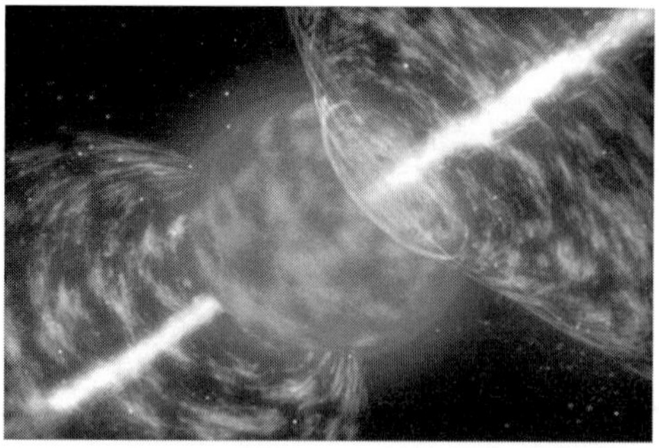

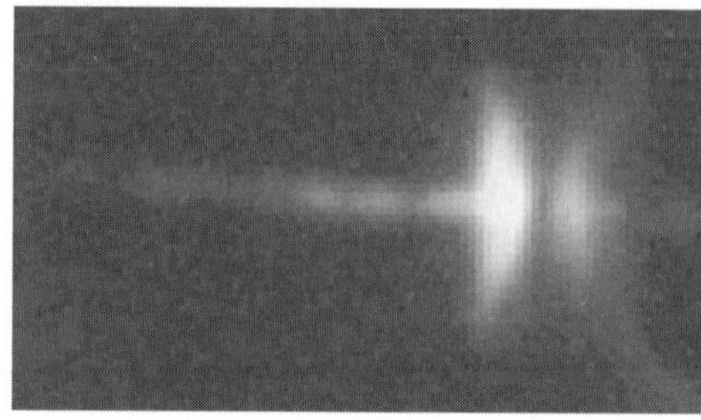

Figure 16.3 Artist's view of a gamma-ray burst (left) and image obtained with the Hubble Space Telescope of a young stellar object (right). A black and white version of this figure will appear in some formats. For the color version please refer to the plate section.

were detected, named at that epoch *quasars*, standing for quasi-stellar objects (see, e.g., Kormendy and Ho, 2013, for a review).

16.2.3 Gamma-ray bursts

Gamma-ray bursts (GRBs), initially detected as flashes of gamma-rays lasting from ten milliseconds to several minutes, are associated with the most energetic explosions in the universe, observed in distant galaxies at cosmological distances billions of light years away from Earth (see Figure 16.3, left). Initial GRBs are usually followed by a longer-lasting "afterglow," emitted at longer wavelengths (from X-ray, to ultraviolet, optical, infrared, and up to radio domain). There are two distinct classes of GRBs: the "short" and the "long" ones. The short GRBs might be the result of the merging of pairs of neutron stars. The "long" GRBs are likely due to a narrow beam of intense radiation released during a supernova or hypernova event, when a rapidly rotating and massive star collapses to form a compact object – either a neutron star or a black hole. While there is no direct observation, models of these events assume that accretion of matter occurs during the formation of the compact object.

It is interesting historically to note that GRBs were first detected in 1967 by the Vela satellites, which had been primarily designed to detect nuclear tests. Probably as many models of GRBs were imagined as the number of GRBs detected, until the detection in 1997 of the first X-ray afterglow by the Beppo-SAX satellite, leading to a direct measurement of the redshift by optical spectroscopy, and thus distance and energy output (see, e.g., Mészáros, 2006, for a review).

16.2.4 Young stellar objects

Young stellar objects (YSOs) are stars in their early stage of stellar evolution, consisting of proto-stars and pre-main-sequence stars. They can be classified in four distinct classes (0, I, II, and III), reflecting an evolutionary sequence, with the deeply embedded Class 0 sources evolving towards a later stage, by dissipating their circumstellar envelopes, and eventually becoming detectable in the optical while becoming pre-main sequence stars. Herbig–Haro objects, a sub-class of YSOs, usually exhibit early stellar evolution phenomena, such as bipolar outflows and jets (see Figure 16.3 right). These phenomena are believed to be caused by the presence of circumstellar accretion disks, at the origin of protoplanetary disks, because they are thought to be the progenitors of planetary systems. In this case, the accreted gas comes from the molecular cloud out of which the star has formed, rather than from a companion star (see, e.g., Feigelson and Montmerle, 1999, for a review).

16.3 What we know about accretion disks

The formation of an accretion disk is a ubiquitous phenomenon in astrophysics: it occurs when gravity forces a flow of matter to be transferred between two objects in an orbital motion, accumulating on a central – usually compact – object, of various nature and mass, including white dwarf, neutron star, and stellar mass – or even supermassive – black hole. Since angular momentum conservation prevents the material from going directly, as a straight flow, from the donor star to the compact object, the material accumulates and spirals inward, creating an accretion disk. Intense frictional forces compress and raise the temperature of the spiralling matter, causing the emission of electromagnetic radiation. The domain where most of the radiation is emitted depends on the mass of the central object: accretion disks of YSOs radiate in the infrared; those of XRBs in X-rays (with temperatures reaching millions of degrees), while those of SMBHs radiate in the ultraviolet domain. The accretion process is one of the most efficient in the universe, converting about 10% of the mass of an object into energy (as compared to $\sim$0.5% for nuclear fusion processes). For a complete and recent review on accretion disks, see, e.g., Abramowicz and Fragile (2013).

Accretion disks very often (always?) give rise to jets coming from the vicinity of the central object. Jets are an efficient way, for a system composed of a star and an accretion disk, to get rid of the angular momentum brought by the matter forming the disk, without losing too much mass. We are able to observe accretion disks surrounding YSOs, XRBs, and SMBHs, although the observation is generally indirect. XRBs exhibiting jets are called "microquasars," because they are the small-scale Galactic analogues of "quasars" hosting stellar-mass black holes, both systems exhibiting disc–jet connection (Chaty, 2007).

We show in Figure 16.4 a schematic sketch of 16 XRBs hosting stellar-mass black holes in the Milky Way, with reasonably accurate dynamical data, obtained through radial velocity detection. The compact object orbiting the companion star creates a periodic perturbation, usually determined via optical spectroscopy, in analogy with the detection of exoplanets around stars, allowing us to get the mass, orbital distance, and inclination of each component of the system. They are long-period systems containing hot supergiants (such as Cyg X-1) and cool red giants (such as GRS 1915+105) and many compact X-ray binaries containing K-dwarf secondaries. GRS 1915+105 is an interesting source, hosting the biggest accretion disk known up to now, which has remained bright in X-rays for more than a decade, since its first known eruption in August 1992 (see, e.g., Rahoui *et al.*, 2010).

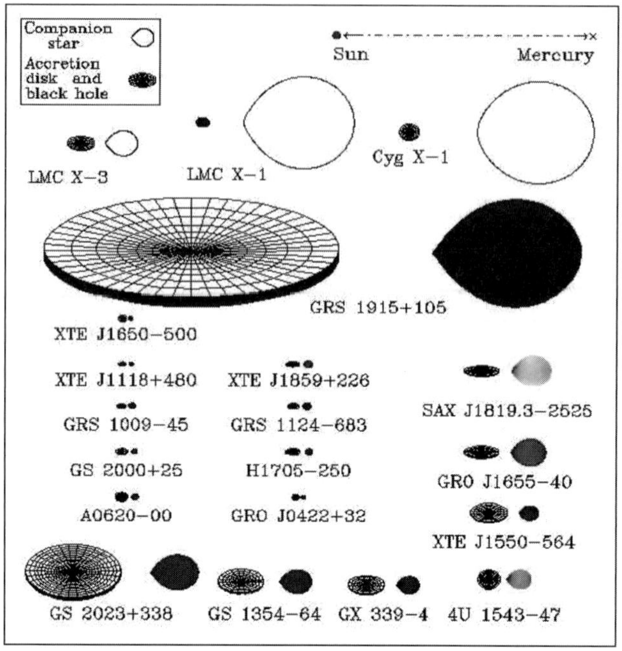

Figure 16.4 Scale drawings of 16 compact binaries hosting a black hole in the Milky Way (the Sun–Mercury distance of 0.4 AU is shown at the top), showing both the accretion disk and the companion star. The estimated binary inclination is indicated by the tilt of the accretion disk, and the color of the companion star roughly indicates its surface temperature. (Courtesy: J. Orosz.) A black and white version of this figure will appear in some formats. For the color version please refer to the plate section.

The best way to study accretion disks is via multi-wavelength observations and modeling of the outbursts exhibited by XRBs, occuring on timescales of 1 to 60 years, and lasting between 20 days and many months. The generally accepted cause of the outburst cycle is an instability arising inside the accretion disk: when the accretion rate from the donor star is too high, leading to a continuous viscous flow towards the compact object which can no longer be supported, matter fills the outer disk until a critical surface density is reached, thus triggering an outburst. This so-called *disc instability model*, initially developed for dwarf novae (hosting a white dwarf; e.g., Lasota, 2001), has since been extended to X-ray novae (see, e.g., Dubus *et al.*, 2001).

16.4 Introduction to diskoseismology

The field of diskoseismology refers to the study of *oscillation modes* in all types of accretion disks, in the general relativity frame. In this context, the fundamental modes (i.e., the g, c and p modes), analogous to internal gravity modes in stars, are trapped as in a resonance cavity, within the inner region of the accretion disks, by the strong gravitational field of a

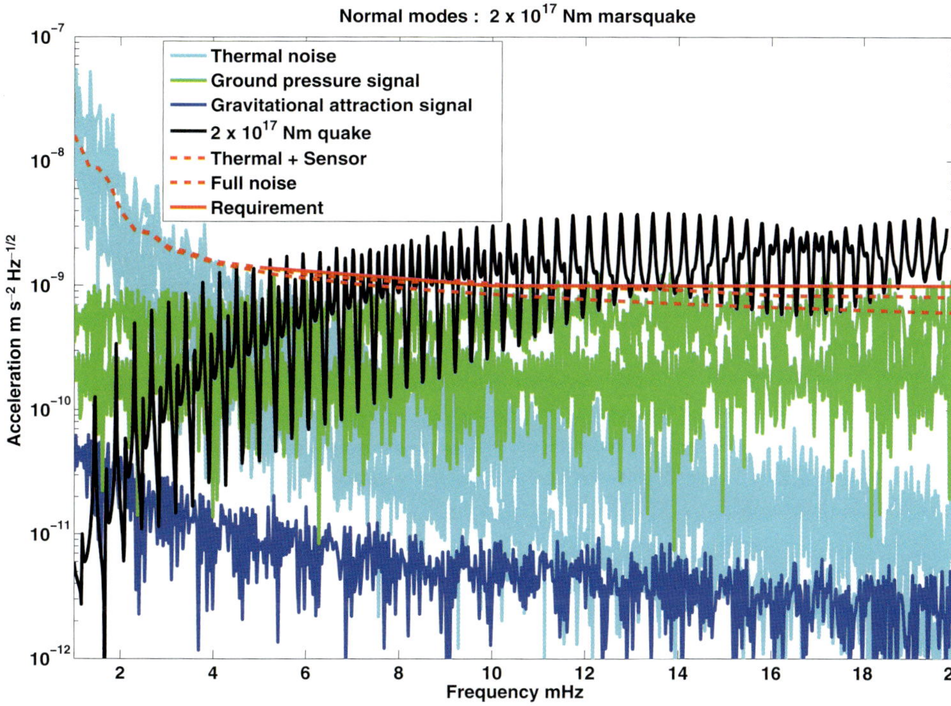

Figure 3.2 The detectability of normal modes for large quakes on Mars is shown in this plot of the calculated acceleration amplitude spectrum (black) for a quake with seismic moment 2×10^{17} Nm compared to SEIS/InSight required (red) and expected (red dashed) sensitivity. Epicentral distance is 90 degrees. Also shown are modeled environmental noise sources included in the red dashed curves: thermal (cyan); atmospheric pressure (green); atmospheric gravity (blue).

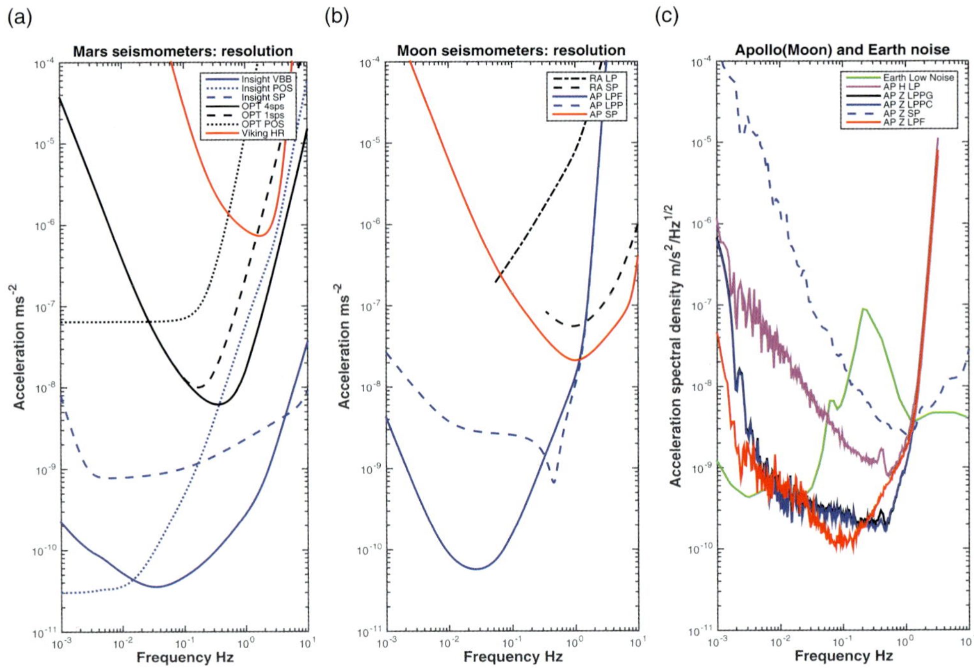

Figure 3.4 Response curves of the past (a) Mars and (b) Moon seismometers. (a) The resolution of the Viking in high data rate mode; the OPTIMISM seismometer in the long-period position mode (POS), in the velocity mode OPTVIT1 at 1 Hz sampling rate, and at 4 Hz (OPT VIT4). The figure also shows the expected resolution of the InSight seismometers, defined here as the RMS in 1/6 of a decade (equal roughly to the product of $0.62 \times \text{ASD} \times \sqrt{f}$ where ASD is the amplitude spectral density in m s^{-2} $Hz^{-1/2}$ and f the frequency). The first two are the velocity flat and acceleration flat outputs of the VBB sensors (VBB and POS) while the third is the velocity flat output of the short-period (SP) instrument. VBB's components are sensitive to acceleration oblique at about 30° with respect to horizontal, while the SP are either vertical or horizontal. At 1 Hz, the expected resolution of the InSight VBB sensor is expected to be 10^3 times better than Viking. (b) The Moon case, with the long (RA LP) and short (RA SP) period analog outputs of the Ranger seismometer, the flat (AP LPF) and peaked (AP LPP) mode of the Apollo LP seismometer, and the short-period seismometer (AP SP). The transfer functions shown are the generic ones. (c) Different sources and levels of seismic noise. The noise levels recorded on the Moon by the different channels of the Apollo seismometers are compared with the Earth-based low noise model of Peterson (1993). Noise levels are from Apollo 12 vertical long-period flat (AP Z LPF) mode and from Apollo 14 vertical short-period (AP Z SP) mode. These noise levels are likely an upper estimate of the Moon noise and may be related to the instruments. For the other components, note that the peaks near 0.45 Hz of the peaked vertical (AP Z LPPG) and horizontal (AP H LP) peaked records might be related to differences between the Apollo 12 transfer function at the time of the recording and the pre-flight generic transfer function. The noise corresponding to the curve AP Z LPPC is obtained by changing the feedback parameters in order to minimize the noise peak (about 10% for the parameter K2 and h, see the Apollo transfer function at http://dartsjaxajp/planet/seismology/apollo/The_Apollo_Seismometer_Responses pdf).

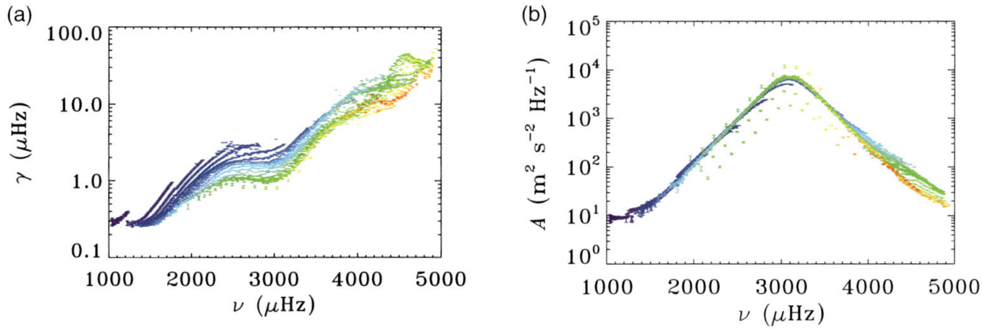

Figure 5.1 Width (a) and amplitude (b) for medium-degree mode fitting to GONG spectra, averaged over 17 years of data. Darker shades represent lower radial orders n.

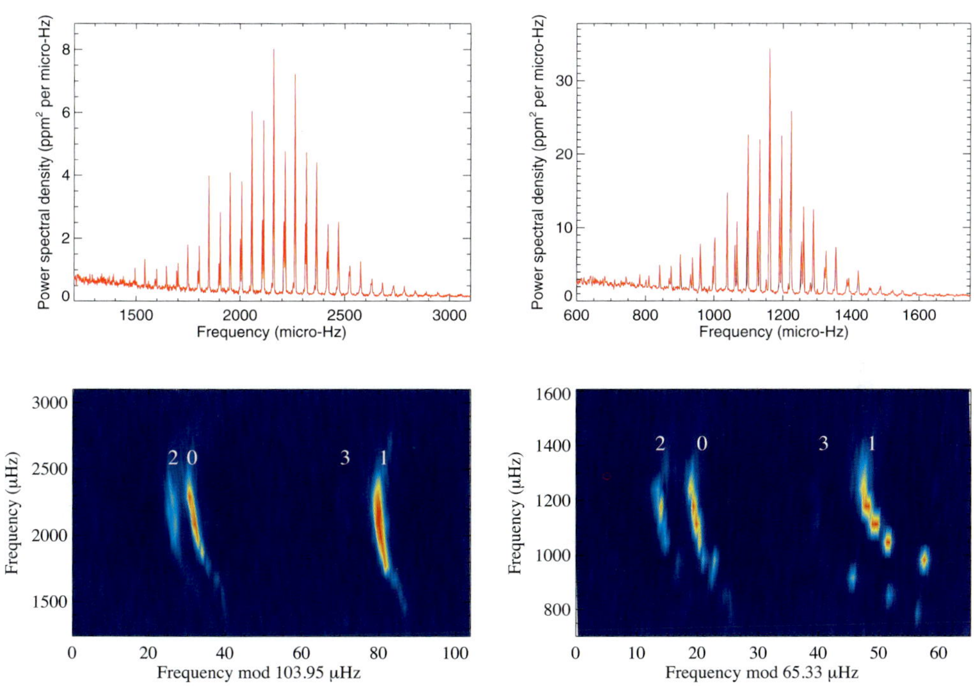

Figure 5.4 Frequency–power spectra (top) and échelle diagrams (bottom) for the G-type main-sequence star 16 Cyg A (HD 186408) (left-hand panels), and the G-type subgiant KIC 6442183 (HD 183159) (right-hand panels). Numbers on the échelle diagrams indicate where the putative ridge of each angular degree lies.

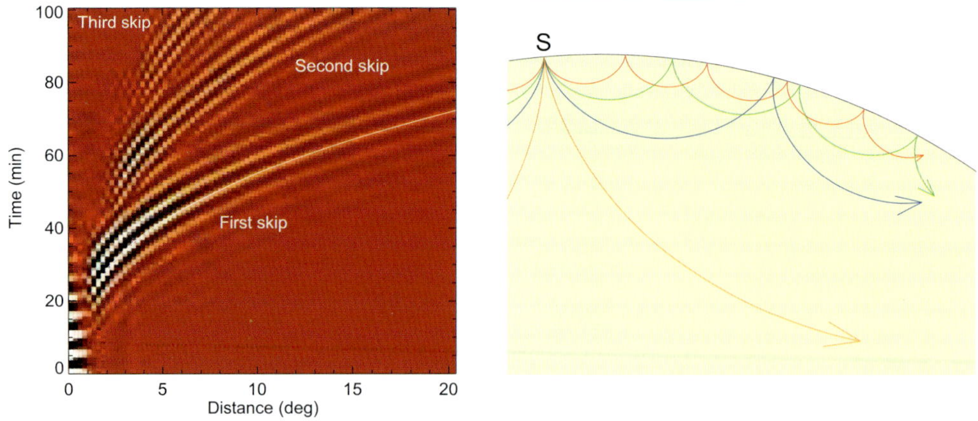

Figure 6.4 (Left) A time–distance cross-covariance function in time (horizontal axis) and between samples of the wavefield spaced different distances (horizontal axis) on the solar surface. The amplitude of the function, modulated by the five-minute solar oscillations, reaches maxima at times corresponding to the travel times of rays corresponding to different acoustic ray paths in the solar interior. (Figure by the author using unpublished cross-covariance measurements provided by T. Duvall, Jr.) (Right) Ray paths of sound waves propagating downward from a specific origin S near the surface, refracting back upward to the surface, and reflecting down again to advance a number of "skips" away. (Figure by the author.)

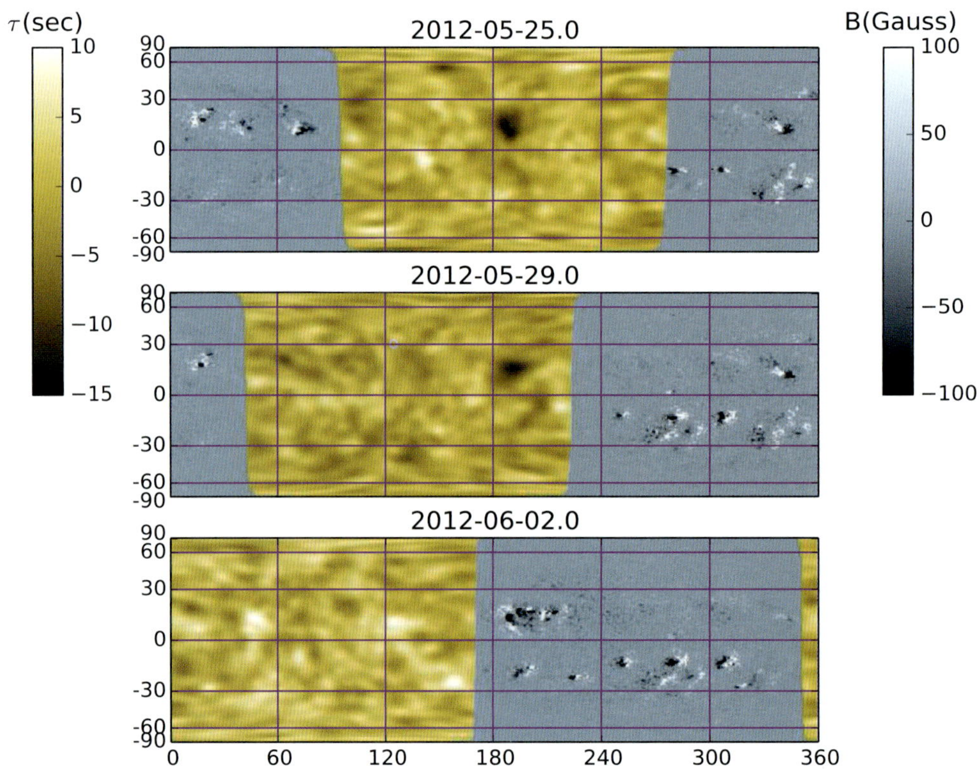

Figure 6.7 Composite farside-seismic (left color scale) and nearside-line-of-sight magnetic maps (right gray scale) show an active region (NOAA AR11498) passing the farside meridian (top panel) and rotating into view from Earth at Carrington coordinate (190W, 13N) in the near hemisphere (bottom) in May–June, 2013. (Figure provided by C. Lindsey.)

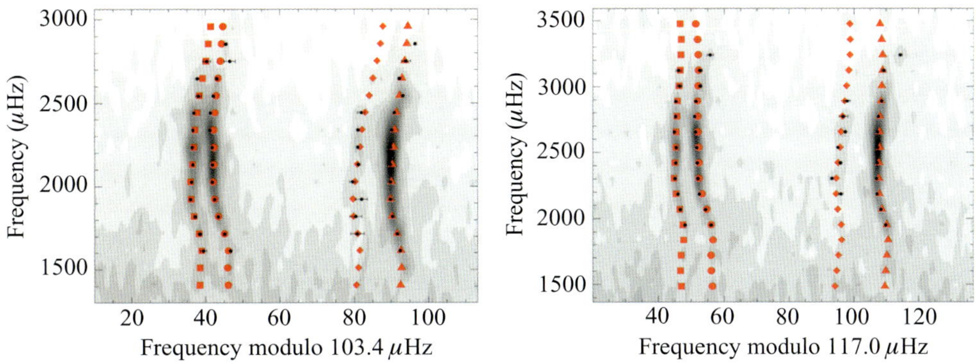

Figure 8.3 Échelle diagrams of 16 Cyg A (left) and B (right) from *Kepler* data. The observed frequencies are indicated as black points with error bars; the frequencies from the best fitting model for each of the two components are indicated with different symbols depending on the mode degree. (Figure reproduced from Metcalfe *et al.*, 2012 with permission.)

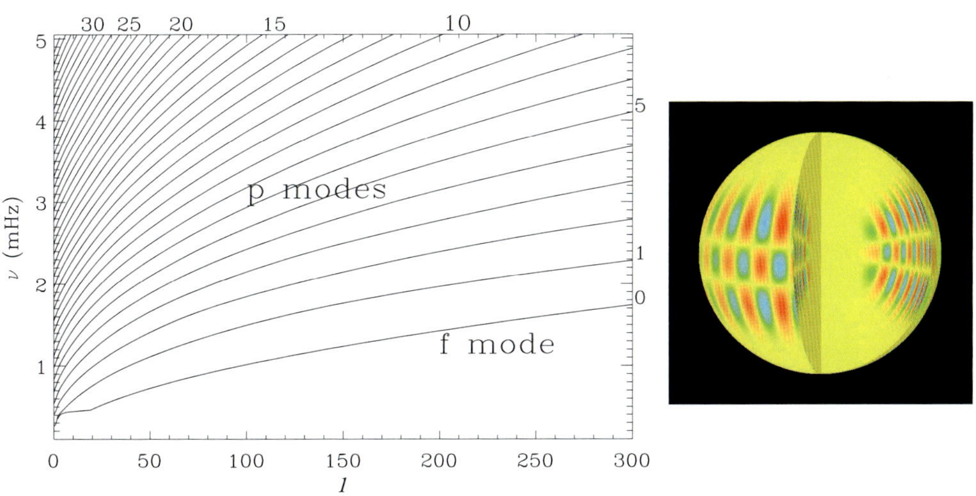

Figure 9.1 Theoretical spectrum of normal modes (p modes and f mode) calculated for a standard solar model (left, model S of Christensen-Dalsgaard *et al.*, 1996), and a normal mode eigenfunction (radial eigenfunction multiplied with the spherical harmonic function) for a p mode of angular degree $l = 20$, azimuthal order $m = 18$, and radial order $n = 14$ (right, Courtesy: A. G. Kosovichev).

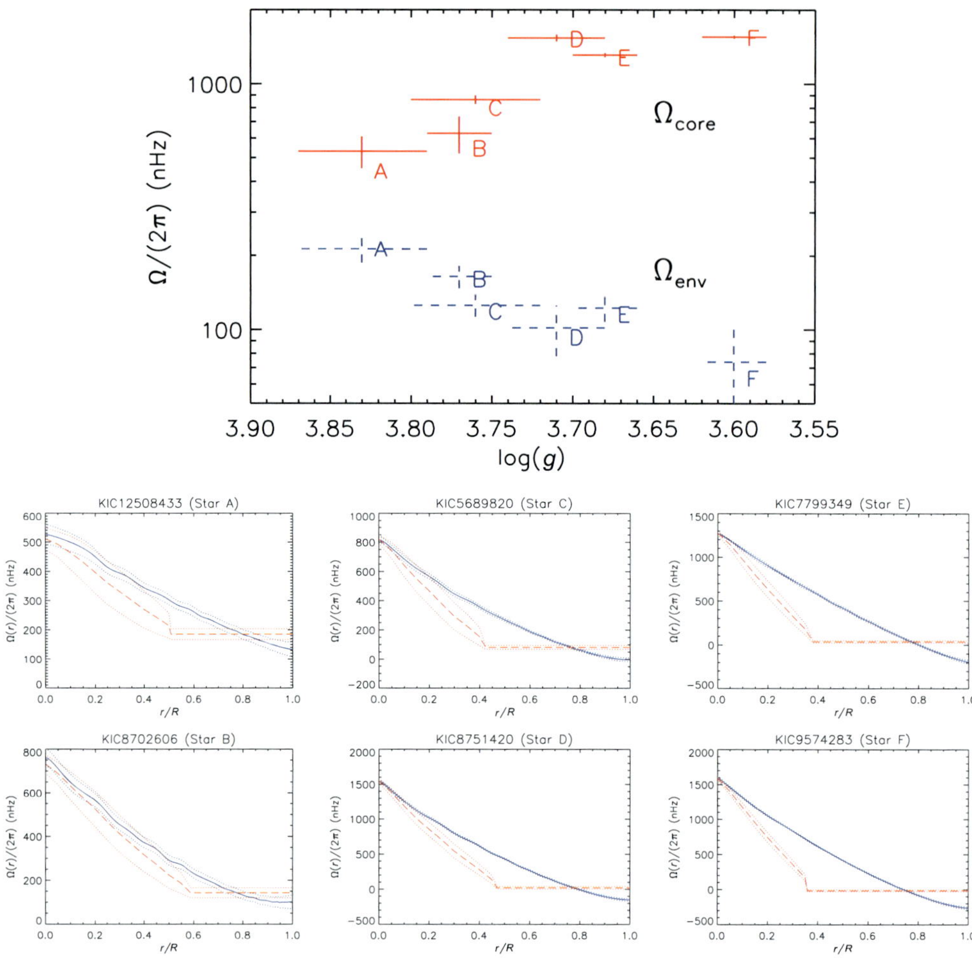

Figure 8.5 (Upper) Core and envelope rotation rates as a function of the surface gravity for six red giants, each specified by a character. The horizontal and vertical lines indicate 1σ error bars. (Middle and lower) Rotation profiles obtained relying on a smoothness condition on the rotation profile for the entire star (blue solid lines) or only in the radiative interior when the convective envelope is assumed to rotate rigidly (red long-dashed lines). The dotted lines indicate the 1σ error bars for both types of inversions. (Reproduced from Deheuvels *et al.*, 2014.)

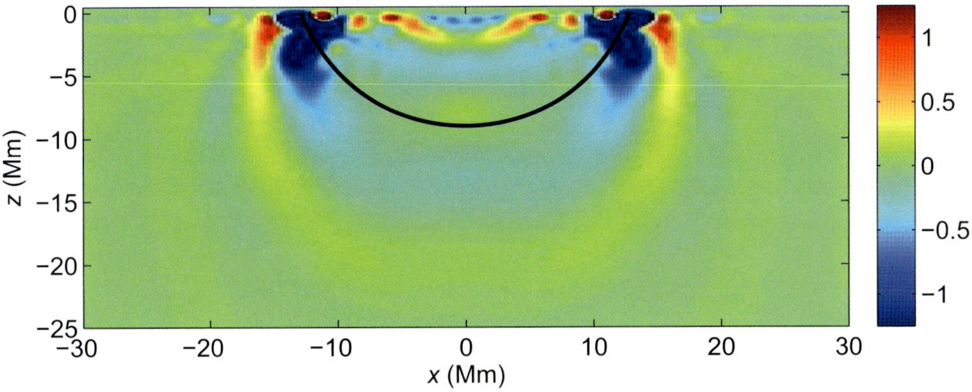

Figure 9.4 A depth slice, in the plane of the ray path, of the 3D Born kernel for sound-speed perturbations on travel times, from the results in Birch *et al.* (2004), for a surface travel distance of 25 Mm. This example shows that the sensitivity is small along the ray path, shown as a black solid curve, and negative in a shell surrounding it, resembling the traditional banana-doughnut shape. (Courtesy: A.C. Birch.)

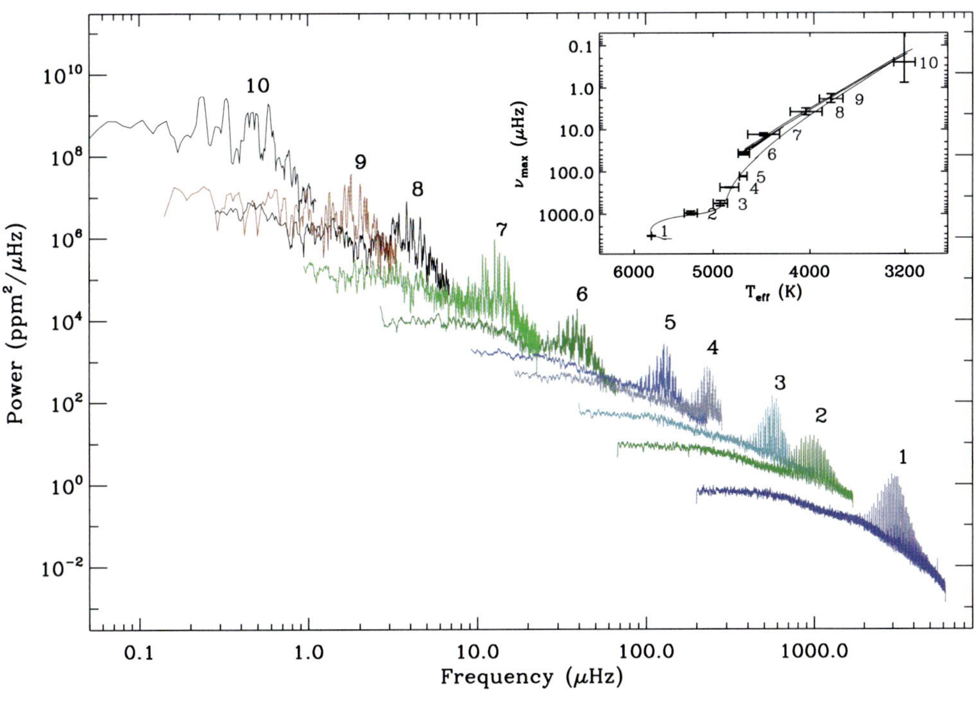

Figure 11.3 Sequence of stellar oscillation spectra starting from a main-sequence star, the Sun (1), to the tip of the red giant branch. Star 6 is a red clump star. Temperatures are from the *Kepler* Input Catalog (Brown *et al.*, 2011) and from Huber *et al.* (2014). The inset represents a seismic HR diagram. The solid line is the evolutionary track of a one-solar-mass stellar model computed using MESA (Paxton *et al.*, 2011, 2013) in which the position of each star is represented. The 1σ error bars in ν_{max} are multiplied by a factor of 10.

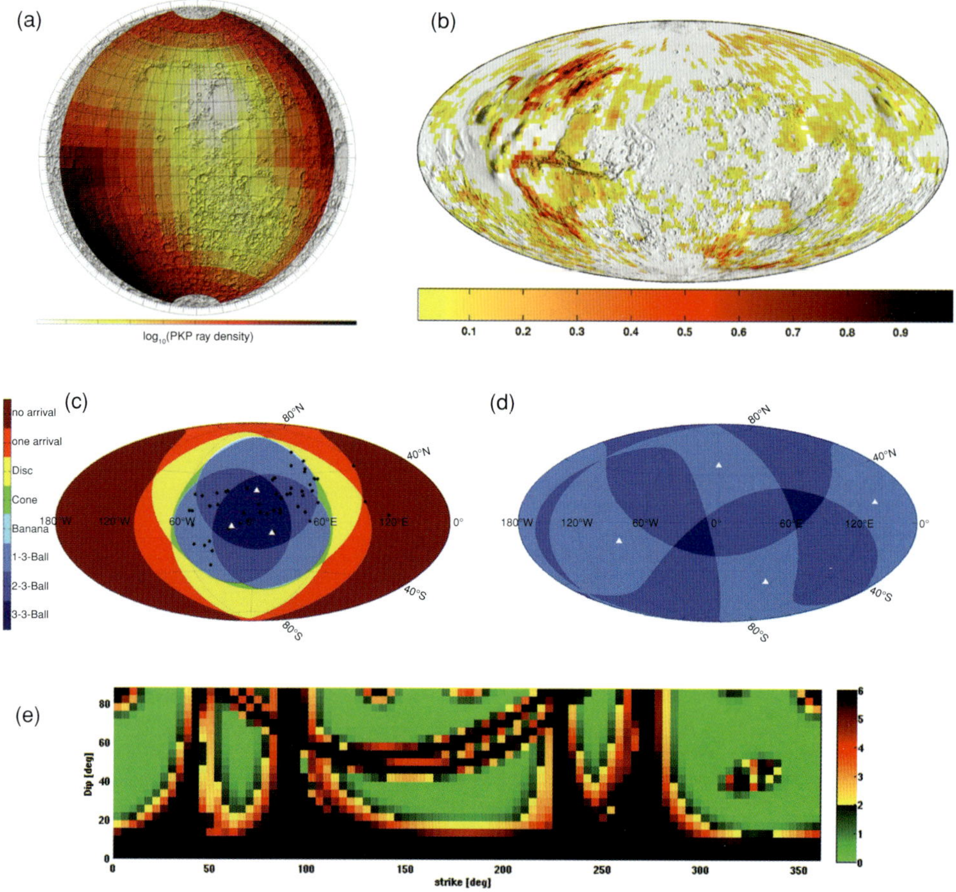

Figure 10.2 (a) Nearside map of PKP ray density from the known distribution of deep moonquake clusters as a function of landing site coordinates, taking predicted cluster occurrence activity and arrival amplitudes into account, shown in logarithmic scale. For this particular phase, landing sites near the limb are favored, particularly in the southwestern quadrant of the Moon, where the likelihood of detecting PKP from the northeastern farside events is greatest. (b) Map showing the probability that a $M_w \leq 4.8$ quake is assigned to 1 km^2 of surface, plotted on a $2.5^\circ \times 2.5^\circ$ grid (Knapmeyer *et al.*, 2006). (c) Location capability of an Apollo-like network. For comparison, real deep moonquake clusters with ball-shaped location clouds are shown (black dots). Triangles are station locations (Hempel *et al.*, 2012). (d) Location capability of a network with four equally distributed, modern seismometers (Hempel *et al.*, 2012). (e) Least-squares misfit between observed and calculated amplitudes on a regular grid of fault orientations averaged over all useable events from the A6 source region. The accepted regions (green) are selected by highlighting fault orientations that predict amplitudes within two standard deviations of the measured values.

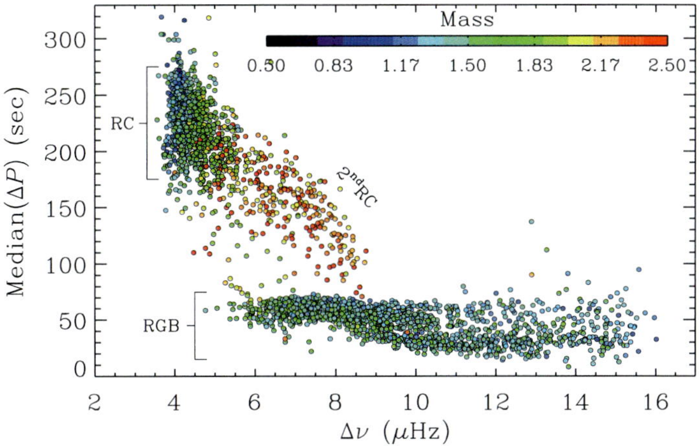

Figure 11.5 Median period spacing of mixed modes for each star analyzed by Stello *et al.* (2013) versus its large frequency separation. Red giant branch (RGB), red clump (RC), and secondary clump (2^{nd}RC) stars are indicated. The masses of the stars obtained from the scaling relations are color coded.

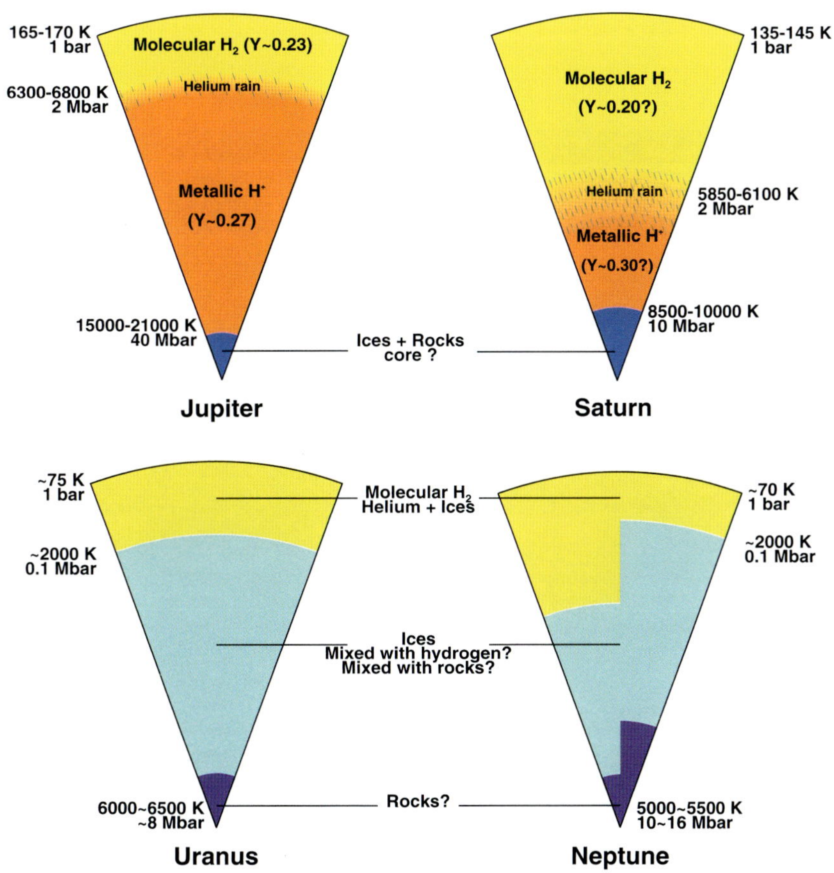

Figure 14.1 Schematic representation of the interiors of Jupiter, Saturn, Uranus, and Neptune. The range of temperatures for Jupiter and Saturn is for models neglecting the presence of the inhomogeneous region (adapted from Guillot, 1999). Helium mass mixing ratios Y are indicated. The size of the central rock and ice cores of Jupiter and Saturn is very uncertain (see text). Similarly, for Uranus and Neptune, considerable uncertainties exist (Nettelmann *et al.*, 2013a). Recent models of Neptune indicate a wider range of possible solutions, of which two representative ones are shown.

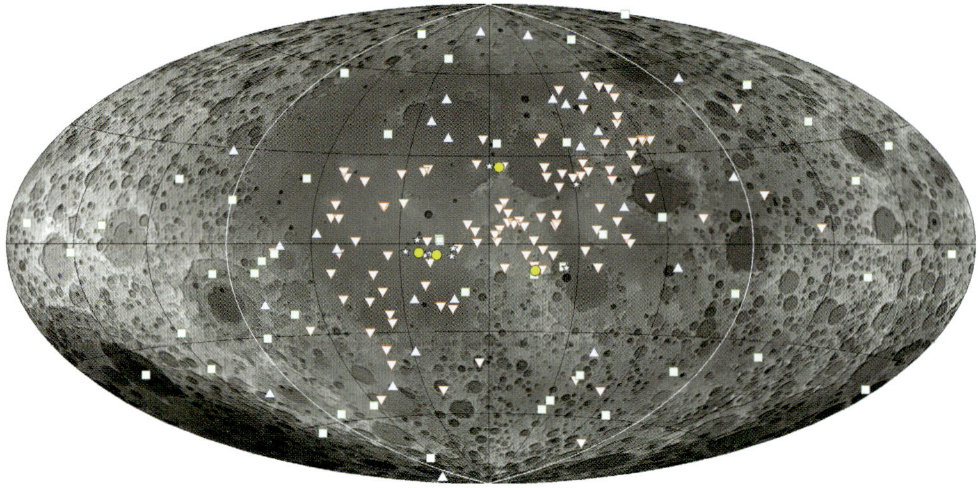

Figure 15.2 Epicenters of deep moonquakes (down-pointing triangles, Nakamura, 2005), shallow moonquakes (up-pointing triangles, Nakamura *et al.*, 1979), the 48 largest meteoroid impacts (squares, Oberst, 1989), and artificial impacts (pentagrams, Toksöz *et al.*, 1974). Circles denote the Apollo seismometer sites. The white meridians separate the near and far sides. Map in equal-area Hammer projection, background topography as in Figure 15.6.

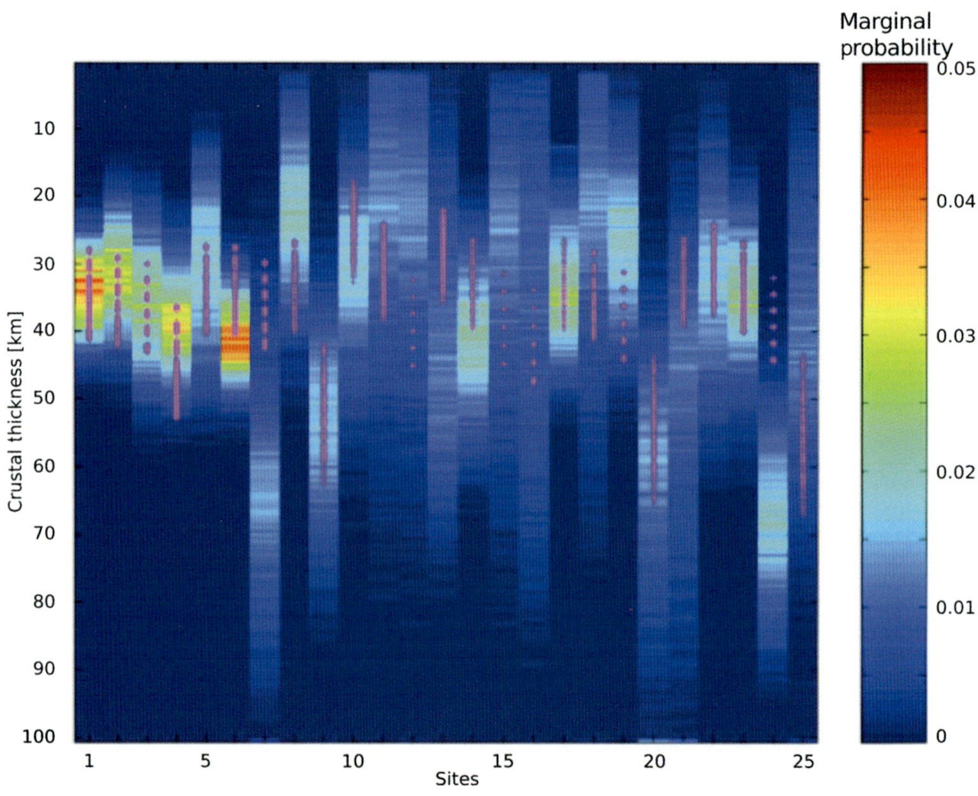

Figure 15.7 Posterior marginal probabilities for the sites investigated by Chenet *et al.* (2006). Sites 1–4 are Apollo sites 12, 14, 15, and 16, Sites 5 and 6 are the impacts of the Apollo 15 and 17 S-IVB stages, respectively. Magenta markers depict possible crustal thicknesses from gravity modeling for each site.

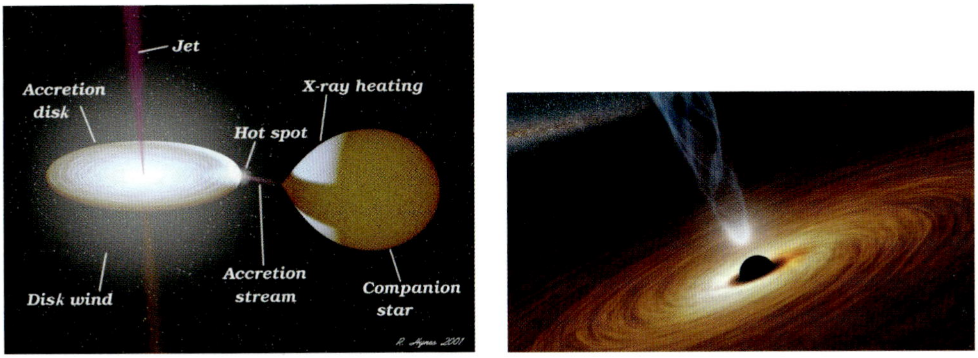

Figure 16.2 Artist's view of a compact X-ray binary (courtesy R. Hynes, left) and of a supermassive black hole (right).

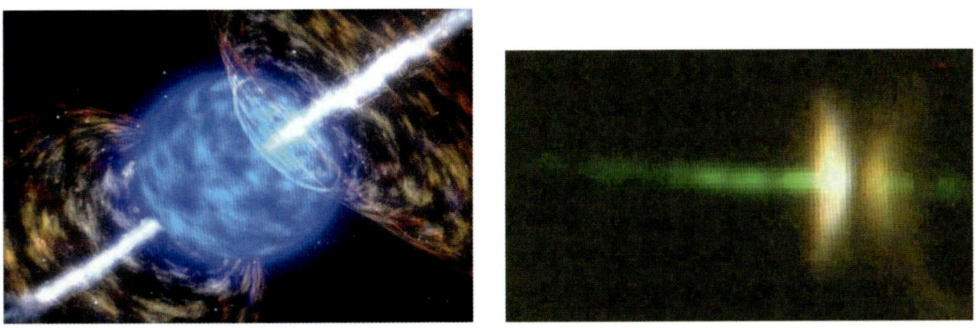

Figure 16.3 Artist's view of a gamma-ray burst (left) and image obtained with the Hubble Space Telescope of a young stellar object (right).

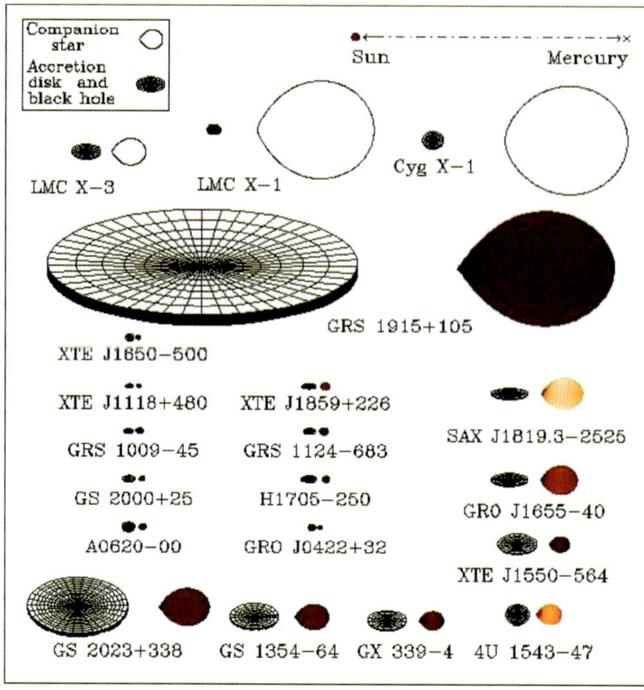

Figure 16.4 Scale drawings of 16 compact binaries hosting a black hole in the Milky Way (the Sun–Mercury distance of 0.4 AU is shown at the top), showing both the accretion disk and the companion star. The estimated binary inclination is indicated by the tilt of the accretion disk, and the color of the companion star roughly indicates its surface temperature. (Courtesy: J. Orosz.)

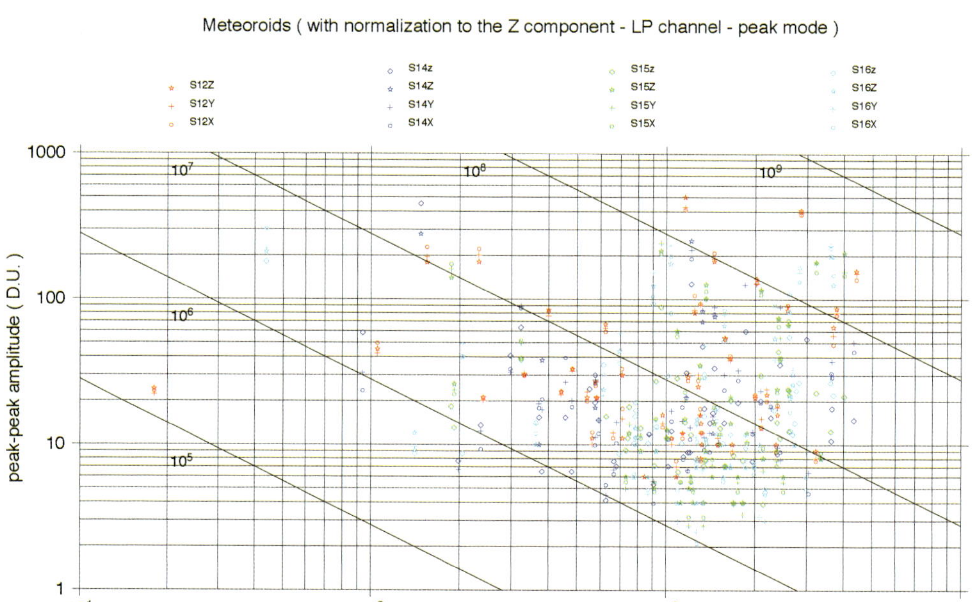

Figure 18.4 Peak-to-peak amplitudes of the impacts detected by the different Apollo channels, in DU, with respect to epicentral distance. The oblique lines are the source amplitudes, from 10^5 Ns up to 10^9 Ns in the y-direction, while the $1/D$ decay of the oblique lines is mainly due to geometrical spreading, as attenuation is very low on the Moon. The moments correspond to impacts from about 3 kg to 30 tons at 20 km/s velocities and with ejecta amplification of 1.5. (Reprint of Figure 9 of Gudkova *et al.*, 2009, with permission of Elsevier.)

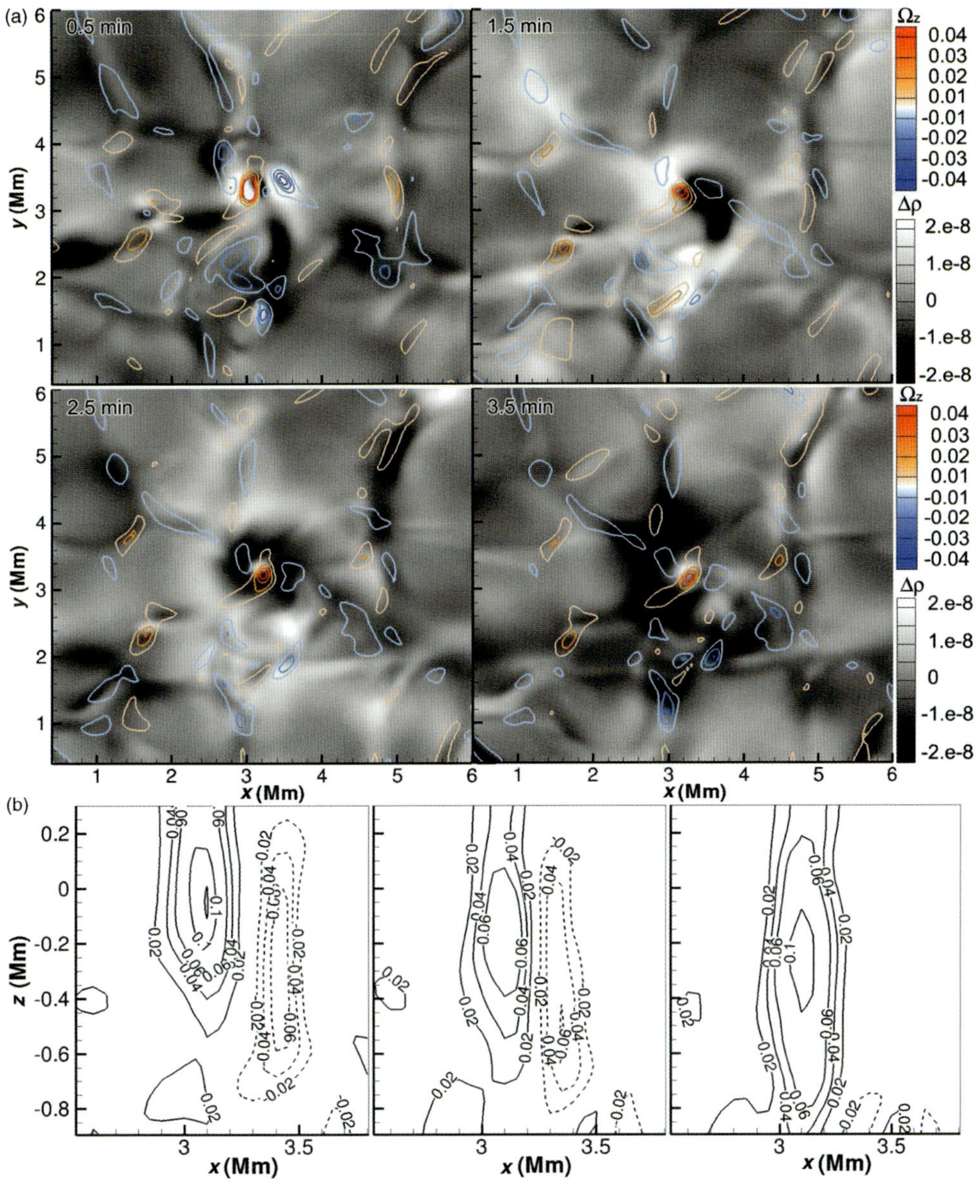

Figure 17.4 (a) Panels displaying the temporal evolution of density fluctuations (calculated as $\rho(t_{i+1}) - \rho(t_i)$) at the solar surface with cadence 1 min illustrate acoustic wave excitation and radial propagation from a vortex source, representing the interaction of two vortices with opposite-sign vertical vorticity. Red and blue lines correspond to the magnitude of the positive (clockwise) and negative (counterclockwise) vertical vorticity. (b) Different stages of the subsurface vortex interaction with a partial annihilation. Solid and dashed isolines show the magnitude of the positive and negative vertical vorticity.

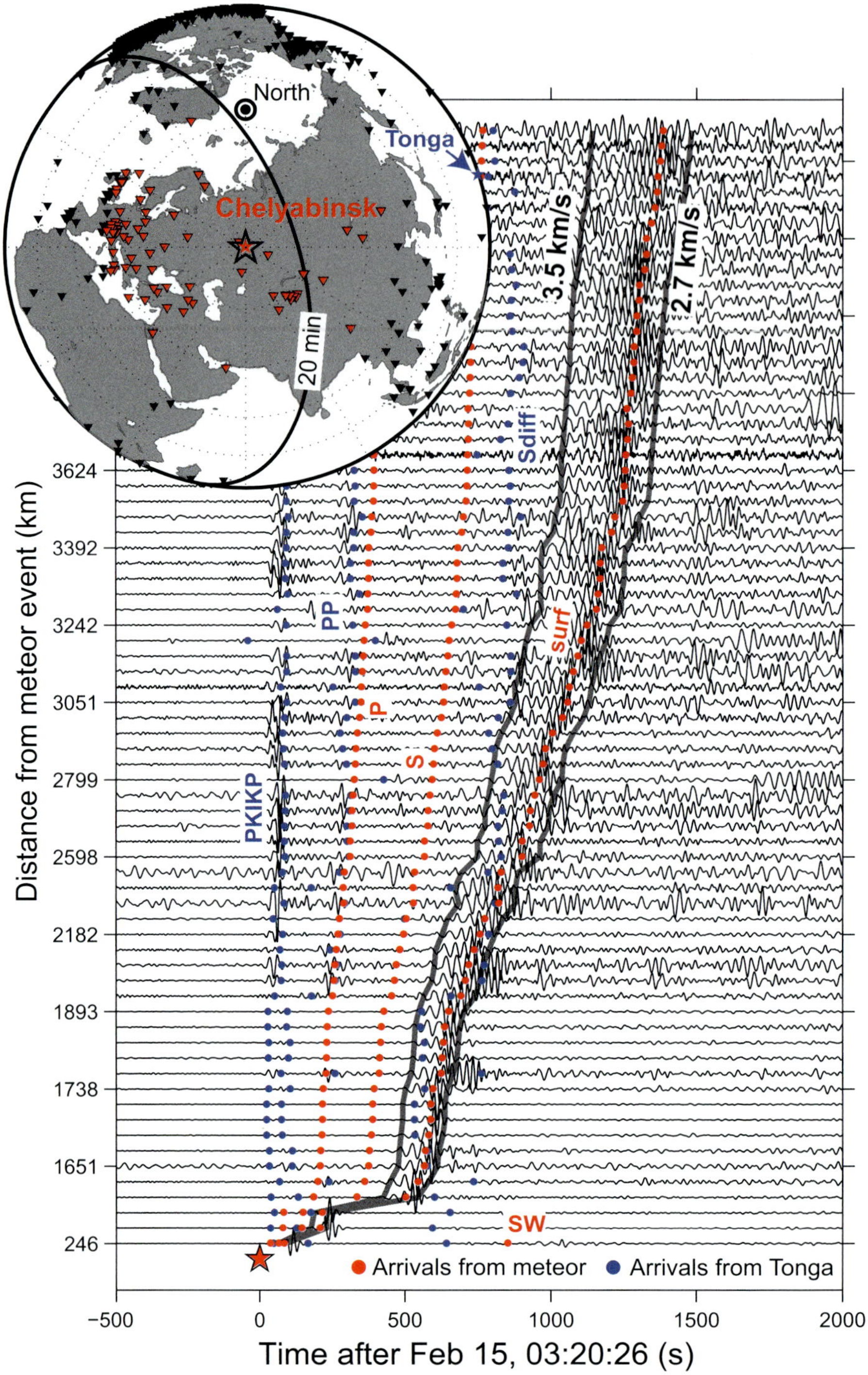

Figure 18.1 Bandpass-filtered vertical seismograms from up to 4000 km of epicentral distance, recorded after the Chelyabinsk blast. The filtering window is between 20 and 60 s and corresponds to the surface wave bandwidth. The blue dots show the arrival of seismic waves associated with an earthquake occurring in the Tonga Islands 20 minutes before the impact. The 20 minute travel-time isocontour for this earthquake is indicated on the Earth's sphere, while an arrow gives its epicenter location. (Reprinted from Tauzin *et al.*, 2013, with permission of AGU.)

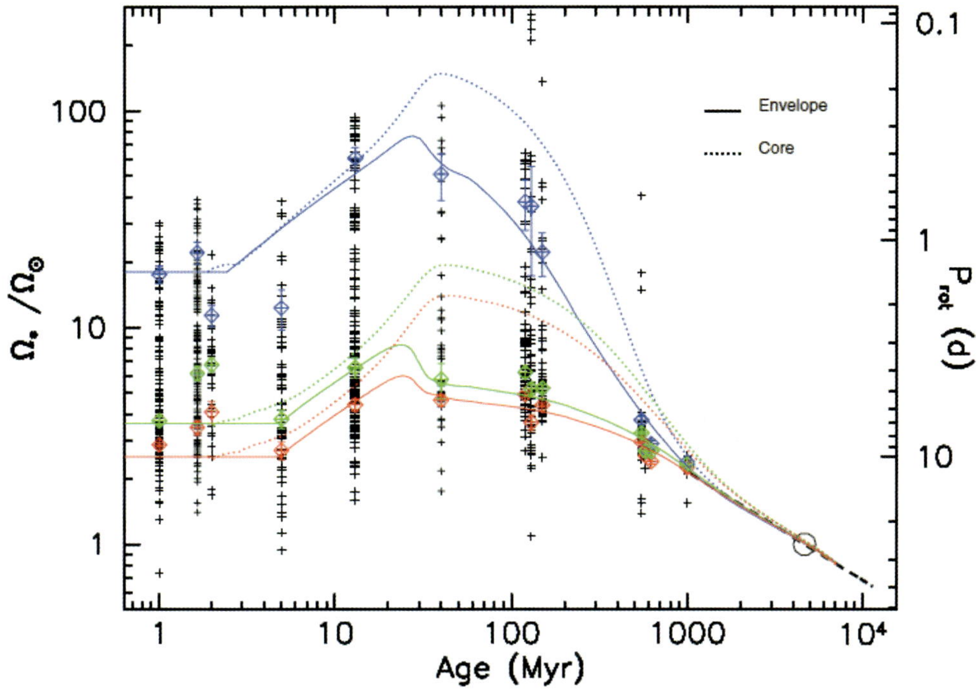

Figure 19.1 Rotation rates of the radiative core (dashed lines) and of the convective envelope (solid lines) of solar-type stars as a function of time for fast (blue), median (green), and slow (red) rotator models. The left axis gives angular velocities normalized to the Sun's. The right axis gives rotational periods (days). The blue, red, and green tilted squares correspond to the 90th percentile, the 25th percentile, and the median, respectively, of the rotational distributions in formation regions and young open clusters. The open circle is the angular velocity of the present Sun and the dashed black line corresponds to the Skumanich relationship (see text). (From Gallet and Bouvier, 2013, courtesy *Astronomy & Astrophysics*.)

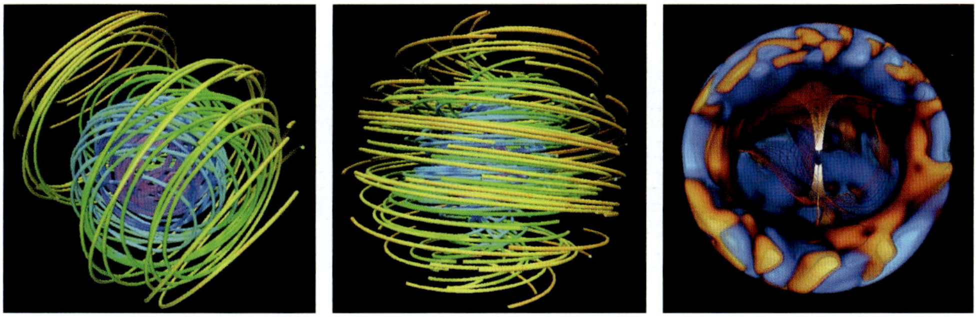

Figure 19.4 (Left) Field configuration resulting from the instability of an initial purely axisymmetric toroidal field (the color scale is a function of the density. (From Duez *et al.*, 2010, courtesy *The Astrophysical Journal*.) (Middle) Stable relaxed large-scale non-force-free twisted fossil field configuration. (From Duez *et al.*, 2010.) (Right) Creation of a toroidal field resulting from the action of the shear of the tachocline on an inital dipolar field in a 3D integrated model of the Sun. (From Strugarek *et al.*, 2011a.) Thin lines are the field lines. In the convective envelope, red indicates positive velocities, and blue, negative velocities.

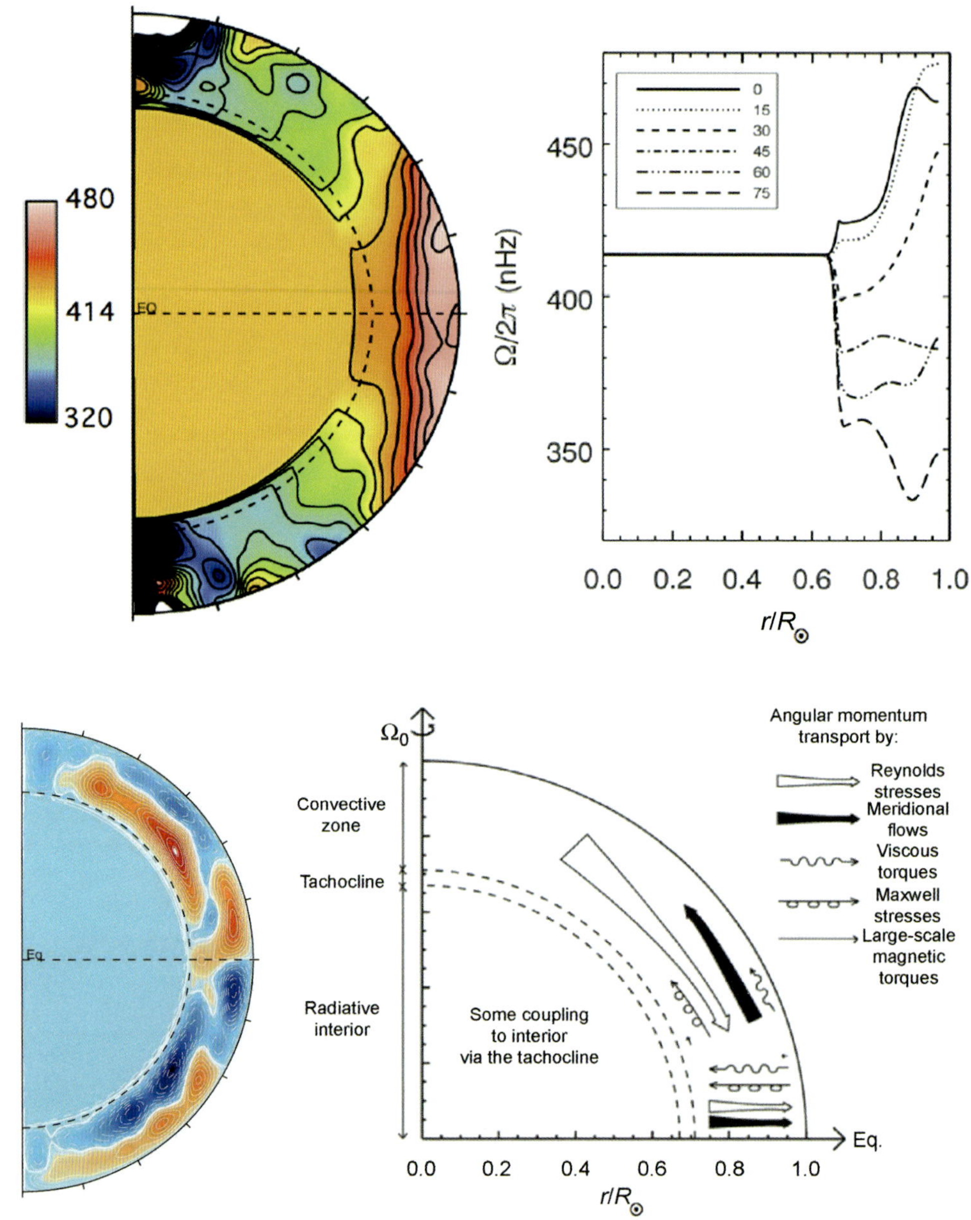

Figure 19.3 (Top) Angular velocity contours for a 3D model of the Sun spanning 97% of the solar radius (left) and latitudinal cuts at indicated latitudes (right), showing clearly a tachocline at the base of the convection zone near 0.7 $R_\odot$ (Brun *et al.*, 2011; Alvan *et al.*, 2014). (Bottom left) Meridional circulation realized in the same 3D solar model (red contours/solid white line correspond to counterclockwise flow). (Bottom right) Summary of angular momentum redistribution in a magnetized turbulent convective shell (Brun, 2004, courtesy *Solar Physics*).

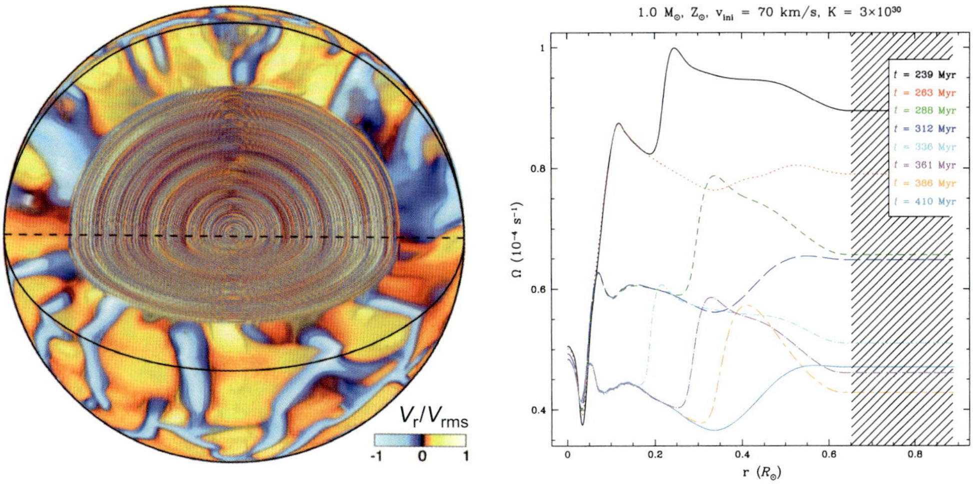

Figure 19.5 (Left) 3D numerical simulation of the full Sun from 0 to $0.97R_{\odot}$. We represent the normalized radial velocity (to the rms one) in the full simulated star. Red and blue correspond to positive and negative velocities respectively. Internal gravity waves excited by turbulent convective plumes are identified in the core. (From Alvan *et al.*, 2014, courtesy *Astronomy & Astrophysics*.) (Right) Evolution of the angular velocity radial profile in a solar-type star along its evolution due to the combined action of internal gravity waves and meridional flows. Hatched region corresponds to the convective envelope. We obtain a spin-down of the core because of the extraction of angular momentum by retrograde waves. (From Alvan *et al.*, 2013, courtesy *Astronomy & Astrophysics*.)

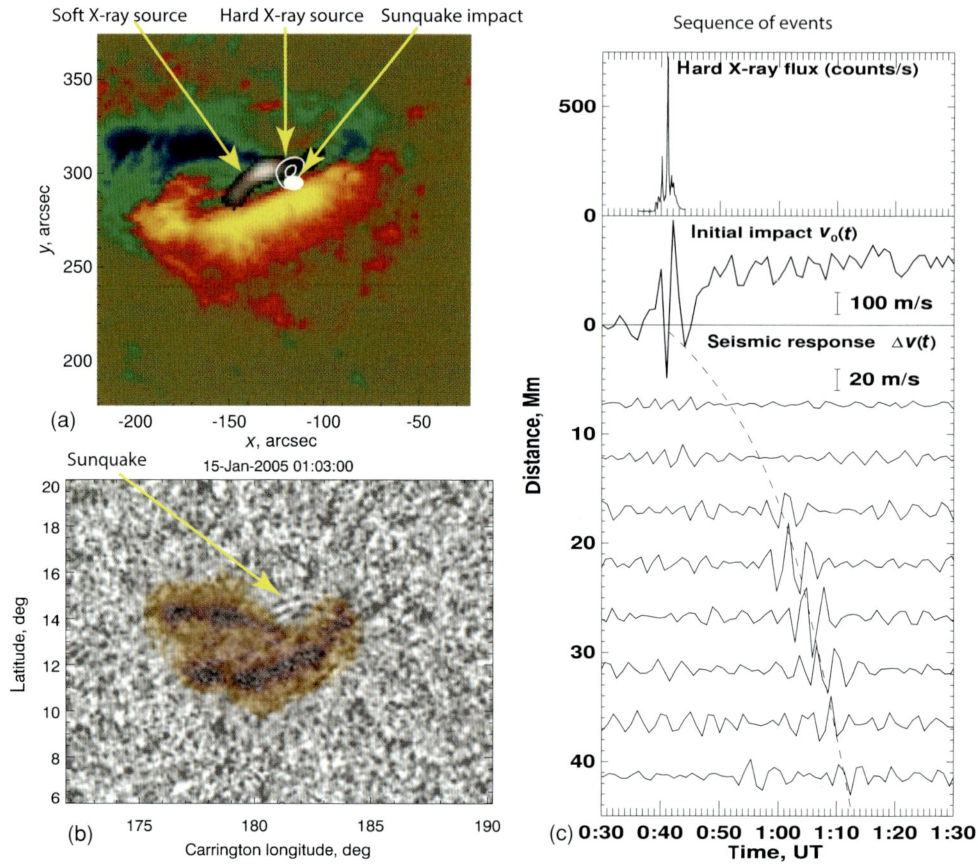

Figure 22.6 Sunquake of X1.2 flare of January 15, 2005: (a) location of the initial sunquake impact (white dot) on the SOHO/MDI magnetogram (color background), soft X-ray (gray-scale structure), and hard X-ray (contour lines) emissions observed by RHESSI; (b) Doppler-velocity image showing the wavefront (gray) overlaid with the corresponding image of sunspots (brown); (c) the sequence of the sunquake events showing the hard X-ray impulse, and Doppler-velocity signals at the initial impact, and at various distances from the impact point. The dashed curve shows the theoretical time–distance relation.

Figure 21.2 Examples of modes in a rotating star ($\Omega = 0.6\Omega_K$). First column: three representative p modes (top to bottom: whispering gallery mode, chaotic mode, island mode). Second column: three representative g modes (top to bottom: mode in the super-inertial domain ($\omega > 2\Omega$) not too affected by rotation, mode in the sub-inertial domain ($\omega < 2\Omega$) affected by a forbidden region in the polar region, a rosette mode in the super-inertial domain. In each panel: on the left-hand side, the variable $p'/\sqrt{\rho}$ is shown in a meridional plane (p' is the pressure fluctuation and ρ the density); on the right-hand side, the kinetic energy of the mode is represented in a meridional plane.

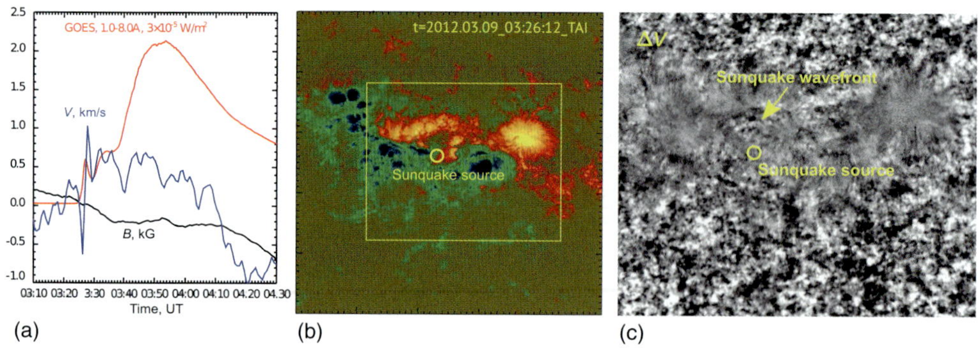

Figure 22.7 Sunquake event produced in the pre-heating phase of the M6.3 flare of March 9, 2012: (a) time profiles of the total soft X-ray flux (1.0–8.0 Å) from the GOES satellite (red), the photospheric velocity (blue) and magnetic field at the flare impact; (b) the impact location on the HMI magnetogram; (c) the Doppler velocity variations showing the sunquake wavefront in the area indicated on the magnetogram by a rectangle.

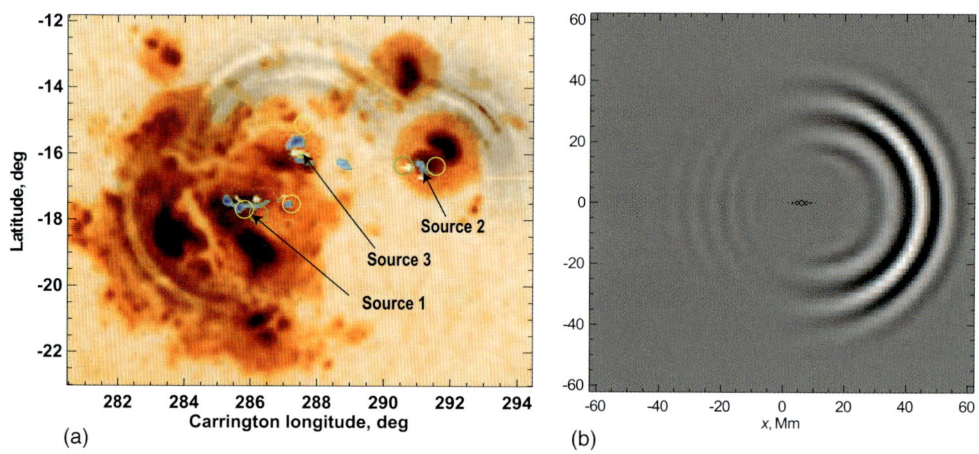

Figure 22.8 (a) Wavefronts of three sunquake events observed by the MDI instrument after the X17 flare of October 28, 2003, at 11:37 UT. The background is the corresponding MDI continuum intensity image; yellow and blue patches show down and up Doppler velocity signals with the amplitude stronger than 1 km/s, observed during the impulsive phase at 11:06 UT; yellow and green circles show locations of the hard X-ray (50–100 keV) and 2.2 MeV gamma-ray sources observed by RHESSI. (b) Theoretical model of sunquakes with a moving impact source, which explains the observed anisotropy of sunquakes. The point impulsive source is moving in the x-direction with the constant speed of 25 km/s. Its strength as a function of time has a Gaussian shape with FWHM of 3 min. The locations of the source are shown by black diamonds at the center, the size of which is proportional to the source strength.

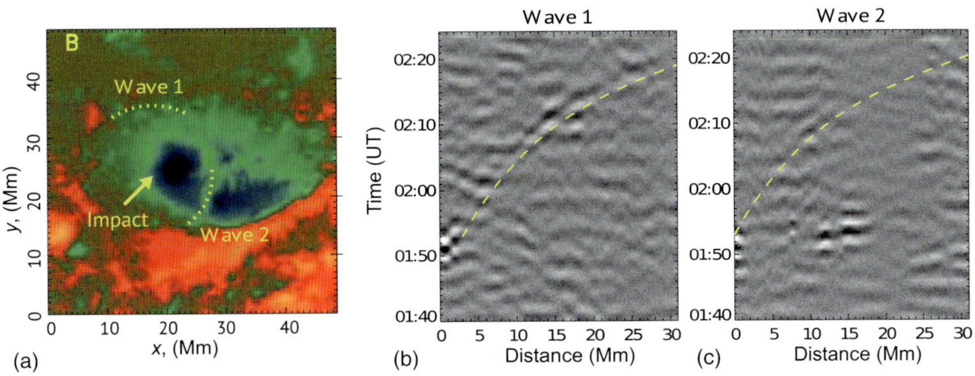

Figure 22.9 Observations of sunquakes of the X2.2 flare of February 15, 2011: (a) SDO/HMI magnetogram of AR 11158 (red, positive polarity; blue, negative polarity) and locations of the initial photospheric impact (arrow), and two wavefronts: Wave 1 traveling through a weak-field region, and Wave 2 traveling through the sunspot umbra with strong magnetic field; panels (b) and (c) show time–distance diagrams of these wavefronts illustrating that Wave 2 traveled slower and had a lower amplitude than Wave 1 (dashed curves show the theoretical time–distance relation).

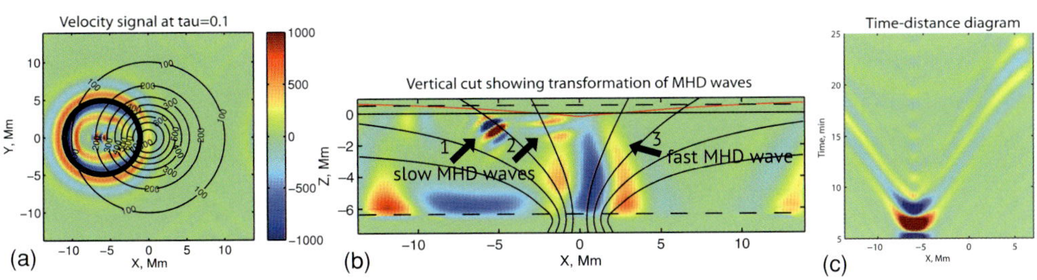

Figure 22.10 Numerical 3D simulations of MHD waves excited by a localized impulsive force and traveling through a sunspot model: (a) snapshot of the z-component of the velocity in the upper photosphere (optical depth $\tau = 0.1$). The circular contours show magnetic field strength in gauss. (b) Vertical cut of the simulated sunquake wave interacting with magnetic field. The helioseismic wave below the surface is a mixture of fast MHD and slow MHD waves. (c) Time–distance sunquake diagram of the photospheric signal.

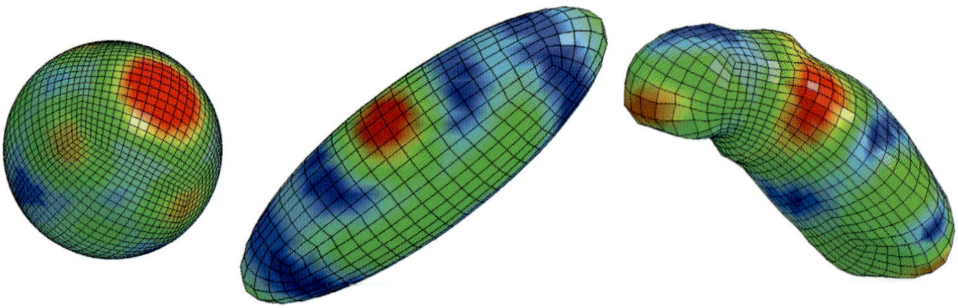

Figure 23.3 The three shapes explored in the normal mode computations: a sphere, an ellipsoid, and Itokawa. The shading is based on the surface acceleration magnitude at 10 s after the blast.

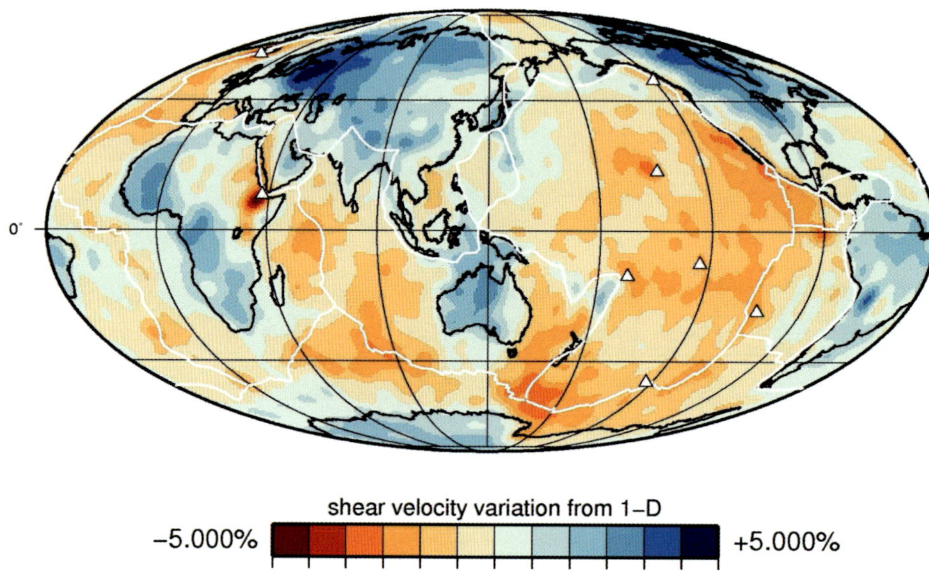

Figure 24.8 A map view of S velocity anomalies (with respect to the PREM model) at a depth of 200 km in global model S40RTS (Ritsema *et al.*, 2011).

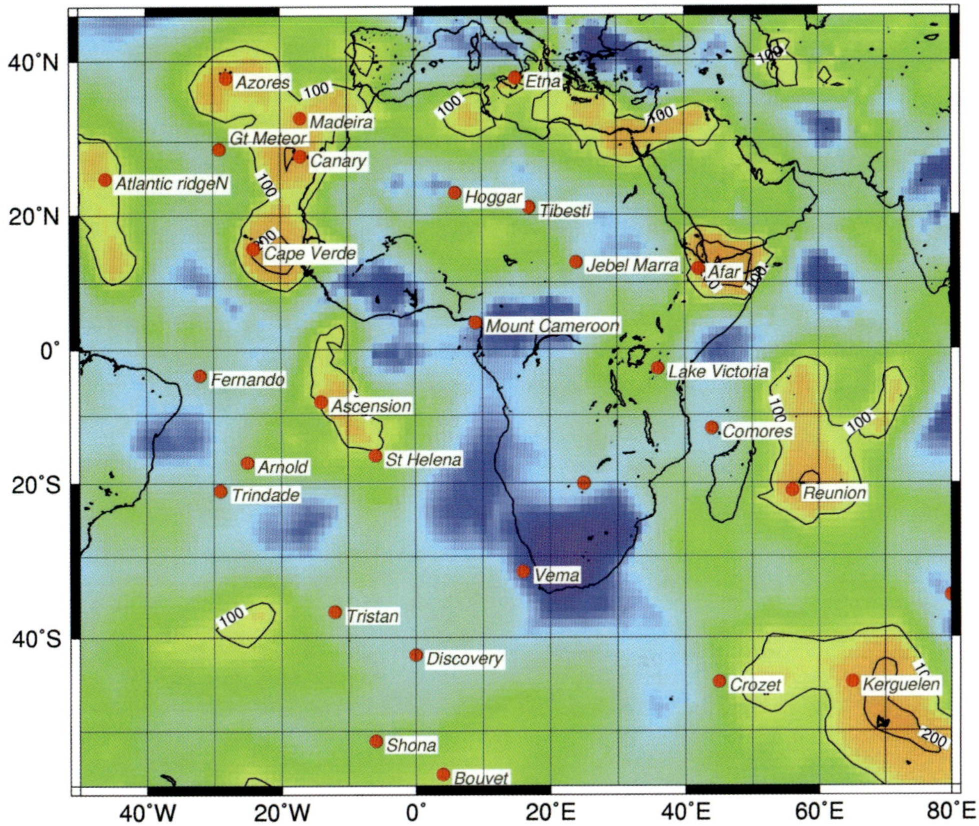

Figure 24.9 A map view of model PRI-P05 (Montelli *et al.*, 2006) converted into a temperature anomaly at a depth of 800 km, showing a correlation between high thermal anomalies of 100–200 K (yellow/orange) and the location of "hotspots" with high observed heat flux at the surface (red dots), indicating plumes rising from the lower mantle.

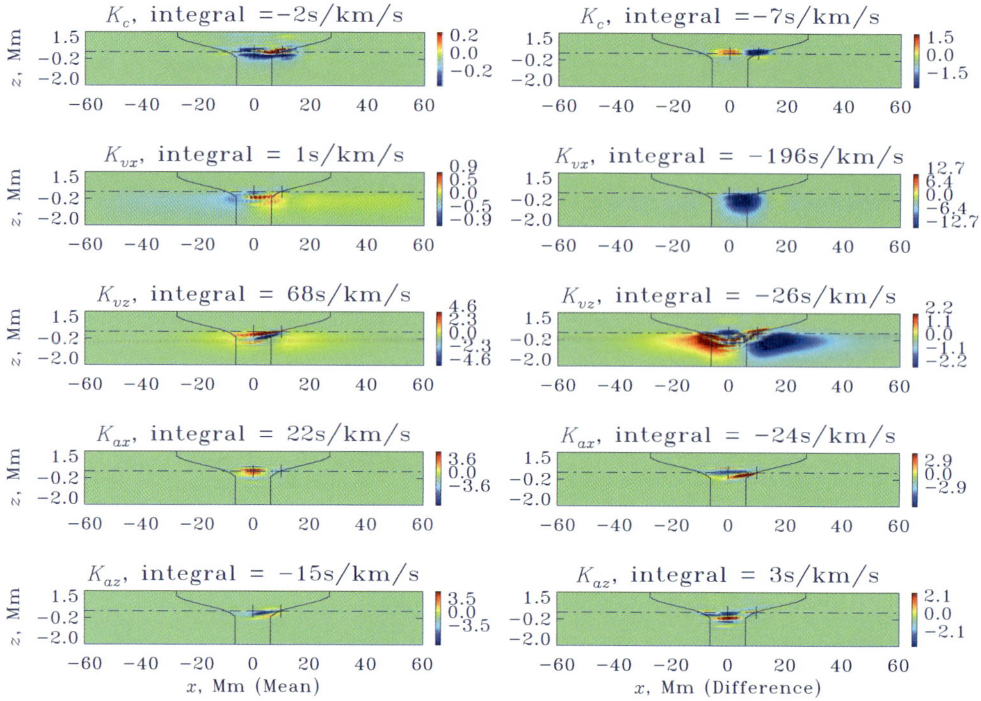

Figure 25.3 f-mode (surface) wavespeed kernels for a point-pair separated by a distance of 10 Mm. The rows display sound speed, flow, and anisotropic (Alfvén) wavespeeds, in that order. Kernels on the left column are for the mean travel time (the average of the negative and positive branch times) between the point pair, whereas the panels on the right column show kernels for the difference travel time (the difference between the negative and positive branch times). The horizontal dot-dash line denotes the height at which observations are made in the quiet Sun and the symbols mark the measurement points. The f-mode is seen to be significantly affected by the spot, as seen in the loss in symmetry of the kernels. Signatures of magnetoacoustic slow and fast modes and hints of conversion to acoustic p_1 may be discerned upon close examination. (Figure taken from Hanasoge *et al.*, 2012a, copyright (2012) by the American Physical Society.)

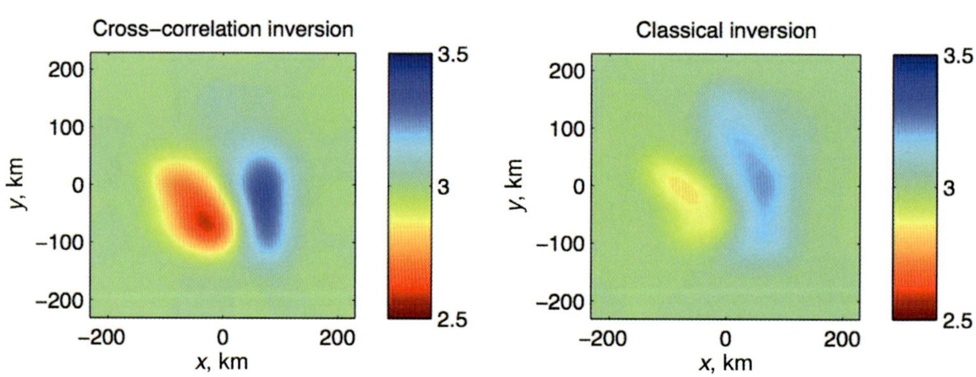

Figure 25.6 Inversions obtained using cross-correlation theory (left panel) and classical source–receiver description (right panel). The true model is shown in Figure 25.5 and "observations" for this model are estimated using the cross-correlation theory described by Hanasoge (2014). While neither inversion is perfect owing to the finite set of measurements, the classical inversion (based on the classical source–receiver description) appears to show larger errors in comparison to the cross-correlation inversion (based on the theory of cross-correlations).

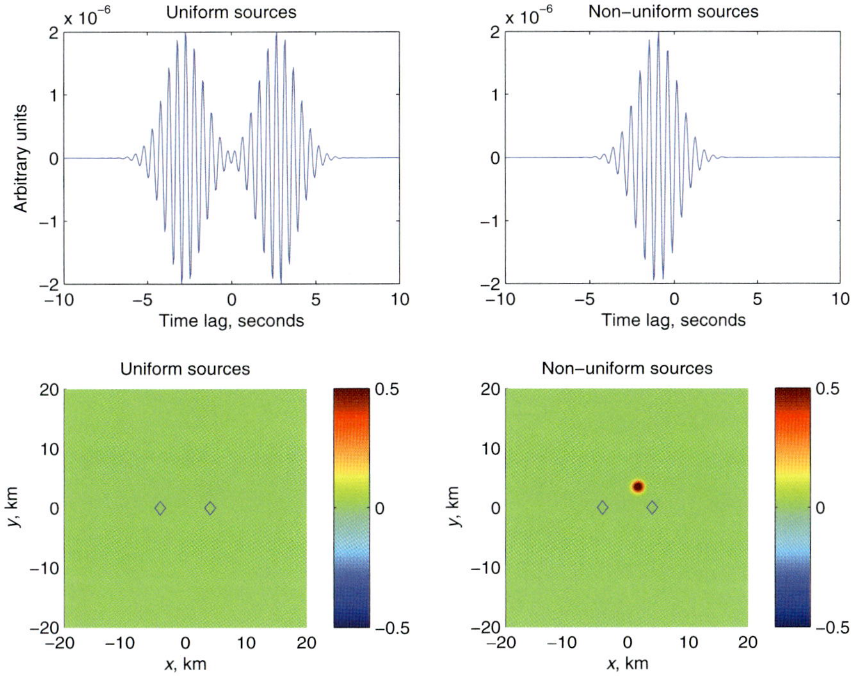

Figure 25.4 Cross-correlation (upper panels) for the source distributions in the corresponding bottom panels with stations marked by the diamonds. For the uniform model of sources, the cross-correlation has significant energy at both positive and negative time lags, peaking approximately at the time taken to travel between the two stations. However, the highly non-uniform distribution on the lower-right panel excites waves that arrive at the two stations nearly simultaneously, i.e., the difference in paths between the spot and the two stations is nearly zero. The cross-correlation is thus almost (but not quite) centered around zero time. These sorts of interpretations are very useful in terrestrial noise tomography where sources are indeed highly non-uniform. (Reproduction of Figure 7 of Hanasoge *et al.*, 2012b, Oxford University Press.)

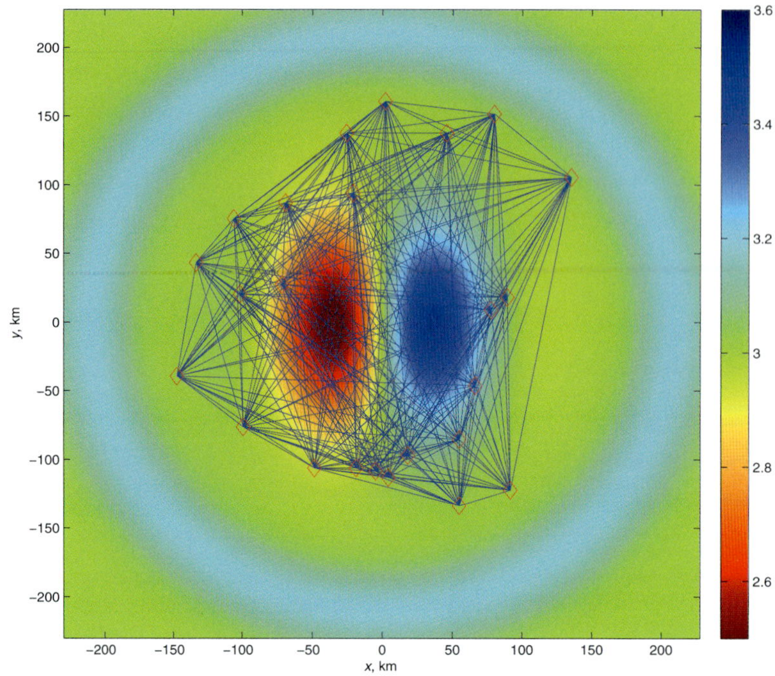

Figure 25.5 A toy problem to investigate the fidelity of cross-correlation tomography in a typical geophysical scenario. The diamonds indicate receivers surrounded by a ring of sources (the light-blue ring). The question is whether the $\pm 20\%$ wavespeed perturbation within the network can be successfully imaged. The ray-path coverage associated with the network is excellent, suggesting that the eventual inversion will likely be good. (Reproduction of Figure 5 of Hanasoge, 2014, Oxford University Press.)

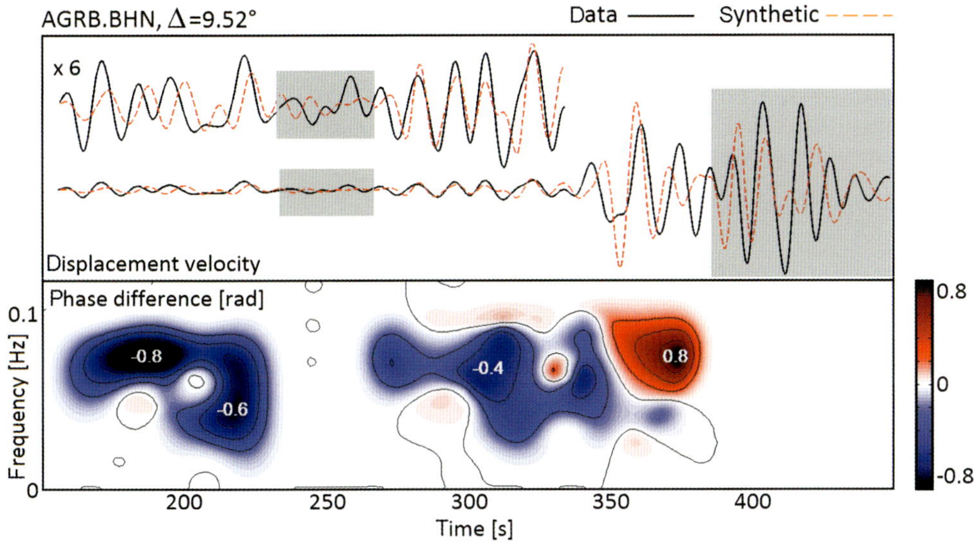

Figure 26.1 Measurement of a time–frequency phase misfit. (Top) Observed (solid) and synthetic (dashed) seismograms for station AGRB in western Turkey (distance to earthquake $\Delta = 9.52°$; see Figure 26.2 for source–receiver geometry). Oscillations prior to 340 s correspond to low-amplitude body waves amplified by a factor of 6 for better visibility. Large-amplitude surface waves appear from 340 to 450 s. Gray shading marks intervals that are excluded from the misfit calculation because waveform differences are too large for phase difference measurements. (Bottom) Phase difference, $\varphi - \varphi_0$, between observations and synthetics as a function of time and frequency. The phase differences between the small-amplitude body waves are as significant as those between large-amplitude surface waves.

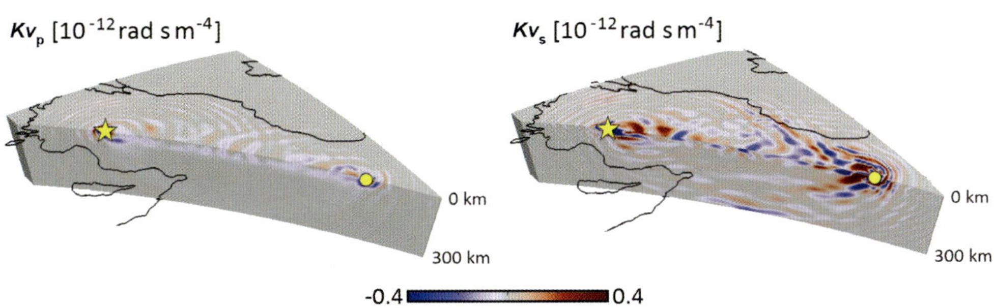

Figure 26.2 Sensitivity kernels $K_{\nu p}$ and $K_{\nu s}$ for the phase difference measurement in Figure 26.1. The positions of the earthquake source and receiver (station ARGB) are indicated by a star and filled circle respectively. Regions that are brighter or darker than average are those where perturbations in υ_p (for the case of $K_{\nu p}$) and υ_s (for the case of $K_{\nu s}$) affect the phase difference measurement. Thus, the kernels indicate where the current model should be modified to reduce the phase misfit.

compact object (neutron star or black hole). The study of these modes would thus constitute our most important source of information on their close surroundings. We will first present the observations of oscillations within the accretion disks, before describing the various models to explain them. One of the constraints of the models based on diskoseismology is due to the determination of the angular momentum of the compact object, testing whether one of the oscillations of the accretion disk corresponds to the fundamental g mode.

16.4.1 Observations of disk oscillations

Time-series observations of compact X-ray binaries, performed by X-ray timing satellites, allowed us to detect rapid variations. The analysis tool commonly used to probe fast variability is the power-density spectrum (PDS, e.g., Leahy *et al.*, 1983). The PDS of compact binaries in the X-ray domain exhibit transient and discrete features, identified as *quasi-periodic oscillations* (QPOs), detected at certain frequencies ranging from 0.01 to 450 Hz. QPOs are generally modeled with Lorentzian profiles, and distinguished from broad power peaks using a coherence parameter: $Q = \nu/\mathrm{FWHM} \geq 2$. The PDS continuum is also interesting for both its shape and integrated amplitude, usually expressed in units of *root mean square* (rms) fluctuations scaled to the mean count rate (for a more complete treatment of QPO definitions and analysis techniques, we refer the reader to, e.g., van der Klis, 2005).

Since QPO phenomena are likely emitted near the inner edge of the accretion disk, probing the innermost regions of strong gravitational fields, they are primordial in the study of accreting compact objects, such as the physical processes related to their X-ray states, the frequencies being linked to their mass, radius, and angular momentum (spin period). They can be used to test both high gravitational fields and fast rotation (with the so-called *Lense–Thirring effect*) in the frame of general relativity theory, by making predictions differing from those of Newtonian gravity. Some of these QPOs might be explained in the context of models based on diskoseismology. Following the literature (such as the excellent review by McClintock and Remillard, 2006), we divide QPOs into low-frequency and high-frequency groups.

Low-frequency QPOs

The QPOs were first identified in compact binaries hosting a white dwarf, and then in binaries with a neutron star (so-called *Z* and *atoll* sources) of unknown spin periods, thought to have relatively low magnetic fields, allowing the accretion disk to get very close to the neutron star before being disrupted by the magnetic field. The spectral variability in these systems was seen to correspond to changes in the QPO, with typical QPO frequencies in the range ~0.1–30 Hz. The fastest oscillations were found in a spectral state called the *horizontal branch*, likely resulting from the combined rotation of both the matter within the accretion disk and the neutron star (as described by the *beat frequency* model). On the other hand, during the *normal* and *flaring branches*, the star approaches its *Eddington luminosity*

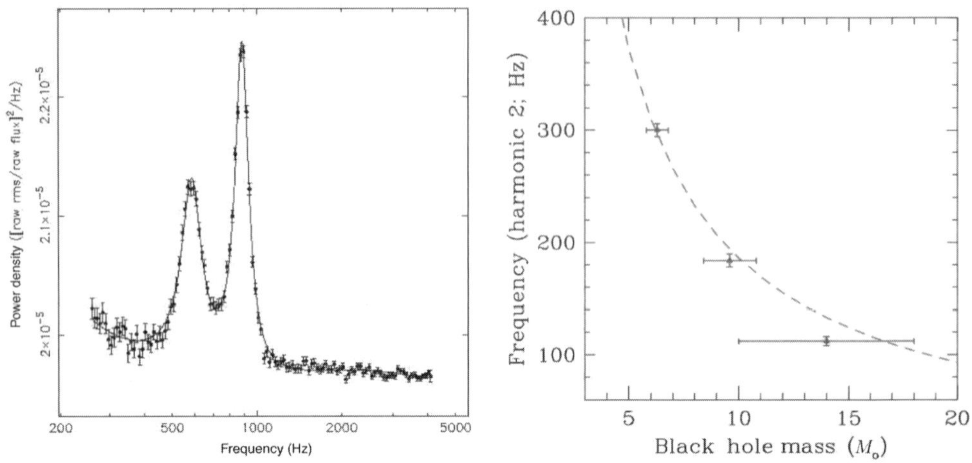

Figure 16.5 (Left) RXTE power spectrum (relative amplitude of oscillation as a function of frequency) of Sco X-1 lightcurve from February, 1996, with a close-up view of the kilohertz twin QPO at ~1000 Hz (van der Klis *et al.*, 1997). (Right) Relationship between QPO frequency and black hole mass for three black hole systems (XTE J1550-564, GRO J1655-40, and GRS 1915+105) displaying a pair of high-frequency QPOs with a 3:2 frequency ratio. The frequencies are plotted for the stronger QPO that represents $2 \times \nu_0$ (the fundamental is generally not seen in the power spectra). The dashed line shows a fit to the data: $\nu_0(\text{Hz}) = 931(M/M_\odot)^{-1}$ (Remillard and McClintock, 2006).

(the maximum X-ray luminosity reached by a compact object of given mass, when the level of radiation becomes so high that it begins to repel the accreted gas).

The origins of these QPOs remain controversial and highly debated. They could be due to an oscillating shot – a sinusoidal variation starting suddenly and decaying exponentially – created by "blobs" of gas in orbit around a rotating, weakly magnetized neutron star, accreted and emitting X-rays while approaching a magnetic pole. Then, once accreted, the blob mass decreases, making the X-rays emission decay. In this context, low-frequency QPOs are more likely related to the rotation of accreted matter than to internal oscillations of the disk.

High-frequency QPOs

Observations, starting in 1996 with the *Rossi X-ray Timing Explorer* (RXTE) satellite, detected fast variability in neutron stars and black holes, with QPO frequencies up to ~450 Hz for black holes and ~2000 Hz for neutron stars respectively. We report for instance in the left panel of Figure 16.5 the power spectrum of Sco X-1, which shows two tiny peaks near ~1000 Hz. These *twin peak* QPOs – two oscillations of roughly the same power – were often detected at high amplitudes in compact X-ray binaries. High-frequency QPOs, with behavior similar to the low-frequency ones, can be used to determine the intrinsic characteristics of compact objects (the equation of state of neutron stars, and mass of black holes).

On one hand, the saturation seen in QPO frequency around neutron stars suggests that rotating blobs have reached the *innermost stable circular orbit* (ISCO) around the compact object, a radius predicted by general relativity, inside which matter must inexorably spiral down to smaller radii. On the other hand, the right panel of Figure 16.5 suggests that the hot gas spirals in toward the black hole event horizon, piles up in the inner part of the accretion disk, radiating X-rays with varying intensity which creates the QPO. The frequency depends on the mass of the compact object, with a higher-frequency QPO being caused by a pile-up zone lying closer to a low-mass black hole, and a lower-frequency QPO being created further away by a higher-mass black hole. High-frequency QPOs might thus be described by diskoseismology.

16.4.2 Models of disk oscillations

A large number of physical mechanisms have been proposed to describe QPOs, and only a few are based on diskoseismology. We briefly discuss here some of these models, referring the reader to McClintock and Remillard (2006) for more details.

Low-frequency QPOs

Models reproducing low-frequency QPOs include global disk oscillations (Titarchuk and Osherovich, 2000), radial oscillations of accretion structures such as shock fronts (Chakrabarti and Manickam, 2000), and oscillations in a transition layer between the disk and a hotter Comptonizing region (Nobili *et al.*, 2000). Alternatively, the accretion–ejection instability model invokes spiral waves in a magnetized disk (Tagger and Pellat, 1999), with a transfer of energy out to the radius where material corotates with the spiral wave. The originality of this model resides in combining magnetic instabilities with Keplerian motion, to explain the observed QPO amplitudes together with their stability.

Low-frequency QPO models have to reproduce both the observed range of frequencies and the fact that oscillations are stronger at photon energies above 6 keV, i.e., where the power law component dominates the accretion disk emission. We point out that the biggest difficulty encountered by these models is to reproduce low-frequency QPOs based on the geometry and flow of the accreted gas, their frequencies being much lower than the Keplerian frequencies, while the expected orbits would be inside the inner part of the accretion disk. Because of the low frequency of these QPOs, all models based on diskoseismology, and therefore related to the inner parts of the accretion disk, do not easily reproduce them, focused instead on the description of the high-frequency QPOs, as discussed below.

High-frequency QPOs

Many models have been created to account for observations of high-frequency oscillations within accretion disks. For instance, the *sonic point model* describes the kHz QPOs taking into account the effects of general relativity, by following the trajectories of blobs of gas,

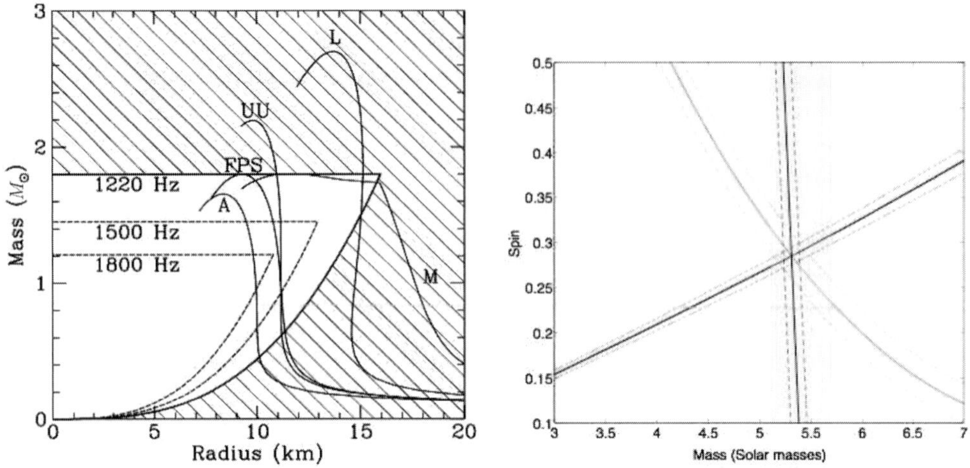

Figure 16.6 (Left) Constraints on the equation of state (EOS) of nuclear matter, brought by the highest observed kHz QPO frequencies of ~1220 Hz. A kHz QPO of 1500 Hz would exclude all current neutron gas EOS (Miller *et al.*, 1998). (Right) Black hole spin as a function of the mass as predicted by the three equations of the relativistic precession model for GRO J1655-40. The derived parameters are $M = 5.31 \pm 0.07\ M_\odot$ and spin $a = 0.290 \pm 0.003$ (Motta *et al.*, 2014).

leaving the inner edge of the accretion disk, and spiraling inward supersonically until they collide with the neutron star surface. This model uses observations of the highest frequency kHz QPOs to relate the pressure of a gas of dense neutrons with its density, and to place constraints on the allowed equation of state by determining the mass–radius equilibrium relation for neutron stars (see Figure 16.6 left; Miller *et al.*, 1998).

Alternatively, since high-frequency QPOs do probe strong gravity fields in accretion disks surrounding *compact* objects, a purely general relativity formulation of diskoseismology, initiated by Kato and Fukue (1980), continued by Nowak and Wagoner (1991, 1992), and then extended by Kato and Nakamura (1998), Wagoner (1999); Wagoner *et al.* (2001), might be necessary to make accurate predictions, and thus correctly describe the observations. These models of "diskoseismic" oscillations, by adopting a global view of the inner disk as a resonance cavity in the general relativity frame, are useful to tentatively explain high-frequency QPOs; however, integral harmonics are not predicted for the three types of diskoseismic modes derived for adiabatic perturbations in a thin accretion disk. Thus, there is clearly a need to further investigate the possibility of "diskoseismic" resonances, since one QPO *has* to be the fundamental g-mode, in order not to rule out the models based on diskoseismology to explain the QPO. By comparing the angular momentum of stellar-mass black holes derived either via continuum and line methods or via diskoseismology, Wagoner (2012) shows that the second highest frequency QPO might be produced by this g-mode, ruling out however the models in which only one QPO frequency corresponds to the ISCO.

In this context, an interesting example is the stellar-mass black hole GRO J1655-40, for which precise mass and spin measurements have been recently obtained, through analysis of a complete set of RXTE X-ray observations, based on frequencies of particle motion in the close vicinity of the ISCO, as predicted by the relativistic precession model (see Figure 16.6, right, Motta *et al.*, 2014). While QPOs keep their mysteries, and diskoseismology has yet to show that it can describe them, they remain the only observables allowing us to test diskoseismology.

16.5 Conclusion

The new field of diskoseismology is a powerful tool to find the mass and angular momentum of compact objects inside a strong gravitational field. The results obtained are currently highly model dependent, and heavily based on general relativity. The new perspectives in this field depend on (i) a deep theoretical understanding of relativistic models in a strong gravity field, (ii) the development of heavy simulations, and (iii) the launch of the next X-ray missions, such as *ATHENA+* (Advanced Telescope for High ENergy Astrophysics), an X-ray observatory to be proposed as a large mission for the ESA (European Space Agency) science program, and also *LOFT* (Large Observatory For X-ray Timing), a proposed ESA Cosmic Vision M3 satellite slated to fly around 2022, and particularly devoted to the timing study of the rapid X-ray variability of compact objects such as neutron stars and black holes.

To conclude, only a broad study based on new observations, simulations, and theoretical developments will allow us to get a better understanding of diskoseismology. This is very promising for understanding the strong gravitational fields of X-ray binaries, supermassive black holes, gamma-ray bursts, and also young stellar objects. As a final note, it is interesting to point out that diskoseismology is a new transverse study, which will ultimately enable us to determine physical characteristics of distinct celestial objects, by using common observational techniques and physical models. In this sense, *diskoseismology is an interdisciplinary field.*

17

Numerical simulations of magnetoconvection and helioseismology

IRINA N. KITIASHVILI

The complex dynamics of highly turbulent solar magnetoconvection is a source of acoustic waves, and also various self-organization processes, observed on different scales on the Sun: from tiny vortex tubes to large-scale convection and active regions. Recent advances in computational capabilities have made it possible to create realistic numerical models, based on first principles, including radiative energy transfer, ionization, and magnetic fields. The simulations have provided important insight into the mechanism of acoustic emission on the Sun, and the physics of wave interactions with flows and magnetic fields. Observations of subsurface dynamics of the Sun by helioseismology are obtained by measuring acoustic travel-time anomalies, and reconstructing variations of the sound speed and behavior of convective flows in the solar interior. However, strong stratification of the turbulent convective medium, inhomogeneity of flow, and complicated topology of the magnetic field create uncertainty in interpretation of helioseismology observations. Because there are no direct subsurface observations, a synergy of the magnetohydrodynamic (MHD) numerical simulations and observations provides a basis for verification and improvement of helioseismology methods.

17.1 Introduction

Understanding and characterization of solar magnetoconvection is a key problem of heliophysics and astrophysics. Solar turbulence driven by convective energy transport determines the dynamical state of the plasma, and leads to excitation of acoustic waves. Observed oscillations contain information about hidden dynamics of the subsurface layers, and are a powerful tool to probe the solar and stellar interior conditions.

Large parallel supercomputers, especially those developed in the last few years, allow us to perform large-scale realistic simulations of MHD processes on the Sun. These simulations have greatly improved our understanding of the multi-scale structure and dynamics of solar

Extraterrestrial Seismology, eds. Vincent C. H. Tong and Rafael A. García. Published by Cambridge University Press.

convection. A characteristic feature of simulations of this type is that they include all essential physics from first principles, and take into account the real-plasma equation of state, radiative transfer, chemical composition, ionization effects, and the effects of magnetic fields, but they do rely on subgrid-scale models of turbulence. Realistic 3D numerical simulations have reproduced and explained many observed effects in sunspots and magnetic active regions (e.g., Schüssler and Vögler, 2006; Cameron *et al.*, 2007; Cheung *et al.*, 2008; Jacoutot *et al.*, 2008b; Kitiashvili *et al.*, 2009, 2010a; Rempel, 2011; Stein and Nordlund, 2012), as well as in the quiet Sun (e.g., Nordlund, 1985; Nordlund *et al.*, 2009; Brandenburg *et al.*, 1996; Carlsson, 1998; Stein and Nordlund, 2001; Wedemeyer *et al.*, 2004; Vögler *et al.*, 2005; Steiner *et al.*, 2010; Kitiashvili *et al.*, 2011a, 2012, 2013; Beck *et al.*, 2013), and allow us to predict new effects which are difficult to detect in observations.

The organization of this chapter is the following. In Section 17.2 I will briefly describe the basic equations, which are solved for modeling of solar convection. Then in Section 17.3, I will discuss oscillatory properties of the solar magnetoconvection, and effects of topology and strength of the magnetic field. The excitation mechanism of oscillations in the quiet-Sun regions is presented in Section 17.4.

17.2 Basic equations and "SolarBox" code

For investigation of the oscillatory properties of solar magnetoconvection and the excitation mechanism of acoustic waves we use the 3D radiative MHD code "SolarBox" (Jacoutot *et al.*, 2008a). The code takes into account fluid flow compressibility in a highly stratified medium, the real-gas equation of state, the standard model of the solar interior, and the OPAL opacity tables (Rogers *et al.*, 1996). The radiative energy is calculated through 3D multi-spectral-bin radiation transfer between fluid elements assuming the local thermodynamic equilibrium.

Because both the hydrodynamic and magnetic Reynolds numbers are very high, an important feature of this code is implementation of various subgrid-scale large eddy simulation (LES) turbulence models (for basic formulation and applications of LES modeling, see Lesieur *et al.*, 2005). Here we adopted the most widely used Smagorinsky model (Smagorinsky, 1963) and its compressible dynamic formulation (Moin *et al.*, 1991; Germano *et al.*, 1991). The turbulent electrical conductivity is calculated by using an extension of the Smagorinsky model for the MHD case (Theobald *et al.*, 1994). The code was verified by comparison with the "Stagger" code (Stein and Nordlund, 2001), and has been used to study solar oscillations (Jacoutot *et al.*, 2008a), magnetoconvection of sunspots (Kitiashvili *et al.*, 2009), and processes of spontaneous formation of magnetic structures (Kitiashvili *et al.*, 2010a).

The governing equations are the grid-cell averaged MHD equations (Priest, 2003) for the conservation of mass (17.1), momentum (17.2), energy (17.3), and magnetic flux (17.4):

$$\frac{\partial \rho}{\partial t} + (\rho u_i)_{,i} = 0, \tag{17.1}$$

$$\frac{\partial \rho u_i}{\partial t} + \left(\rho u_i u_j + (P_{ij} + \rho \tau_{ij})\right)_{,j} = -\rho \phi_{,i}, \tag{17.2}$$

$$\frac{\partial E}{\partial t} + \Bigg[E u_i + (P_{ij} + \rho \tau_{ij}) u_j - (\kappa + \kappa_T) T_{,i} + \left(\frac{c}{4\pi}\right)^2 \frac{1}{\sigma + \sigma_T} \left(B_{i,j} - B_{j,i}\right) B_j + F_i^{rad} \Bigg]_{,i} = 0, \tag{17.3}$$

$$\frac{\partial B_i}{\partial t} + \left[u_j B_i - u_i B_j - \frac{c^2}{4\pi(\sigma + \sigma_T)} \left(B_{i,j} - B_{j,i}\right) \right]_{,j} = 0, \tag{17.4}$$

where ρ is the average mass density, u_i is the Favre-averaged velocity (density weighted) (Favre, 1969), B_i is the magnetic field, and E is the average total energy density, $E = \frac{1}{2}\rho u_i u_i + \rho e + \rho \phi + \frac{1}{8\pi} B_i B_i$, where ϕ is the gravitational potential and e is the Favre-averaged internal energy density per unit mass. F_i^{rad} is the radiative flux, which is calculated by solving the radiative transfer equation, and P_{ij} is the averaged stress tensor, $P_{ij} = \left(p + \frac{2}{3}\mu u_{k,k} + \frac{1}{8\pi} B_k B_k\right) \delta_{ij} - \mu \left(u_{i,j} + u_{j,i}\right) - \frac{1}{4\pi} B_i B_j$, where μ is the viscosity. The gas pressure p is a function of e and ρ calculated through a tabulated equation of state (Rogers *et al.*, 1996); τ_{ij} is the Reynolds stress, κ is the molecular thermal conductivity, κ_T is the turbulent thermal conductivity, σ is the molecular electrical conductivity, and σ_T is the turbulent electrical conductivity.

17.3 Oscillations in the solar magnetoconvection

There is no doubt that there is strong coupling between oscillations of the solar surface and turbulent convection, which is affected by magnetic fields generated locally in subsurface layers or emerging from deeper layers of the convection zone. Studying the interaction of highly turbulent near-surface convection and magnetic fields allows us to determine links between properties of magnetoconvection and oscillations.

17.3.1 Effects of magnetic fields on properties of magnetoconvection

It is well known from ground and space observations that dynamical properties of the solar surface are oscillatory and highly inhomogeneous, and usually divided between conditions of quiet-Sun and active regions. Such division also reflects a multi-scale distribution of magnetic fields, from subgranular to global. Because the oscillations are produced by turbulent flows, such inhomogeneity, and also complicated magnetic topology, can have significant influence on oscillation modes. Properties of solar oscillations have been investigated using 3D numerical simulations by many authors (see, e.g., Stein and Nordlund, 2001; Georgobiani *et al.*, 2000; Nordlund and Stein, 2001; Jacoutot *et al.*, 2008a,b; Kitiashvili *et al.*, 2011b). The results have provided important insights into the excitation mechanism (e.g., Stein and Nordlund, 2001), and interpretation of the mode line asymmetry (Georgobiani *et al.*, 2003). The simulations have been also used for testing the local correlation tracking technique (Georgobiani *et al.*, 2007) and time–distance helioseismology (e.g., Zhao *et al.*, 2007).

Analyzing the MHD simulations with initially vertical magnetic fields of various strengths, Jacoutot *et al.* (2008b) found enhanced excitation of high-frequency "pseudo-modes," which reached a maximum amplitude for a moderate field strength of ~600 G. It was suggested that this effect may qualitatively explain the phenomenon of "acoustic halos" observed around sunspots and active regions in the frequency range of 5.5–7.5 mHz (see, e.g., Braun *et al.*, 1992; Jain and Haber, 2002).

To simplify the modeling of the topologically complicated and strongly varying structure of magnetic fields on the solar surface, it is instructive to consider small regions with simple magnetic topology. This approach allows us to assume that the mean magnetic field in such regions has constant inclination and strength. However, no restrictions on the local topological structure and dynamics are imposed (Kitiashvili *et al.*, 2009). To investigate the effects of magnetic field strength and topology we perform a series of simulations varying the magnetic field strength from 0 to 3000 G, and the inclination angle from 0° (vertical to the photosphere) to 85° (almost horizontal). Such a simulation setup is used for studying basic features of the Evershed flow and filamentary structures of sunspot umbra and penumbra, including the "sea-serpent" behavior of magnetic field lines outside the penumbra (Kitiashvili *et al.*, 2010b).

Magnetic fields have strong influence on convective motions (Figure 17.1). For instance, in regions with a strong vertical magnetic field (see Figure 17.1a,c,e), such as the sunspot umbra, convection is largely suppressed. The umbra convection is characterized by horizontally small and vertically long convective cells, and slow flows (Schüssler and Vögler, 2006; Kitiashvili *et al.*, 2011b). Such convective structures are observed as small-scale regions of enhanced brightness, "umbral dots" (e.g., Ortiz *et al.*, 2010). The horizontal magnetic fields tend to stretch convective granules along the field lines. Increasing the magnetic field inclination causes stretching of convective granules and formation of filamentary penumbra-like structure (Figure 17.1b,d,f). In the inclined field ($\alpha = 85°$) of the same strength, the magnetic field structure is very different and forms arch-like structures below the surface and almost horizontally elongated vortex tubes at the surface (Kitiashvili *et al.*, 2011b). The magnetic field inclination breaks the horizontal homogeneity of convection and leads to formation of a mean shear flow (Evershed flow) along the magnetic field lines. The effect becomes stronger with increasing field inclination and strength (Kitiashvili *et al.*, 2009).

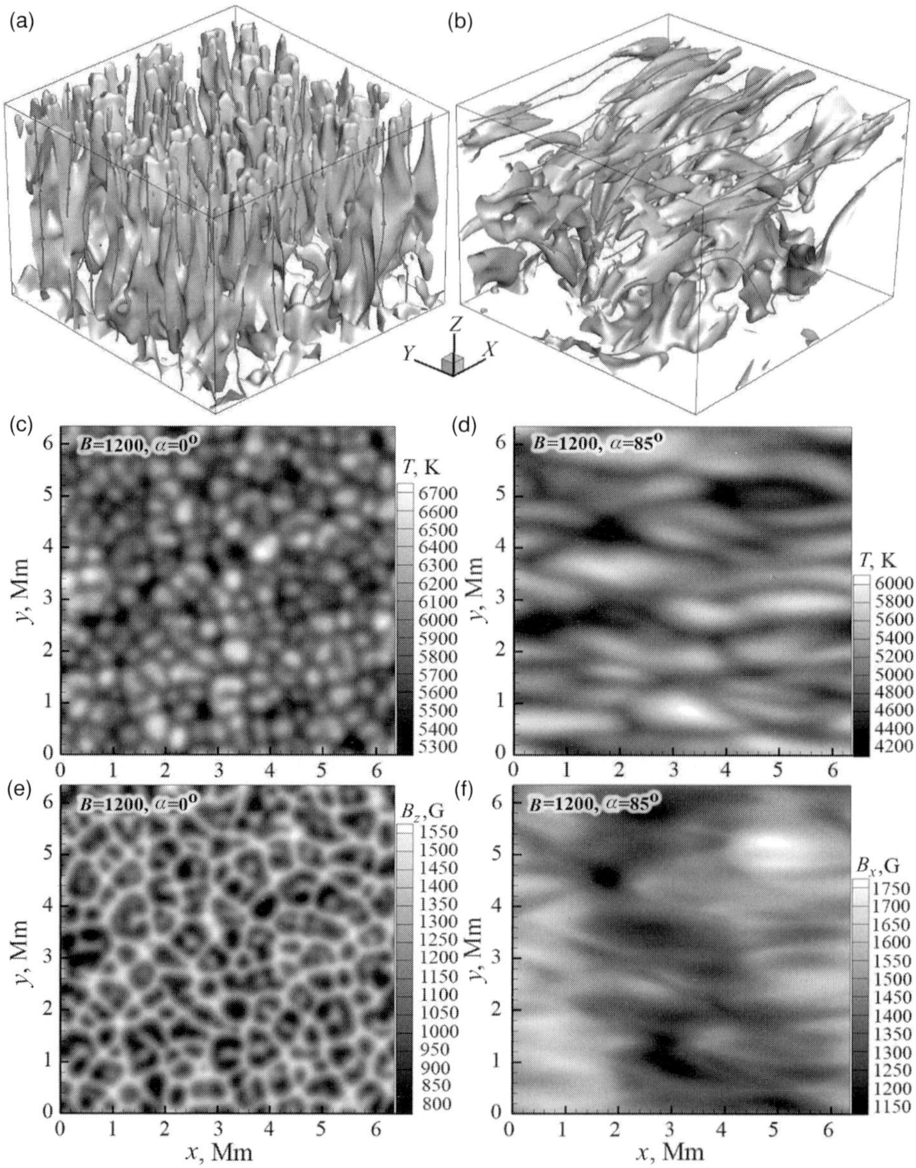

Figure 17.1 Effect of magnetic field inclination on the photospheric structure of convection for the mean magnetic field of 1200 G. Panels a, c, and e correspond to simulations with initially uniform vertical magnetic field ($\alpha = 0^\circ$), and panels b, d, and f show the results for the inclined field $\alpha = 85^\circ$. Panels a and b illustrate the 3D structure of the magnetoconvection, where isosurfaces correspond to the field strength of 1450 G and 1700 G respectively. Curves illustrate the topology of magnetic field lines. Panels c and d show the temperature distribution, and panels e and f show the vertical magnetic field distribution at the photosphere.

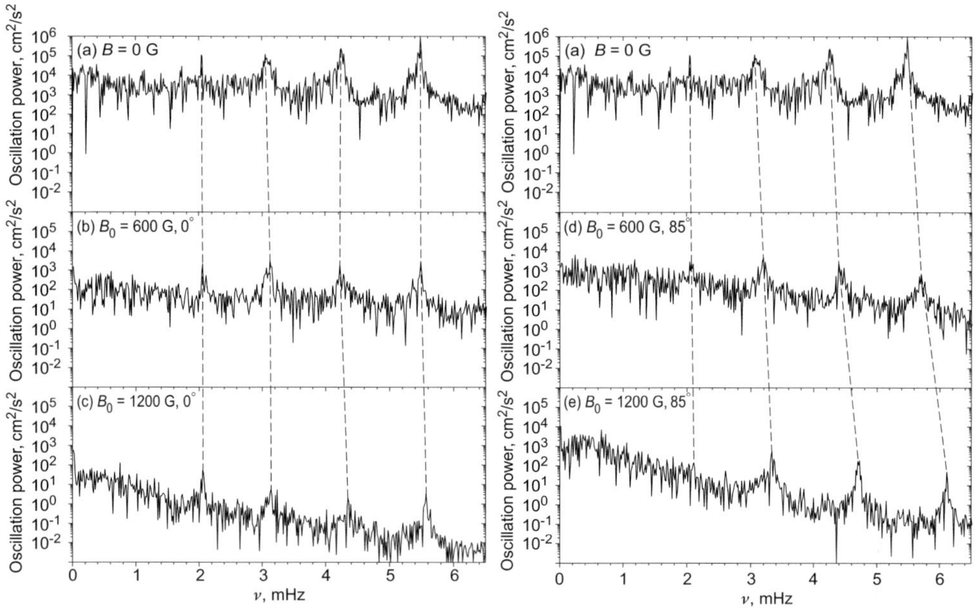

Figure 17.2 Oscillation power spectra of the vertical velocity for different magnetic field strengths and inclinations: (a) $B_0 = 0$ G; (b) $B_0 = 600$ G, $\alpha = 0°$; (c) $B_0 = 1200$ G, $\alpha = 0°$; (d) $B_0 = 600$ G, $\alpha = 85°$; and (e) $B_0 = 1200$ G, $\alpha = 85°$.

17.3.2 Frequency shift of the radial oscillations

Because both the magnetic field strength and inclination affect the dynamical and structural properties of magnetoconvection, it is interesting to consider effects of magnetic fields on the properties of solar oscillations in local regions. To investigate radial or large-scale oscillations under different conditions it is natural to calculate the power spectral density for the vertical velocity fluctuations, averaged over the whole horizontal domain corresponding to the photospheric level. This corresponds to considering radial oscillations or oscillations of a low angular degree with the horizontal wavelength much larger than the size of the computational domain (6 Mm). Because of the small domain of the simulations, no attempt is made to extract high-degree modes. The power spectra are calculated using 20 hours of the simulated vertical velocity with a cadence of 30 seconds, after the magnetoconvection reaches a stationary state.

Figures 17.2b,c illustrate how the power spectrum changes with increasing vertical magnetic field strength, for values of 0, 600, and 1200 G. Figures 17.2d,e show the power spectra for the highly inclined fields of the same strengths. The two identical top panels (Figure 17.2a) show the spectrum without a magnetic field, for comparison. In the frequency range of 0–6 mHz, the power spectra display four mode peaks, the frequencies of which are determined by the resonant conditions between the bottom boundary conditions of our

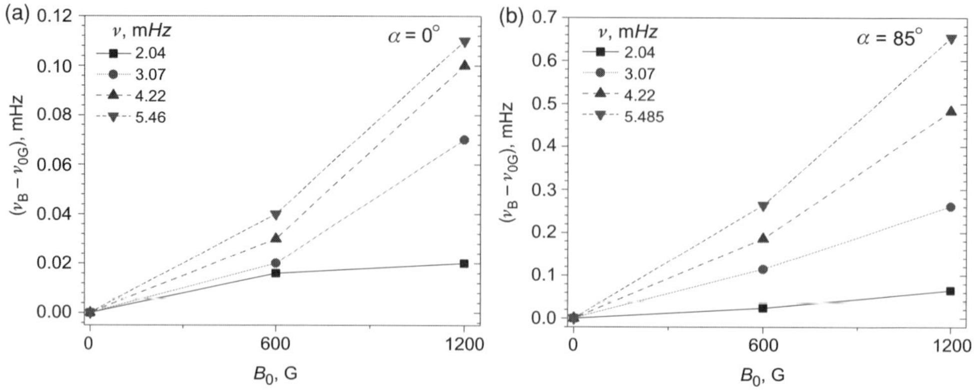

Figure 17.3 Frequency shifts of the oscillation modes (indicated by different symbols) relative to the case of non-magnetic convection for the vertical ($\alpha = 0°$, panel a) and almost horizontal ($\alpha = 85°$, panel b) magnetic fields.

domain and the near-surface reflective layer. In this simulation, the high-frequency spectra are not considered because the top boundary condition is reflecting.

Comparison of the oscillation spectra in Figure 17.2 shows three main dependencies: (i) suppression of the power by the magnetic field, (ii) shift of the mode peaks to higher frequencies, which is greater for the stronger field, and (iii) increase of the width of the resonant peaks with the magnetic field. The frequency shift increases with the mode frequency. It is interesting that the power suppression is stronger for the vertical magnetic field than for the inclined field, but the frequency shift is significantly stronger for the inclined field than for the vertical field. Also, the frequency dependence of the shift is stronger for the inclined magnetic field. The frequency shifts estimated from the positions of the peak maxima are shown in Figure 17.3.

Understanding of the effects of different conditions of magnetoconvection is closely related to the mechanism of acoustic wave excitation. In the next section I describe one of the most prominent sources of acoustic oscillations, caused by interaction of vortex tubes in subsurface layers of the convection zone.

17.4 Acoustic wave excitation on the Sun

17.4.1 Vortices in the solar photosphere

It is known that turbulent flows are a primary source of observed oscillations in the photosphere. Numerical models support the observational evidence of near-surface acoustic sources located in the intergranular lanes or in their vicinity (Goode *et al.*, 1992) and allow us to identify an important class of acoustic sources associated with interactions of turbulent vortex tubes. Because helical motions represent intrinsic properties of turbulent flows, it is not surprising that swirling motions are found in observations of the highly turbulent solar magnetoconvection. Small-scale vortex flows in the photosphere ($\sim$3 Mm in diameter) were initially detected by Brandt *et al.* (1988), and small-scale swirling flows ($\sim$0.5 Mm) were observed by Wang *et al.* (1995). Observations have shown that vortices are ubiquitous in non-magnetic quiet-Sun regions (Pötzi and Brandt, 2005; Bonet *et al.*, 2008, 2010). These whirlpools are associated with intergranular lanes. They are $\leq$0.5 Mm in diameter, and their lifetime varies from 5 min to $\geq$20 min. Such swirling motions correspond to vertically oriented vortex tube structures that have been found in numerical simulations (e.g., Brandenburg *et al.*, 1996; Kitiashvili *et al.*, 2011a).

17.4.2 Mechanism of acoustic wave excitation due to vortex tube interactions

The vortices frequently interact with each other leading to the appearance of concentric waves, excited approximately at a "join" center of the vortex cores (Figure 17.4). The wavefronts have the form of the circle or a sector, depending on the source structure, the interference of waves from other sources, and on the dynamic of surrounding flows. Figure 17.4 shows the process of acoustic wave excitation in a sequence of the density difference with a cadence of 30 s, in a horizontal slice corresponding to the surface. This process of acoustic wave excitation is very common in the simulations. However, identification of individual excitation events is difficult because the wave amplitude is of the same order of magnitude as the amplitude of acoustic waves coming from other random acoustic sources and the turbulent convection. Nevertheless, many excitation events with circular-shaped wave flows, similar to those shown in Figure 17.4, can be identified and examined.

The propagating wavefronts can also be shown on time–distance diagrams, obtained by plotting the density perturbations averaged over a range of angles for different distances from a vortex location. Figure 17.5 shows two examples of acoustic wave propagation in such time–distance diagrams. The diagrams show normalized density fluctuations. The light inclined ridges correspond to the acoustic waves. The wave speed is on average 8–10 km/s, but may vary probably due to background convective flows. In some cases, sequences of acoustic waves (Figure 17.5b), when several wavefronts are produced at $\sim$2–2.5 min intervals, can be seen.

Figure 17.6 shows snapshots of the surface layer for the temperature, density, vertical velocity, and the magnitude of horizontal velocities in a hydrodynamic run. The white squares indicate several vortex structures (whirlpools). The detailed structure of a single vortex is shown in the bottom left corner of each panel in Figure 17.6 (Kitiashvili *et al.*, 2011a). Structure of the vortices can be described by the following properties: (i) low-temperature core; (ii) dramatically decreased density and gas pressure; (iii) strong (up to 9 km/s) down flows; and (iv) high-speed, often supersonic, horizontal velocities up to 11 km/s.

The vertical vorticity distribution reveals a correlation between the excitation events of the acoustic waves and the interaction of vortices of opposite vorticity (Figure 17.4).

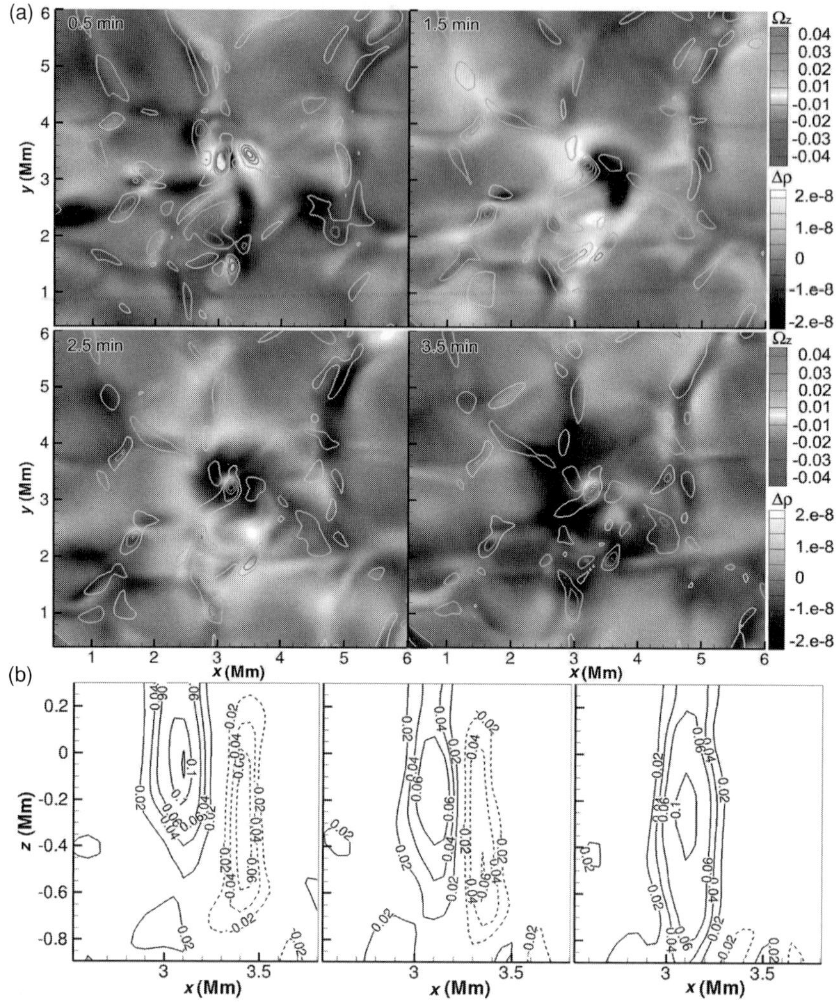

Figure 17.4 (a) Panels displaying the temporal evolution of density fluctuations (calculated as $\rho(t_{i+1}) - \rho(t_i)$) at the solar surface with cadence 1 min illustrate acoustic wave excitation and radial propagation from a vortex source, representing the interaction of two vortices with opposite-sign vertical vorticity. Red and blue lines correspond to the magnitude of the positive (clockwise) and negative (counterclockwise) vertical vorticity. (b) Different stages of the subsurface vortex interaction with a partial annihilation. Solid and dashed isolines show the magnitude of the positive and negative vertical vorticity. A black and white version of this figure will appear in some formats. For the color version please refer to the plate section.

In this particular example, two local concentrations of the vertical vorticity with opposite sign move close together and partially annihilate under the surface, resulting in a partial cancelation (reduction of the magnitude) of the negative vorticity. In general, the vortex interaction is a complex process, in which the turbulent flows deform the shape of the

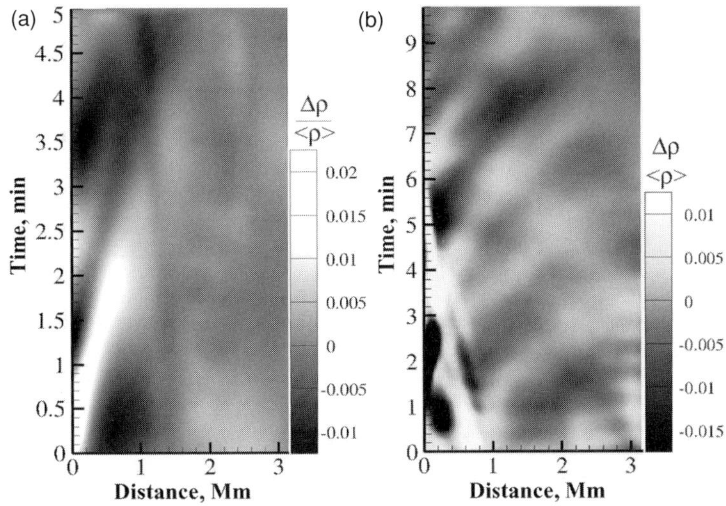

Figure 17.5 Time–space diagrams of normalized density fluctuations for (a) a single wave excitation event and (b) a series of acoustic wave excitation events. The slope of the ridges corresponds to a mean speed of 8–10 km/s.

swirls and also stretch smaller vortices around bigger "main" ones. The depth of the wave excitation events (or the vortex annihilation process) is within the top 0.5 Mm layer. But more detailed investigation is needed, because the structure and dynamics of the vortex tubes result in very complicated wavefronts.

17.5 Conclusions

Realistic numerical simulations have provided us with important insights into the structure and dynamics of solar convection and excitation of acoustic oscillations on the Sun, as well as solar-type stars (Stein and Nordlund, 2001; Kitiashvili *et al.*, 2011a). The simulations represent an important tool for understanding many observed phenomena, verification and validation of theoretical models, and interpretations of observations. Radiative MHD simulations of magnetoconvection in the upper turbulent convective layer and the low atmosphere of the Sun allow us to investigate changes in the oscillation power spectrum in local regions with different magnetic field strengths and topology. The oscillations in these simulations are naturally excited by the turbulent convection.

The results illustrated in this chapter show two basic effects in magnetic field regions: power suppression and shifts of the mode peaks to higher frequencies. The power suppression is stronger for the vertical magnetic field, but the frequency shift is stronger for the inclined magnetic field. The stronger frequency shift in the inclined magnetic field can be understood in terms of physical properties of fast MHD waves. The speed of these waves is higher when they travel across the magnetic field lines, and this leads to higher modal

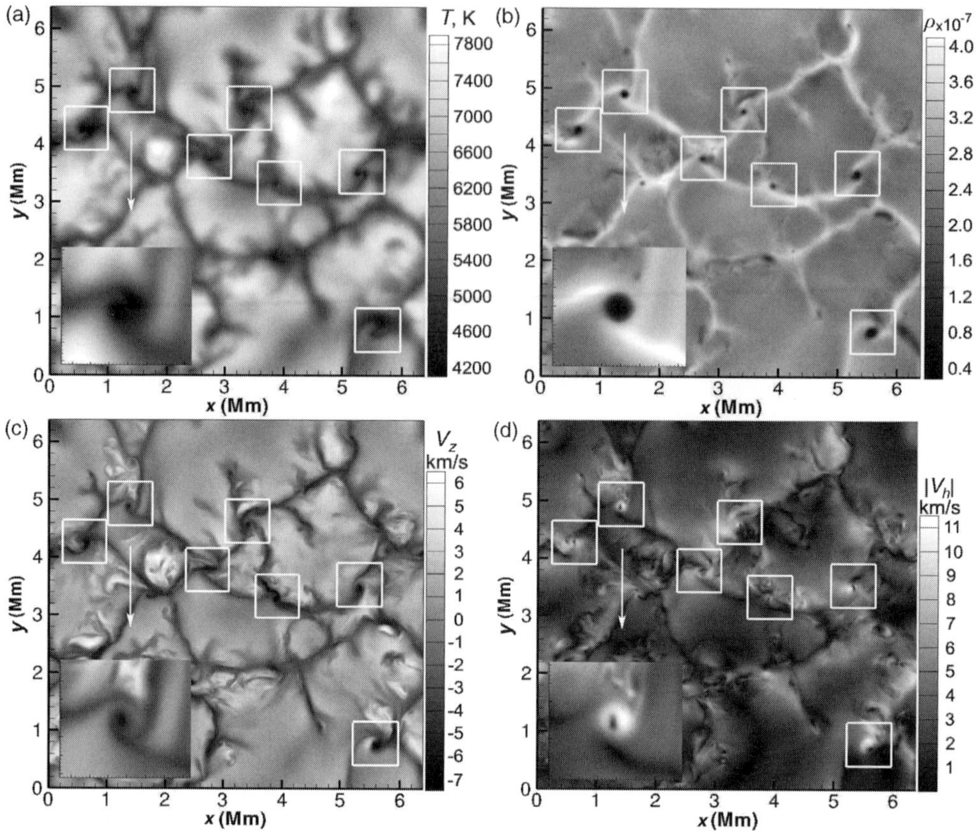

Figure 17.6 Snapshots of the simulated solar surface for (a) temperature, (b) density, (c) vertical velocity, and (d) magnitude of the horizontal velocities. The white squares indicate some of the largest whirlpools in the snapshots. The whirlpools are located in the intergranular lanes. A magnified whirlpool is shown in the left bottom corner of each panel.

frequencies. These results may have important implications for local helioseismology analysis and for comparison with linear modeling of oscillations in magnetic fields (e.g., Khomenko *et al.*, 2009; Parchevsky and Kosovichev, 2009b). In particular, the results indicate that the frequency shifts measured in sunspot regions using the ring-diagram technique of local helioseismology without discriminating the sunspot umbra and penumbra regions may come mostly from the penumbra, because the contributions of various parts of the analyzed area are weighted by the local oscillation power of these parts.

Currently, the excitation mechanisms are still not sufficiently well understood, particularly when oscillations are strongly coupled to the subsurface turbulent convection. This coupling limits interpretation of observations and development of advanced approaches in data analysis, in particular in helioseismology. The stochastic wave excitation in the turbulent convective medium and frequent interference of the wavefronts make the problem of solar oscillations very complicated. The results of our simulations reveal an important source of solar oscillations due to strong interacting vortices in the top layers of the convection zone. The vortices are primarily generated in the intergranular lanes (Kitiashvili *et al.*, 2012). They are numerous and interact with each other. The process of interaction of vortices with opposite-sign vorticity can lead to their partial annihilation. This can produce strong density perturbations and result in excitation of acoustic waves. The dynamics of the vortices and waves is complicated and depends on the structure of surrounding convective motions.

Such a mechanism of acoustic wave excitation due to the interaction of vortex tubes is very common in solar convection simulations. However, it is not unique, e.g., excitation may also be due to collapse of granules (Skartlien *et al.*, 2000). Thus, the understanding of the primary sources of solar oscillations and their interaction with turbulent flows and magnetic fields requires further detailed investigations using high-resolution numerical models and observations.

18

Impact seismology on terrestrial and giant planets

PHILIPPE LOGNONNÉ AND TAICHI KAWAMURA

18.1 Short review of impact seismic records in the Solar System

Impact structures affect all planetary surfaces in the Solar System and especially those not resurfaced by recent tectonic processes or fast erosion. Planetary seismology and remote sensing are the tools used to monitor the dynamics of large impact processes and associated shock waves, in a variety of conditions, from the airless Moon to the atmosphere-protected Earth.

18.1.1 Meteorites and bolides on Earth

The largest impact ever instrumentally recorded occurred in the early days of seismology. This was the famous great Siberian meteor (Ben-Menahem, 1975) with an energy estimated to be about 12.5 megatons (1 ton of TNT $= 4.185 \times 10^9$ J). It was recorded by two Russian seismic stations at about 1000 km and 5000 km, in addition to pressure records.

The development of worldwide infrasound and seismic networks today allows the detection of much smaller impacts, down to kilogram size in mass. Most of these impacts are detected by their generated airwaves (see Edwards *et al.*, 2008; Edwards, 2008 for a review), either with infrasound sensors or by seismometers detecting the surface displacement generated by the airwave. This acoustic detection can be done not only locally but also at larger distances, thanks to the tropospheric waveguide (Edwards, 2008).

Airwaves are also generating seismic waves through conversion processes at the Earth's surface. This is illustrated by Figure 18.1, which shows the recent seismic observations made by Tauzin *et al.* (2013) following the Chelyabinsk meteor blast of February 15, 2013. This event generated large Rayleigh surface waves propagating at between 2.7 and 3.5 km/s, with amplitudes corresponding to an event with surface wave magnitude $M_s \sim 3.7$. Comparable precursor Rayleigh waves, arriving well before the atmospheric air wave, are also found in many of the impacts reported by Edwards *et al.* (2008), but of course with smaller amplitudes.

Extraterrestrial Seismology, eds. Vincent C. H. Tong and Rafael A. García. Published by Cambridge University Press.

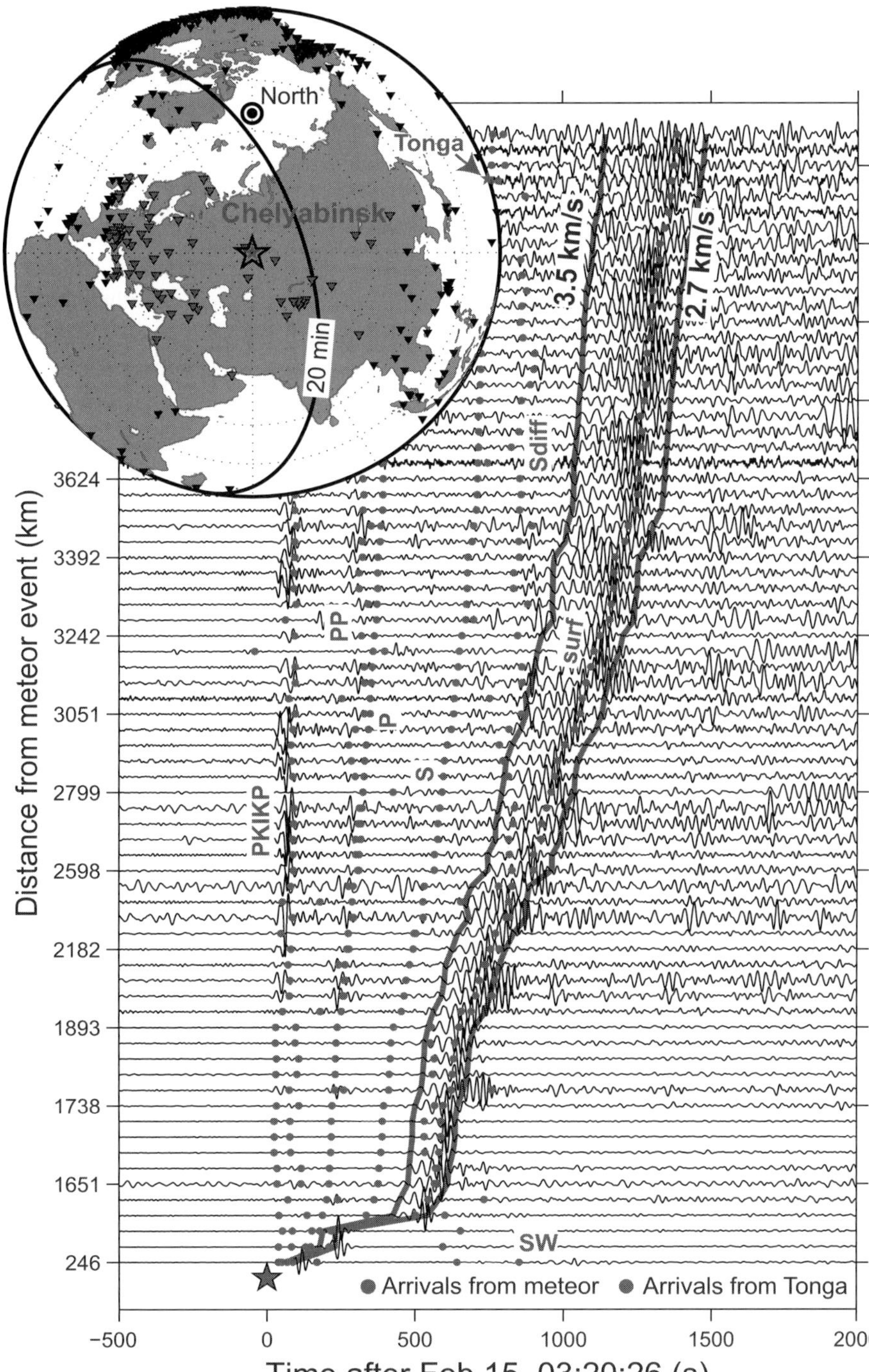

Figure 18.1 Bandpass-filtered vertical seismograms from up to 4000 km of epicentral distance, recorded after the Chelyabinsk blast. The filtering window is between 20 and 60 s and corresponds to the surface wave bandwidth. The blue dots (light gray dots in the black and white figure) show the arrival of seismic waves associated with an earthquake occurring in the Tonga Islands 20 minutes before the impact. The 20 minute travel-time isocontour for this earthquake is indicated on the Earth's sphere, while an arrow gives its epicenter location. (Reprinted from Tauzin *et al.*, 2013, with permission of AGU.) A black and white version of this figure will appear in some formats. For the color version please refer to the plate section.

18.1.2 The Moon and Apollo

No atmospheric shielding or coupling occurs on the airless Moon, so the seismic signals are directly related to the impact on the ground. Natural or artificial impacts therefore constituted an important fraction of the observed sources, and about one-fifth of the seismic signals detected (1753 of the 9442 identified and classified events) by the Apollo seismic network were impacts.

Artificial impacts were made through the collision on the Moon of either the Saturn IVB (SIVB) upper-stage or the Lunar Module (LM) (Figure 18.2). The 14 230 ± 260 kg SIVBs impacted at 2.56 ± 0.02 km/s while the LMs were smaller seismic sources, with 2320 ± 60 kg impacting at 1.68 ± 0.02 km/s. The size of the generated crater's rim diameter predicted by scaling laws is about 30 m for the SIVB and 6.5 m for the LM (Lognonné *et al.*, 2009). These diameters are comparable to observations made by the imager of NASA's Lunar Reconnaissance Orbiter (LRO) spacecraft (Robinson *et al.*, 2010). Most of these impacts generated large signal-to-noise seismic signals, even at large distances, and were crucial in the determination of the crustal thickness, as the location and precise time of the impacts were known. The impact time of the Apollo 16 SIVB, due to failure of the tracking, is however unknown, while its location has been recovered by LRO data.

Natural impacts were typically detected at a rate of about 150 per year with different detection rates for the Apollo stations resulting from differences in local (site) amplification (Lognonné *et al.*, 2009). This much higher rate as compared to the Earth is of course directly related to the lack of atmosphere on the Moon and therefore the lack of shielding. Natural impacts were also important for interior structure analysis, as only the geographical location (latitude and longitude) and impact time had to be determined in the seismological inversions.

Figure 18.3 shows typical records of a large natural impact (with a 60–70 m estimated crater diameter, Gudkova *et al.*, 2011) while Figure 18.4 provides the typical distances and amplitudes of impacts located with the long-period Apollo channels. As we will see later, only the momentum impulse is constrained by the measurement of the seismic signals. Assuming a 20 km/s impact velocity, Figure 18.4 therefore shows that the typical mass of these impacts ranged from tens of kg to a few thousand kg. Many impacts were also detected on the Apollo short-period channel, but most of the time at only one seismic station.

18.1.3 Jupiter and Shoemaker–Levy 9

The impacts, in July 1994, of the disrupted pieces of the Shoemaker–Levy 9 (SL9) comet were considered to be large enough to generate waves potentially detectable from Earth through remote sensing. The amplitudes of signals were theoretically predicted for different masses of impactors (e.g., Lognonné *et al.*, 1994). For an impact with energy greater than 10^{21} J, peak-to-peak temperature fluctuations greater than $\sim$0.01 K were expected for 10 mHz frequency P-waves, while surface waves below 3 mHz were expected to generate fluctuations in excess of 0.01 K for impacts greater than 2×10^{21} J (Lognonné *et al.*, 1994).

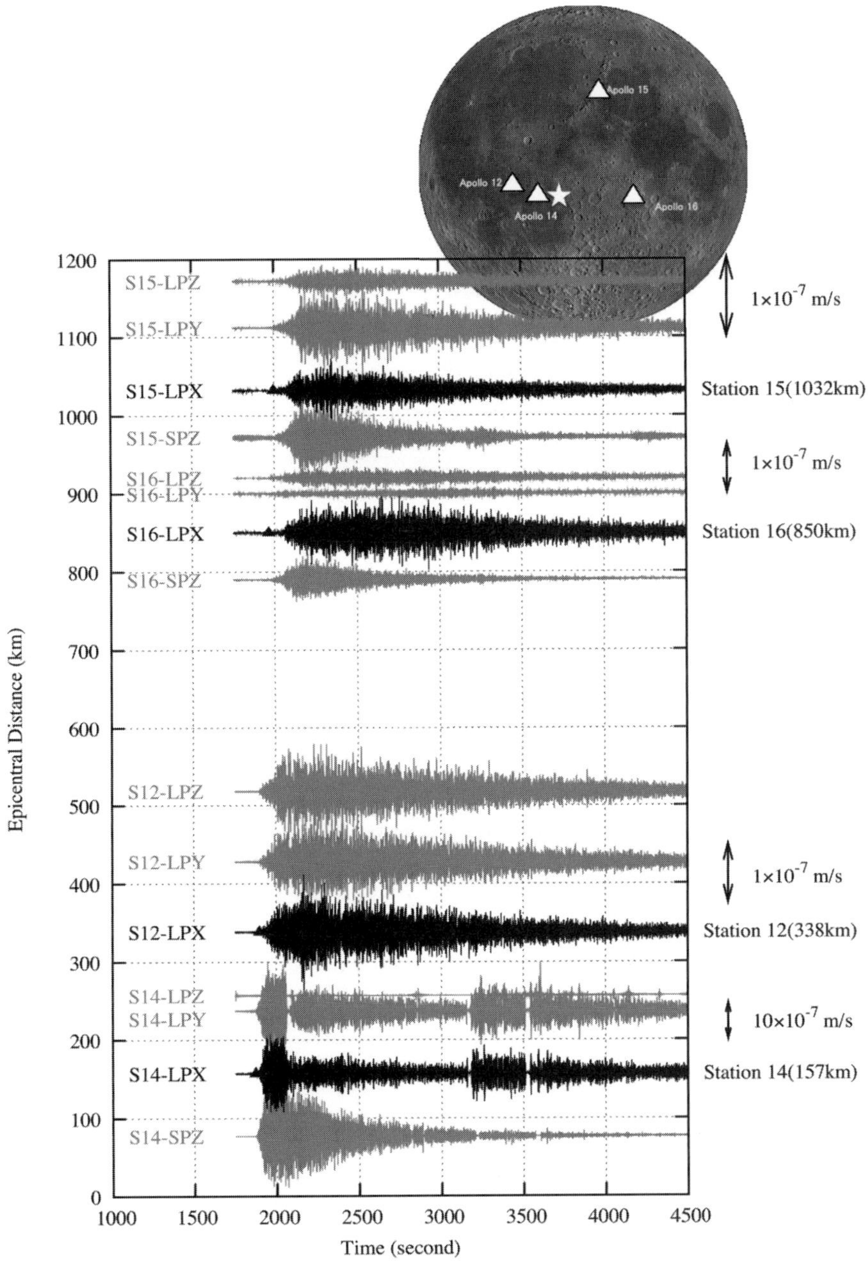

Figure 18.2 Ground velocity records from the Apollo seismic network, of the impact of the Apollo 17 Saturn V upper stage (Saturn IVB) on the Moon on December 10, 1972 (at distances of 338, 157, 1032, and 850 km from the Apollo 12, 14, 15, and 16 stations, respectively). Data are shown for the long-period seismometer (LPX, Y, Z) plus the vertical axis of the short-period seismometer (SPZ). Amplitudes at the Apollo 14 station, 157 km from impact, are saturated mainly due to S-waves trapped in the regolith. The first P arrival is typically 10 times smaller. Note the 10-db gain change at the middle of the LPX and LPY records of station 14, which saturated for the large signal. The amplitude arrow for this station is for the amplitudes before this gain change. (The lunar globe was taken from LROC observation by NASA (http://photojournal.jpl.nasa.gov/catalog/PIA14011). Apollo stations and deep moonquake nests were added by the authors.)

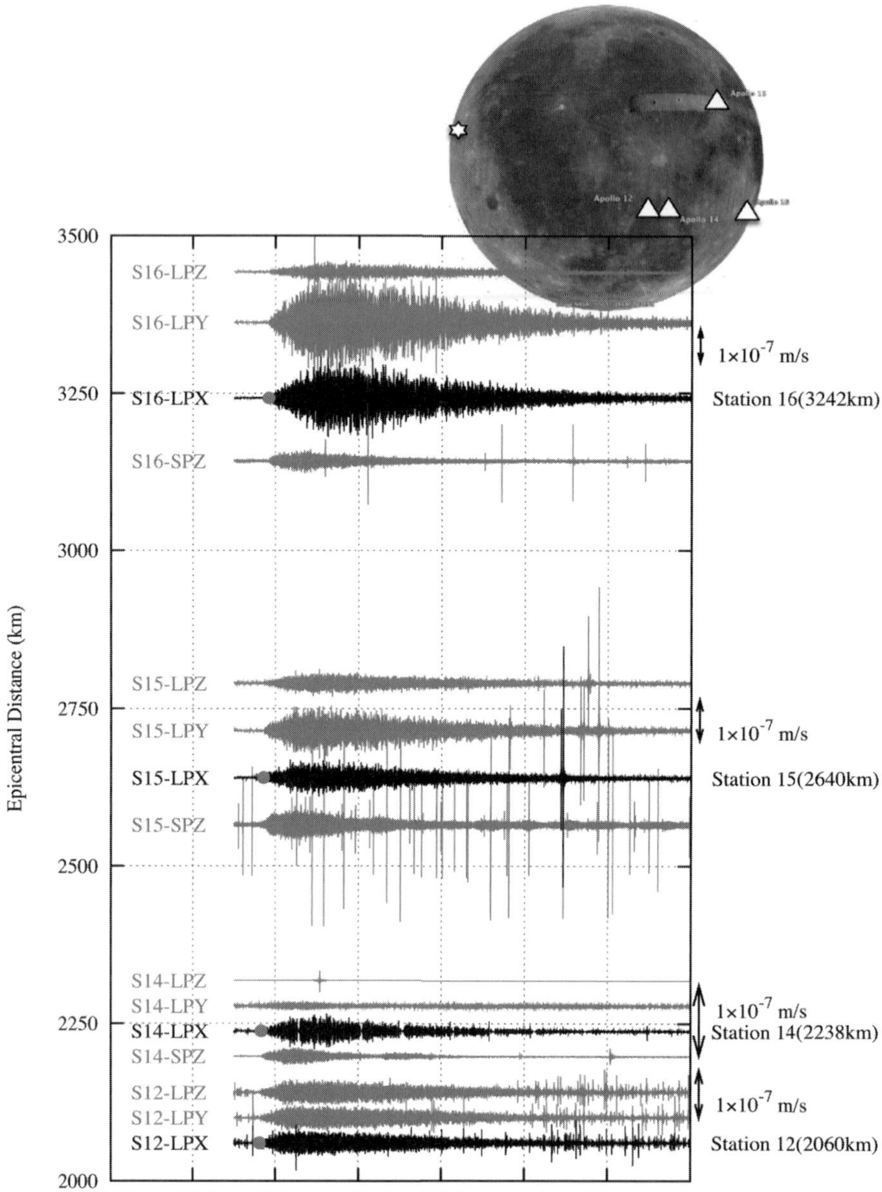

Figure 18.3 Typical records of the large natural impact occurring on November 14, 1976, recorded at the Apollo stations. The same processing as for Figure 18.2 was done. Note that amplitudes at a similar epicentral distance are larger than those of Figure 18.1. The mass of the impact has been estimated at about 25–35 tons assuming an impact velocity of 20 km/s. (The lunar globe was taken from LROC observation by NASA (http://photojournal.jpl.nasa.gov/catalog/PIA14011). Apollo stations and deep moonquake nests were added by the authors.)

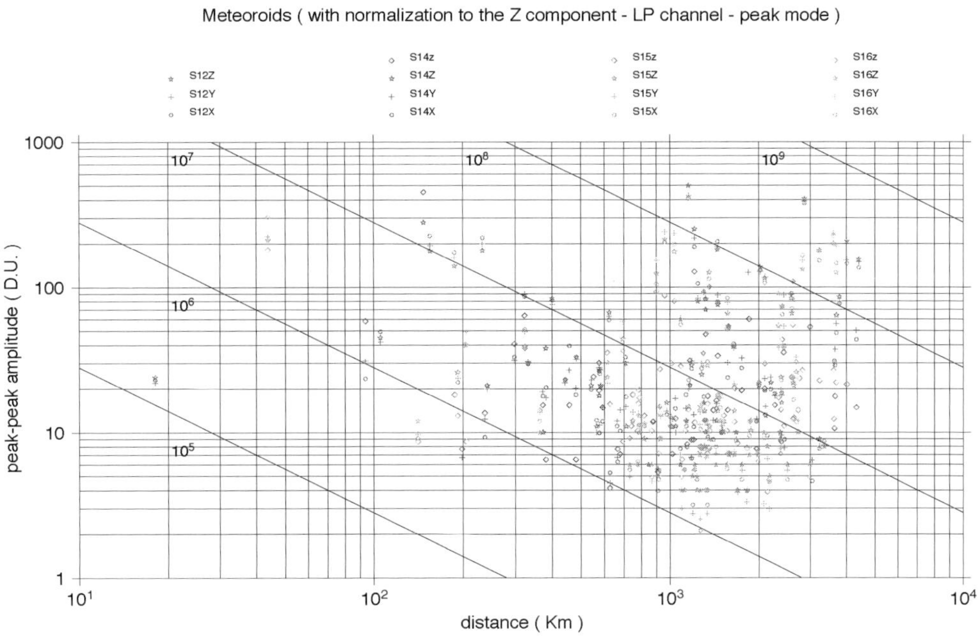

Figure 18.4 Peak-to-peak amplitudes of the impacts detected by the different Apollo channels, in DU, with respect to epicentral distance. The oblique lines are the source amplitudes, from 10^5 Ns up to 10^9 Ns in the *y*-direction, while the $1/D$ decay of the oblique lines is mainly due to geometrical spreading, as attenuation is very low on the Moon. The moments correspond to impacts from about 3 kg to 30 tons at 20 km/s velocities and with ejecta amplification of 1.5. (Reprint of Figure 9 of Gudkova *et al.*, 2009, with permission of Elsevier.) A black and white version of this figure will appear in some formats. For the color version please refer to the plate section.

No seismic wave observations were however reported for the SL9 impacts (e.g., Mosser *et al.*, 1996), therefore putting upper limits on the impact energy of $1–2\times10^{21}$ J. Though the impacts did not generate detectable body waves, they did however produce a ring-like pattern in the Jovian atmosphere, with two rings propagating at 210 m/s and 450 m/s (Figure 18.5 after Hammel *et al.*, 1995). These waves were probably gravity waves propagating either in the stratosphere (Walterscheid *et al.*, 2000) or in a deeper layer. For Jupiter and the SL9 observations, an enhancement by a factor 10 of the water content at the 10 bar depth (about 80 km below the 1 bar level) is necessary to explain the observed wavefront speed (e.g., Kanamori, 2004).

18.2 The impact seismic source

18.2.1 Seismic source on airless planets and small bodies

On airless bodies, the impactor reaches the surface without any deceleration, releasing both its momentum and kinetic energy to the planetary surface. Impact velocities from

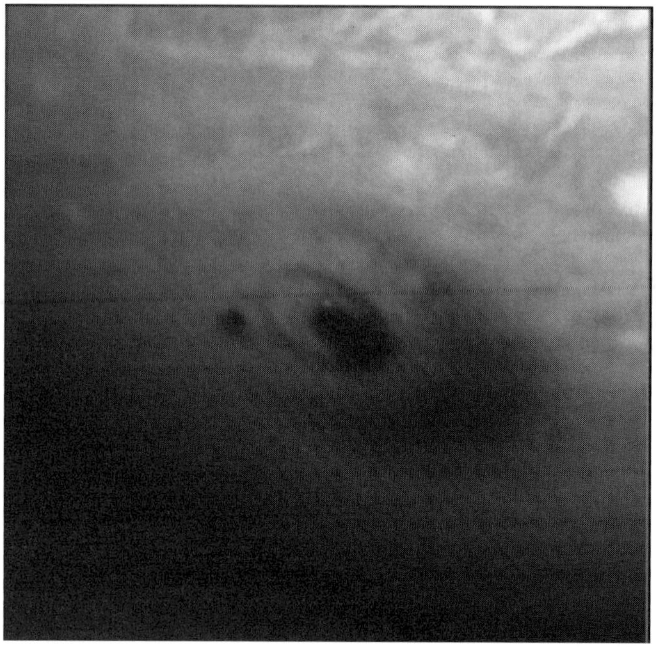

Figure 18.5 Wave features generated by the impact G of the Shoemaker-Levy 9. (After Hammel *et al.*, 1995, and from NASA Hubble Space Telescope WFP camera.)

prograde objects decrease with the distance from the planet to the Sun and range from 5 to 50 km/s. While the mean impact velocity on the Moon is ~20 km/s, the ratios with those on Mercury, Venus, Earth, and Mars are 2.16, 1.28, 1.04, and 0.54 respectively (LeFeuvre and Wieczorek, 2008). A few impactors, especially those related to comets or cometary swarms, have retrograde motions and can reach the planets with much higher velocities. In all cases, both the generated shock wave and momentum carried by ejecta must be accounted for in a rigorous modeling of the seismic source.

The first way to model this seismic source is to quantify the fraction of the kinetic energy transferred into seismic wave energy, defined as the seismic efficiency. The seismic efficiency of meteorite impacts has been widely discussed, either for impacts on the Moon (Laster and Press, 1968; McGarr *et al.*, 1969; Latham *et al.*, 1970; Lognonné *et al.*, 2009), on Mars (Davis, 1993; Teanby and Wookey, 2011), or on asteroids (e.g., Richardson *et al.*, 2004). Large uncertainties are found, with values ranging from 10^{-6} to 10^{-4}, which leads to one order of magnitude uncertainty in the amplitude of the generated seismic waves.

The second approach is to focus on long-period seismic waves, for which the area affected by the shock waves is small enough compared to the wave's wavelength to be considered as a point source. A point-force seismic source (McGarr *et al.*, 1969;

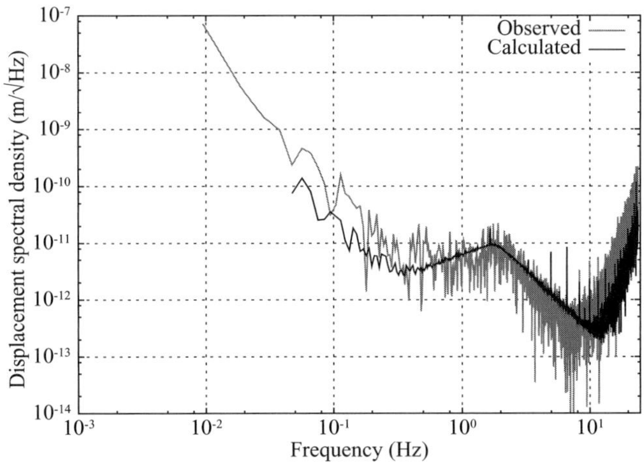

Figure 18.6 Displacement spectrum of the Apollo 17 SIVB impact recorded at the Apollo station 16. The spectrum is a composite spectrum of the long period and short-period vertical channels both corrected from the instruments transfer functions. The corner frequency is at about 2 Hz and corresponds to a maximum of amplitude. The instrument noise increases below 0.2 Hz and above 10 Hz.

Lognonné *et al.*, 2009) can then be used, and its time dependence can furthermore be assumed as instantaneous for periods much larger than the typical duration of the non-linear regime of the shock wave. The seismic source is then very similar to a percussion force and can be expressed as $\mathbf{f}(x, t) = \mathbf{n}_0\, \delta(\mathbf{x} - \mathbf{x}_s)\, \Phi(t)$, where $\mathbf{n}_0$ is the point-force unit vector and $\delta(\mathbf{x} - \mathbf{x}_s)$ is the 3D Dirac space function at the source location. For an instantaneous impact releasing a momentum $p = mv$ to the ground, $\Phi(t) = p\delta(t)$, where $\delta(t)$ is the Dirac delta function. For a homogeneous medium, the displacement amplitude of the P or S body waves in the far field is then proportional to $\Phi(t - r/\alpha)/[\rho r \alpha^2]$, where r is the distance, ρ the density, and α the P or S seismic velocity (Aki and Richards, 2002). Note here that in contrast to quakes, this displacement is not proportional to a seismic moment (in Nm), but to an impulse (in Ns).

Analysis of the seismic records of the Apollo artificial impacts has confirmed this simple approach based on linear momentum conservation: for low velocity impacts, the amplitude is proportional to the linear momentum of the impactor, while for larger velocities it is amplified by up to a factor of 2 due to impact-generated ejecta (Lognonné *et al.*, 2009). While a clear corner frequency followed by a rapid ω^{-2} decrease of the spectrum is found in most of the impact spectra (see Figure 18.6 for a typical spectrum associated with the record of an Apollo artificial impact), a slight increasing trend with respect to the flat displacement spectrum (up to ω, see Gudkova *et al.*, 2011) is found before the corner frequency, which is not predicted by a simple point force. This has been found to be related to shock wave blast in the subsurface (Gudkova *et al.*, 2014).

18.2.2 Seismic sources on planets with atmospheres

The recent deployment of infrasound sensors, in the framework of the Comprehensive Test Ban Treaty, has multiplied observations of sonic booms associated with the impact of small (1 kg) asteroids in Earth's atmosphere. Earth's typical rate of 3×10^{-5}–2×10^{-4} km^2/yr for 1 kg objects leads indeed to approximately 4–25 detectable events per year within 200 km of the receiver. When detected by an infrasound sensor at regional (<200 km) distances, the associated signals are short-duration (~0.1 s) pressure pulses, with typical amplitudes ranging from 0.01 to 0.1 Pa for the most common events, and 0.1 to 1 Pa for the largest ones (Edwards, 2008).

These shock waves are generated by the deceleration of the impactor in the atmosphere. Due to the high Mach numbers the atmospheric source can be considered as a line source, radiating acoustic energy cylindrically outward along the impact trajectory (ReVelle, 1976). In some cases, a final disruption/explosion generates a second seismic source, radiating spherical shock waves. Last but not least, but only in a very few reported cases on Earth, the impactor reaches the surface with a terminal velocity and mass large enough to generate direct seismic waves, as for the Moon and airless bodies. Larger impacts on Earth can also be detected at larger distances, thanks to the tropospheric waveguide (Edwards, 2011).

Waves in the atmosphere transition from the shock wave regime to the weakly nonlinear and finally linear regimes. For impacts with sources at about 80 km altitude, the transition to the linear regime is at Earth altitudes between 30 km and 40 km. Below this the pressure wave sees little attenuation. The geometrical decay of $D^{1/2}$ for cylindrical sources generates an attenuation of $\sim 3/4$ at the ground. This is counterbalanced by the pressure increase toward the ground with amplitude increasing by the square root ratio of the ground to 35 km pressure. This provides an amplification of about 12 at the ground. Blast pressure amplitudes are therefore about one order of magnitude larger at the Earth's surface than at 35 km altitude.

Figure 18.7a–d illustrates the comparisons between acoustic propagation conditions on Earth and on Mars. If similarities between Mars (from the surface up to 50 km altitude) and Earth (from 35 to 80 km altitude) are found for the density and pressure, large differences in terms of acoustic velocities (200–230 m/s for Mars, 310–380 m/s for Earth) and especially attenuation factor appear. Attenuation, as noted by Bass and Chambers (2001) and Williams (2001), is several orders of magnitude larger due to the specific properties of carbon dioxide and its molecular relaxation, which generates an absorption peak at about 100 Hz at the Martian surface, which is also shifted toward lower frequencies when the pressure decreases with altitude.

This absorption makes propagation of the high-frequency content (e.g., for frequencies above a few Hz) of the blast extremely difficult, and the only signals expected to propagate through the atmosphere will be at periods of a few tens of seconds (Figure 18.7d). These periods correspond to those of Rayleigh surface waves, very commonly excited

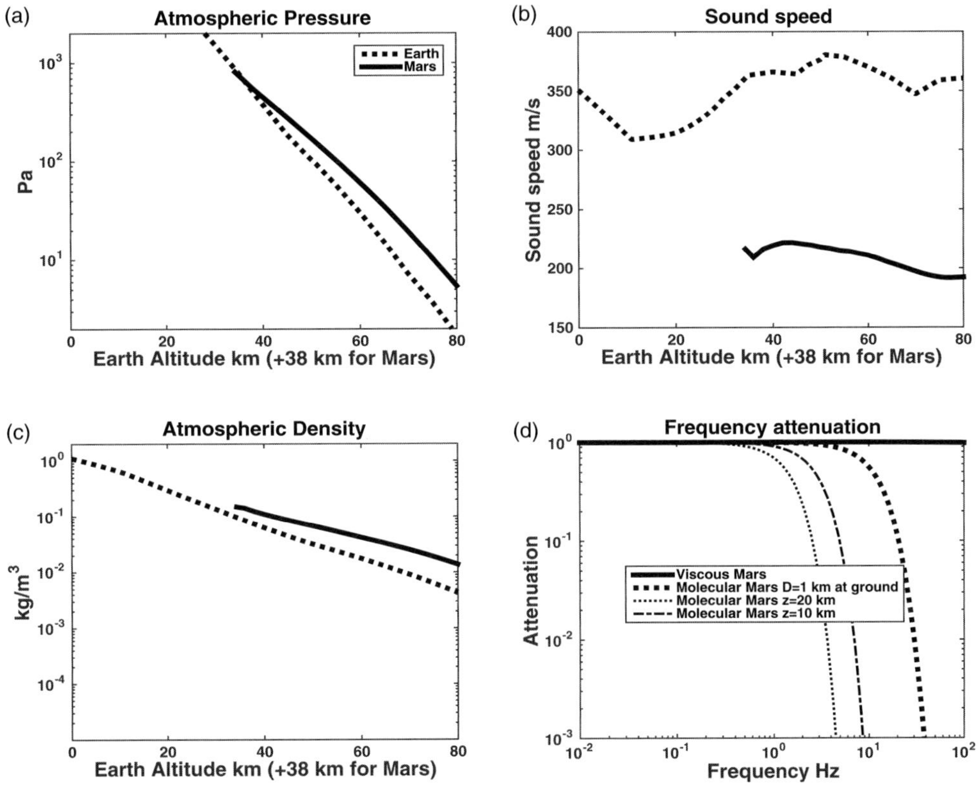

Figure 18.7 Comparison of the acoustic conditions in Mars and Earth upper atmospheres. Mars' surface pressure is approximately equal to Earth's at 38 km altitude, and altitudes for Mars on all curves are therefore shifted by this value. Atmospheric pressure, (b) sound speed, (c) atmospheric density and (d) attenuation at different frequencies (attenuation is the exponential decay of the amplitude with distance due only to viscosity and relaxation). The atmospheric properties are expected to experience daily and geographical variations. From right to left and for the attenuation: the first two lines are the attenuation due to viscosity and to molecular relaxation for 1 km of propagation at the ground, showing the major impact of molecular relaxation. The long dashed and dotted line are the attenuation for sources at 10 km and 20 km altitude respectively, showing that only the long period acoustic waves can reach the surface for blasts originating from more than 20 km altitudes.

by the blasts on Earth, and which propagates faster than the direct acoustic waves. As a consequence, they arrive as precursors waves on Earth's records, prior to the direct acoustic wave arrival. Such waves might therefore be excited too on Mars, but likely with tiny amplitudes as this excitation process will be reduced not only by the atmospheric attenuation but also by the lower amplitude of the pressure waves on the ground as compared to Earth.

18.3 Impact frequency–amplitude relation

18.3.1 Lunar impact detection rate and impact hum

The impact rate on the Moon has been studied extensively by Oberst and Nakamura (1987, 1991). They proposed an annual rate of impacts given by the following relation:

$$\log_{10} N = -0.99 \log_{10} E + 11.4,$$

where N is the number of events per year, and E is the kinetic energy in J, assuming an impact velocity of 14 km/s and a seismic coupling efficiency of 10^{-6}. This gives an energy of $10^{11.5}$ J for the largest yearly impact, equivalent to a mass of about 3500 kg. For the 7.7 years of operations of the Apollo seismic network, this puts an upper bound of 2.6×10^{12} J (about 0.62 kilotons of TNT) and a mass of about 26 tons for the largest recorded impact. The sources of these large impacts were recently estimated by Gudkova *et al.* (2011). The impulses found were in the range of 5×10^8–10^9 N s. Assuming an ejecta amplification factor of 1.5 and taking the same 14 km/s impact velocity, this leads to comparable estimates for the largest impacts in the range of 2.3–4.6 10^{12} J. Many uncertainties remain in both the propagation condition and the seismic efficiency/source model.

For smaller impacts of about 1 kg (about 10^8 J) the Oberst and Nakamura rate suggests about 3000/yr. As recently shown by Lognonné *et al.* (2009), such a rate is well retrieved by a Monte Carlo simulation of impacts constrained by the impact flux rate determined by other techniques, including those classified by the detection of impacts in the Earth's upper atmosphere using early warning satellites (e.g., Brown *et al.*, 2002). Here again, an uncertainty of a factor of 2 remains, due to the unknowns in the propagation properties used for the Moon, including attenuation.

Finally, impacts are probably the source of the background seismic noise on airless bodies such as the Moon. Such bodies are expected to have a relatively low seismic noise, as their surfaces are only subject to long-period, diurnal temperature changes. However, impacts (including micro-impacts) occur continuously. Lognonné *et al.* (2009) suggested impacts as a primary source of micro-seismic noise, with a proposed background noise ranging between 2/1000 and 2/100 of the Apollo resolution around 0.5 Hz. This might be an ultimate limit for future lunar seismometers, although very challenging, as this noise level corresponds to ground accelerations in the range of only 10^{-12}–10^{-11} m/s^2.

18.3.2 Expected detection rate on Mars

On Mars, both the velocity and the mass of the impactors are reduced during atmospheric entry, which can be assessed by integrating the impactor trajectory equation in the atmosphere (e.g., Lognonné and Johnson, 2007). The atmospheric shielding effect on Mars is small, comparable to the Earth at altitudes of 30 km or higher (ReVelle, 1976). Figure 18.8 illustrates both this atmospheric shielding effect and the typical rate of events occurring during two Earth years, by using a random simulation based the impactor rate model of LeFeuvre and Wieczorek (2008). The entry flux of meteorites is 2.6 times larger than for

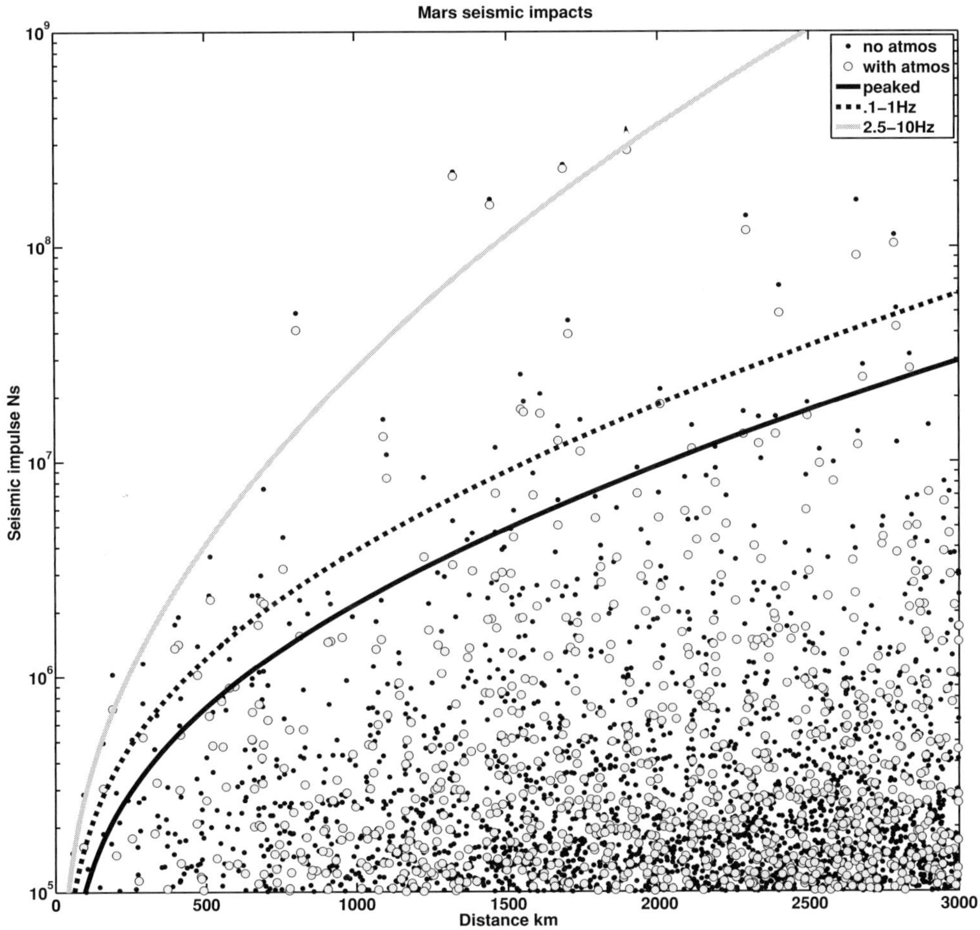

Figure 18.8 Random simulation of impact events to be detected by a Mars mission during two Earth years of operation. The distribution of impactors has mass and velocities given by LeFeuvre and Wieczorek (2008) probabilities. The seismic amplitudes are estimated following the methods described by Lognonné *et al.* (2009) and calibrated by using artificial Moon impact data corrected for Mars condition in the crust (Q of 600 and shear velocity of 4.5 km/s). Small dots are impact impulses without atmospheric ablation, while circles are with an atmosphere for the same impactor populations. Note that the atmospheric shielding is significant for the small ones, but marginal for the large. Lines are for detection thresholds corresponding to signal to noise of 3 and flat noise levels of 5×10^{-10} m s^{-2} Hz$^{-1/2}$ and 10^{-8} m s^{-2} Hz$^{-1/2}$ in the 0.1–1 Hz and 2.5–10 Hz bandwidths respectively.

the Earth–Moon due to the proximity of the asteroid belt (e.g., Davis, 1993), but it is balanced by the impact velocity, which is about half that for the Earth–Moon. The number of impacts per year detected by a seismic station on Mars might be about 10 seismic events per year with amplitudes larger than 5×10^{-9} m/s^2. These would be associated with impacts of objects with masses of several hundred kilograms. These flux predictions might, however, be overestimated, as recent observations of the impact rate from orbiters suggest that the impact rate is 4–5 times smaller (Daubar *et al.*, 2013). Most of the difference between Mars and the Moon is expected to be associated with differences in crustal seismic attenuation. This attenuation is unknown for Mars, but is likely to be less than on Earth.

18.3.3 Temporal and spatial variations

As impact rates and the strength of the associated seismic source depend on the distributions of the impactors with respect to the orbit of the studied body and of their terminal impact velocity, temporal variations of the impact rates as well as differences with respect to latitude and/or angular position with respect to the orbital plane are expected. On the Moon, additional effects are related to the orbit of the Moon with respect to Earth, which generates a leading-trailing asymmetry observed in the density of craters. This enhancement on the leading side has been detected by Kawamura *et al.* (2011) in the collection of lunar impacts detected by Apollo. On Mars, 20% more impacts are expected at high latitude as compared to the equator, and temporal variations are expected with the variation of the distance between Mars and the asteroid belt, up to a factor of 4–5 between the maximum flux and the minimum one during a Mars year (LeFeuvre and Wieczorek, 2008).

18.4 Perspectives and future

We can expect that both the future NASA InSight mission to Mars in 2016 and further planetary seismology projects on the Moon and Mars in the 2020s decade will provide new seismic observations of impacts. SL9-type impacts on Jupiter were a unique observation: Even if new smaller impacts were detected on Jupiter (e.g., Hammel *et al.*, 2010), the recurrence time of events large enough for remote sensing of seismic waves has been estimated to be 500 years or more (Roulston and Ahrens, 1997).

Future planetary seismic missions can leverage the possibility of observing surface impacts with both seismometers and remote sensing techniques. On the Moon, as impacts are known to generate a flash when hitting the lunar surface (Ortiz *et al.*, 2006), both time and location might be obtained. On Mars, however, only the location will be obtained through the differential analysis of high-resolution imaging data taken before and after the impact (e.g., Daubar *et al.*, 2013). Complete datasets will then be obtained, with location, crater size and morphology, and seismic wave records for both Mars and the Moon, with for the Moon the additional measurement of the impact time and flash emission spectrum. The expected number of impacts to be detected and located on Mars through seismic waves remains modest (one to five per year depending on instrument noise) but comparable to the artificial impact performed during Apollo. A much higher rate of observations will be possible on the Moon due to the much lower attenuation (Yamada *et al.*, 2011). High-resolution seismic tomography of the lunar crust can therefore be imagined, when seismometers will be deployed again on the Moon and will be coupled with lunar flash monitoring from Earth or from lunar high-altitude orbit.

19

Angular momentum transport in stars: From short to long time scales

ALLAN SACHA BRUN AND STÉPHANE MATHIS

In this chapter we briefly summarize how angular momentum is being transported and exchanged between convective and radiative zones in stars. We discuss what physical processes influence the internal rotation history of stars on short to long (secular) time scales.

19.1 The astrophysical context

Stars are rotating magnetic bodies with complex internal and external dynamics. Observations using helioseismology (e.g., García *et al.*, 2007), asteroseismology (e.g., Deheuvels *et al.*, 2014), and spectropolarimetry (e.g., Donati and Landstreet, 2009) techniques put more and more constraints on this intricate dynamics. To get a complete and coherent picture of dynamical processes in stars and of the associated transport of angular momentum that goes beyond the "standard" modeling of stellar structure and evolution (Maeder, 2009) one needs to develop new models by introducing an improved physical description of these time-dependent processes. However, to simulate such processes in a star in full detail requires treating spatial and temporal scales spanning about 10 orders of magnitude. This is clearly not yet feasible, even with the most powerful computers available today. Therefore, one can choose to describe what occurs on a dynamical time scale (such as a convective turnover time or stellar magnetic cycles) or on the long-term evolution where the typical characteristic time scale is the dominant nuclear reactions. The same applies for spatial scales. One has to choose which relevant scale one needs to model in order to accurately describe the spatial dependence of the physical processes (convection motions, MHD instabilities, transport and mixing processes, surface dynamics).This is the reason why it is nowadays necessary to use and couple 1D, 2D, and 3D models to get a global picture of macroscopic MHD transport processes in stars over short to secular time scales.

In this chapter, we report on the state of the art of the modeling of the transport of angular momentum in stars both in convection and in radiation zones and we present our main contributions to this field of research. Our main objectives are to study stars' angular

Extraterrestrial Seismology, eds. Vincent C. H. Tong and Rafael A. García. Published by Cambridge University Press.

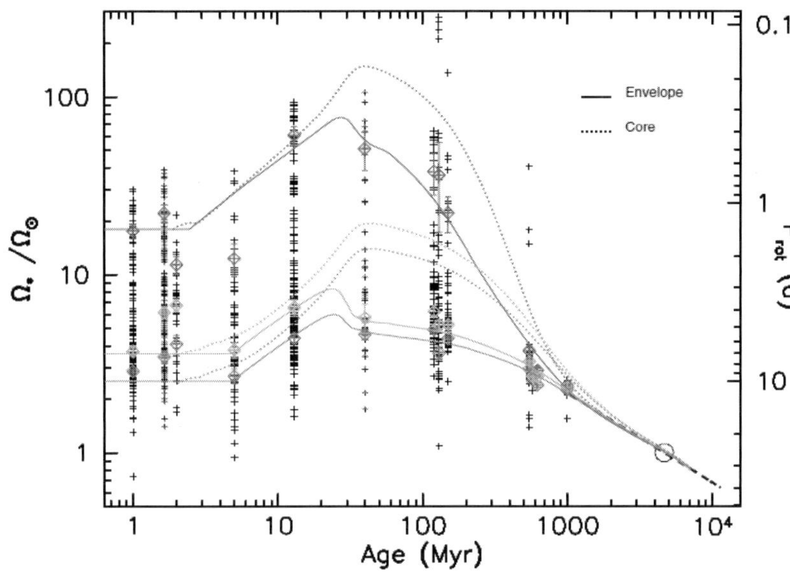

Figure 19.1 Rotation rates of the radiative core (dashed lines) and of the convective envelope (solid lines) of solar-type stars as a function of time for fast (blue), median (green), and slow (red) rotator models. The left axis gives angular velocities normalized to the Sun's. The right axis gives rotational periods (days). The blue, red, and green tilted squares correspond to the 90th percentile, the 25th percentile, and the median, respectively, of the rotational distributions in formation regions and young open clusters. The open circle is the angular velocity of the present Sun and the dashed black line corresponds to the Skumanich relationship (see text). (From Gallet and Bouvier, 2013, courtesy *Astronomy & Astrophysics*.) A black and white version of this figure will appear in some formats. For the color version please refer to the plate section.

momentum history (see, e.g., Bouvier *et al.*, 1997; Bouvier, 2013; Gallet and Bouvier, 2013, and Figure 19.1) and the related profile of differential rotation (García *et al.*, 2007; Deheuvels *et al.*, 2014).

On Figure 19.1, we show the rotational history of a sample of stars covering the pre-main-sequence (PMS) up to the late evolution of main-sequence (MS) solar-like stars, e.g., 10 Gyr (Gallet and Bouvier, 2013). Briefly summarized, following the formation of a proto-star through the collapse of a molecular cloud, stars interact with their surrounding accretion disk and maintain a constant angular velocity (horizontal plateau) as long as the disk locking is operating (Zanni and Ferreira, 2013). Then they contract and speed up until they reach the zero-age main-sequence (ZAMS) (tip of the color curves) when they start burning their hydrogen. Through the action of a stellar wind (Kawaler, 1988; Matt *et al.*, 2012), stars then brake on a time scale of a few billion years with a law such that $\Omega(t) \propto t^{-1/2}$ (Skumanich, 1972). As can be seen, the braking is more efficient for faster rotating stars such that all stars converge to the same rotation rate by about the age of the Sun (small black circle on the figure). This tight relationship between stellar age and rotation has been named *gyrochronology* by Barnes (2003).

In Section 19.2, we recall the definition of the angular momentum in stars and we discuss the simplified bi-zone models used to study stellar rotation history (MacGregor and Brenner, 1991; Allain, 1998). In Section 19.3, we focus on the dynamics of convective regions, discussing couplings between differential rotation, meridional circulation, turbulence, and magnetic field and the related transport of angular momentum. Then, in Section 19.4, we discuss transport and mixing processes operating on secular time scales in stellar radiation zones. In Section 19.5, we conclude and discuss the future developments.

19.2 Angular momentum in stars

The concept of angular momentum is fundamental to understanding the dynamics of rotating bodies. A particle with respect to an origin possesses an angular momentum $\mathbf{J}$ equal to

$$\mathbf{J} = \mathbf{r} \times \mathbf{p} = \mathbf{r} \times m\mathbf{v}, \tag{19.1}$$

with $\mathbf{r}$ the position vector relative to the origin, $\mathbf{p}$ the particle impulsion, m its mass and $\mathbf{v}$ its velocity. Note that in the literature the term "specific angular momentum" is often used and is defined as $\mathbf{J}_{spec} = \mathbf{r} \times \mathbf{v}$. One important application of the concept of angular momentum is its conservation when no net external torques are applied to the rotating object. Indeed, if we take the time derivative of Equation (19.1), we get

$$\frac{d\mathbf{J}}{dt} = \frac{d(\mathbf{r} \times m\mathbf{v})}{dt} = \mathbf{r} \times m\mathbf{a} = \mathbf{r} \times \mathbf{F} = \boldsymbol{\tau} \tag{19.2}$$

with $\mathbf{a}$ the acceleration, $\mathbf{F}$ the net force (i.e., the vector sum of all forces acting on the object), and $\boldsymbol{\tau} = \mathbf{r} \times \mathbf{F}$ the net torque. If no net torques are applied, then $\frac{d\mathbf{J}}{dt} = 0$ and the angular momentum of the object is conserved.

On Figure 19.2 (left panel) we show a simple situation of a solid disk of mass m and radius r, rotating around an axis at angular velocity $\boldsymbol{\Omega}$. The angular momentum can be shown to be $\mathbf{J} = I\boldsymbol{\Omega}$, with I the moment of inertia, which in the case of a thin solid disk is given by $I = mR^2/2$.

Let us now apply the concept of angular momentum exchange and conservation to a star like our Sun. For simplicity we model the star with two zones, a core and an envelope (see Figure 19.2, right panel) rotating around one axis along the z-direction. Each zone possesses its own moment of inertia I and angular velocity Ω. The angular momentum is, respectively, $J_{core} = I_{core}\Omega_{core}$ and $J_{env} = I_{env}\Omega_{env}$. Let us now find the exact quantity of angular momentum exchange ΔJ between the two zones required to have them rotate uniformly at the rotation rate Ω. The initial state is $J^c_{init} = J_{core}$ and $J^e_{init} = J_{env}$. Applying total angular momentum conservation, i.e., $J^c_{init} + J^e_{init} = J^c_{final} + J^e_{final}$, implies that the final state is $J^c_{final} = J_{core} - \Delta J = I_{core}\Omega$ and $J^e_{final} = J_{env} + \Delta J = I_{env}\Omega$. A simple substitution

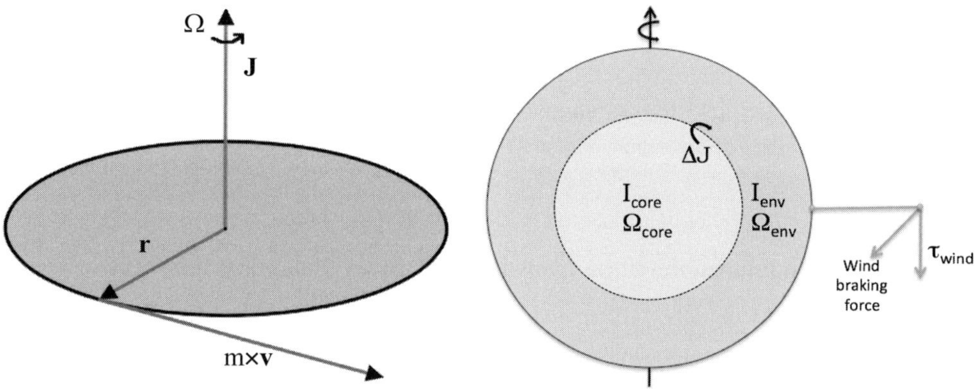

Figure 19.2 (Left) Cartoon representation of a rotating solid disk of mass m and the relevant physical quantities: angular velocity $\mathbf{\Omega}$, angular momentum $\mathbf{J}$, distance to the axis $\mathbf{r}$, and velocity $\mathbf{v}$. (Right) The two zones model of a star, showing the exchange of angular momentum ΔJ between a core and an envelope rotating at different rates and the braking action of a stellar wind.

yields

$$\Delta J = \frac{I_{env} J_{core} - I_{core} J_{env}}{I_{core} + I_{env}}. \tag{19.3}$$

If an external torque, such as the one coming from a stellar wind τ_{wind}, is applied to the star, the angular momentum evolution of the convective envelope and radiative interior can be expressed as (MacGregor and Brenner, 1991)

$$\frac{dJ_{core}}{dt} = -\frac{\Delta J}{t_c}, \tag{19.4}$$

$$\frac{dJ_{env}}{dt} = \frac{\Delta J}{t_c} - \frac{J_{env}}{t_w}, \tag{19.5}$$

with t_c the coupling time scale between the two zones due to the combined action of turbulence, waves, magnetic fields, and viscous stresses and t_w the wind braking time scale. We further assume that the convective envelope instantaneously transmits the applied surface torque to the base of the convective envelope. Many authors have studied the behavior of the two zones model under various prescriptions of the stellar wind torque (Kawaler, 1988; Matt *et al.*, 2012), determining what coupling time scale is required to reproduce the rotation distribution observed in stellar open clusters (MacGregor and Brenner, 1991; Bouvier *et al.*, 1997; Allain, 1998; Barnes, 2003; Denissenkov *et al.*, 2010; Gallet and Bouvier, 2013; see also Figure 19.1). In order to reproduce those observations with the two zones model, a value of t_c between 1 and 100 Myr is required. We refer to Matt *et al.* (2012), Reiners and Mohanty (2012), Zanni and Ferreira (2013), and Reville *et al.* (2014) for recent discussion on the sensitivity of the stellar wind torque to stellar parameters such as stellar rotation, magnetic field amplitude, and topology.

Such a two zones model can be improved by resolving self-consistently in one, two, or three dimensions the internal structure, rotational, chemical, and magnetic evolution of stars on secular time scales. We refer to Section 19.4 for a specific discussion on this fast developing field. We now discuss angular momentum transport on short dynamical time scales.

19.3 Angular momentum transport on a short dynamical time scale

Stars are dynamical objects, which display on their surface a large range of processes varying on short to long time scales. Solar-like stars with their surface convective envelope are typically turbulent and magnetic and as such exhibit properties that vary on very short (minutes) time scales, such as granulation or small-scale magnetic structures, to month-long variations (rotation, active regions, coronal holes), to decadal time scales (activity cycles). How such turbulent upper layers establish some order (large-scale differential rotation, 11-year cycle) in what seems like chaos is puzzling. Currently, we believe that in order to explain the large range of observed surface phenomena, a non-linear dynamo is operating in and at the base of the convective envelope (Parker, 1993; Brun *et al.*, 2004, 2013; Charbonneau, 2010). In order to understand the non-linear operation of the solar dynamo, theoretical considerations supported by state-of-the-art 3D time-dependent dynamo simulations are required.

We here discuss briefly how global dynamo numerical simulations help to explain the observed (helioseismically inferred) differential rotation and meridional circulation and the establishment of an 11-year cycle. We show on Figure 19.3 the rotation and meridional flow profiles achieved in a model of the Sun going from $r = 0$ to $r = 0.97R_\odot$. We note the prograde differential rotation at the equator and the slow poles, with a sharp transition at the base of the convection zone to a solid body rotation, as is clearly apparent in the top right panel on Figure 19.3. This shear layer at the base of the convection zone is called the tachocline (Spiegel and Zahn, 1992a; Strugarek *et al.*, 2011a). The meridional circulation is complex, possessing several large-scale cells per hemisphere and is of order 20 m/s (Brun *et al.*, 2011; Alvan *et al.*, 2014). In order to characterize how the large-scale differential rotation profile in magnetized convective shells representative of stellar convection zones is achieved, one must identify the main physical processes responsible for redistributing angular momentum. With the proper choice of stress-free and potential-field or perfect-conductor boundary conditions at the top and bottom of the domain, no net external torque is applied on the system, and thus angular momentum is conserved. We further assume that over the time scale considered here, a decade to a few centuries, the stellar radius and rotation rate $\mathbf{\Omega}_0$ are assumed constant in time and the rotation axis is aligned along the z-axis, i.e., $\mathbf{\Omega}_0 = \Omega_0\,\hat{e}_z$. Assuming a constant rotation rate over the solar cycle is well justified by observations, and so seems to be the case for the solar radius. Of course, if mass loss through a magnetized stellar wind is considered, the angular momentum of the star evolves on the stellar wind braking time scale (see introduction above).

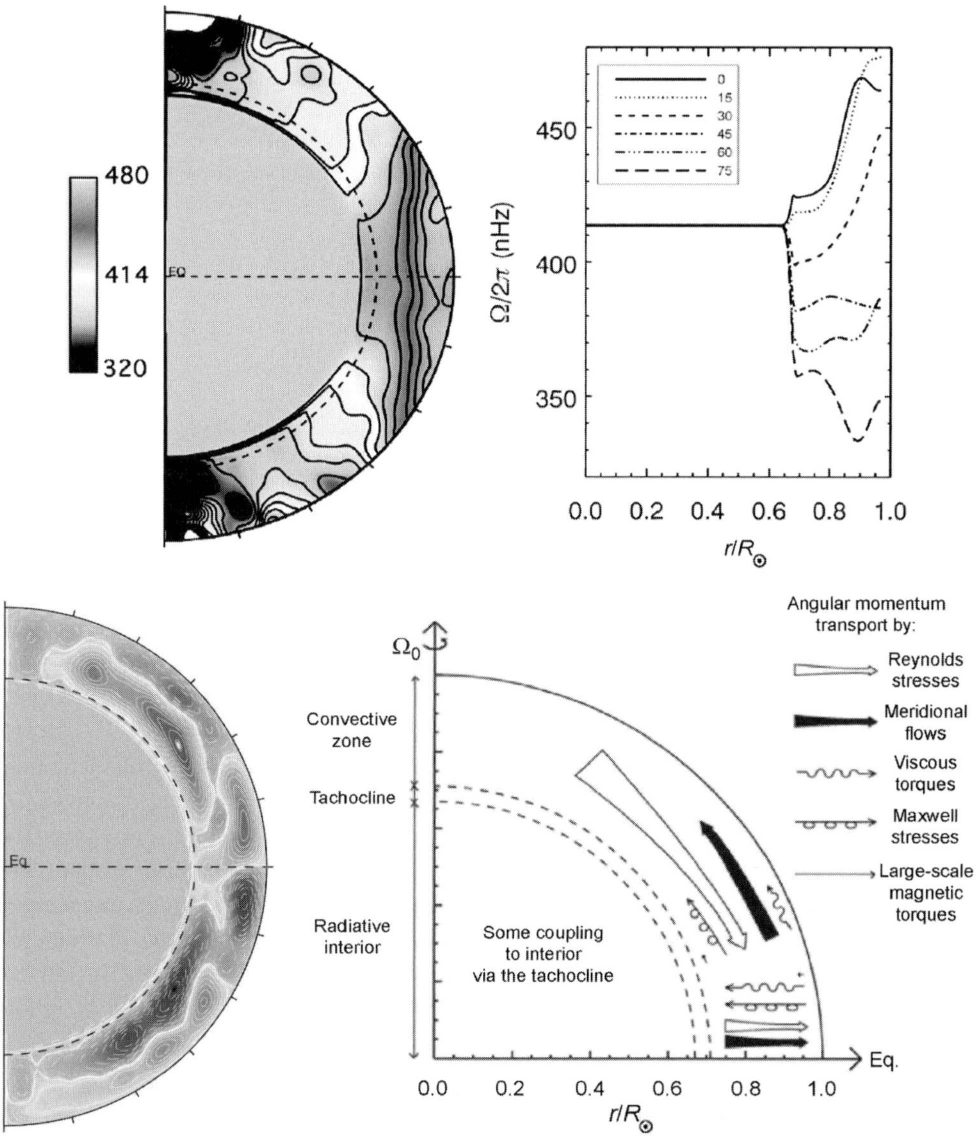

Figure 19.3 (Top) Angular velocity contours for a 3D model of the Sun spanning 97% of the solar radius (left) and latitudinal cuts at indicated latitudes (right), showing clearly a tachocline at the base of the convection zone near 0.7 $R_\odot$ (Brun *et al.*, 2011; Alvan *et al.*, 2014). (Bottom left) Meridional circulation realized in the same 3D solar model (red contours/solid white line correspond to counterclockwise flow). (Bottom right) Summary of angular momentum redistribution in a magnetized turbulent convective shell (Brun, 2004, courtesy *Solar Physics*). A black and white version of this figure will appear in some formats. For the color version please refer to the plate section.

We may assess the physical processes responsible for the transport of the specific angular momentum, $\mathcal{L} = r^2 \sin^2\theta\,\Omega$, within these systems by considering the ϕ-component of the momentum equation expressed in conservative form and averaged in time and longitude (as in Brun and Toomre, 2002; Brun *et al.*, 2004; Miesch, 2005):

$$\bar{\rho}\,\partial\mathcal{L}/\partial t = -\nabla\cdot\mathcal{F}, \quad (19.6)$$

where $\mathcal{F}$ is the flux of angular momentum. It simplifies in the stationary case in the spherical coordinates system to

$$\frac{1}{r^2}\frac{\partial(r^2\mathcal{F}_r)}{\partial r} + \frac{1}{r\sin\theta}\frac{\partial(\sin\theta\mathcal{F}_\theta)}{\partial\theta} = 0, \quad (19.7)$$

where the mean radial ($\mathcal{F}_r$) and latitudinal ($\mathcal{F}_\theta$) angular momentum fluxes:

$$\mathcal{F}_{r,\theta} = \mathcal{F}_{r,\theta}^{VD} + \mathcal{F}_{r,\theta}^{RS} + \mathcal{F}_{r,\theta}^{MC} + \mathcal{F}_{r,\theta}^{MS} + \mathcal{F}_{r,\theta}^{MT} \quad (19.8)$$

are the sum of angular momentum transport by viscous diffusion (which we denote as $\mathcal{F}_r^{VD}$ and $\mathcal{F}_\theta^{VD}$), Reynolds stresses ($\mathcal{F}_r^{RS}$ and $\mathcal{F}_\theta^{RS}$), meridional circulation ($\mathcal{F}_r^{MC}$ and $\mathcal{F}_\theta^{MC}$), Maxwell stresses ($\mathcal{F}_r^{MS}$ and $\mathcal{F}_\theta^{MS}$), and large-scale magnetic torques ($\mathcal{F}_r^{MT}$ and $\mathcal{F}_\theta^{MT}$) (cf. Brun *et al.*, 2004). The Reynolds stresses are associated with correlations of the fluctuating velocity components that arise from organized tilts within the convective structures, especially in the downflow plumes (e.g., Brummell *et al.*, 1998; Miesch *et al.*, 2000). In the same spirit, the Maxwell stresses are associated with correlations of the fluctuating magnetic field components that arise from tilt and twist within the magnetic structures. In Figure 19.3 (lower right panel) we show the overall balance of the five terms in a turbulent dynamo model representative of the Sun (Brun, 2004). We see that the differential rotation is prograde at the equator because the Reynolds stresses transport angular momentum equatorward, being opposed in the latitudinal direction by the other processes. In the radial direction the balance of terms is more equilibrated, with the meridional circulation and the Reynolds stresses being outward on average. When we consider the non-linear coupling between the convective envelope and the stably stratified radiative interior in 3D models, we see that it influences the profile of the differential rotation through the retroaction of the thermal wind at the base of the convection zones. It also allow us to study in detail the interface between the two zones and the amount of angular momentum transferred (Gilman *et al.*, 1989). We can see that the meridional circulation only penetrates to about 0.02 $R_\odot$ below the convective envelope and that the penetration of convective motions overall is limited by the stiff stratification. Yet these penetrative motions trigger internal waves with a large spectrum of wavenumbers, as we will see below. The interaction of an inner fossil field with the convective envelope and hence the magnetic stresses linking the two zones can also be quantified in the presence or not of a dynamo-generated field (Forgács-Dajka and Petrovay, 2002; Strugarek *et al.*, 2011a). We now discuss in more detail the angular momentum exchange over longer time scales.

19.4 Angular momentum transport on a long dynamical time scale

In addition to transport of angular momentum on a short dynamical time scale, secular transport of angular momentum occurs in stellar radiation zones. In these regions, differential rotation induces transport of angular momentum, heat, and chemicals that modify a star's evolution, lifetime, late stages of evolution, and nucleosynthetic properties (see, e.g., Maeder, 2009). Moreover, since very strong constraints have been obtained on the Sun's internal rotation (e.g., García *et al.*, 2007, 2011) and on more evolved low-mass stars (e.g., Deheuvels *et al.*, 2012; Mosser *et al.*, 2012b), it becomes necessary to build stellar models that also account for MHD processes on secular time scales. In this section, we therefore summarize the most recent progress that has been achieved for the modeling of secular exchanges of angular momentum in the Sun and low-mass stellar interiors.

19.4.1 Differential rotation and associated large-scale meridional circulation and shear-induced turbulence

The first mechanisms that transport angular momentum in stellar radiation zones are the large-scale meridional circulation and turbulence induced by differential rotation (e.g., Zahn, 1992; Mathis and Zahn, 2004). A coherent picture of their interactions has been given in Decressin *et al.* (2009). The viscous stresses of the shear-induced turbulence ($\mathcal{F}^{VD}_{r,\theta}$), the torque applied by stellar winds, and structural adjustments during stellar evolution drive the meridional circulation. This flow advects heat and the radiation zone relaxes to a new thermal state. Because of the latitudinal gradient of entropy, a new differential rotation is built because of the thermal-wind balance. However, these two mechanisms, if applied to the Sun and low-mass stars, are unable to reproduce the angular velocity profile of the radiative core of the present Sun (Pinsonneault *et al.*, 1989), the light-element mixing in solar-type stars on the main sequence (e.g., Talon and Charbonnel, 2005) and the rotation profiles in the central radiation zone of low-mass subgiant and red giant stars observed by *Kepler* (Eggenberger *et al.*, 2012b; Marques *et al.*, 2013; Ceillier *et al.*, 2013). Therefore, other processes such as magnetic fields (e.g., Rudiger and Kitchatinov, 1997; Gough and McIntyre, 1998; Strugarek *et al.*, 2011a,b) or internal waves excited by penetrative convection (e.g., Schatzman, 1993; Zahn *et al.*, 1997) must be studied.

19.4.2 Magnetic fields

The admitted paradigm for the origin of magnetic fields in the solar and low-mass star radiative cores is the fossil field hypothesis (Mestel, 1965). Fossil fields are believed to originate from the trapping of the interstellar magnetic flux during star formation or/and to be a remnant of the field generated by the convective dynamo during the pre-main sequence (e.g. Zanni and Ferreira, 2013, and references therein). If the initial field is a small-scale turbulent field, we have to predict the topology of the resulting field after the birth of the radiative region. Indeed, the field geometry directly impacts the angular momentum

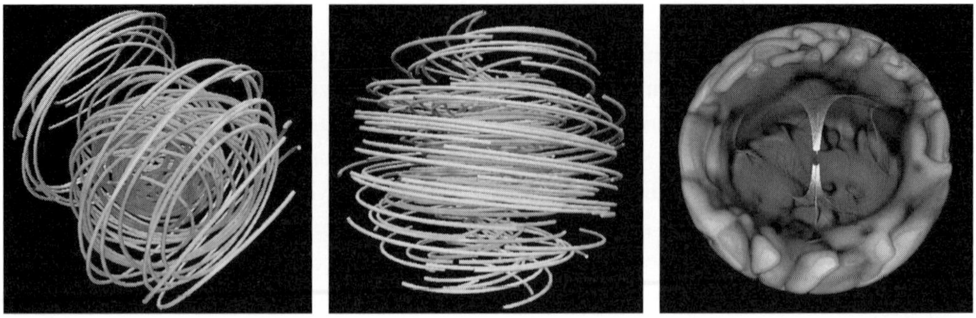

Figure 19.4 (Left) Field configuration resulting from the instability of an initial purely axisymmetric toroidal field (the color scale is a function of the density. (From Duez *et al.*, 2010, courtesy *The Astrophysical Journal*.) (Middle) Stable relaxed large-scale non-force-free twisted fossil field configuration. (From Duez *et al.*, 2010.) (Right) Creation of a toroidal field resulting from the action of the shear of the tachocline on an inital dipolar field in a 3D integrated model of the Sun. (From Strugarek *et al.*, 2011a.) Thin lines are the field lines. In the convective envelope, red indicates positive velocities, and blue, negative velocities. A black and white version of this figure will appear in some formats. For the color version please refer to the plate section.

transport by magnetic torques ($\mathcal{F}_{r,\theta}^{MT}$ and $\mathcal{F}_{r,\theta}^{MS}$) and the resulting differential rotation profile. Understanding the initial configuration of fossil fields is linked to MHD relaxation processes, in which an initial turbulent magnetic field is converted into a large-scale configuration thanks to a selective decay of the ideal MHD invariants such as magnetic helicity ($\mathcal{H} = \int_V (\nabla \times \mathbf{B}) \cdot \mathbf{B}\,\mathrm{d}V$ with $\mathbf{B}$ the magnetic field and V the volume) and energy. The case of the self-gravitating plasma of stellar radiation zones, where the gaseous pressure dominates, has been studied both theoretically (Duez and Mathis, 2010) and using numerical simulations (Braithwaite and Spruit, 2004). Resulting non-force-free relaxed states minimize the total energy for given magnetic helicity and fluid invariants due to the stable stratification of these regions. Both axisymmetric and non-axisymmetric configurations can be obtained (Braithwaite, 2008; Duez and Mathis, 2010; and Figure 19.4). The field is organized on large-scale non-force-free configurations that are initially stable (see Figure 19.4), because of their mixed geometry (poloidal and toroidal), confirming the pioneering work of Tayler (1973) and Markey and Tayler (1973), who demonstrated that purely toroidal and poloidal fields are unstable (see Figure 19.4).

This initial stable situation could be modified by differential rotation (Ω-effect) and magnetic instabilities. Due to large-scale shear, the toroidal field can be amplified to a point where the field will lose its stable mixed character and become unstable. If the instablity is able to create an efficient regeneration of the initial field from the correlations of the fluctuations of the velocity and of the field due to the instability (the so-called α-effect), it could trigger a dynamo; this possibility is still, however, strongly debated (Spruit, 2002; Braithwaite, 2006; Zahn *et al.*, 2007; Rüdiger *et al.*, 2012).

Once the initial (axi- or non-axisymmetric) non-force-free magnetic relaxed configuration has been established, the large-scale magnetic torque ($\mathcal{F}_{r,\theta}^{MT}$) and Maxwell stresses

($\mathcal{F}^{MT}_{r,\theta}$) transport angular momentum. This leads to strong coupling between the core and the envelope and to two possible final rotational states.

- In the case where the field is axisymmetric, a Ferraro state is established by magnetic local torques. In this state, the differential rotation is damped on poloidal (i.e., meridional) magnetic surfaces. The resulting angular velocity is frozen in the poloidal field lines, i.e., $(\mathbf{B}_p \cdot \nabla)\Omega = 0$, where $\mathbf{B}_p$ is the poloidal field component. This prediction has been confirmed by 2D and 3D numerical simulations of the dynamical evolution of magnetized stellar radiation zones (e.g., Charbonneau and MacGregor, 1993; Brun and Zahn, 2006; Spada *et al.*, 2010; Strugarek *et al.*, 2011a). Note also that the final differential rotation profile depends on the penetration of the fossil field lines in the envelope that transmit its differential rotation to the core (Charbonneau and MacGregor, 1993).
- In the case of strong non-axisymmetric magnetic fields, a uniform rotation is obtained as in the case of oblique rotators (Moss, 1992). In the case of weak non-axisymmetric fields, Spruit (1999) proposed that they evolve to an axisymmetric configuration because of rotational smoothing processes, leading finally to the Ferraro state. However, this scenario has not been confirmed by the first 3D numerical simulations for an initial oblique dipole by Strugarek *et al.* (2011a), and more complete studies must be achieved in the near future.

Finally, in the magnetized case, a meridional circulation could also modify the internal rotation configuration of the star. As in the hydrodynamical case, meridional flows in these regions are driven by internal and applied torques: the Lorentz torque ($\mathcal{F}^{MT}_{r,\theta}$), turbulent Maxwell ($\mathcal{F}^{MS}_{r,\theta}$) and Reynolds stresses ($\mathcal{F}^{RS}_{r,\theta}$), the torque applied by stellar winds, and structural adjustments during stellar evolution. In the case where all the sources vanish, the meridional circulation dies after an Eddington–Sweet time and the star settles into a baroclinic state described by the magnetic thermal-wind equation. We thus understand that the meridional circulation, if we consider a star without structural adjustments and applied torques, will be driven by the residual magnetic torque and Maxwell stresses until the star reaches a torque-free state (Mestel *et al.*, 1988; Mathis and Zahn, 2005). Further, the interactions between an inner fossil field and downwelling meridional circulations coming from the convective envelope have also been studied. Gough and McIntyre (1998) (see also Acevedo-Arreguin *et al.*, 2013) proposed that such flows can confine the field to the radiative core and prevent the transmission of the differential rotation of the convective envelope. But this scenario is still not verified in the case of 3D non-axisymmetric simulations of the Sun (Strugarek *et al.*, 2011b) where the field lines of the fossil field penetrate the convection zone which leads to a non-uniform rotation of the solar core.

19.4.3 Internal waves

The last mechanism able to transport angular momentum in the stably stratified radiative cores of the Sun and solar-type stars is the action of internal waves that are excited by turbulent convective downflows, also called plumes, at the radiation/convection

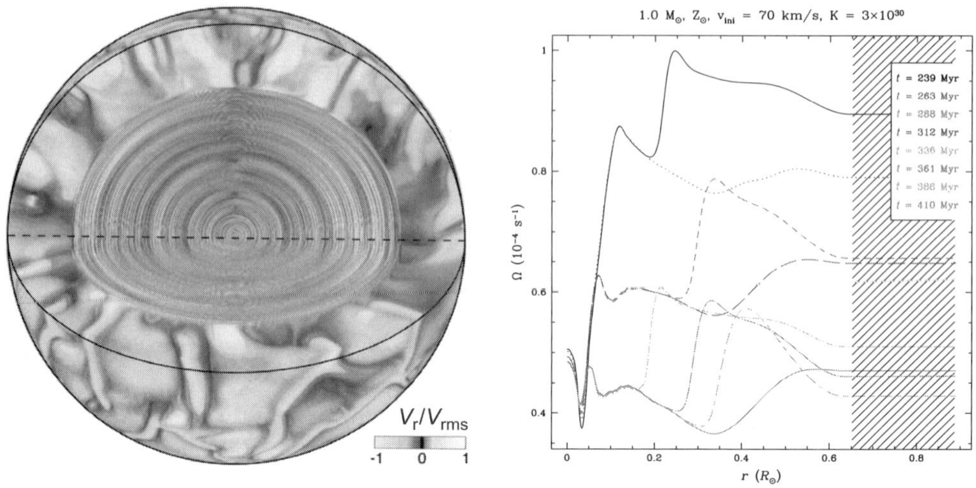

Figure 19.5 (Left) 3D numerical simulation of the full Sun from 0 to $0.97R_\odot$. We represent the normalized radial velocity (to the rms one) in the full simulated star. Red and blue correspond to positive and negative velocities respectively. Internal gravity waves excited by turbulent convective plumes are identified in the core. (From Alvan *et al.*, 2014, courtesy *Astronomy & Astrophysics.*) (Right) Evolution of the angular velocity radial profile in a solar-type star along its evolution due to the combined action of internal gravity waves and meridional flows. Hatched region corresponds to the convective envelope. We obtain a spin-down of the core because of the extraction of angular momentum by retrograde waves. (From Alvan *et al.*, 2013, courtesy *Astronomy & Astrophysics.*) A black and white version of this figure will appear in some formats. For the color version please refer to the plate section.

borders (Dintrans *et al.*, 2005; Rogers and Glatzmaier, 2005; Brun *et al.*, 2011; Alvan *et al.*, 2014, see left panel in Figure 19.5). These waves transport angular momentum that is deposited/extracted where they are dissipated because of heat and viscous diffusion (e.g., Zahn *et al.*, 1997) or where they reach critical layers where their frequency is proportional to the local angular velocity (Goldreich and Nicholson, 1989; Alvan *et al.*, 2013). The momentum flux, carried by Reynolds stresses ($\mathcal{F}_{r,\theta}^{RS}$) of the waves, is of the same (respectively of opposite) sign as the energy flux for prograde (retrograde) waves. This leads to a net deposit (extraction) of angular momentum by prograde (retrograde) waves in the case of the Sun and solar-type stars. If differential rotation is initially present, a net flux of angular momentum is transported because of the Doppler effect that differentiates the respective damping rates of prograde and retrograde waves. In the case of a core rotating faster than the envelope, prograde waves are damped closer to the excitation region than retrograde ones that propagate deeper into the star. A net extraction of angular momentum is obtained through different fronts that propagate from the interior to the exterior during the evolution and drive an efficient spin-down of the core (see right panel in Figure 19.5). This net extraction, combined with the strong horizontal shear-induced turbulence due to the stable stratification, is therefore a candidate to explain the observed quasi-uniform rotation rate down to 0.2 $R_\odot$ of the solar radiative core as well as the mixing in low-mass stars (Charbonnel and Talon, 2005). Internal gravity waves can be strongly modified by rotation and by magnetic fields through Coriolis acceleration (Mathis *et al.*, 2008; Mathis, 2009) and the Lorentz tension force (Mathis and de Brye, 2011, 2012) respectively. In this case, internal gravity waves become magneto-gravito-inertial waves. Their propagation is modified with possible vertical and latitudinal trappings. For example, gravito-inertial waves (without a magnetic field) with frequencies below 2Ω are trapped in an equatorial belt. Moreover, their radiative damping is increased. Therefore, the transmission of energy from convective plumes to the waves is modified as well as the efficiency of the transport of angular momentum. In the case of solar-type stars, Mathis and de Brye (2012) have demonstrated that the net bias in favor of retrograde waves and the corresponding extraction of angular momentum is conserved in the case of magneto-gravito-inertial waves.

19.5 Future prospects

As we have shown, our understanding of the key magnetohydrodynamical processes acting inside and at the surface of stars has made important progress recently. We are starting to have a clearer picture of what are the physical processes transporting angular momentum in stars, how the radiative interior and convective envelope in solar-like stars non-linearly interact, and how a stellar wind can brake a star over a long period of time. Of course these processes are inter-related; for instance, dynamo action in a convective envelope depends on the rotation rate of the stars whose braking via a magnetized wind depends on dynamo action! Hence, developing a global picture taking into account all the feedbacks is difficult, but starting to be reachable, with global 3D simulation coupling a convective dynamo to a stably stratified interior now being possible thanks to the recent advances of supercomputers. To improve the modeling of angular momentum exchanges on secular time scales, efforts to build new generations of dynamical stellar evolution codes will be continued. First, 1D stellar evolution codes will strongly benefit from couplings with systems of 2D equations for the transport of angular momentum, magnetic field, heat, and chemical elements and for large-scale meridional flows (Mathis and Zahn, 2004, 2005). To treat the case of rapidly rotating stars, corresponding 2D dynamical stellar evolution codes that also take into account the flattening of stellar structure because of centrifugal acceleration are also being developed (Espinosa Lara and Rieutord, 2013). In both cases, all the non-linear and non-axisymmetric phenomena such as waves, instabilities and turbulence will be treated using prescriptions and scaling laws coming from 3D numerical simulations on dynamical time scales, asymptotic theoretical studies, and laboratory experiments. To guide these theoretical improvements, constraints on both surface dynamics and magnetism via an activity proxy such as Ca II H and K lines or spectropolarimetry (ESPaDOnS, SPIRou) and stellar interiors via space-borne asteroseismology experiments (with CoRoT, *Kepler*, and in the future the selected mission PLATO) are our best guide.

20

From the Sun to the distant stars, and back: Future research directions

JØRGEN CHRISTENSEN-DALSGAARD

20.1 Introduction

Until recently our knowledge of stellar structure and evolution was based almost exclusively on observations of the superficial characteristics of stars, i.e., their surface temperature, gravity, and composition, in some cases combined with the surface luminosity, and in relatively rare cases observational determinations of radii and masses, with observations of eclipsing binaries providing accurate information (e.g., Andersen, 1991). This was combined with modeling, allowing us to infer from these surface properties the structure of the interior of a star, including its evolutionary state and hence age. In a few binary systems it is possible to determine the apsidal motion, which provides a measure of the internal mass distribution (e.g., Claret and Giménez, 2010). Also, in the unique case, amongst non-exploding stars, of the Sun, observations of neutrinos provide information about the nuclear reactions in the stellar interior (for a recent review, see Turck-Chièze and Couvidat, 2011), although these were for a long time at variance with the observations.

Thus, observations only provide somewhat indirect information on which to base a test of our modeling and understanding of stellar evolution. This is a serious concern. Stars are evidently very complex systems, controlled by atomic processes and interactions in the thermodynamic properties of stellar matter, interactions between matter and radiation, and nuclear energy generation that are far from understood or yet reproduced in a controlled manner in the laboratory. Also, they have hydrodynamic instabilities, including those potentially associated with rotation, which are at best treated in a rudimentary fashion in the modeling, but which probably have substantial effects on the evolution of stars. Magnetic fields, which are present on large scales in some stars and which are evident in solar and stellar activity, add another layer of complexity in the physics of stars that is rarely even considered in the modeling. Given our present rudimentary representation in the modeling of the physics of stellar interiors, one might doubt whether the results of the modeling can be used reliably in other areas of astrophysics, such as the stellar age determinations that play a central role in the study of the evolution of the Galaxy, or the understanding of the

Extraterrestrial Seismology, eds. Vincent C. H. Tong and Rafael A. García. Published by Cambridge University Press.

chemical evolution of the universe. Observations providing more direct information about stellar interiors would obviously be very valuable to test the modeling and possibly indicate how it should be improved. Helioseismic and asteroseismic analyses of observations of stellar oscillations provide such more direct inferences.

It is evident that the seismic studies of the Sun yield far more detail than what can be obtained for the distant stars. However, the Sun is just a single star in a relatively simple stage of evolution. Thus it is crucial that seismic data, although more limited than for the Sun, are now available for stars spanning a very broad range in mass, evolutionary state, and composition. By exploiting the complementary nature of the very detailed information for a single star and the limited information for many stars, we may hope to obtain a more complete view of stellar evolution and stellar internal physics, thus also improving our understanding of the Sun.

Concerning the Sun, I here concentrate on global helioseismology, which is of most relevance to the study of stellar evolution; however, I note that local helioseismology may provide information of great value to the understanding of near-surface properties of stars. Asteroseismology covers a huge range of stellar properties, from red giants to white dwarfs. In the present discussion of the interplay between helio- and asteroseismology, the stars of most relevance are probably those that show solar-like oscillations, and hence these are the ones that I consider in the following.

As a background, a few words on the properties of the oscillations may be helpful; further details are provided by, for example, Aerts *et al.* (2010). Most oscillations observed in the Sun are acoustic modes, i.e., essentially standing sound waves, of generally rather high radial order n, and with maximum power at a cyclic frequency $\nu_{\max} \simeq 3000\ \mu\text{Hz}$, corresponding to a period near 5 minutes. The variation of the modes across the stellar surface is characterized by their degree l determining the total number of nodal lines, and the azimuthal order m determining the number of nodal lines crossing the equator. In the Sun, oscillations are observed over a very large range of degrees, from $l = 0$ (corresponding to spherically symmetric, or *radial*, oscillations) to degrees of several thousand. In unresolved observations of distant stars, however, only the lowest degrees are observable, typically up to $l = 2$ or 3; the signal from modes of higher degree is suppressed by cancellation across the stellar surface. Such low-degree acoustic modes approximately satisfy the relation $\nu_{nl} \simeq \Delta\nu(n + l/2 + \epsilon)$, where ν_{nl} is the cyclic frequency of a mode of radial order n and degree l, and ϵ is a phase predominantly determined by the near-surface layers of the star. The so-called *large-frequency separation* $\Delta\nu$ is given by

$$\Delta\nu = \left(2\int_0^R \frac{dr}{c}\right)^{-1}, \tag{20.1}$$

where c is the adiabatic sound speed, r is the distance to the center of the star and R is its surface radius.

Detection of solar-like oscillations in distant stars is typically characterized by the identification of a localized enhancement of power in the power spectrum of the observations and the presence of a uniformly spaced set of peaks in power. Thus, the first quantities to be determined are the frequency ν_{max} at maximum power and the large frequency separation $\Delta\nu$. These *global seismic parameters* already have substantial diagnostic potential. From the basic scaling relations of stellar structure it follows that $\Delta\nu$ scales as the square root of the mean density of the star, i.e.,

$$\Delta\nu \propto \left(\frac{M}{R^3}\right)^{1/2} , \tag{20.2}$$

where M is the mass of the star. Also, it appears that ν_{max} scales as the so-called acoustic cut-off frequency of the stellar atmosphere, leading to

$$\nu_{max} \propto \frac{M}{R^2} T_{eff}^{-1/2} , \tag{20.3}$$

where T_{eff} is the effective temperature. From observed values of $\Delta\nu$ and ν_{max}, and a determination of T_{eff}, the mass and radius of the star can be estimated, either from a direct analysis of these relations (e.g., Kallinger *et al.*, 2010b) or by fitting to grids of stellar models (Stello *et al.*, 2009a; Basu *et al.*, 2010; Quirion *et al.*, 2010).

In the Sun, spatially resolved observations have allowed the detection of a huge number of acoustic modes, up to high degree. Their characteristics are well described by their asymptotic properties, according to which the modes are confined outside a distance r_t from the center satisfying

$$\frac{c(r_t)}{r_t} = \frac{2\pi\nu}{\sqrt{l(l+1)}} , \tag{20.4}$$

for $l \geq 1$; radial modes essentially reach the center. Thus, low-degree modes, such as those that are observed in distant stars, reach the stellar core, whereas the modes of highest degree observed in the Sun are confined to a thin region near the solar surface. This property is of huge importance for our ability to carry out helioseismic inversion: the variation in the depth of penetration allows us to separate the contributions to the frequencies from the different layers of the Sun. In particular, it is used in the simple, but surprisingly successful, asymptotic techniques for helioseismic inversion (e.g., Gough, 1984; Christensen-Dalsgaard *et al.*, 1989; Vorontsov, 1989).

In evolved stars, a second class of oscillation modes become relevant, the internal standing gravity waves, whose restoring force is buoyancy. When a star exhausts its central hydrogen and leaves the main sequence it develops a compact helium core. As a result, the local gravitational acceleration in the deep interior of the star becomes very high, pushing up the frequencies of the gravity waves to the level of the acoustic modes and hence giving rise to *mixed modes*, which have both buoyancy and acoustic character. Their frequencies depend on the detailed structure of the deep interior of the star, including the composition profile, and hence they have very high diagnostic potential. As a star evolves, the acoustic-mode frequencies decrease and the gravity-mode frequencies increase, leading to *avoided crossings*, where the frequencies may approach quite closely without actually crossing (e.g., Osaki, 1975). In the extreme case of red-giant stars, this leads to a dense spectrum of buoyancy-dominated modes. These also satisfy an asymptotic relation, most simply expressed in terms of the oscillation periods Π_{nl}:

$$\Pi_{nl} = \frac{\Pi_0}{\sqrt{l(l+1)}}(n + \epsilon_g) ; \tag{20.5}$$

here

$$\Pi_0 = 2\pi^2 \left(\int N \frac{dr}{r} \right)^{-1} , \tag{20.6}$$

where N is the so-called buoyancy frequency, i.e., the characteristic frequency of internal gravity waves, and ϵ_g is a phase depending on the properties near the boundaries of the region where the gravity waves propagate.

For a spherically symmetric star the frequencies are independent on m. This degeneracy is lifted by rotation, which induces a frequency splitting depending on the internal rotation rate and the properties of the modes.

20.2 Helioseismic successes and challenges

A discussion of the impressive results of helioseismology has already been provided in previous chapters of this book. Here I give a brief summary and consider some of the outstanding issues, with some emphasis on placing the Sun in the broader general stellar context. For a recent review of helioseismology, see Gough (2013).

It is remarkable that the solar data have proven sufficiently extensive and accurate to allow inversion for structure, most commonly represented by sound speed, and rotation in most of the solar interior. The most extensive results on solar structure have been obtained using different versions of optimally localized averaging inversion (Gough *et al.*, 1996; Basu *et al.*, 1997; Kosovichev *et al.*, 1997) to determine the difference in sound speed between the Sun and various models. In early results of such detailed analyses, using models that included microscopic diffusion and settling, the sound-speed difference was dominated by a localized region just beneath the solar convection zone, where the solar sound speed exceeded that of the model by up to 0.2%, at a highly significant level. This discrepancy could in part be eliminated by invoking weak mixing of the otherwise rather steep gradient in the helium abundance established by settling, hence decreasing the mean molecular weight in the critical region and increasing the sound speed to match that of the Sun (e.g., Brun *et al.*, 1999; Christensen-Dalsgaard and Di Mauro, 2007). Such mixing could reasonably be assumed to be caused by penetration of convection beyond the limit of instability and/or through instabilities in the tachocline, the region of strong rotational shear inferred at the base of the convection zone.

Recent revisions of the determination of the solar surface composition (Asplund *et al.*, 2009) have substantially reduced the inferred abundances of, in particular, carbon, nitrogen, and oxygen, and consequently the opacity in the radiative interior of solar models. The resulting models differ from the solar sound speed by up to around 1%, with the difference gradually decreasing from near the base of the convection zone to the solar center (for a review, see Basu and Antia, 2008). It remains unclear how this difference can be eliminated, assuming that the revised abundance determinations are reliable. An obvious, but hardly satisfactory, explanation might be that there are compensating errors in the opacity tables (e.g., Bahcall *et al.*, 2005a; Christensen-Dalsgaard *et al.*, 2009), although the magnitude of the required corrections is well beyond the estimated uncertainties in opacity calculations.

Although the heavy-element abundance mainly affects the opacity, it also has a small, but in principle detectable, influence on the equation of state, particularly through the adiabatic compressibility Γ_1, which directly affects the sound speed. This was used in an analysis by Vorontsov *et al.* (2013), who carried out an inverse analysis partly based on the asymptotic properties of the acoustic modes, concentrating on the structure of the convection zone. This analysis indicated a heavy-element abundance in the convection zone even lower than what was obtained by Asplund *et al.*, hence further aggravating the problems with the modeling of the radiative interior. Thus, although the helioseismic analysis undoubtedly provides an accurate determination of the sound speed in most of the solar interior, the understanding of this structure in terms of solar modeling is still far from complete.

A major uncertainty in the modeling of cool stars with an outer convection zone, including the Sun, is the treatment of the outermost parts of the convection zone. Here convection is substantially superadiabatic and hence the structure is affected by the detailed treatment of convection, often, but certainly not satisfactorily, dealt with in terms of some version of a mixing-length formulation. Also, the momentum transport by convection, acting as a turbulent pressure, is generally neglected in the calculation of stellar models or at best treated in a fashion that is hardly consistent. The calculations of stellar oscillations are generally carried out in the adiabatic approximation, neglecting the perturbation to the turbulent pressure, and when non-adiabatic calculations are attempted, the treatment of the perturbation to the convective flux remains uncertain. These different deficiencies in the modeling of cool stars are collectively referred to as "the surface effect" and contribute to the difference between the computed and observed frequencies. They generally dominate the differences between observed solar oscillation frequencies and those of up-to-date solar models (e.g., Christensen-Dalsgaard *et al.*, 1996). Since the effects are concentrated very near the surface, the influence on the frequencies is essentially just a function of frequency. Given this property, the observed set of modes in the solar case is sufficiently rich to allow the near-surface contribution to the frequency differences to be isolated from differences arising from structure differences in the rest of the solar interior, allowing reliable inversions. However, in the stellar case this effect remains a major concern, as discussed in Section 20.3.

Although a major strength of helioseismology is the possibility, thanks to the richness of the observed modes, to determine the properties of the solar interior as a function of position through inverse analysis, this can usefully be combined with other types of analysis to investigate specific properties. An important example is the study of *acoustic glitches*, i.e., features in solar structure that vary on a scale smaller than the local wavelength of the modes (Gough, 1990). These include the rapid variation in the sound-speed gradient at the edges of convective regions and the variation in sound speed resulting from the localized decrease in

Γ_1 in the second helium ionization zone. The latter effect has been used to infer the helium abundance in the solar convection zone (e.g., Vorontsov *et al.*, 1991; Pérez Hernández and Christensen-Dalsgaard, 1994; Basu and Antia, 1995). Analysis of the glitch at the base of the solar convective envelope has been used to put stringent constraints on the extent of overshoot beneath the convection zone (Basu *et al.*, 1994; Monteiro *et al.*, 1994; Roxburgh and Vorontsov, 1994). The analysis shows that the transition to the radiative region is smoother than predicted by solar models including helium settling (Christensen-Dalsgaard *et al.*, 2011), indicating the presence of mixing processes smoothing the composition profile in this region. Analysis of the glitch at the base of the convective envelope also provides information about the depth of the convection zone, in terms of the *acoustic depth*, i.e., the sound travel time from the surface. A careful analysis of the acoustic glitches in the Sun, based just on the low-degree modes that can be observed in distant stars, was carried out by Houdek and Gough (2007).

Extensive observations of rotational splittings for a broad range of modes have enabled inversions providing detailed inferences of solar internal rotation, as functions of latitude and distance to the center, over most of the solar interior (for reviews, see Thompson *et al.*, 2003; Howe, 2009). In the convection zone, the angular velocity is predominantly a function of latitude and differs strongly from a dependence solely on the distance to the rotation axis ("rotation on cylinders") predicted by simple models of convection-zone dynamics. The bulk of the radiative interior rotates essentially as a solid body, at a rate corresponding to the convection-zone rotation at an intermediate latitude; in the transition between the convection zone and the radiative interior there is therefore a steep radial gradient of angular velocity in the tachocline, of thickness probably around or less than $0.04R_\odot$ (Antia *et al.*, 1998; Charbonneau *et al.*, 1999). The angular velocity of the solar core, within $0.2R_\odot$, is uncertain, although the indications, within considerable error bars, are for a continuing solid-body rotation (e.g., Chaplin *et al.*, 1999; García *et al.*, 2004; Eff-Darwich and Korzennik, 2013).

The origin of this rotation pattern is poorly understood. Hydrodynamical modeling, although far from resolving the relevant scales, can to some extent reproduce the rotation profile in the convection zone (Miesch *et al.*, 2008). However, the origin of the near-constant rotation of the radiative interior and of the detailed properties of the tachocline is still unclear. I discuss this further in Section 20.4.

The frequencies of solar oscillation vary with solar activity, on average by about 0.4 μHz between solar minimum and maximum (Woodard and Noyes, 1985). The dependence of the variation on frequency and degree strongly indicates that it is dominated by the near-surface region (Libbrecht and Woodard, 1990), and indeed the frequency variations are very closely correlated with the surface magnetic activity (e.g., Bachmann and Brown, 1993; Chaplin *et al.*, 2001). It is not clear whether the dominant cause of these variations is the direct effect of the magnetic field on the oscillations, or whether the related changes in solar structure dominate. Interestingly, analysis of low-degree modes has shown a 2-year cycle in the frequency variations overlying the more dominant 11-year cycle, leading to speculations about the presence of two separate dynamo mechanisms in the Sun

(Broomhall *et al.*, 2012b). Solar activity also affects the physics of the modes: Chaplin *et al.* (2000) found that the damping rate increases with increasing activity and the amplitudes decrease, whereas the energy input to the modes stays approximately unchanged.

20.3 Asteroseismic successes and challenges

The recent development of asteroseismology has been dominated by space-based photometric observations. Following the serendipitous use of a star tracker on the WIRE spacecraft (Buzasi *et al.*, 2000), the Canadian MOST mission (Walker, 2003) was the first mission dedicated to photometric observations of stellar variability. It has made many valuable contributions to the study of stellar oscillations but does not quite reach the sensitivity required for detailed studies of solar-like oscillations in main-sequence stars. This level has been dramatically achieved by the CoRoT (Baglin *et al.*, 2009) and *Kepler* (Borucki *et al.*, 2010) missions, both having the combined goal of detecting extra-solar planets and carrying out asteroseismology. An overview of the asteroseismic potential of *Kepler* was provided by Gilliland *et al.* (2010); the activities have been organized in the *Kepler Asteroseismic Science Consortium* (KASC; Kjeldsen *et al.*, 2010).

As far as the solar-like oscillations are concerned, it is obvious that valuable experience is obtained from the analysis of low-degree solar oscillations, such as obtained from the BiSON network and the GOLF and in particular VIRGO instruments on SOHO; clearly the photometric data from VIRGO share many properties with the observations by CoRoT and *Kepler*. In preparation for the missions, this transfer of experience was formalized in the AsteroFLAG effort (Chaplin *et al.*, 2008), developing and testing with artificial data techniques for asteroseismic analysis, on the basis of similar efforts in the solar FLAG project (Chaplin *et al.*, 2006).

A particular strength of *Kepler* has been the possibility to observe a sample of several hundred stars showing solar-like oscillations, on and near the main sequence. These were observed for one month in the early phases of the mission, as part of a survey to identify targets for more extensive observation. An initial analysis of these data based on the global asteroseismic parameters $\Delta\nu$ and $\nu_{\rm max}$ provided determinations of masses and radii of a large sample of stars (Chaplin *et al.*, 2011a). After correction for possible biases a comparison was made with the predicted distribution of stellar masses and radii in a Galactic model, indicating good agreement between the inferred and modeled radius distributions, whereas the inferred mass distribution showed a surplus of low-mass stars and a deficiency of high-mass stars compared with the model. Although certainly preliminary, these results are a strong indication of the power of asteroseismic ensemble studies in investigations of Galactic structure and evolution.

Studies of global stellar properties have been made for thousands of red giants, greatly extending the potential of asteroseismic probes of Galactic properties, particularly given the intrinsic brightness of these stars, allowing them to be observed over large distances. From the inferred radius, the observed effective temperature and the apparent brightness, the distance to the red giant can be estimated, while the mass provides a measure of the age, given that most of the life of the star is spent on the main sequence. It is important to note, however, that such age inferences are closely related to a better understanding of stellar evolution.

From the point of view of relations to helioseismology, the main interest is obviously to investigate the internal properties of the stars and hence extend the seismic investigations of stellar interiors beyond the Sun. It is probably fair to say that these detailed studies based on the analysis of the individual observed frequencies, or suitable combinations of them, are still in their infancy. A limiting factor is the errors in the modeling of the near-surface layers (cf. Section 20.2) which, unlike in the solar case, cannot be fully eliminated in the analysis of the frequencies. By assuming a functional form of the required frequency correction modeled on the solar correction, procedures have been developed that allow the stellar correction to be estimated (Kjeldsen *et al.*, 2008a; Christensen-Dalsgaard, 2012); however, such an assumption can obviously be questioned, particularly for stars whose properties differ substantially from those of the Sun. Thus, some have considered it preferable to regard the near-surface effect as a source of systematic error, included in the weights used when fitting model frequencies to the observations. This was formalized in a Bayesian sense by Gruberbauer *et al.* (2012).

A powerful alternative is to base the analysis on frequency combinations that are essentially insensitive to the surface errors. Particularly important are ratios between frequency separations such as $r_{02}(n) = (\nu_{n0} - \nu_{n-1\,2})/(\nu_{n1} - \nu_{n-1\,1})$ (Roxburgh and Vorontsov, 2003, 2013; Otí Floranes *et al.*, 2005). These quantities provide information about the properties of the stellar core and hence the age of the star. They can be combined with the lowest observed frequencies that on very general grounds are expected to be insensitive to the near-surface layers (Christensen-Dalsgaard and Thompson, 1997).

At the present level of the asteroseismic observations, inverse analysis for stellar structure is probably infeasible. Thus, the analyses have employed what in the helioseismic context has been called the *forward method*, where models are adjusted to fit the observed frequencies, in general by minimizing a measure of the difference between the observed and the model observables, as a function of a few global model parameters. An important aspect is the comparison of results obtained with different evolution codes, to obtain an estimate of the systematic uncertainty resulting from problems with the stellar modeling.

The search for the best-fitting model is complicated by the existence of local minima in the measure of the goodness of the fit, which may lead to the identification of a model that is not optimal. To minimize this effect, one possible solution is the use of a genetic algorithm as implemented in the *Asteroseismic Modeling Portal* (AMP; Metcalfe *et al.*, 2009), a user-friendly tool for the computation of stellar models and fitting to observational data. Even though the technique is relatively efficient, the fit to a single set of observations still requires the computation of a huge set of stellar evolution tracks, handled by an underlying supercomputer grid.

A reliable statistical characterization of the results of asteroseismic analyses is still a topic that needs considerable further work. It is an essential requirement to ascertain whether the models used in the fits are consistent with the data, or whether (more interestingly) improvements are required in the model physics. The statistical analysis obviously involves both the determination of the oscillation frequencies and the analysis of the frequencies to investigate the stellar properties. In the former case, substantial progress is being made through the use of Bayesian methods such as the Markov Chain Monte Carlo (MCMC; Handberg and Campante, 2011) analysis, which yields an unbiased estimate of the probability density function of the frequencies. MCMC analysis has also been applied to stellar-model fitting (Bazot *et al.*, 2012); like the analysis using the genetic algorithm, it is extremely computationally demanding, but it certainly deserves to be further explored.

An early analysis of individual oscillation frequencies was made for the star KIC 11026764 by Metcalfe *et al.* (2010a), based on 33 days of *Kepler* data. Different techniques were applied for the frequency analysis and modeling, to test the sensitivity of the results to systematic effects. The star is on the subgiant branch, with clear identification of mixed modes, whose frequencies vary rapidly with evolution. Thus, for a given model calculation, the fit yields a very precise determination of stellar parameters, particularly the radius and age. Different calculations were found to give rather different results, with masses roughly speaking clustering around 1.13 and 1.23 $M_\odot$, and correspondingly higher (lower) initial helium abundance for the lower-(higher-)mass stars. This anticorrelation between mass and initial helium was also noted by, for example, Fernandes and Monteiro (2003) and Metcalfe *et al.* (2009); it is a natural consequence of the homology scaling of stellar properties (e.g., Kippenhahn and Weigert, 1990). The degeneracy between mass and helium has emerged in the asteroseismic investigations as a significant issue, which needs further work.

The diagnostic potential of mixed modes was also utilized by Deheuvels and Michel (2011) in the analysis of CoRoT observations of the star HD 49385. They made a simultaneous fit to models having the observed large frequency separation $\Delta\nu$ and the observed frequency of the avoided crossing between the acoustic and gravity-wave dominated modes. As an alternative, Deheuvels *et al.* (2012) made a simultaneous fit to $\Delta\nu$ and the asymptotic period spacing $\Delta\Pi_0$ (cf. Equation (20.6)). Such fits determine the region in parameter space of the best-fitting model, and they can subsequently be refined through a finer search and fits to the individual frequencies.

Mathur *et al.* (2012) carried out extensive fits to 22 stars observed during the early operation of *Kepler*. They compared results of several independent fits to the global seismic parameters with fits to individual frequencies using AMP. The agreement between the different techniques on the determination of mass and radius was relatively good, with differences around 4–6% and less than 2%, respectively. This provides support for the simpler global fits in the analysis of large numbers of stars. The AMP analysis also estimated a substantial number of parameters and properties of the best-fitting models, including the age with apparently high precision, although not including possible systematic errors from the modeling. A somewhat worrying result was that the best-fitting model in several cases had an initial helium abundance below the abundance obtained from Big Bang nucleosynthesis, which obviously seems unrealistic. This tendency, which has also been found in other recent fits to asteroseismic data, is probably related to the helium–mass degeneracy mentioned above. To resolve it, better techniques for the analysis of the asteroseismic data, or other observational constraints, will probably be needed.

An extensive analysis was carried out by Silva Aguirre *et al.* (2013) of two main-sequence stars observed by *Kepler* for about 270 days: KIC 6106415 (given the informal name Perky according to the custom in the KASC to name the stars after team-members' cats) and KIC 12009504 (named Dushera). Except for two models with anomalously low initial helium abundance (and hence high mass), the mass of Perky was found to be in the range 1.05–1.13 $M_\odot$, with ages between 3.8 and 5.3 Gyr. For Dushera the masses obtained were between 1.12 and 1.26 $M_\odot$, with ages between 3.1 and 4.3 Gyr, with no cases of very low initial helium abundance. All models of Dushera had a convective core, generally with a strong indication that mixing beyond the convective core was needed, in the form of convective overshoot or enhanced diffusion. This is the first time that asteroseismic evidence for additional core mixing has been obtained for solar-like stars and clearly indicates the potential for investigating such processes with the available data. For Perky, all models except one had a radiative core. The fact that one model showed preference for core convection might indicate that the star is close to the limit of stability and hence could offer further constraints on this process.

Main-sequence stars showing solar-like oscillations typically have masses below around 1.5 $M_\odot$. If such a star has a convective core, the mass contained in the core generally grows with age, as a result of the increasing importance in the energy generation of the CNO cycle. This results in a (nearly) discontinuous composition profile at the edge of the core, with a corresponding discontinuity in the sound speed. Such sharp sound-speed features, or *acoustic glitches*, have a potentially observable effect on the oscillation frequencies. Cunha and Metcalfe (2007) analyzed the asymptotic behavior of the oscillations in such models and developed a frequency combination that specifically depends on the jump in sound speed and hence is sensitive to the age of the star and/or effects of additional mixing beyond the convective core. This was further analyzed by Brandão *et al.* (2014), who determined the dependence of this and similar diagnostics on stellar properties. The results indicated that a significant determination of the discontinuity in sound speed would be possible from such data, for a substantial selection of stars.

As discussed in Section 20.2, the edge of convective cores is just one example of an acoustic glitch. The glitch associated with the second helium ionization zone has a potential for determining the helium abundance in the stellar envelope, while the glitch at the base of the convective envelope may provide information about the depth of the convection zone and, possibly, about the structure of this region. The potential for such asteroseismic investigations was analyzed by Pérez Hernández and Christensen-Dalsgaard (1998), Ballot *et al.* (2004), and Mazumdar (2005). Evidence for the effect of the helium glitch was found in CoRoT observations of a red giant by Miglio *et al.* (2010), and later in a solar-like star by Mazumdar *et al.* (2012). Furthermore, Mazumdar *et al.* (2014) studied a few *Kepler* stars with several methods obtaining very consistent results, demonstrating the potential of this technique to obtain reliable inferences about the stellar interior. Given the mass–helium degeneracy mentioned above, a constraint on the envelope helium abundance would be highly desirable.

Observation of rotational splitting in main-sequence solar-like stars is complicated because the splitting is typically comparable with the line width of the observed peaks. In the solar case it is only through observations over many years that enough information has been obtained to provide reliable constraints on rotation of the deep interior (see also García *et al.*, 2004; Broomhall *et al.*, 2012a). Thus, for main-sequence stars the information achieved has generally been just an average rotational splitting, mostly in connection with asteroseismic analysis of observations of exoplanet hosts (see below). In more evolved stars, on the other hand, the presence of mixed modes has provided much more detailed information. Beck *et al.* (2012) first, and Deheuvels *et al.* (2012, 2014) later, determined rotational splittings of mixed modes in several subgiant and red-giant stars, and on this basis deduced a core rotation exceeding the rotation of the surface. This was confirmed and greatly extended by Mosser *et al.* (2012b) in an analysis of around 300 stars. These included both stars on the ascending red-giant branch, with a hydrogen-burning shell around the helium core, and stars in the "red clump" where in addition there is central helium burning. In the former case there seemed to be an increase of core rotation rate with evolution, while the helium-burning stars showed slower rotation, undoubtedly as a result of core expansion following helium ignition and hence increase of the moment of inertia of the core. Thus we are now beginning to get a picture of the evolution of rotation with age, at least in the later stages of the life of solar-like stars. There is no doubt that the available *Kepler* data will allow such analysis for many more stars.

Kepler asteroseismology has made important contributions to the mission through the determination of properties of stars found to host exoplanets. The characterization of the planets requires precise information about stellar properties, in particular radius and mass, which are difficult to determine from "classical" astrophysical observations. Also, asteroseismology may allow determination of the age of the host star, and hence the planetary system, of great importance to the study of the evolution of planetary systems. A limitation has been that most of the exoplanet candidate hosts (the so-called Kepler objects of interest, or KOIs) found by *Kepler* are too faint for asteroseismic investigation. Even so, Huber *et al.* (2013a) provided estimates of radius and mass for 66 *Kepler* planet-candidate host stars, based on analysis of the global seismic observables $\Delta\nu$ and $\nu_{\rm max}$. More detailed analyses, based on individual frequencies, are restricted to a smaller number of stars, although additional cases are under study. An interesting example is Kepler-10b, the first rocky planet detected by *Kepler* (Batalha *et al.*, 2011). This was recently re-analyzed by Fogtmann-Schulz *et al.* (2014), who confirmed the earlier results, although with substantially higher precision. In particular, the age was determined as 10.4 ± 1.4 Gyr, making this one of the oldest known exoplanetary systems. Furthermore, the analysis of KOIs has substantially increased the number of stars with asteroseismic data having masses below solar; the *Kepler* targets selected specifically for asteroseismology included few such stars.

An important part of the asteroseismic investigations of exoplanet hosts is the determination of the stellar rotation rate and, in particular, the orientation of the pulsation axis, assumed to coincide with the rotation axis. Stochastically excited oscillations in main-sequence stars are expected to reach an amplitude that on average does not depend on the azimuthal order of the mode. However, as discussed by Gizon and Solanki (2003), the *observed* amplitudes depend on the geometry of the modes. In particular, the relative amplitudes of the different rotationally split components depend on the inclination i of the rotation axis relative to the line of sight: If the axis points towards the observer only modes with $m = 0$ are observed, with the axis in the plane of the sky only modes for which $l + m$ is even are visible, and in intermediate cases the relative amplitudes provide a measure of the inclination of the axis. An interesting analysis of the diagnostic potential of such observations, based on artificial data, was carried out by Ballot *et al.* (2006). They found a strong correlation between the rotational splitting $\delta_{\rm rot}\nu$ and the inclination i, such that for the best-fitting solutions $\delta_{\rm rot}\nu \sin i$ was approximately constant. The resulting shape of the probability density function in a $(\sin i, \delta_{\rm rot}\nu)$ diagram has led to such diagrams being known as "banana plots."

The inclination of the rotation axis is closely related to the properties of planetary systems detected by the transit method, where the plane of the orbits must be such that the planets pass in front of the star. It is normally assumed that planets form in a disk left behind from the formation of the star from a rotating cloud, spun up during contraction because of angular-momentum conservation; hence the disk very likely is in the equatorial plane of the star. Thus one would expect that for stars with transiting planets the rotation axis is in the plane of the sky, and departures from this would be an indication that the system has somehow been perturbed. Chaplin *et al.* (2013a) analyzed *Kepler* data for two main-sequence stars hosting two and three transiting planets, respectively. In both cases the inferred position of the rotation axis was consistent with being in the plane of the sky, as naively expected. I also note that Gizon *et al.* (2013) used CoRoT data to determine the average rotation rate for a main-sequence star hosting a non-transiting exoplanet, found from radial-velocity observations.

A more surprising case was analyzed by Huber *et al.* (2013b). This is a relatively unevolved red giant with two transiting planets in co-planar orbits, and with a rich spectrum of mixed modes. The amplitude pattern of the rotationally split modes was clearly inconsistent with a rotation axis in the plane of the sky, given that all three components of the $l = 1$ modes were observed. The analysis determined an inclination of the rotation axis from the line of sight of $47^\circ \pm 6^\circ$; reassuringly, inclinations determined separately from pressure-dominated and gravity-dominated mixed modes showed no significant difference between the inclinations in the core and outer layers of the star. The system contains a third orbiting but non-transiting object; Huber *et al.* (2013b) suggested that the torques from this object are responsible for the inclination of the planetary orbits relative to the equatorial plane of the star.

The observation of solar-like oscillations in a broad range of stars provides very valuable constraints on the modeling of the physics of the modes, such as their damping rates and amplitudes. Chaplin *et al.* (2009) determined damping rates from ground-based observations of solar-like oscillations and compared with model predictions based on the formulation of Balmforth (1992b) (see also Houdek *et al.*, 1999). They inferred a fairly steep increase in the damping rate with increasing effective temperature. Appourchaux *et al.* (2012b) determined average linewidths for a substantial number of main-sequence and subgiant stars from *Kepler* observations, indicating an even steeper dependence on effective temperature. These and other observations were compared with computed linewidths by Belkacem *et al.* (2012), showing generally good agreement with the observations. These results are preliminary but show strong promise for a better understanding of the damping and excitation of the modes. Further progress will result from studies under way of the frequency dependence of the damping rates and amplitudes. This may also provide a stronger basis for the scaling relation (Equation (20.3)) for the frequency $\nu_{\rm max}$ at maximum power. Despite its widespread use in analyses of the observations, and some limited theoretical underpinning (Belkacem *et al.*, 2011), this relation is mainly based on empirical evidence, leaving some concern regarding its general validity.

Given the variations of solar oscillation properties with the phase in the solar cycle (see Section 20.2), similar variations might be expected in other Sun-like stars. This does indeed seem to be the case, although the studies of the relation between stellar activity and oscillations are still in their infancy. The first indication of activity-related changes in stellar oscillation frequencies was found by Metcalfe *et al.* (2007) for β Hydri, combining ground-based oscillation observations with activity measurements from the IUE satellite. However, in a major breakthrough, García *et al.* (2010) detected anticorrelated variations in oscillation frequencies and amplitudes from CoRoT observations of the star HD 49933, which is somewhat more massive and hotter than the Sun; the variations were apparently correlated with activity variations as inferred from the photometric variability presumably linked to the presence of starspots. The cycle period has not yet been determined for this star, but it appears to be at most around one year. Further analysis by Salabert *et al.* (2011) showed that the frequency shifts increased in magnitude with frequency, as observed in the Sun and indeed expected from effects localized in the near-surface layers of the star (cf. the discussion of the near-surface effects in Section 20.2).

With the extended observations of a substantial number of stars, the *Kepler* data offer very good prospects for further detection of stellar-cycle variations, and searches for such variations are now under way (e.g., Karoff *et al.*, 2013; Mathur *et al.*, 2014). In addition, it is important to carry out observations of variability in stellar activity for the relevant *Kepler* targets, e.g., by monitoring the CaH and K lines, even though these observations will not be contemporaneous with the *Kepler* observations.

20.4 Combining helio- and asteroseismology

From the point of view of studies of the Sun, the major aspects of asteroseismology of solar-like stars are to place the Sun in a broader context and to investigate the past history and future evolution of the Sun. An initial effort in this direction was made on the basis of early *Kepler* data by Silva Aguirre *et al.* (2011), who determined a sequence of stars with masses close to 1 $M_\odot$ at varying stages of evolution, based on the global seismic parameters. As a result of selection towards higher-amplitude targets, these stars were all at a later stage of evolution than the Sun. Also, the lack of detailed determinations of the composition of the stars limited the potential for analysis. There should clearly be a much stronger basis for this type of investigation given the available data, including also analysis of individual frequencies.

Helioseismology has provided important information about the physics of stellar interiors, particularly the equation of state (Elliott and Kosovichev, 1998; Basu *et al.*, 1999a; Baturin *et al.*, 2000). One of the goals of asteroseismology has been to extend such information to the various physical conditions found in other stars. However, up to now the available data and/or techniques for analyzing these data have not been sufficient to meet this goal. An important limitation is the lack of accurate independent constraints on stellar global parameters, such as mass, radius, and luminosity. An issue that may be amenable to further progress concerns the revision of the solar composition (Asplund *et al.*, 2009) and the resulting difficulties for the solar models to match the helioseismic inferences. Since stellar composition is typically determined relative to the Sun, the change in the inferred solar composition directly affects the compositions assumed for stars. So far, it seems that no systematic investigation of the resulting effects on the asteroseismic analysis has been carried out. However, I note that VandenBerg *et al.* (2007) investigated the effect on the modeling of the open cluster M67 (see also Magic *et al.*, 2010). Obviously, asteroseismology has a unique potential for studying phenomena that are not relevant for the present Sun, such as convective-core overshoot (Silva Aguirre *et al.*, 2013) or effects of rapid rotation. Also, the detection of acoustic glitches associated with the base of the convective envelope (Mazumdar *et al.*, 2014) holds a promise of studies of the details of this important region in a variety of stars, thus extending the solar investigations.

Mode physics is an area of potentially strong interaction between helio- and asteroseismology. Extensive determinations of oscillation amplitudes and damping rates are becoming available from the CoRoT and *Kepler* data, extending from the main sequence to the red-giant branch. The dependence of power on frequency strongly indicates that common damping and excitation mechanisms are at work, the latter inferred to be stochastic excitation by near-surface convection of otherwise damped modes. Strikingly, this mechanism also seems to be responsible for oscillations in very luminous red giants observed by OGLE (Soszyński *et al.*, 2009), with a continuous variation of behavior all along the sequence from the Sun to the tip of the red-giant branch (Mosser *et al.*, 2013). Indeed, stochastic excitation was inferred by Christensen-Dalsgaard *et al.* (2001) for semiregular variables with periods of several months. Also, improvements in the modeling of the asteroseismic data may help constraining the origin and properties of the near-surface corrections that have to be applied to the individual frequencies; an early attempt in this direction was made by Mathur *et al.* (2012). These observations provide the possibility of studying the detailed interaction between convection and pulsations over a broad range of stellar parameters and hence for improving our understanding of these intricate processes, perhaps, in return, improving our ability to analyze the helioseismic data.

A crucial issue displaying the complementarity between helio- and asteroseismology is the evolution of stellar rotation, which remains a major uncertain aspect of stellar evolution. It is generally assumed that stars, including the Sun, were formed spinning quite rapidly, following the contraction of an interstellar cloud; this is confirmed observationally by the typically fast rotation of young stars (e.g., Barnes, 2003). In the case of stars like the Sun, angular momentum is lost through a magnetically confined stellar wind, presumably linked to the convection zone. However, the evolution of the rotation in stellar interiors is very uncertain. With no transport of angular momentum the gradual contraction of the stellar core with evolution would lead to an increased rotation rate. The nearly uniform rotation, at a rate comparable with the surface rotation, found by helioseismology shows the action of transport processes in the Sun, but the nature of these has not been identified with any certainty.

It is obvious that asteroseismology has a huge potential for improving our knowledge about stellar rotation and hence, one might hope, our understanding of its evolution. In sun-like main-sequence stars this will likely be restricted to determining an average rotation rate, heavily weighted towards the outer layers of the star. This can help constrain the evolution of rotation with age, ideally also for stars younger than the Sun. Also, if an independent measure of the surface rotation rate could be obtained, some information could be available about the variation of rotation with depth. The analysis of rotational splittings in red giants discussed above is a very important contribution. This shows an increase of the rotation rate of the core with evolution as qualitatively expected from the contraction of the core, but still implying angular-momentum transport that is far more efficient than predicted by existing models (Eggenberger *et al.*, 2012a; Marques *et al.*, 2013).

The asteroseismic investigation of stellar activity holds very considerable promise, starting with the early analysis by García *et al.* (2010) and extending the studies in the solar case of the relation between activity and oscillation properties. It is unlikely in the foreseeable future that we shall have asteroseismic observations extending over the 11-year duration of (half) a solar cycle. Thus, it is interesting that García *et al.* (2010) found a cycle length of at most a year in the F5V star HD 49933, while cycles of 2 and 3 years were found in two exoplanet host stars by Metcalfe *et al.* (2010b, 2013). This gives hope that the existing *Kepler* data may hold cases of detectable activity variation over their 3–4 year span, and searches for such cases are under way, including also determinations of rotation periods and activity variation from the photometric variability of the stars (García *et al.*, 2014b). Combining such analyses with asteroseismic determination of the depth of the convective envelope and perhaps some constraint on internal rotation would be extremely important as tests of the ongoing attempts to model stellar cycles, including detailed magnetohydrodynamical modeling (for a review, see Toomre *et al.*, 2012). The results could greatly improve our ability to understand and perhaps predict the solar magnetic variability, a rare case of asteroseismology potentially having substantial practical and societal importance.

20.5 Prospects for future observations

Both helio- and asteroseismology are ultimately data driven. In the case of helioseismology, a wealth of data has been obtained over the past two decades from a range of ground- and space-based facilities. Of these, the GOLF and VIRGO instruments are still operational on the SOHO spacecraft, while the HMI instrument on SDO is providing data with high spatial resolution. Barring technical failure, these instruments should remain operational for several years, assuming the availability of the required financial resources. However, for now no follow-up helioseismic space mission has been approved. An interesting possibility would be to use the GOLF-NG, extending GOLF by carrying out observations at several depths in the sodium lines (Turck-Chièze *et al.*, 2012). On the ground, the BiSON and GONG networks can be maintained and hence, in principle, be kept operating indefinitely, but of course critically depending on funding. The importance of long-term continuity in helioseismic observations is particularly evident in the study of the effects of solar-cycle variations, where spatially resolved observations have not yet covered a full magnetic cycle, and not least in the light of the peculiar behavior of the present cycle (Howe *et al.*, 2009; Basu *et al.*, 2012).

The huge development of asteroseismology over the past decade has to an overwhelming extent been driven by the CoRoT and *Kepler* missions. CoRoT has unfortunately been terminated due to a failure in the data chains, although after operating for almost twice its nominal lifetime. Also, *Kepler* in 2013 suffered loss of two of the four momentum wheels, making stable observations of the original field impossible. However, a proposal was made to continue *Kepler* in the so-called K2 mission, observing selected fields along the ecliptic for up to 3 months each; here the remaining two wheels suffice to keep stable pointing, with intermittent attitude correction using thrusters to compensate for solar radiation pressure, which is the main perturbing force (Howell *et al.*, 2014). This was approved by a NASA Senior Review in May 2014, granting *Kepler* a 2-year extension in this mode. The shift in operations trades continuity and long time series for a much broader selection of targets. The long continuous observations of a single field in the initial phase of the *Kepler* mission have been remarkably successful in providing high-quality data on solar-like oscillations in main-sequence and subgiant stars, and particularly important in allowing detailed investigations of the mixed modes in red giants. Indeed, it seems unlikely that similarly long-duration high-quality data will be obtained in the foreseeable future; clearly we need to make the most of the data already obtained with *Kepler*. On the other hand, the restriction to a single field meant observation of only a limited number of rare but very interesting targets. In contrast, K2 could provide outstanding possibilities for the study of targets such as rapidly oscillating Ap stars, pulsating white dwarfs and other compact hot stars, hot massive stars, as well as clusters where main-sequence and subgiant stars are available for asteroseismology.

Further missions are on the way for space photometric asteroseismology. The TESS mission (Transiting Exoplanet Survey Satellite; Ricker *et al.*, 2010), selected by NASA for a launch in 2017, will provide an all-sky survey, with each field being observed for at least one month, concentrating on relatively bright and hence nearby stars. This provides far better possibilities for follow-up observations than in the cases of CoRoT and *Kepler*. The driving goal of the mission is the search for exoplanets, but with an observing cadence of at least once a minute, the mission will provide a remarkably rich basis for asteroseismology, including for solar-like stars. Recently ESA selected the PLATO mission for launch in 2024 (Rauer *et al.*, 2013), combining research on exoplanets with asteroseismology in several very large fields, in some cases observed for 2–3 years. PLATO will concentrate on relatively bright and nearby stars, and a large fraction of the stars found to host exoplanets will be characterized using asteroseismology.

Despite the remarkable successes of space-based photometric asteroseismology, there certainly remains important potential for ground-based asteroseismology, observing the oscillations in radial velocity. It was noted by Harvey (1988) that velocity observations provide a substantially better signal-to-noise ratio than do observations in intensity, when the "noise" background is dominated by stellar phenomena, such as granulation. This has been confirmed by the observations of solar oscillations in photometry and velocity from the SOHO satellite (Grundahl *et al.*, 2007). As a result, velocity observations allow detection of modes over a broader frequency range, particularly towards lower frequency, where the mode lifetimes are longer and the frequency precision consequently higher. Also, velocity observations are more sensitive to modes of degree $l = 3$, further extending the asteroseismic diagnostics. The potential of velocity observations has been demonstrated by observations using existing telescopes with very stable spectrographs, although only for relatively short observing periods (e.g., Kjeldsen *et al.*, 2005). To obtain extended observations is the goal of the SONG (Stellar Observations Network Group) project (Grundahl *et al.*, 2011) to establish a network of seven or eight telescopes with a suitable geographical distribution to ensure nearly continuous observations over the entire sky. By optimizing the design it is expected to reach the required sensitivity with 1 m telescopes to observe oscillations in sun-like stars brighter than magnitude 6. The first telescope is essentially operational on Tenerife, while a second telescope is undergoing commissioning in China. Funding for further telescopes will be sought through international collaboration.

The Sun is unique amongst seismic targets in having very well-determined basic properties, such as mass, radius, age, and luminosity. For other stars the seismic analysis depends critically on the availability of supplementary spectroscopic or photometric observations that provide additional constraints on stellar properties. Planning and carrying out such observations have been important parts of the space missions (for *Kepler*, see Molenda-Żakowicz *et al.*, 2010; Uytterhoeven *et al.*, 2010); in particular, the spectroscopic observations are time-consuming, owing to the relative faintness of typical targets, and a great deal of work is still required. A very promising effort is the APOKASC collaboration, combining the APOGEE (the Sloan Digital Sky Survey III (SDSS-III) Apache Point Galactic Evolution Experiment; Mészáros *et al.*, 2013) near-infrared spectroscopic survey with the activities of KASC, with the main emphasis on red giants. A major improvement will result in a few years time from the ESA Gaia mission (Eyer *et al.*, 2013), launched in December 2013, which will provide accurate parallaxes and other stellar parameters for essentially all asteroseismically relevant targets. Studies should be started now on how these data can be optimally used in the asteroseismic analysis. Further strong constraints can be obtained by including interferometric radius measurements, as discussed by Cunha *et al.* (2007) and Creevey *et al.* (2007), who also presented a powerful technique for analyzing the improvement in parameter determination resulting from inclusion of additional observables. Such further constraints on stellar properties will undoubtedly help in breaking the degeneracy between the mass and initial helium abundance of the star, discussed above.

21

Stellar dynamics: Rotation, convection, and magnetic fields

SAVITA MATHUR, JÉRÔME BALLOT, AND RAFAEL A. GARCÍA

21.1 Introduction

Stars are changing entities in a constant evolution during their lives. At non-secular time scales – from seconds to years – the effect of dynamical processes such as convection, rotation, and magnetic fields can modify the stellar oscillations. Convection excites acoustic modes in solar-like stars, while rotation and magnetic fields can perturb the oscillation frequencies, lifting the degeneracy in the azimuthal component m of the eigenfrequencies (see Chapter 9 for the case in which rotation is slow and first-order perturbative theory can be used). Moreover, the interaction between rotation, convection, and magnetic fields can produce magnetic dynamos, which sometimes yield to regular magnetic activity cycles.

In this chapter we review how stellar dynamics can be studied and explain what long-term seismic observations can bring to the understanding of this field. Thus, we show how we can study some properties of the convective time scales operating in a star like the Sun. We also compare the stratified information we can obtain on the internal (radial) differential rotation from main-sequence solar-like stars to the Sun, and to more evolved subgiants and giants. We complement this information on the internal rotation with the determination of the surface (latitudinal differential) rotation obtained directly from the lightcurves. Indeed, when stars are active there can be spots on their surfaces dimming the light emitted. When the star rotates, the emitted light will be modulated by the presence of these spots with a period corresponding to the rotation rate at the active latitudes (where the spots develop). We finally give a brief summary of stellar magnetic studies based on spectroscopic observations and then we discuss the use of seismology to better understand the stellar magnetism of solar-like stars and the existence of possible magnetic cycles. We conclude this chapter by discussing the seismology of fast rotating stars and, from a theoretical point of view, what are the current challenges to infer properties of the internal structure and dynamics of intermediate- and high-mass stars.

For stars harboring planets, activity is an important factor needed to define the habitable zone – at least for carbon-based life – as it is known on Earth. If a star is very active, the energetic particles reaching the planets located in the habitable zone can destroy the

Extraterrestrial Seismology, eds. Vincent C. H. Tong and Rafael A. García. Published by Cambridge University Press.

elements necessary for the development of life unless the planets develop a protection like the Earth's magnetosphere, preserving our planet from these particles. Therefore, stronger constraints are imposed on the planets in order to be good candidates for the development of life.

Understanding those magnetic activity cycles is necessary for us to properly predict them. Unfortunately, we are far from a complete understanding of the physical processes responsible for these magnetic cycles. For the first time in history, we are able to access the internal structure of many stars while, at the same time, putting observational constraints on the convection properties, rotation, and magnetic fields for stars other than the Sun in many different configurations of the parameter space. Asteroseismology is changing our preconceived thoughts on this topic and what we can expect in the coming years.

21.2 Constraining convection

Stars like the Sun have an outer convection zone where the energy is transported by convective motions. We have seen that the turbulence in the convection zone is responsible for the excitation of acoustic modes. It also plays a crucial role in the stellar magnetic activity. In a star, depending on the effective temperature and its evolutionary stage, the depth of the convective zone can vary. For main-sequence stars of spectral type F ($T_{\rm eff}$ between 6000 K and 7000 K), the convection zone is much thinner than in the Sun and can be as shallow as only a small percentage of the stellar radius for the hottest stars, while in the Sun the convection zone represents $\sim$30% of the solar radius. Stars cooler than the Sun, of spectral types K and M, have deeper convective envelopes than the Sun, and the coolest M-type dwarfs are even fully convective. When a solar-like star evolves and becomes a red giant, its core contracts and its outer envelope expands, and the convection zone represents more than 50% of the stellar radius. One way of determining the depth of the convection zone is to model the star and try to match the spectroscopic parameters ($T_{\rm eff}$, Fe/H ...) and the frequencies of acoustic modes, as seen in Chapter 9. The best-fit model that is obtained provides the complete structure of the star and in particular the depth of the convection zone (e.g., Mathur *et al.*, 2012). Of course, we have to remember that this model is obtained with our current knowledge of the physics of stellar evolution, which is not yet complete.

We can also study the pattern of the acoustic-mode frequencies, which allows us to define the mean large frequency separation and the small separation. However, the repetition of the pattern is not completely perfect. The transition from the radiative zone to the convection zone, which is a sharp change in the thermal gradient, affects the modes and thus their frequencies. By studying the small deviation of the mode frequencies from the regular pattern, we observe an oscillation whose period is related to the size of the convection zone. We call such oscillations "acoustic glitches." This method has been applied to the Sun (Christensen-Dalsgaard *et al.*, 1991) and also to some solar-like stars (Mazumdar *et al.*, 2014). For a few tens of stars, the determination of the depth of the convection zone with stellar models and from the analysis of the acoustic glitches show a rather good agreement.

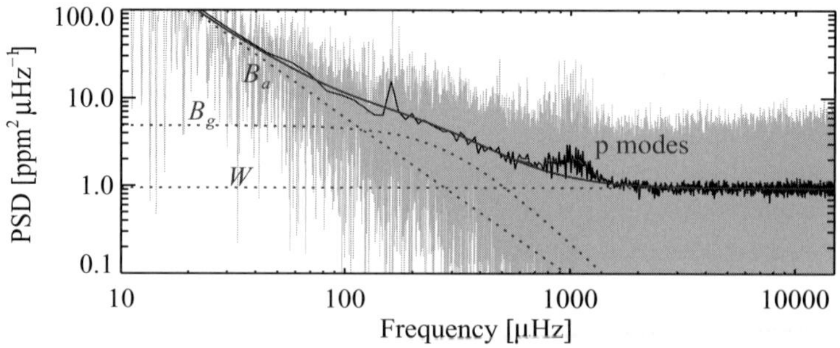

Figure 21.1 Power spectrum density of the solar-like star HD 169392 (gray) and smoothed one (black) observed by the CoRoT satellite. The background is represented by white noise (W), the granulation (B_g), and a power law to reproduce the magnetic activity (B_a). (Adapted from Mathur *et al.*, 2013.)

Finally, one of the manifestations of the convection is the granulation observed on the stellar surface where hot plasma rises towards the stellar surface, and after cooling down descends back towards the interior. This phenomenon can be studied in the power spectrum computed with photometric observations for instance. We call it the stellar background, on top of which are observed the acoustic modes of the star. By fitting the stellar background following the formulation described by Harvey (1985), we can measure the time scale of the granulation, $\tau_{\rm gran}$:

$$P_{\rm H}(\nu) = \frac{4\sigma^2 \tau_{\rm gran}}{1 + (2\pi \nu \tau_{\rm gran})^{\alpha}}, \tag{21.1}$$

where $P_{\rm H}(\nu)$ is the total power at frequency ν, σ is the characteristic amplitude of the granulation, and α is a positive parameter characterizing the slope of the decay.

First applied to the Sun, this formula has been tested on a large number of solar-like stars and red giants (e.g., Mathur *et al.*, 2011a; Samadi *et al.*, 2013) showing the validity of the model. An example is shown in Figure 21.1. The analysis of almost 1000 red giants showed that this time scale is inversely proportional to the mean large separation and to $\nu_{\rm max}$ (Mathur *et al.*, 2011a). The scaling relations seen in Chapter 11 allow us to relate these two global parameters to the surface gravity of the star. Thus there is an anticorrelation between the granulation time scale and $\log g$. The granulation time scale can also be studied in the time domain by analyzing the standard deviation of the lightcurve on time scales shorter than 8 hours. This quantity, called "flicker," has been studied by Bastien *et al.* (2013) and its relation to the granulation has been shown by Cranmer *et al.* (2014).

21.3 Constraining rotation

The transport of angular momentum is very important in understanding the rotation history of stars. When they are born, their surface velocities depend on the conditions of the protostellar disk, and hence a variety of rotation periods are observed at the beginning of the main sequence. At this evolutionary stage, the existence of two different regimes delimited by the so-called Kraft break has been determined (Kraft, 1967). Stars with masses larger than 1.3 $M_\odot$ have such thin convection zones that the generation of magnetic winds does not happen, thus the rotation periods of these stars barely evolve on the main sequence. Stars below this mass limit lose angular momentum very rapidly via the magnetized stellar winds. The loss laws are of the form $dJ/d\omega$ (Kawaler, 1988; Krishnamurthi *et al.*, 1997), which means that fast rotators slow down more than slow rotators. This leads to a very narrow range of rotation periods at the end of the main sequence, allowing the determination of gyrochronology relations (e.g., Barnes, 2003). Finally, when the stars leave the main sequence, because of the expansion of the envelope, both regimes of stars undergo a spin-down of their envelope. Models on the transport of angular momentum have been developed to understand the coupling that takes place in these different stages and regimes of the stars (e.g., van Saders and Pinsonneault, 2013; Epstein and Pinsonneault, 2014).

Most of the knowledge we have on stellar interiors is obtained from the calculation of stellar structure and evolutionary models (see Chapter 8). Those models are driven by many different physical mechanisms that are not perfectly understood. What is the effect of internal rotation in stars? First, centrifugal acceleration reduces local gravity and models of rotating stars behave as if they had lower masses. Second, rotation induces transport of chemical elements and angular momentum by the meridional circulation (Eddington, 1925) and shear mixing (e.g., Zahn, 1992; Maeder and Zahn, 1998; Mathis and Zahn, 2004), which modifies the global evolution of a star (Maeder, 2009; Eggenberger *et al.*, 2010; Marques *et al.*, 2013).

Surface abundances of light elements can be used as indirect diagnostics of the efficiency of internal rotational mixing (Talon and Charbonnel, 2010) including, in particular, the use of lithium depletion (Eggenberger *et al.*, 2012b). However, it is much more effective to directly compare the rotation rate inferred from the models with the observed one. This has been made possible thanks to the determination of the internal rotation profile obtained with seismic techniques applied to the Sun and other more distant stars.

In the case of the Sun, we have thousands of high-degree p modes available, allowing us to infer a very precise 2D rotation map in the outer convection zone (see Figure 12.6). However, the number of acoustic modes probing the rotation rate in the radiative regions towards the center of the Sun reduces drastically and only low-degree modes provide reliable information in these deep layers. Chaplin *et al.* (2004) studied the potential of low-degree p modes to measure rotation gradients between 0.1 and 0.3 $R_\odot$. They concluded that changes of at least 110 nHz would be needed to detect a gradient in the core rotation. Indeed, the rotation profile in the Sun is only well constrained using p modes down to ~ 0.2 $R_\odot$ (Couvidat *et al.*, 2003; García *et al.*, 2008a; Eff-Darwich and Korzennik, 2013). To go beyond this limit it is necessary to measure mixed modes such as the $l = 2$ and $l = 3$ modes with $n = 1$ (see for example Figure 2 in García *et al.*, 2004) or individual gravity modes (Mathur *et al.*, 2007, 2008). Unfortunately, those modes have tiny amplitudes at the surface of the Sun because they are evanescent in the convection region, which extends over the outer 30% of the solar radius (Appourchaux *et al.*, 2010). However, by measuring the asymptotic spacing of dipole gravity modes in the Sun – in a similar way to what is now commonly done in the analysis of red giants (see Chapter 11) – it has been possible to determine that the core of the Sun below 0.2 $R_\odot$ could rotate faster than the rest of the radiative region (García *et al.*, 2007, 2008b). However, it would be necessary to measure individual splittings of those gravity modes to perform a proper inversion of the deepest solar layers to confirm this result.

In the stellar case the situation is different. For main-sequence solar-like stars, there are only a few dozen low-degree p modes that can be observed. Their propagation cavities are very similar and we can only obtain an average rotation rate of the interior (Gizon *et al.*, 2013) weighted towards the surface. Moreover, due to the width of the modes (inversely proportional to their lifetimes, see Chapter 5) and the correlation between rotational splitting and inclination angle, the extraction of reliable splittings is only possible for stars with high inclination angles and long-lifetime modes (e.g., Ballot *et al.*, 2006). In any case, we hope that in young F-stars with shallow convective envelopes, pure g modes could be observed as they will not be completely damped in this zone, allowing us to have a direct view of the conditions in the deep interior.

The situation drastically changes when stars are more evolved. In the subgiant and early red giant branch, when the hydrogen in the core is exhausted and stars only burn hydrogen in a shell, there are a few measurable mixed modes that are very sensitive to the deepest radiative interior, in particular they provide invaluable information about the internal rotation (Deheuvels *et al.*, 2012, 2014). In this case, combining the information carried by acoustic and mixed modes, we can obtain the rotation at two distinct radii: one weighted towards the inner radiative interior and the other one weighted towards the external convective envelope.

Finally, in the case of red giants (burning hydrogen in a shell or after the ignition of the helium in the core), the measurement of many mixed modes (typically more than 10) allows us to have a good determination of the rotation in the radiative cores of these stars (Beck *et al.*, 2012; Goupil *et al.*, 2013).

The ensemble analysis of the rotation of hundreds of red giants (Mosser *et al.*, 2012b) confirms an increase in the rotation of the inner contracting radiative core compared to the expanding outer envelope, but much less steep than what we would expect as a result of internal angular momentum distribution (Sills and Pinsonneault, 2000). This is a similar result to what we already knew from the analysis of the solar internal rotation profile (Turck-Chièze *et al.*, 2010). This slower rotation rate in the core of red giants allows a better agreement with the results found in low-mass white dwarfs exhibiting slower rotation rates than what we expect for conservation of angular momentum (Kawaler *et al.*, 1999).

These results confirm the necessity of invoking extra physical processes in the stellar evolution codes that could help to evacuate angular moment from the inner core towards the outer envelope (Eggenberger *et al.*, 2012a; Ceillier *et al.*, 2013). Two possible mechanisms have been proposed. The first one invokes the existence of a fossil magnetic field in the stellar interior that was trapped during the first stages of the stellar evolution (Gough and McIntyre, 1998). The second one implies the transport of angular momentum by internal gravity waves excited in the convective to radiative transition region (Talon and Charbonnel, 2005).

To complement the information on stellar internal differential rotation, the long and uninterrupted photometric time series required for asteroseismic studies are excellent for measuring any modulation in the photometry induced by stellar spots crossing the visible stellar disk (García *et al.*, 2013; Lanza, 2013), which is proportional to the surface rotation rate at the active latitudes. The information on the surface rotation can be extracted by performing detailed spot modeling (Mosser *et al.*, 2009; Mathur *et al.*, 2010; Fröhlich *et al.*, 2012) or by more general treatments using automatic analysis of the cross-correlation of the temporal signal (McQuillan *et al.*, 2013), the analysis of the low-frequency part of the periodogram (Reinhold *et al.*, 2013; Nielsen *et al.*, 2013), or by using time–frequency analysis coupled to the other two methods (Mathur *et al.*, 2014; García *et al.*, 2014b). The determination of surface rotation rates allows a natural link with classical analysis based on surface rotation using spectroscopy. The rates impose strong constraints on the rotation of the external layers for subgiants and giants, while also providing information about the existence of surface magnetic fields at these stages of evolution. Asteroseismology, on the other hand, provides information on the rotation profile in deeper layers of the stars, which is crucial in understanding the transport of angular momentum.

21.4 Constraining magnetic fields

Different techniques are used to measure magnetic fields and magnetic activity in stars. Classical observations are based on the measurement of optical, ultraviolet, or X-ray emission from the magnetically heated gas. Starspots and flares that are associated with magnetic fields are also used to infer their magnetic activity.

Spectroscopic surveys such as the one at the Mount Wilson Observatory (Wilson, 1978) consist of measuring changes in spectroscopic lines (such as Ca HK and Hα) due to the magnetic field present in the stars. These observations, which are done for several years on a large number of stars, have allowed us to observe and discover a broad variety of magnetic activity behavior during recent decades for stars with different fundamental properties (rotation, mass, spectral type, evolutionary stage, etc.). These studies suggested a relation between the length of the magnetic cycles and the surface rotation periods (Saar and Brandenburg, 2002; Böhm-Vitense, 2007). This also points to the existence of two different branches called the "inactive" and the "active" branch. No clear explanation of the differences between the two branches has been given so far, but the most common one implies a different location of the dynamo process more or less deep in the stars.

Theory shows that the rotation and the magnetic cycle periods are related to the Rossby number $\mathrm{Ro} = \tau_c / P_{\mathrm{rot}}$, where τ_c is the convective turnover time and P_{rot} is the surface rotation period. In order to understand the detailed mechanisms responsible for the existence of magnetic activity cycles, it is crucial to have precise knowledge of the internal stellar structure and rotation. From so many decades of observations of the Sun, we believe that for a star like the Sun, with an outer convective zone, the magnetic activity results from a dynamo process where rotation, convection, and magnetic fields are interacting all the time. From this picture, we would then expect fast rotators to have shorter magnetic activity cycles which can be observed with missions such as *Kepler* or PLATO. Finally, spectroscopic observations by Reiners (2006) suggest that faster rotating F-type stars with masses larger than $1.2\,M_\odot$ have a stronger differential rotation, which is an important ingredient in the magnetic activity and makes these stars interesting study cases.

In the last few years, cycle lengths shorter than a few years have been observed with spectroscopic and spectropolarimetric observations. By analyzing the electromagnetic spectrum, we can study the variation of some absorption lines that are sensitive to the magnetic field. A classical proxy of the magnetic activity is the S-index, based on Ca II H and K lines, which is sensitive to the magnetic field in the photosphere and the chromosphere. For the Sun, the average S-index is around 0.171. A good example of such spectroscopic study is the F-type star HD17051 (also known as ι Horologii) which was monitored in Ca II H and K with the SMARTS 1.5 m telescope for 5 years. This study suggested a cycle period of 1.6 years for a surface rotation of 8 days (Metcalfe *et al.*, 2010b). However, longer datasets did not show another cycle but a chaotic variation of the S-index (Sanz-Forcada *et al.*, 2013).

Spectropolarimetric observations consist of observing the light spectrum and also measuring the polarization of the light over the whole spectral range, which allows us to study large-scale magnetic fields in the stars. Because a star rotates, at different rotational phases, we can observe different distributions of the magnetic field. Polarized spectra are collected by spectropolarimeters such as ESPaDOnS at the CFHT (Hawaii), NARVAL on TBL (France), and HARPSpol (Chile). Some very interesting results have been obtained for solar-like stars in recent years. Fares *et al.* (2009) detected a change of polarity in the magnetic field of the F-type star τ Boo, the star with the shortest cycle period observed. The Bcool project detected magnetic fields in 67 stars through their spectropolarimetric survey of 170 stars. This study shows that magnetic mapping can be done for a variety of moderately active stars (see Marsden *et al.*, 2013, for more details).

With the long and uninterrupted asteroseismic measurements collected by the CoRoT and *Kepler* missions, we have a unique database of lightcurves to study activity cycles in other stars for which we can probe the internal layers as well. We have already talked about using time–frequency analysis to measure the surface rotation period. As this measurement is linked to the presence of spots and thus magnetic activity, such time–frequency analysis detects signatures of magnetic activity cycles. Mathur *et al.* (2014) analyzed 22 F-type stars observed for around 1400 days with a time–frequency method, leading to the detection of magnetic activity signatures for two stars. This study should be extended to more stars with different spectral types and rotation periods to have more statistics on the existence of cycles in the stars and to check the accuracy of the rotation–activity–age relation.

Seismology also allows us to look for hints of magnetic activity. Indeed, for the Sun, we know that when the magnetic activity increases, the acoustic modes have smaller amplitudes while their frequencies shift towards higher values (Salabert *et al.*, 2009b). This can be explained by the fact that there is a change in the effective temperature affecting the sound speed (Goldreich *et al.*, 1991) and thus the frequencies of the modes. Do we see the same pattern in other stars? García *et al.* (2010) studied the CoRoT target, HD 49933, which is an F-type star with a rotation period of 3.4 days. Similarly to what is seen in the Sun, they detected the same anticorrelation in the acoustic-mode parameters suggesting that the star was going through a magnetic cycle (see Figure 1 of García *et al.*, 2010), which was confirmed later by spectroscopic observations. Furthermore, a deeper analysis of this star showed that the shift in frequency is larger for the higher frequency modes (Salabert *et al.*, 2011). This is also what we observe in the Sun, and it suggests that the magnetic changes happen closer to the surface as the high-frequency modes are more sensitive to the outer layers. In the case of the existence of a magnetic field in the stellar deep interior, for example of fossil origin, modes propagating in these regions would be split into different components separated by a magnetic splitting that would be proportional to the magnetic field strength (Goode and Thompson, 1992; Rashba *et al.*, 2007). However, a first analysis of the fine structure of mixed modes in 300 red giant stars, performed by Mosser *et al.* (2012b), showed that all of the measured mixed modes could be interpreted as rotationally split. If a magnetic field exists in those stars, it must be confined very deep or its strength must be too small to be measured with the current observations.

21.5 Seismology of rotating stars

Rotation has a direct impact on the oscillation modes and their frequencies. Some of those effects have already been discussed in Chapter 12, but they are only valid as long as the rotation is slow. Before going further, we should clarify what is a "slow rotation." The rotation is slow when the effects of both the Coriolis force and the centrifugal distortion are small. Denoting the rotation rate by Ω, this leads to the following conditions: $\Omega \ll \omega$ and $\Omega \ll \Omega_K$, where ω is the oscillation frequency, while $\Omega_K = \sqrt{GM/R^3}$ is the Keplerian break-up rotation rate, with G, M, R the gravitational constant, the mass and the equatorial radius of the star, respectively. When a star spins at its Keplerian break-up rotation rate, the centrifugal force at the surface, along the equator, is exactly balanced by the gravity; hence it corresponds to the maximal rotation rate above which the star would break up.

When these two conditions ($\Omega \ll \omega$ and $\Omega \ll \Omega_K$) are fulfilled, it is possible to treat the effects of the rotation as a small perturbation. In such a case, the first step is to solve the oscillation equations neglecting the rotation. We then obtain a set of eigenmodes characterized by their quantum numbers (n, l, m) associated with eigenfrequencies $\omega_{n,l}$ which are independent of m. The first-order correction in Ω does not affect the modes, only the frequencies are modified. We can identify two contributions to this first-order term: (1) the waves are advected by the rotation, then prograde (retrograde, respectively) modes have slightly higher (lower) frequencies than zonal ($m = 0$) modes, which stay unaffected.

Prograde (retrograde) modes are modes with $m < 0$ ($m > 0$), i.e., modes propagating around the axis in the same direction as the rotation (the opposite direction). (2) Coriolis force also modifies the frequency, but this is generally very weak for p modes, only g modes are affected.

If we need to take the second-order effects ($O(\Omega^2)$) into account, the centrifugal distortion must be considered, and both the frequencies and the spatial distributions of modes are perturbed. Methods that take into account the first-order effects of rotation were developed in Cowling and Newing (1949) and Ledoux (1951), second-order corrections were proposed for example by Saio (1981), Dziembowski and Goode (1992), and Suárez *et al.* (2006).

Perturbative approaches are the easiest way to treat slow rotation. However, for g modes – which have long periods – the condition $\Omega \ll \omega$ is generally not fulfilled. More general methods should then be used in such cases; see, for example, the work of Dziembowski and Kosovichev (1987). However, the most classical approach, developed and commonly used in geophysics (Eckart, 1960), is the so-called traditional approximation. Within this approximation, the oscillation equations are rewritten, leading to a problem that is also separable in r (radius) and θ (co-latitude). However, the horizontal dependency of modes is no longer described with spherical harmonics, as in the non-rotating case, but with Hough functions (e.g., Unno *et al.*, 1989). This traditional approximation has been applied to compute g modes in various stars (see, for example, Berthomieu *et al.*, 1978; Lee *et al.*, 1992; Townsend, 1997; Savonije, 2005).

In practice, the second assumption ($\Omega \ll \Omega_K$) is no longer valid for $\Omega \geq 0.1\Omega_K$ (Reese *et al.*, 2006; Ballot *et al.*, 2010). Above this limit, perturbative approaches fail in recovering correct mode frequencies. For a typical A-type star, this would correspond to an equatorial velocity $v \geq 50$ km s^{-1}. Unfortunately, such a situation is very common for intermediate- and high-mass main-sequence stars (e.g., Royer *et al.*, 2007). As a consequence, rotation may strongly impact the frequencies of p modes, especially in δ Scuti and β Cephei stars, and g modes, especially in γ Doradus and slowly pulsating B-type (SPB) stars. Complete computations of oscillations must then be performed to solve this problem.

The first challenge is to model a distorted star. Taking advantage of the spherical symmetry, non-rotating stars are modeled as 1D objects (the structure depending only on the radius r). However, rapidly rotating stars must be considered as 2D objects, for which physical quantities are function of r and θ. New generations of more realistic stellar models are now emerging(e.g., Deupree, 2011; Roxburgh, 2006; MacGregor *et al.*, 2007; Rieutord and Espinosa Lara, 2013).

The second challenge is to solve the system of equations governing the oscillations in the presence of rotation. In the non-rotating case, the oscillation problem is separable with respect to the radial and horizontal variables: the horizontal component of a mode is described with a spherical harmonic of a given degree l (and azimuthal order m); thus, for a given l, we only have to solve a 1D problem. This is not true for a rotating star: the problem is no longer separable with respect to r and θ; as a consequence, one mode cannot be described by a unique spherical harmonic.

Two-dimensional oscillation codes have been developed to solve this problem (e.g., Clement, 1984; Lee and Baraffe, 1995; Savonije *et al.*, 1995; Reese *et al.*, 2006; Ouazzani, 2011). More details and discussions on technical aspects can be found in Ballot *et al.* (2013). Recent investigations with accurate codes, such as the TOP code (Reese *et al.*, 2006, 2009), allowed us to better understand how the oscillation spectrum is reorganized in fast spinning stars. To illustrate this, we compute oscillations in a simplified stellar structure consisting of a polytropic model with a polytropic index $\mu = 3$ (i.e., a model where pressure p and density ρ are linked through the relation $p \propto \rho^{1+1/\mu}$), which mimics a radiative star. Typical modes are shown in Figure 21.2 for a uniform rotation rate $\Omega = 0.6\Omega_{\rm K}$.

If we consider p modes, we notice that they present very different spatial structures. A ray theory for rapidly rotating stars is very helpful to understand these differences. We note a first attempt by Perdang (1988) in this direction. The recent work of Lignières and Georgeot (2008, 2009) allowed us to properly classify the modes. For a given m, we can distinguish between three families of modes. First, the whispering gallery modes are similar to the modes found in non-rotating stars. At high rotation rates, they have a large number of latitudinal nodes. Second, we have the chaotic modes, so called because they are associated with chaotic trajectories in the ray dynamics. The modes themselves do not have any chaotic behavior. They are characterized by irregular nodal lines and they propagate down to the stellar center. Third, the island modes are associated with stable periodic trajectories in the ray dynamics. The energy of the mode is concentrated around this stable trajectory. We can count the number of nodes along this trajectory (denoted $\tilde{n}$) and in the transversal direction (denoted $\tilde{l}$). For a given m, these two numbers ($\tilde{n}$,$\tilde{l}$) fully characterize an island mode. For example, the island mode plotted in Figure 21.2 is characterized by $\tilde{l} = 0$ and $\tilde{n} = 20$ (and $m = 3$). Island modes correspond to the low-degree modes of the non-rotating case.

The p-mode spectrum of a rotating star is then the superimposition of subspectra corresponding to these different families. Thus, the global spectrum seems more complex, and the regularity of the spectrum, described at $\Omega = 0$ by Tassoul's relation (Tassoul, 1980), $\omega_{n,l} = \Delta_0(n + l/2 + \alpha)$, is apparently lost. However, new regularities appear inside the families. The regularities of the island modes have been extensively studied numerically by Lignières *et al.* (2006) and Reese *et al.* (2008, 2009), and theoretically by Pasek *et al.* (2011, 2012). It has been shown that, for island modes, $\omega_{\tilde{n},\tilde{l},m} = \tilde{n}\tilde{\Delta}_n + \tilde{l}\tilde{\Delta}_l + |m|\,\tilde{\Delta}_m - m\Omega + \tilde{\alpha}^{\pm}$ where $\tilde{\Delta}_n$, $\tilde{\Delta}_l$, $\tilde{\Delta}_m$, and $\tilde{\alpha}^{\pm}$ are parameters that are related to the stellar structure. Recovering regularities is a crucial question because regularities *are* very helpful in identifying modes in observations, and mode identification is a prerequisite for any sensible inference on internal structure and dynamics.

For g modes, the spatial structure will first depend on its frequency ω, expressed in the co-rotating frame (not the observer's one). Generally, when a mode stays in the super-inertial regime ($\omega > 2\Omega$), its structure stays similar to that of a g mode in a non-rotating star and is only slightly deformed. However, for a mode in the sub-inertial domain ($\omega < 2\Omega$) a more visible change is observed due to the appearance of a polar, approximately conical, forbidden region where waves cannot propagate (see Dintrans and Rieutord, 2000). The critical angle defining this cone is close to $\arccos(\omega/2\Omega)$. Modes are then more and more

Figure 21.2 Examples of modes in a rotating star ($\Omega = 0.6\Omega_K$). First column: three representative p modes (top to bottom: whispering gallery mode, chaotic mode, island mode). Second column: three representative g modes (top to bottom: mode in the super-inertial domain ($\omega > 2\Omega$) not too affected by rotation, mode in the sub-inertial domain ($\omega < 2\Omega$) affected by a forbidden region in the polar region, a rosette mode in the super-inertial domain. In each panel: on the left-hand side, the variable $p'/\sqrt{\rho}$ is shown in a meridional plane (p' is the pressure fluctuation and ρ the density); on the right-hand side, the kinetic energy of the mode is represented in a meridional plane. A black and white version of this figure will appear in some formats. For the color version please refer to the plate section.

confined around the equator plane when ω/Ω decreases. A new family of modes has been discovered recently, the so-called rosette modes (Ballot *et al.*, 2012). These modes have been found in the super-inertial regime and exhibit a drastic change in their structure. For these modes, the kinetic energy is focused around a set of loops, forming a "rosette" pattern. They appear for rotation rates as low as $\Omega \approx 0.1\Omega_K$. The formation of the rosette modes can be understood with two different approaches: (1) Similarly to the island p modes, they can be associated with stable periodic trajectories in the ray dynamics (Ballot *et al.*, 2012); (2) Their formation can also be explained as an interaction among g modes having close frequencies (Takata and Saio, 2013).

22

Sunquakes: Helioseismic response to solar flares

ALEXANDER G. KOSOVICHEV

22.1 Introduction and overview

Sunquakes observed in the form of expanding wave ripples on the surface of the Sun during solar flares represent packets of acoustic waves excited by flare impacts and traveling through the solar interior. The excitation impacts strongly correlate with the impulsive flare phase, and are caused by the energy and momentum transported from the energy release sites. The flare energy is released in the form of energetic particles, waves, mass motions, and radiation. However, the exact mechanism of the localized hydrodynamic impacts that generate sunquakes is unknown. Solving the problem of the sunquake mechanism will substantially improve our understanding of the flare physics. In addition, sunquakes offer a unique opportunity for studying the interaction of acoustic waves with magnetic fields and flows in flaring active regions, and for developing new approaches to helioseismic acoustic tomography.

Solar flares represent a process of rapid transformation of the magnetic energy of active regions into the kinetic energy of charged particles, plasma flows, and heating of the solar atmosphere and corona. The primary energy release during the flares is believed to occur in the corona as a result of magnetic reconnection. It is generally believed that most of the energy released by the reconnection goes directly and indirectly (via plasma waves) to acceleration of electrons and protons which are injected into flaring magnetic loops (Figure 22.1a). Most of the observed radiation is produced either directly by these particles or indirectly through energization of the background plasma.

It was suggested long ago (Wolff, 1972) that flares may cause acoustic waves traveling through the solar interior, similar to the seismic waves in the Earth. Because the sound speed increases with depth, the waves are refracted in the deep layers of the Sun (Figure 22.1b), and then appear on the surface, forming expanding rings, similar to ripples on a water surface. Theoretical modeling (Kosovichev and Zharkova, 1995) predicted that the speed of the expanding seismic ripples increases with distance because the distant waves propagate into the deeper interior where the sound speed is higher.

Extraterrestrial Seismology, eds. Vincent C. H. Tong and Rafael A. García. Published by Cambridge University Press.

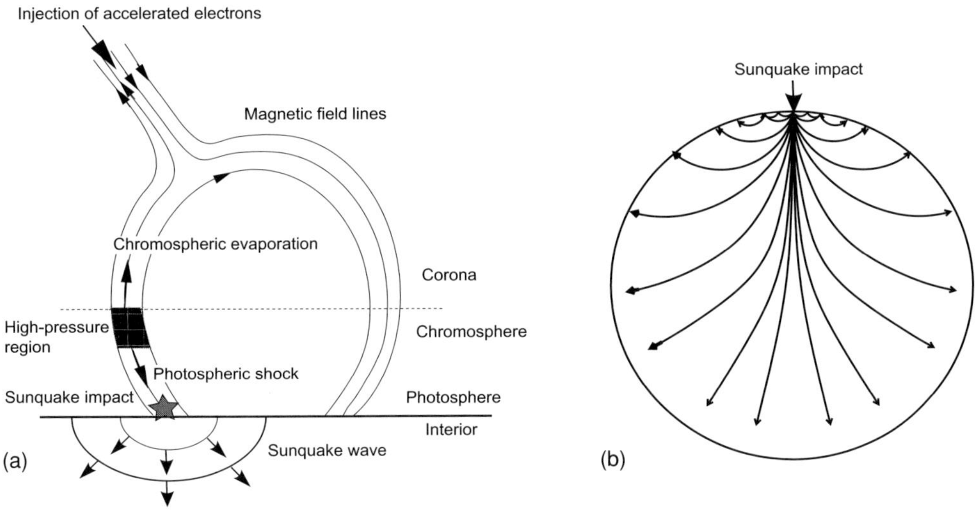

Figure 22.1 (a) Illustration of a sunquake mechanism in the "thick-target" model of solar flares: high-energy particles injected into a magnetic flux tube heat an upper chromospheric layer, creating a high-pressure region which produces "chromospheric evaporation" and a shock traveling to the photosphere and causing a sunquake impact. (b) Illustration of ray paths of acoustic waves excited by an impulsive source on the solar surface. The waves travel into the deep interior and are reflected back to the surface, producing sunquake ripples.

First observations of the seismic waves caused by the X2.6 flare of July 9, 1996, proved these predictions (Figure 22.2; Kosovichev and Zharkova, 1998). These observations also showed that the source of the seismic response was a strong shock-like compression wave propagating downwards in the photosphere. This led to a suggestion that the seismic response can be explained in terms of the so-called hydrodynamic "thick-target" model (Figure 22.1a). In this model (e.g., Kostiuk and Pikelner, 1975; Livshits *et al.*, 1981; Fisher *et al.*, 1985; Kosovichev, 1986), a beam of high-energy particles, accelerated in the corona, heats the solar chromosphere, resulting in a strong compression of the lower chromosphere. This compression produces a chromospheric eruption ("evaporation") and a downward-propagating shock wave which hits the solar surface and causes a seismic response. This shock was observed in the Dopplergrams by the Michelson Doppler Imager (MDI) instrument on the NASA-ESA mission SOHO as a localized large-amplitude velocity impulse of about 1 km/s, immediately after the hard X-ray impulse produced by high-energy electrons (Figure 22.3a). This velocity impulse represents the initial hydrodynamic impact exciting the seismic waves. In addition, it was found that the seismic waves had a significant anisotropic quadrupole component. However, the seismic wavefront was almost circular (Figure 22.2).

Figure 22.3b shows a time–distance diagram for the axi-symmetrical component of the sunquake wave. The diagram is obtained by azimuthally averaging the Doppler velocity signal centered at the initial impact location. In this diagram, the initial impact

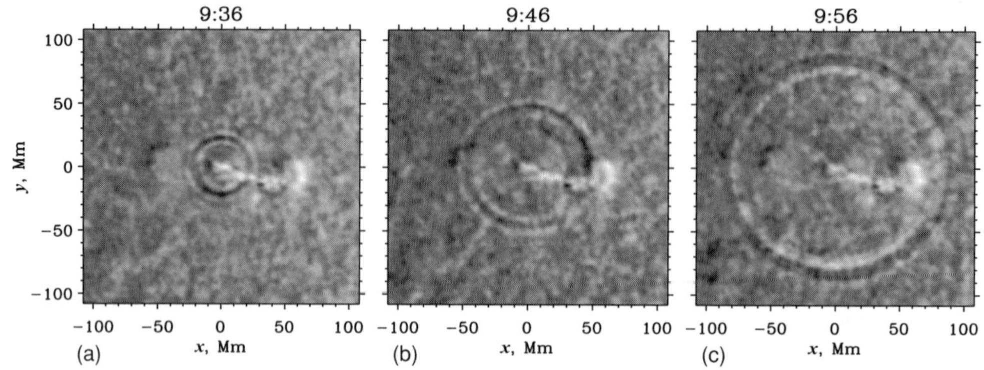

Figure 22.2 A sequence of enhanced Doppler-velocity images of the first sunquake observed on July 9, 1996, by the Michelson Doppler Imager on the SOHO spacecraft. The sunquake is observed on the solar surface as a circular expanding wave originating from an impact of the X2.6 solar flare. Dark shading shows upflows, and white shows downflows. The noisy background is caused by solar granulation. Bright permanent features are flows associated with magnetic fields of the active region. The sunquake wave signal in these images is enhanced by a factor of 4.

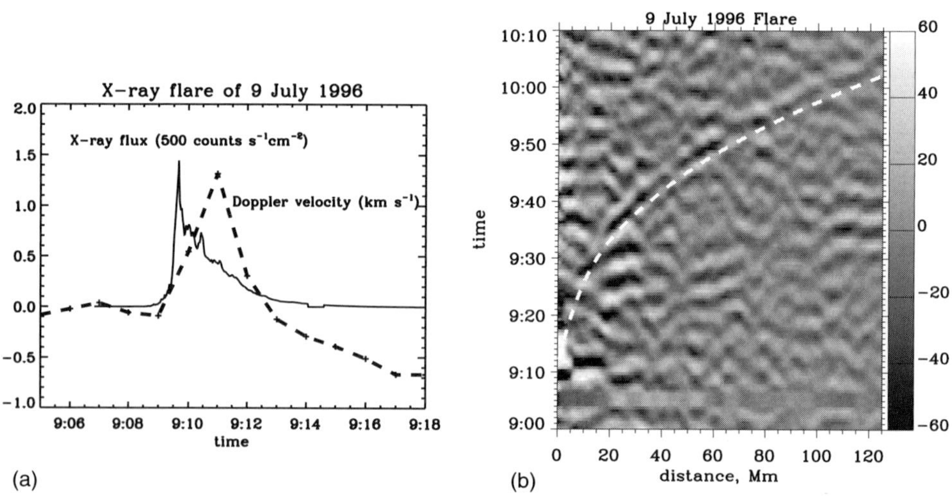

Figure 22.3 (a) The hard X-ray flux of the July 9, 1996, flare (solid) and the Doppler velocity signal of the localized flare impact that generated a sunquake; (b) a time–distance diagram of the sunquake wave, obtained by azimuthally averaging and stacking the Doppler-velocity images with 1-min cadence. The strong variations of velocity at about 9:10 UT represent the initial impact. The white dashed curve shows a theoretical time–distance relation of acoustic waves, calculated in a ray-path approximation.

started at about 9:10 UT, and the seismic wave is displayed as a narrow ridge with a characteristic slope, which follows the theoretical time–distance relations for acoustic waves traveling in the unperturbed conditions of the quiet Sun. In this case, the theoretical relation is calculated for a standard solar model using the acoustic ray-path approximation (e.g., Kosovichev, 2011). While the appearance of sunquakes is similar to water ripples, their physical properties are quite different. In particular, as follows from Figure 22.3b the surface speed of sunquake waves rapidly increases with the travel distance. Immediately after initiation the wave speed is about 10 km/s, which is close to the surface speed of sound, and then it rapidly increases to $\sim$100 km/s. The apparent surface speed increases because the acoustic waves travel through the solar interior where the sound speed is much higher than on the surface, and more distant waves travel through deeper interior (see Figure 22.1b).

Subsequent observations of solar flares made by SOHO/MDI (Scherrer *et al.*, 1995) did not show noticeable sunquake signals even for X-class flares until the big events of October 2003. This search was carried out by calculating an "egression" power for high-frequency acoustic waves during the flares (Donea *et al.*, 1999). Thus, it was suggested that sunquakes are a rather rare phenomenon on the Sun, which occur only under some special conditions. Surprisingly, 7 years after the first event several flares did show strong "egression" signals indicating new potential sunquakes (Donea and Lindsey, 2005). Of course, detection of sunquakes depends on their amplitude relative to the background solar noise. Presumably, all flares generate some seismic response, but if the amplitude is not high enough the waves may be lost in the background noise.

The helioseismology observations with the MDI discovered several powerful events in 2003–2005 during the declining phase of solar cycle 22 (Kosovichev, 2006b; Donea, 2011). Perhaps the most powerful event of the cycle was on January 15, 2005 (Section 22.3.1). The observations revealed that in most events the helioseismic response is anisotropic, and that the wavefront shape may significantly deviate from circular. These effects are mostly associated with the sunquake source extension along footpoints of magnetic arcades (observed in the form of bright "flare ribbons"), and also with the source motion, which can reach supersonic speeds (Kosovichev, 2006c). The source motion is often related to expansion of the flare ribbons, and probably reflects properties of magnetic reconnection and energy release in the solar atmosphere (Sturrock, 1966). Recent observations with the Helioseismic and Magnetic Imager (HMI) instrument on board the Solar Dynamics Observatory (SDO) (Scherrer *et al.*, 2012) show that sunquakes are much more common events than was thought previously, and that they can be frequently observed for moderate-class flares (Kosovichev, 2012b).

22.2 Theory of the helioseismic response

Generally, the helioseismic response is caused by the energy and momentum transfer from a place of magnetic energy release in the solar atmosphere to the surface. Observations of Doppler velocities clearly reveal localized impacts in the low atmosphere and photosphere, which represent sources of sunquake events. Without such impacts sunquakes are not observed. Therefore, a theory of sunquakes must include two parts: (1) calculations of the hydrodynamic impact for a flare model assuming some mechanism of the energy release; (2) calculations of the helioseismic waves produced by this impact. So far, the theory has

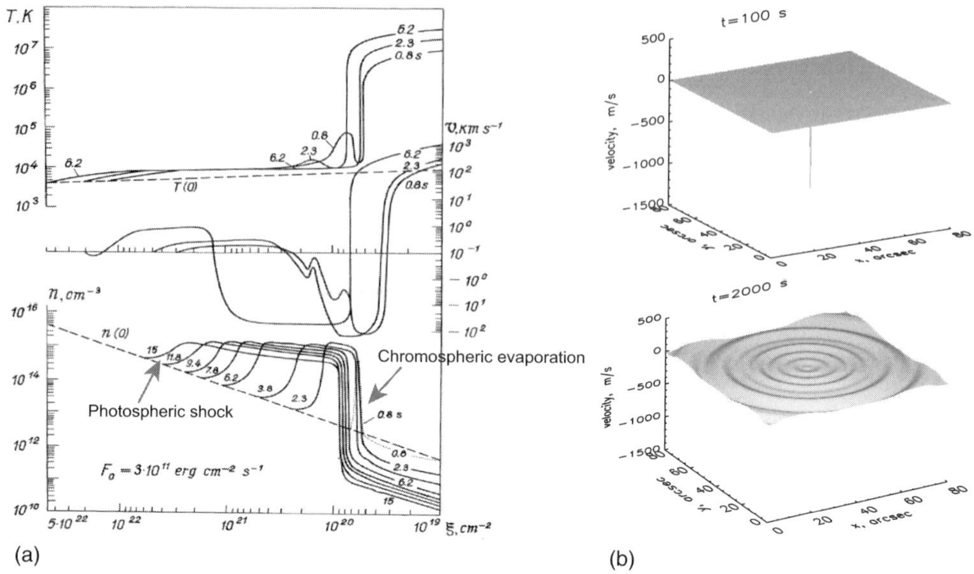

Figure 22.4 (a) Numerical simulations of the thick-target model of solar flares in a vertical magnetic flux tube with a constant cross-section: temperature, velocity, and density profiles for different moments of time. The chromospheric heating and dynamics were initiated by a 10 s injection of high-energy electrons at the top of the tube with the maximum energy flux of 3×10^{11} erg cm^{-2} s^{-1} with a power-law energy spectrum with the low-energy cut-off at 20 keV. The plasma compression behind the shock moving towards the photosphere can reach a factor of ~100 because of the radiative energy losses. This region can be a source of white-light emission. (b) Theoretical model of sunquakes showing surface velocity perturbations at $t = 200$ and 2000 s, produced by a point momentum source initiated at $t = 0$.

been developed for a relatively simple hydrodynamic "thick-target" flare model, and the helioseismic response is calculated in the framework of a standard solar model and also for magnetostatic sunspot models.

The hydrodynamic "thick-target" model was first developed by Kostiuk and Pikelner (1975), and calculated in varying detail by many authors. In this model (schematically illustrated in Figure 22.1a) high-energy particles accelerated in the corona are injected into a magnetic flux tube. They heat the upper chromospheric layers to high temperature, creating a high-pressure region which expands producing "chromospheric evaporation" and a shock traveling towards the photosphere. Numerical simulations of the flare dynamics in a one-dimensional magnetic flux tube model revealed that, due to the intense radiative energy losses, this shock can compress the plasma by a factor of ~100 to very high density (Figure 22.4a; Kosovichev, 1986), and thus transfer substantial momentum into the lower atmosphere. The compressed relaxation zone behind the shock front can also be a source of the impulsive continuum emission observed in flares (Livshits *et al.*, 1981). Quantitative details of the hydrodynamic "thick-target" model are still being developed (Allred *et al.*,

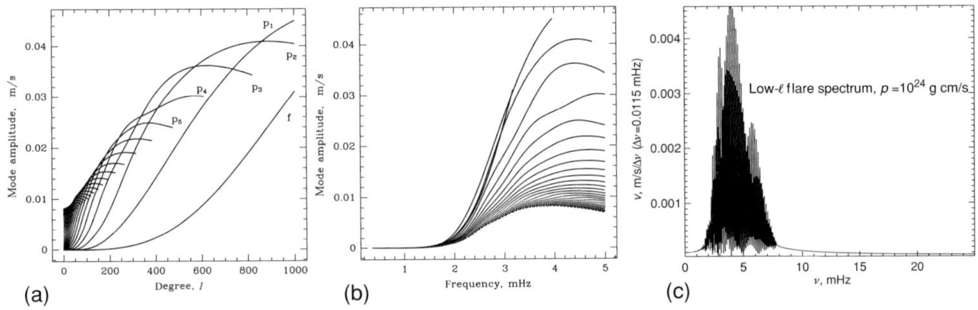

Figure 22.5 (a) Theoretical amplitudes of solar acoustic (p) and surface gravity (f) modes of varying angular degree ℓ, excited by the point source; (b) the mode amplitudes as a function of oscillation frequency; (c) the spectrum of low-degree modes ($\ell = 0, 1, 2$) for observations of the Sun as a star. Most of the oscillation power of sunquakes is concentrated around the acoustic cut-off frequency at $\sim$5 mHz.

2005; Rubio da Costa *et al.*, 2011), and currently it is unclear if this shock can explain the impact sources observed in the Doppler shift of photospheric absorption lines.

The helioseismic response to a localized source can be calculated by using the standard stellar pulsation theory (e.g., Unno *et al.*, 1989). A solution of the non-radial stellar pulsation equation written in symbolic form for a displacement vector $\boldsymbol{\xi}$,

$$\frac{\partial^2 \boldsymbol{\xi}}{\partial t^2} + \mathcal{L}\boldsymbol{\xi} = 0, \tag{22.1}$$

for initial conditions $\boldsymbol{v}(r, \theta, \phi, 0) = \boldsymbol{v}_0(r, \theta, \phi)$, can be obtained in terms of normal mode eigenfunctions, $\boldsymbol{\xi}_{nlm}$:

$$\boldsymbol{v}(r, \theta, \phi, t) = \sum_{nlm} \frac{\langle \boldsymbol{\xi}^*_{nlm} \cdot \boldsymbol{v}_0 \rangle}{\langle \boldsymbol{\xi}^*_{nlm} \cdot \boldsymbol{\xi}_{nlm} \rangle} \boldsymbol{\xi}_{nlm} \cos(\omega_{nlm} t) e^{-\gamma_{nlm} t}, \tag{22.2}$$

where the angle brackets mean the integration over the solar mass. Assuming that the impact is localized in a very small volume at the surface and calculating the integrals, we obtain for the radial component of the oscillation velocity at the surface:

$$v_r(R, \theta, \phi, t) = \sum_{nlm} \frac{P_0}{M_\odot I_{nl}} (2\ell + 1) P_\ell(\cos\theta) \cos(\omega_{nl} t) e^{-\gamma_{nl} t}, \tag{22.3}$$

where P_0 is the total momentum of the impact, $M_\odot$ is the solar mass, I_{nl} is the mode inertia, P_ℓ is the Legendre polynomial, ω_{nl} are the mode eigenfrequencies, and γ_{nl} are their damping times. The denominator, $M_\odot I_{nl}$, is often called the "mode mass." Thus, the mode amplitude excited by the impulsive point source is equal to the total momentum divided by the mode mass and multiplied by a geometrical factor $(2\ell + 1)P_\ell$. The theoretical sunquake surface velocity signals are illustrated in Figure 22.4b.

The spectrum of the excited oscillation modes is shown Figure 22.5a,b. For the total momentum 10^{24} g cm/s, the calculated total amplitude corresponds to the maximum amplitude observed in the sunquake events. This means that the total momentum of the flare impact does not exceed 10^{24} g cm/s. Then, using the same solution, we calculate the amplitudes of low-ℓ modes excited by this impact. The velocity spectrum of $\ell = 0$–2 modes is shown in Figure 22.5c. The maximum amplitude of these global modes does not exceed 0.4 cm/s (Kosovichev, 2009). This is about 100 times smaller than the amplitude of stochastically excited low-ℓ modes. The maximum amplitude of the flare-excited modes is in the frequency range of 4–5 mHz. These model results are inconsistent with the suggestion of Karoff and Kjeldsen (2008) that they had observed global sunquake oscillations.

The helioseismic response in inhomogeneous sunspot regions can be studied only numerically. An example of such numerical simulations is presented in Section 22.3.3.

22.3 Properties of sunquake sources

22.3.1 Comparison observations with the thick-target flare model

Simultaneous observations of sunquake events using Doppler shift data and X-ray and gamma-ray flare emissions are used for investigation of the relationship between the flare-accelerated particles and the sunquake impacts in the photosphere. In particular, observations of the X1.2 flare of January 15, 2005, showed a good agreement with the "thick-target" flare model (Kosovichev, 2006a,b). Figure 22.6a shows the soft (gray loop-like structure) and hard X-ray (white contours) emissions, as well as the location of the initial photospheric impact (white spot) overlaid on the magnetogram. Figure 22.6b shows a superposition of the Doppler velocity and continuum intensity images. Evidently, this event provides a nice example of the thick-target model configuration: The hard X-ray emission is observed at the footpoint of a magnetic loop filled with dense high-temperature plasma, and the sunquake impact is observed just beneath the hard X-ray source. However, the sequence of events (Figure 22.6c) casts doubts in this model scenario. The photospheric impact started simultaneously with or even earlier than the hard X-ray peak (top two panels), whereas the thick-target model predicts a delay between the hard X-ray emission and the photospheric impact. In this model, the hard X-ray emission (bremsstrahlung) is produced by Coulomb scattering of high-energy electrons in the upper chromosphere. The upper chromosphere is heated during this process, and then expands causing the chromospheric evaporation and the photospheric shock. This means that there must be a significant delay between the hard X-ray impulse and the photospheric signal, roughly corresponding to a hydrodynamic response time of about 100–200 s. However, the observations do not show this delay.

The discrepancy between the model predictions and observations became particularly apparent with the start of sunquake observations from the SDO. The first sunquake event detected by the HMI instrument on the SDO showed that the hydrodynamic impact occurred at the very beginning of the flare impulsive phase and that the hard X-ray flux was rather low (Kosovichev, 2011). This discrepancy was particularly pronounced in the M6.3 flare of March 9, 2012, in which the photospheric impact at $\sim$3:27 UT was more than 10 min prior the main impulsive phase, and was associated with a small peak of the soft X-ray

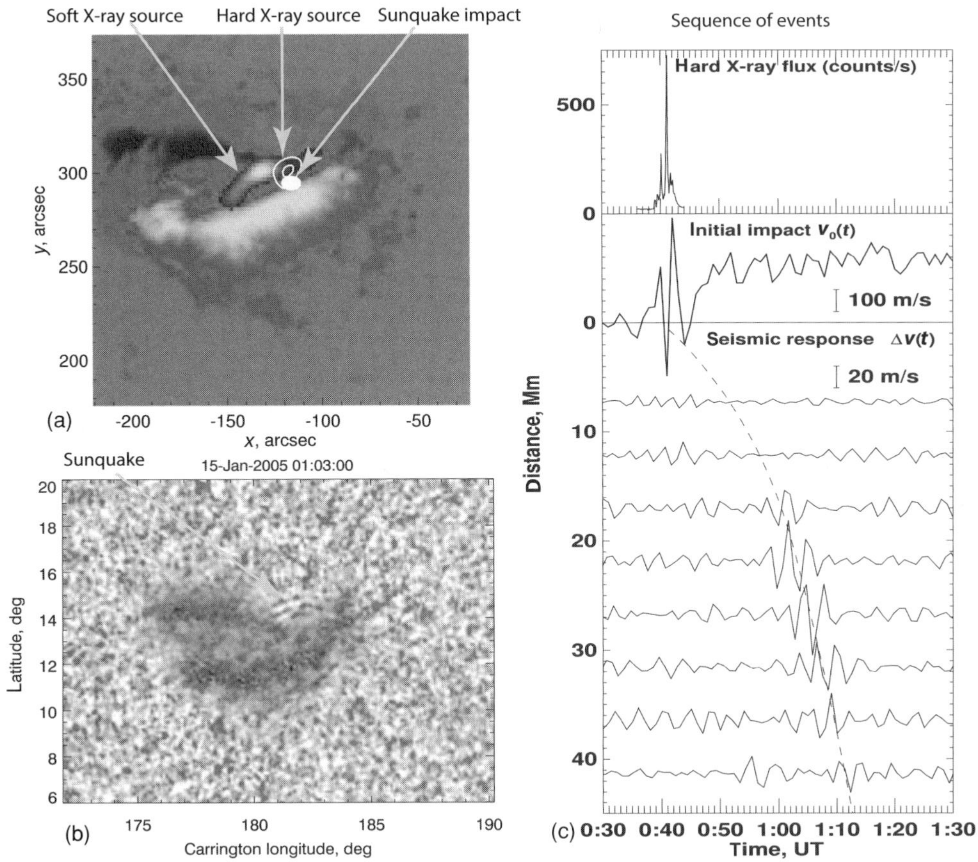

Figure 22.6 Sunquake of X1.2 flare of January 15, 2005: (a) location of the initial sunquake impact (white dot) on the SOHO/MDI magnetogram (color background), soft X-ray (gray-scale structure), and hard X-ray (contour lines) emissions observed by RHESSI; (b) Doppler-velocity image showing the wavefront (gray) overlaid with the corresponding image of sunspots (brown); (c) the sequence of the sunquake events showing the hard X-ray impulse, and Doppler-velocity signals at the initial impact, and at various distances from the impact point. The dashed curve shows the theoretical time–distance relation. A black and white version of this figure will appear in some formats. For the color version please refer to the plate section.

emission (Figure 22.7a Kosovichev, 2012b). Analysis of the HMI Doppler images and magnetograms showed that the impact was located in a region of relatively weak magnetic field, near a magnetic neutral line (Figure 22.7b). Variations of the magnetic field were observed after the velocity impact. This rules out the hypothesis of a magnetic origin for the impact due to an impulsive Lorentz force (so-called "McClymont jerk") (Hudson *et al.*, 2008). Observations of this flare with the Extreme-ultraviolet Imaging Spectrometer (EIS) on the Hinode spacecraft revealed strong chromospheric evaporation characterized by 150–200 km/s upflows observed at the same time as the photospheric impact in multiple

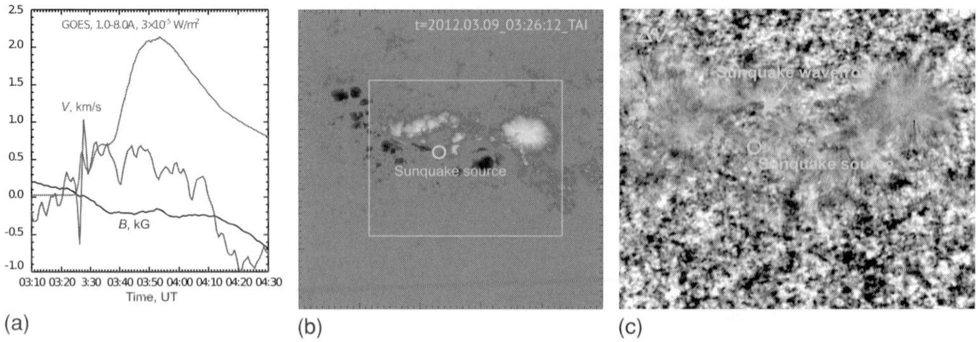

Figure 22.7 Sunquake event produced in the pre-heating phase of the M6.3 flare of March 9, 2012: (a) time profiles of the total soft X-ray flux (1.0–8.0 Å) from the GOES satellite (red), the photospheric velocity (blue) and magnetic field at the flare impact; (b) the impact location on the HMI magnetogram; (c) the Doppler velocity variations showing the sunquake wavefront in the area indicated on the magnetogram by a rectangle. A black and white version of this figure will appear in some formats. For the color version please refer to the plate section.

locations in multi-million degree spectral lines of flare ions (Doschek *et al.*, 2013). The detection of the chromospheric evaporation is consistent with the "thick-target" model scenario. However, the early timing of the impact indicates that the photospheric impact was probably caused by energy release in the low atmosphere, and not by energy transport from the high corona as suggested by the model. These observations may lead to paradigm change in the theory of solar flares.

22.3.2 Anisotropy of wavefront and source motion

A characteristic feature of the seismic response in this and most other flares is anisotropy of the wavefront: the observed wave amplitude is much stronger in one direction than in the others. In particular, all three seismic waves excited during the X17 flare of October 28, 2003 (Figure 22.8a), had the greatest amplitude in the direction of moving impact sources located in the bright "flare ribbons" (Kosovichev, 2006b,c). The motion of flare ribbons is often interpreted as a result of the magnetic reconnection processes in the corona. When the reconnection region moves up it involves higher magnetic loops, the footpoints of which are located further apart. A sequence of energy injection events in the reconnecting loops creates the effect of apparent source motion. Of course, there might be other reasons for the wavefront anisotropy, such as inhomogeneities in temperature, magnetic field, and plasma flows. However, the source motion seems to be the primary factor.

In such cases, the seismic wave is generated not by a single impulse but by a series of impulses moving on the solar surface. The seismic effect of the moving source can be easily calculated by convolving the wave Green function with a time-dependent moving source function. The results of these calculations show a strong anisotropic wavefront (Figure 22.8b), qualitatively similar to the observations by Kosovichev (2007). Curiously,

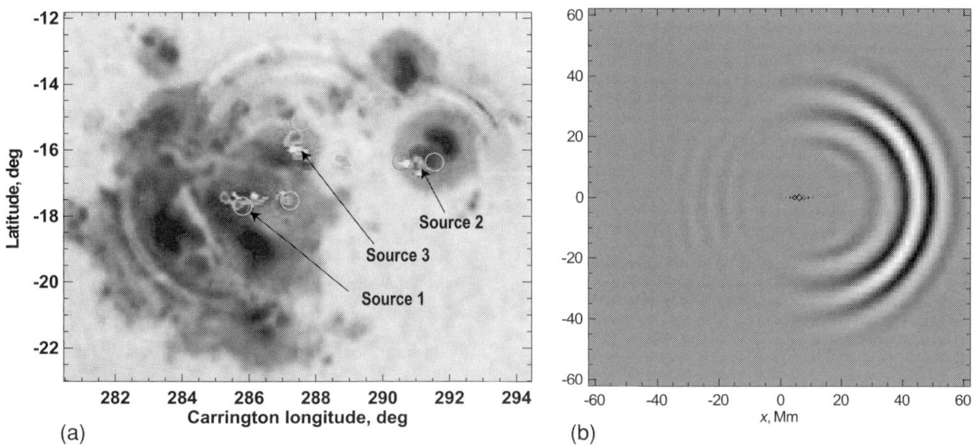

Figure 22.8 (a) Wavefronts of three sunquake events observed by the MDI instrument after the X17 flare of October 28, 2003, at 11:37 UT. The background is the corresponding MDI continuum intensity image; yellow and blue patches show down and up Doppler velocity signals with the amplitude stronger than 1 km/s, observed during the impulsive phase at 11:06 UT; yellow and green circles show locations of the hard X-ray (50–100 keV) and 2.2 MeV gamma-ray sources observed by RHESSI. (b) Theoretical model of sunquakes with a moving impact source, which explains the observed anisotropy of sunquakes. The point impulsive source is moving in the x-direction with the constant speed of 25 km/s. Its strength as a function of time has a Gaussian shape with FWHM of 3 min. The locations of the source are shown by black diamonds at the center, the size of which is proportional to the source strength. A black and white version of this figure will appear in some formats. For the color version please refer to the plate section.

this effect is quite similar to the anisotropy of seismic waves on Earth, when the earthquake rupture moves along the fault. Thus, taking into account the effects of multiple injections of accelerated particles in a realistic 3D magnetic geometry is very important for sunquake theories.

22.3.3 Interaction of sunquake waves with magnetic field of sunspots

Understanding of magnetic field effects on propagation of acoustic waves is of particular interest for helioseismology because this opens new diagnostics of subsurface structure and dynamics of sunspots and active regions (Kosovichev, 2012a). Local helioseismology techniques are based on measurements of perturbations of wave travel times and oscillation frequency shifts, which require long time series (typically 8–24 hours) because of the stochastic nature of solar oscillations. Also, because the distribution of wave excitation sources in active regions is not uniform, the interpretation of cross-covariance functions and power spectra is not straightforward (e.g., Zhao *et al.*, 2012a). Sunquakes provide a direct view of the wavefronts traveling through various areas of active regions with different magnetic properties and flow patterns. In combination with numerical modeling of MHD waves in sunspot models, this provides valuable information about wave interactions with

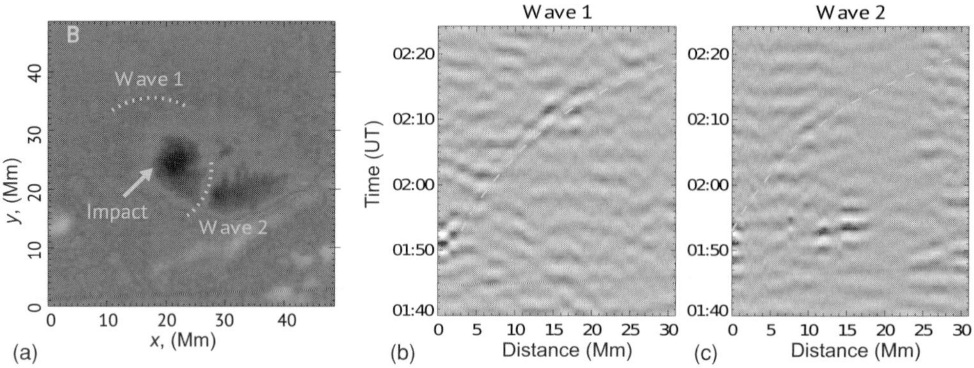

Figure 22.9 Observations of sunquakes of the X2.2 flare of February 15, 2011: (a) SDO/HMI magnetogram of AR 11158 (red, positive polarity; blue, negative polarity) and locations of the initial photospheric impact (arrow), and two wavefronts: Wave 1 traveling through a weak-field region, and Wave 2 traveling through the sunspot umbra with strong magnetic field; panels (b) and (c) show time–distance diagrams of these wavefronts illustrating that Wave 2 traveled slower and had a lower amplitude than Wave 1 (dashed curves show the theoretical time–distance relation). A black and white version of this figure will appear in some formats. For the color version please refer to the plate section.

magnetic fields. These studies are in an initial stage, and here I present two observational and theoretical examples.

Figure 22.9a illustrates propagation of sunquake waves during the X2.2 flare of February 15, 2011 (Kosovichev, 2011). The sunquake impact source was located in the sunspot penumbra close to the umbra–penumbra boundary. One part of the sunquake (Wave 1) traveled outside the spot through the penumbra, whereas the other part (Wave 2) traveled through the sunspot umbra (dark blue region on the magnetogram). The time–distance diagrams of these waves (Figure 22.9b,c) show that Wave 2 traveled slower and had lower amplitude than Wave 1. This travel-time delay for waves traveling through sunspots confirms the previous results obtained by time–distance helioseismology for relatively short travel distances (Kosovichev *et al.*, 2000). However, it is important to note that the travel-time variations strongly depend on the frequency contents (bandwidth and mean frequency) of the acoustic wave-packets.

The origin of the amplitude decrease in sunspots is currently unknown. This can be due to several effects, including absorption of wave energy, wave scattering, transformation into different modes, changes in the acoustic cut-off frequency, etc. Numerical simulations of sunquakes for sunspot models with different physical properties can help us to disentangle these effects and investigate their role. An example of such simulations (Parchevsky and Kosovichev, 2009a; Parchevsky *et al.*, 2010) performed for the self-consistent magnetostatic sunspot model of Khomenko and Collados (2008) is shown in Figure 22.10. In these simulations the flare impact was modeled by a localized point-like impulsive vertical force at the outer edge of the sunspot model. The time profile of the impulse was chosen to model a frequency spectrum similar to the solar spectrum. One part of the excited wave traveled outside the magnetic structure, and the other part traveled through the structure. The

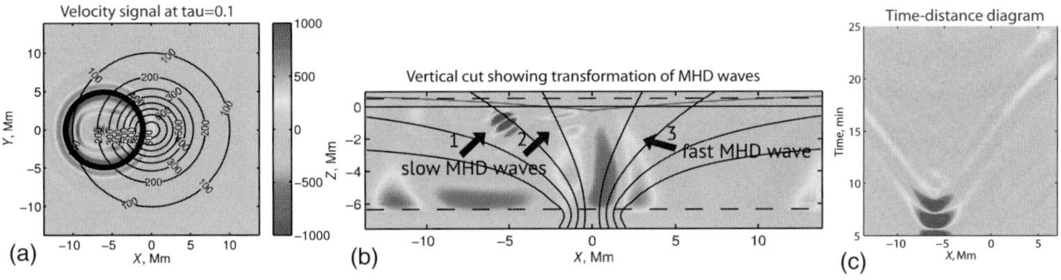

Figure 22.10 Numerical 3D simulations of MHD waves excited by a localized impulsive force and traveling through a sunspot model: (a) snapshot of the z-component of the velocity in the upper photosphere (optical depth $\tau = 0.1$). The circular contours show magnetic field strength in gauss. (b) Vertical cut of the simulated sunquake wave interacting with magnetic field. The helioseismic wave below the surface is a mixture of fast MHD and slow MHD waves. (c) Time–distance sunquake diagram of the photospheric signal. A black and white version of this figure will appear in some formats. For the color version please refer to the plate section.

amplitude reduction of the second wavefront is apparent. The vertical cut in Figure 22.10b shows that the impact force excited two types of MHD waves: a fast MHD wave (arrow 3) which formed circular sunquake-type ripples at the surface, and a slow MHD wave (left arrow 1) which traveled along the magnetic field lines into the interior, and thus did not affect the surface wave pattern. Similarly, our simulations showed that when the slow MHD wave (indicated by arrow 2) is excited because of the wave transformation at the level where plasma $\beta \equiv P/8\pi B^2 = 1$ it travels along the field lines into the interior. This allows us to conclude that the observed sunquake phenomenon in magnetic field regions represents fast MHD waves, and that the wave transformation does not play a significant role. The simulations reproduce the amplitude reduction when the waves travel through the strong field region. It seems that this effect is related to the wave reflection from the $\beta = 1$ layer, which in sunspots is below the photosphere level. After the wave passes through the spot its amplitude is restored to the quiet-Sun values, as illustrated by the time–distance diagram in Figure 22.10c. This means that the amplitude reduction in sunspots is probably not due to wave absorption, but rather due to changes in the surface reflection properties.

22.3.4 Comparison of impact sources with holography reconstruction

There are two approaches for detection and investigation of sunquakes. The first approach is based on visual detection of the initial impacts and traveling sunquake waves (which are usually most visible ~20 min after the impact) in a series of images of Doppler-shift running differences processed through a high-pass filter in order to partially separate sunquake signals from convective noise, and also on the time–distance diagrams constructed for the impact locations. Generally, strong initial impacts in the photosphere with a speed of 1 km/s or greater are good indications of sunquakes. Such impacts are easily observed in Dopplergrams. The second approach (called "acoustic holography") attempts to

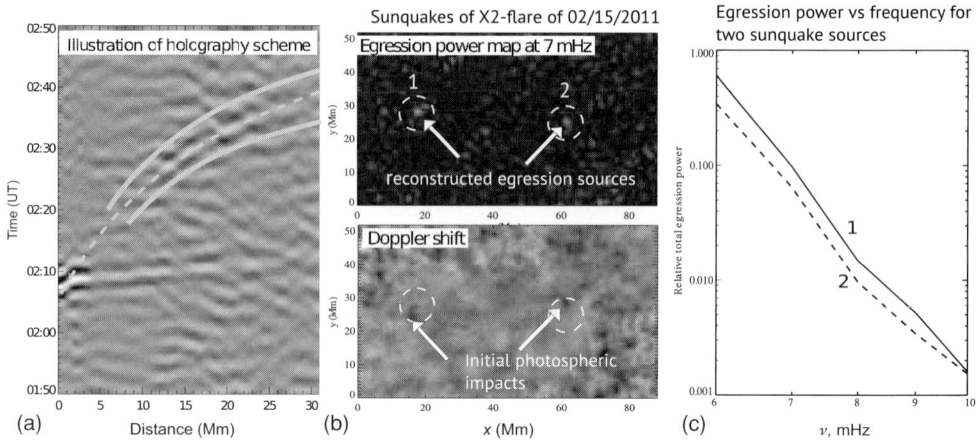

Figure 22.11 Analysis of sunquakes using the holography approach (Donea *et al.*, 1999): (a) illustration of the holography scheme, which integrates the wavefront signal and extrapolates it back in time to the origin assuming a standard time–distance relation (dashed curve) for solar acoustic waves; (b) top panel: egression power image of two sunquake events of the X2.2-class flare of February 15, 2011 (dashed circles indicate the locations of the egression source), bottom panel: Doppler-velocity image showing the initial sunquake impacts as black point sources near the edges of the dashed circles; (c) the egression power as a function of frequency for the two events of the X2.2 flare (marked 1 and 2).

reconstruct the initial sources by tracing the sunquake wave signals back in time to their origins, assuming that the wave propagation follows theoretical Green's function relations for point acoustic sources in the horizontally uniform quiet-Sun models (see Figure 22.11a). The reconstruction is performed for different frequency filters for every point in the flaring region, and the results are plotted as "egression power maps." Then these maps are visually inspected for spikes that are considered as candidates for sunquake sources. When these spikes coincide with localized flare indicators as sources of continuum emission, they are identified as sunquakes. For strong events, both approaches are equally reliable. However, when the flare perturbations are weak, the time–distance approach may miss weak sunquake events, while the holography reconstructions may lead to false positive identifications. In addition, in the real Sun conditions the impact sources and the wave properties may be quite different from the theoretical Green function. Therefore, the location and times of the reconstructed sources may differ from the initial flare impacts (Kosovichev and Zhao, in preparation). This discrepancy is illustrated in Figure 22.11b, in which the top panel shows the locations of two sunquake sources identified in the egression power map at 7 mHz, and the bottom panel presents a Doppler shift map, processed for the time–distance analysis, showing locations of the actual flare impacts. Evidently, there is a shift of about 5 Mm in these locations. This discrepancy has to be taken into account in sunquake studies. The egression power rapidly decreases with frequency (Figure 22.11c).

22.4 Challenges in understanding sunquakes

Since the first detection in 1996, sunquakes have been extensively studied using data from the SOHO and SDO space missions, and also from the ground-based network GONG. However, the mechanism of these events and their physical properties are still mysterious. Here, I summarize some of the primary questions, potential answers, problems and challenges.

- *Why are sunquake events observed only in some flares?*
New observations with the HMI instrument on the SDO reveal that sunquakes are more common than was thought previously. These observations showed that sunquake events may occur even for relatively weak flares of X-ray class as low as M1, whereas in some strong X-class flares the photospheric impacts are almost absent, and helioseismic response is not detectable. A full statistical analysis of the HMI observations is not completed yet, and I can only make some preliminary conclusions. It appears that the helioseismic response is a characteristic of compact energy release events in the low atmosphere. Such energy release events usually are not accompanied by coronal mass ejections (CMEs), but generate coronal shocks (often observed by the SDO/AIA instrument). In flares accompanied (and perhaps triggered) by CMEs the energy is often released in the high corona. Such flares may have a high X-ray class, but produce virtually no photospheric impact. My current conjecture is that sunquakes are more common for "compact" ("confined") flares than for "eruptive" ("dynamic") flares. This flare classification introduced by Pallavicini *et al.* (1977) and Svestka (1986) is based on characteristics of X-ray and optical emissions and dynamics. Complex flares (such as the X2.2 flare of February 15, 2011) may have both "compact" and "eruptive" components. The relationship between sunquakes and the two types of flares requires detailed statistical studies, and is certainly very important for our understanding of the basic mechanisms of energy release in solar flares.
- *Why are the initial photospheric impacts observed in the early impulsive phase?*
The photospheric impacts observed at the very beginning of the flare impulsive phase or even during the pre-heating phase represent one of the greatest puzzles of sunquakes and challenge the standard flare model. The standard thick-target model (Section 22.3.1) predicts that the photospheric impact is a result of expansion (accompanied by shocks) of the upper chromosphere region heated by high-energy electrons accelerated in the corona. In this scenario, the photospheric impact should follow the hard X-ray impulse with a time delay of $\sim 10^2$ s, which is needed for the perturbations and shocks to propagate from the upper chromosphere to the photosphere. However, in many events the photospheric impact is observed at the beginning of the impulsive phase before the hard X-ray impulse reaches the maximum. This can be explained in two ways: (1) the low-chromosphere plasma is directly heated by deeply penetrating particles, electrons or protons, which do not produce significant X-ray or gamma-ray emission; (2) the initial sunquake impacts are due to the flare-related dynamics, e.g., caused by the Lorentz force or by a flux-rope eruption, rather than high-energy particles. The first scenario seems to be more likely

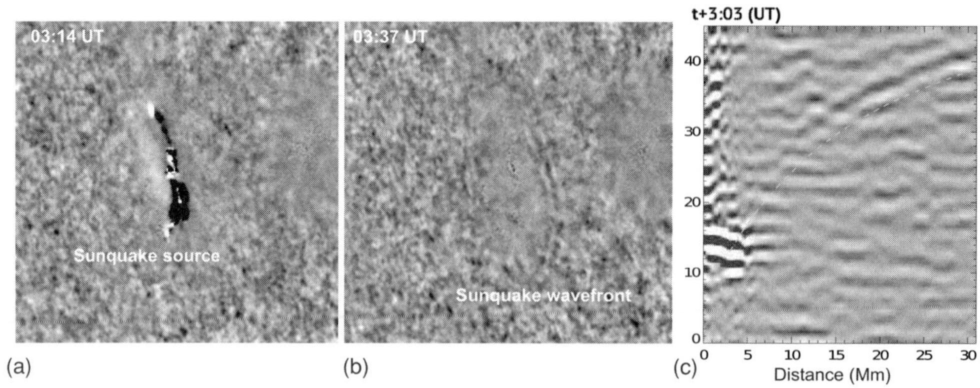

Figure 22.12 One of the strongest events of Solar Cycle 24 observed during the X1.8 flare of October 23, 2012: (a) Doppler velocity variations at 03:14 UT, showing an unusually long impact source extended along a flare ribbon; (b) the sunquake wavefront at 03:37 UT; (c) the sunquake time–distance diagram, showing the source motion at the beginning of the flare and the sunquake ridge (dashed curve is the theoretical relation).

because the initial sunquake impacts are often synchronized over very large distances, which can be achieved only by relativistic particles, and also because the speed of apparent motion of sunquake sources on the surface is often significantly higher than the local Alfvén speed or sound speed.

- *What is the mechanism of the sunquake anisotropy and variations of wave amplitude?* Observations show that most sunquakes are highly anisotropic. The anisotropy is usually observed as a significant enhancement of the wave amplitude in one direction compared to the amplitude in other directions. The extent of the amplitude enhancement can be very narrow or can extend over almost half of the circle. Also, the wavefront can be far from circular. For instance, in the greatest sunquake event observed so far in solar cycle 24 on October 23, 2012, during the X1.8 flare, due to the almost linear source (Figure 22.12a) the sunquake wavefront had a very elongated elliptical shape (Figure 22.12b), with the strongest amplitude on one side of the ellipse. (The sunquake time–distance diagram is shown in Figure 22.12c.) The linear source structure is caused by a series of almost simultaneous impacts located in a flare ribbon parallel to the magnetic neutral line. Such extended sources are not uncommon, but this was, perhaps, the longest source observed with the space MDI and HMI instruments. Generally, there are two main sources of wavefront anisotropy: (1) wave propagation through the inhomogeneous structure and flows of active regions; (2) rapid motion of the excitation source. Typically, when the waves travel through regions of strong magnetic fields of sunspots their amplitude is reduced, but when the wave re-emerges from sunspots the amplitude is restored to the quiet-Sun values. The numerical simulations presented in Section 22.3.3 show that impulsive force or momentum impacts in magnetic regions excite fast and slow MHD waves. However, the slow wave travels along the magnetic field lines into the deep interior, and does not appear in the surface signal. Also, when the fast MHD wave reaches the region of plasma $\beta = 1$, it is partially transformed into a slow MHD wave, which also travels into the interior and does not affect the observed wavefronts. When the wave traveling through a sunspot emerges on the other side its amplitude is restored to the quiet-Sun values. The rapid source motion is a typical feature of sunquakes, and, likely, the primary cause of the wave amplification and anisotropy due to the interference effect (Section 22.3.2).

- *What causes the rapid motion of sunquake sources?*
  The sunquake source motion is observed as a sequence of individual point-like impacts (Kosovichev, 2006c) or as a wave-like motion (Kosovichev, 2011). In the first case, the apparent speed can reach 50–60 km/s, substantially exceeding the speed of sound or Alfvén waves. The impact sources are usually located in footpoints of magnetic loops or arcades (Figure 22.1a), observed as bright flare ribbons, and this motion corresponds to the ribbon expansion, but sometimes the impact sources can move along the ribbons (Kosovichev, 2007). The cause of the ribbon expansion is not fully understood. The current paradigm, based on the thick-target model, is that the flare ribbons represent a response of the low atmosphere to the precipitation of energetic particles accelerated in a magnetic reconnection region, and that the ribbons expand as the reconnection process moves higher up in the corona. The rapid source motion, particularly in the form of a sequence of discrete impacts, clearly indicates that the sunquakes most likely represent a helioseismic response to a series of impacts of flare-accelerated particles.
- *What is the relationship between the flare-accelerated particles and sunquake initiation?*
  Perhaps this is the central question, which represents the importance of sunquakes for our understanding of solar flares. The spatial and temporal properties of the sunquake sources clearly indicate a close association with flare-accelerated particles, however, this relationship is not as straightforward as it follows from the flare thick-target model (Figure 22.1a). It is particularly puzzling that the sunquake impacts are observed at the very beginning of the impulsive phase, and that it seems that there is no clear correlation between the hard X-ray and gamma-ray emission of solar flares. One idea to solve this puzzle is that the particles that transport the flare energy into the low chromosphere are of relatively low energy and do not produce significant X-ray and gamma-ray emissions. Perhaps these particles are accelerated in the low chromosphere, or penetrate through low-density magnetic flux tubes. These and other possibilities require detailed theoretical modeling and comparison with other observational effects, such as the transient variations of the photospheric line profiles, including rapid variations of magnetogram signals, and continuum emission (e.g., Kosovichev and Zharkova, 2001).
- *Can sunquakes be caused by variations of magnetic field (Lorentz force) or by filament eruptions?*
  Alternative ideas to explain the localized impacts in the low atmosphere include effects of the Lorentz force due to the rapid restructuring of the magnetic field configuration, resulting from magnetic reconnection and dissipation (e.g., Hudson *et al.*, 2008). However, the observations do not show a correlation between the magnetic field variations

and the sunquake impact sources (e.g., Kosovichev, 2011). Also, quite often the impacts are observed in weak-field regions (e.g., see Figure 22.7). It was also suggested that the back reaction of erupting filaments (flux ropes) or CMEs might generate sunquake events (e.g., Zharkov *et al.*, 2011). In this case the synchronization of distant impact sources has to be with the Alfvén speed, which is quite slow in the low atmosphere. This contradicts the observations of almost simultaneous impacts over long distances (e.g., see Figure 22.12a). Also, in many sunquake cases, flux-rope eruptions are not observed. Therefore, it seems unlikely that the sunquake photospheric impacts and helioseismic waves are caused by these mechanisms.

- *Can the seismic response flares be observed in global oscillation modes and on other stars?*
The expanding sunquake waves represent high-degree p modes, however, the theory (Section 22.2) predicts that global low-degree modes are also excited. Their detection has been reported from a statistical analysis of the correlation between flare soft X-ray signals (observed by the GOES satellites) and total solar irradiance measurements from the space observatory SOHO (Karoff and Kjeldsen, 2008). However, there was no unambiguous detection of the whole-Sun oscillations caused by individual flare events. Indeed, the theory predicts that the amplitude of the low-degree modes of sunquakes is significantly lower than the amplitude of stochastically excited oscillations (Figure 22.5c) (Kosovichev, 2009). However, in the case of significantly more powerful stellar flares, the impact on the star's surface can be much greater. This may lead to excitation of the global low-degree oscillations to significantly higher amplitudes than on the Sun. A preliminary study of the Kepler short-cadence data provides indications of such "starquakes" (Kosovichev, 2014). However, a statistical study for a large sample of stars and longer observing intervals are needed for investigation of the "starquakes."

23

Seismic waves in small bodies: Sources and propagation

JAMES D. WALKER, WALTER F. HUEBNER, SIDNEY CHOCRON, RORY P. BIGGER, AND TRENT KIRCHDOERFER

Our Solar System is full of small bodies. The planets occupy an obvious role as the beautiful heavyweights, but smaller bodies are found throughout the Solar System in the form of asteroids, comet nuclei, and moons of Mars and the outer giants. Even Pluto has multiple moons. In spite of their abundance, there is not a large amount of mass in small bodies. An estimate for the mass of all the objects in the main belt puts it at around 4% the mass of the Moon (Krasinsky *et al.*, 2002). Though many of their properties can be understood by looking at the surface and from surface samples, there are many questions that require looking inside the body.

For example, a central question of the bodies' origin and evolution is whether the asteroid or comet is an accretion of small objects into something larger, or is it a small piece of an originally larger object. An important clue to answering this question is the interior structure of the asteroid or comet nuclei. Being able to look inside would allow us to see the structure and whether the object is one solid piece, a small number of large pieces, or a lot of small objects held loosely together (the oft referred to "rubble pile"). In addition, understanding the interior structure is important if we want to move an asteroid or a comet nucleus, either for engineering reasons or to prevent one from striking Earth.

There are two ways of looking inside an object. Electromagnetic radiation or high-energy particles is one approach, which will not be discussed here save to mention that the power requirements are large and penetration depths of things such as "ground-penetrating radar" are not large. The second approach is to use mechanical waves propagating through the body, i.e., seismology.

The many successes of seismology have been described in the chapters in this book. Here we focus on some of the unique opportunities and challenges posed by performing seismology on small bodies. Given the extensive use of seismology in studying the Earth and in resource exploration there, it should be kept in mind that a major difference between Earth seismology and extraterrestrial seismology is that in the latter we expect few sources and few seismometers, due to cost, weight, and size constraints in delivering them to a

Extraterrestrial Seismology, eds. Vincent C. H. Tong and Rafael A. García. Published by Cambridge University Press.

celestial body. This lack of spatially distributed data collection points precludes use of most of the advances made in the last few decades in tomography.

23.1 Material properties: Density, wave speeds, and strength

It turns out that only two independent values are necessary to describe an isotropic elastic body. For seismological purposes, two independent properties that can be directly inferred are the longitudinal sound speed c_L, sometimes denoted α, representing a wave where the material motion is in the same direction as the propagation of the wave, and the shear wave speed c_S, sometimes denoted β, representing a wave where the material motion is transverse to the direction of propagation of the wave. In seismology, the longitudinal wave is referred to as a primary wave or P wave, since it is faster and arrives at the seismometer first ($c_L > c_S$), and the shear wave is referred to as a secondary wave or S wave, since it arrives second. In a shear wave, the transverse motion occurs in the plane normal to the propagation direction; when further clarification is desired, the component of the shear motion that lies in the plane defined by the propagation direction and the normal to the surface is called SV (for vertical) and the transverse motion normal to that is called SH (for horizontal). Given the two wave speeds and the density, it is possible to explicitly compute the two elastic constants, for example, Young's modulus and Poisson's ratio.

Currently, some asteroid bulk (global) densities have been measured or inferred. Densities are based on mass estimates divided by volume estimates. Mass is always estimated by the deflection of another body, either an asteroid's moon in a binary system, a spacecraft, other asteroids, or Mars. The asteroid's gravitational attraction alters the other body's trajectory, which is measured, thereby revealing the mass of the asteroid. Images of the asteroid taken by spacecraft or ground- or space-based telescopes, radar images during Earth flybys, or shapes inferred by stellar occultations make it possible to estimate an asteroid's volume. Shapes and densities of selected asteroids are shown in Table 23.1 (the last column in the table is discussed later). The table comprises values in three compilations (Britt *et al.*, 2002; Hilton, 2002; Baer, 2010) and original sources. Densities of solid rocks found on Earth typically lie in the range of 3 to 5 g/cm^3. The upper lunar surface density is around 2.7 g/cm^3. The estimated asteroid densities, ranging from 0.8 to 6.7 g/cm^3, imply for the most part a high degree of porosity, either micro- or macro-porosity. Micro-porosity is small pores in the rock, while macro-porosity is spaces between blocks of materials. Presumably macro-porosity is generated during aggregation formation or post-collision processes in asteroids, where cavities form during reassembly. These blocks may be held together only by mutual gravitational attraction and minor adhesive forces.

In general, the composition of asteroids is based on remote sensing of broad-band spectra from their surfaces, but surface compositions may differ from the interior. To date we have no directly measured sound speeds of asteroid materials. However, we do have directly measured sounds speeds on the surface of the Moon. The regolith layer at the Apollo 14 and Apollo 16 landing sites was measured to have speeds of 104 m/s and 114 m/s,

Table 23.1 *Estimated bulk densities for several asteroids. Taxonomy (Tholen classification based on eight color broad-band spectra between 0.31 μm and 1.06 μm): B, F, G, C, major groups of dark carbonaceous asteroids; S, major group of silicaceous (stony) asteroids; M, E, P, major groups of metallic asteroids; A, D, T, R, Q, V, minor groups of asteroids; double letters indicate a combination of two types.*

Asteroid (Taxonomy)	Size or diameter (km)	Density (g/cm^3)	Estimated max modal period (s)
1 Ceres (G)	$975\times975\times909\pm3.6^{c}$	2.08 ± 0.08^{ab}	930
2 Pallas (B)	$582\times556\times500\pm18^{h}$	2.8 ± 0.4^{ab}	415
4 Vesta (V)	$572.6\times557.2\times446.4\pm0.2^{\ell}$	$3.456\pm0.03^{\ell}$	330
10 Hygiea (C)	$530\times407\times370\pm3.4^{c}$	2.10 ± 0.11^{c}	505
11 Parthenope (S)	153 ± 1.55^{c}	2.72 ± 0.12^{a}	120
15 Eunomia (S)	$357\times255\times212\pm7.5^{c}$	3.13 ± 0.53^{c}	230
16 Psyche (M)	$240\times185\times145\pm14^{c}$	6.73 ± 3.05^{c}	80
20 Massalia (S)	$160\times145\times132\pm4.65^{c}$	3 ± 0.8^{ab}	105
21 Lutetia (M)	$121\times101\times75\pm1\times1\times13^{i}$	3.4 ± 0.3^{j}	72
45 Eugenia (F)	$232\times193\times161\pm8^{c}$	$1.2+0.6-0.3^{ab}$	385
87 Sylvia	$384\times262\times232\pm5^{c}$	1.62 ± 0.3^{ab}	640
90 Antiope (C)	$93\times87\times83.6\pm0.5^{c}$	1.8 ± 0.8^{b}	125
121 Hermione (C)	187 ± 3^{c}	1.37 ± 0.14^{c}	265
243 Ida (S)	$59.8\times25.4\times18.6\pm0.6^{c}$	2.6 ± 0.5^{ab}	46
253 Mathilde (C)	$66\times48\times46^{k}$	1.3 ± 0.2^{k}	100
433 Eros (S)	$34.4\times11.2\times11.2\pm0.1^{c}$	2.67 ± 0.03^{d}	26
617 Patroclus (P)	121.8 ± 3.2^{e}	$0.8+0.2-0.1^{e}$	305
704 Interamnia (F)	$349\times339\times274\pm11^{c}$	2.3 ± 0.5^{c}	305
762 Pulcova (FC)	137 ± 1.6^{c}	0.9 ± 0.1^{c}	305
804 Hispania (P)	158 ± 2.9^{c}	1.98 ± 0.44^{c}	160
854 Frostia (V)	$14\times10\times10\pm5.5\times4\times4^{f}$	0.89 ± 0.14^{f}	31
25143 Itokawa (S)	$0.535\times0.294\times0.209\pm0.001^{g}$	1.9 ± 0.13^{g}	0.56
1999 KW4	$1.532\times1.495\times1.347\pm0.04^{c}$	2.39 ± 0.9^{ab},	1.5
1089 Tama (S)	$11\times7.5\times7.5\pm1^{f}$	2.52 ± 0.30^{f}	8.7
1313 Berna	$15\times12\times12\pm6\times5\times5^{f}$	1.22 ± 0.15^{f}	25
2000 DP 107 (C)	0.8 ± 0.08^{c}	$1.62+1.2-0.9^{ab}$	1
2000 UG11 (QR)	0.23 ± 0.02^{c}	$1.47+0.6-1.3^{ab}$	0.31
4492 Debussy	$10\times6.5\times6.5\pm4\times2.5\times2.5^{f}$	0.91 ± 0.10^{f}	22

[a]Britt *et al.* (2002), [b]Hilton (2002), [c]Baer (2010), [d]Veverka *et al.* (2000); Yeomans *et al.* (2000), [e]Marchis *et al.* (2006), [f]Behrend *et al.* (2006), [g]Ostro *et al.* (2004); Fujiwara *et al.* (2006), [h]Schmidt *et al.* (2008), [i]Sierks *et al.* (2011), [j]Pätzold *et al.* (2011), [k]Veverka *et al.* (1999), [ℓ]Russell *et al.* (2012).

Table 23.2 *Density and sound speed (using 1 MHz transducers) ranges for chondrite meteorites*

Meteorite group	ρ (g/cm^3)	c_L (m/s)	c_S (m/s)	Poisson's ratio
Ordinary chondrites (Jones, 2009)				
H-chondrites	2.52–3.49	1753–5376	786–2998	0.11–0.44
L-chondrites	3.08–3.42	1578–6327	792–3390	0.04–0.39
LL-chondrites	2.9–3.50	1466–4954	875–3401	0.05–0.26
Carbanaceous chondrites (Ibrahim, 2012)				
CM-chondrites	1.96–2.55	1542–3472	956–2235	0.03–0.28
CO-chondrites	2.40–3.36	1747–3088	1051–1982	0.09–0.24
CV-chondrites	2.65–3.26	1016–4203	665–2618	0.06–0.20

respectively, with respective thicknesses of 8.5 m and 12.2 m. Below that was a 299 m/s and 250 m/s layer, respectively. At the Apollo 17 landing site, the 250 m/s layer was estimated to have a thickness of 248 m and below it was a 1200 m/s layer. References (Kovach *et al.*, 1971, 1972, 1973; Kovach and Watkins, 1973b) describe the active seismometry experiments that were performed.

Otherwise, the only notions we have for sound speeds related to asteroids are those inferred from meteorites. Meteorites are classified in groups, with the goal of grouping meteorites according to a common origin, i.e., a single parent body such as an asteroid. Modern meteorite classification relies on placing specimens into groups in which members share certain key physical, chemical, and mineralogical properties consistent with a common origin. The main categories of these groups are stony meteorites (including chondrites and achondrites), stony–iron meteorites (including pallasites and mesosiderites), and iron meteorites (including magmatic irons and non-magmatic or primitive irons). These groups are divided into many subgroups based on their mineralogy, bulk chemical composition, and oxygen isotope abundances.

With the sample return in 2010 by Hyabusa from 25143 Itokawa, the surface dust has been matched with LL chondrite meteorite material (Nakamura *et al.*, 2011). Thus, it is now confirmed that ordinary chondritic meteorites match at least one asteroid, and we can presume there are other matches. Our main purpose here is simply to list sound speeds in recovered meteorites, as they give a range of sound speeds we might see in asteroid seismology. Table 23.2 shows recent measurements of some chondrites, including measurements of density. The values within these ranges are not entirely random, in that increasing density typically implies increasing sound speeds, and increasing porosity in the rock implies decreasing sound speeds, but there is a lot of scatter. A wide range of sound speeds is exhibited. Poisson's ratio is computed from $\nu = (r/2 - 1)/(r - 1)$, where $r = (c_L/c_S)^2$.

Material strength is also of interest, but since it involves destructively testing samples there is little data available on meteorite strength. Petrovic (2001) notes the absence of data, but points out that compressive strength of stony meteorites ranges from 150 to 450 MPa, with tensile strength being roughly 10% of compressive strength (a value similar to that for rock and other brittle materials such as ceramics and glass). Iron meteorites have measured strengths ranging from 350 to 700 MPa. It should be kept in mind that these meteorites survived entry into Earth's atmosphere as well as an impact with the ground, and thus may not be representative of asteroid material strength.

Additional geophysical data of interest for asteroids and comets can be found in Greenberg and Huebner (2002), Huebner *et al.* (2001), and neodata.space.swri.edu.

23.2 Seismic wave paths and travel times

Since small bodies are small, it is possible to produce sound waves that travel through the whole body with relatively small sources. On Earth it requires large earthquakes or nuclear explosions to produce seismic signals that can be sensed around the planet, but on small bodies relatively small seismic sources will be sufficient. Obvious seismic sources are explosives and impactors. Both of these are being considered for spacecraft missions.

A not so obvious seismic source is sunlight producing thermal stresses that can lead to motion (seismometers on the Moon detected seismic events during daylight; Duennebier and Sutton, 1974). Relative internal motion for a small, rotating fractured rock or rubble pile could also produce noise that is measurable on the surface. These small bodies may have a lot of inherent noise, all just waiting to be recorded and analyzed. In a similar vein, a solar-powered hammer (thumper) that attaches to the surface and then provides regular, weak seismic signals could be utilized. Planned signals from explosives and impacts are easiest to analyze, since the magnitude, location, and time of the seismic impulse are known.

The most direct way of analyzing seismic signals is ray tracing – that is, the signal source is known and seismic wave transit paths to the seismometer are drawn through the body and, based on first arrival, a sound speed is determined for the path. That is how the above-quoted lunar surface information was obtained. Ray tracing can help distinguish interior structure as internal reflections are inferred, indicating changes in interior composition or large structural flaws. There will also be multiple paths due to reflection off near and far free surfaces, and the ray-tracing technique builds a model of internal structure that is consistent with the observed surface motions at the seismometer, based on various wave speeds, wave types, and paths. In this approach, it is straightforward to estimate the required temporal resolution of the recording devices. For example, if the material has $c_L = 5000$ m/s and a travel path of $L = 200$ m (either a reflection off the far side at 100 m or a direct distance between source and seismometer), and it is desired to know the sound speed to 1% ($\Delta c_L = 50$ m/s), then sampling needs to occur with $\Delta t = L/c_L - L/(c_L + \Delta c_L) \approx (L/c_L)(\Delta c_L/c_L) = 4$ milliseconds, i.e., a rate of $1/\Delta t = 250$ Hz (1 Hertz = 1/second).

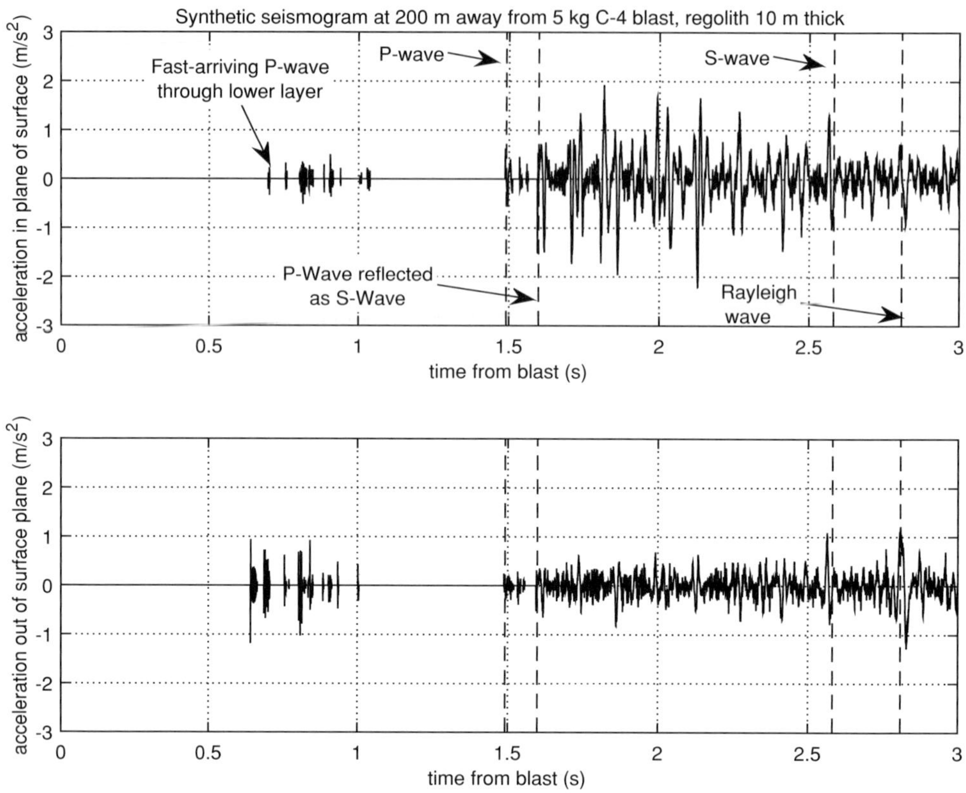

Figure 23.1 Example of a synthetic seismogram showing first arrivals 200 m from the source, where there is a $c_L = 134$ m/s, 10 m thick regolith layer on top of a $c_L = 335$ m/s layer. Initial arrivals are P waves traveling in the lower layer. The other marked waves are direct wave arrivals traveling in the upper layer, save the marked arrival of a P wave reflecting as an S wave at the layer interface. The computation was performed using CTH (McGlaun *et al.*, 1989).

The longitudinal (P) and shear (S) waves travel through the body. There are also surface waves that travel near the surface. The Rayleigh wave (a combination of P and SV motion) and the Love wave (SH motion) travel along the surface, the latter when the surface is layered. These waves typically have larger material ground motion associated with them than the longitudinal and shear waves. Arrival times provide information about the longitudinal and shear wave speeds of the material. Figure 23.1 shows a synthetic seismogram for a layered surface structure, showing accelerations both in the surface plane and out of plane. For this asteroid computation, the top 10 meters of material is a low sound speed regolith. The arrivals of the various waves are marked in the seismogram. The first arrivals are due to waves that travel through the regolith layer, then through the lower, faster layer, and then travel back up through the regolith layer. For such a wave, a use of calculus shows that the travel time is

$$t = \left(x + 2y\sqrt{(c_{L2}/c_{L1})^2 - 1}\right) / c_{L2} = 0.73 \text{ s}, \tag{23.1}$$

where x is the distance between the source and the seismometer, y is the thickness of the surface layer, subscript 1 refers to the upper slower layer and subscript 2 to the lower faster layer. In the figure, the earliest arrival time is slightly earlier than the estimated time due to the finite size of the explosive and crater-opening dynamics. The P and S waves that travel directly near the surface are explicitly marked. For a material with Poisson's ratio 0.25, the Rayleigh wave speed is given by $c_R = 0.53c_L$. There is also a conspicuous arrival of a P wave reflecting off the layer interface as a shear wave, whose arrival time is computed by the minimum time it takes for a wave to travel that path:

$$t = \min_z \left(\sqrt{z^2 + y^2}/c_{L1} + \sqrt{(x-z)^2 + y^2}/c_{S1}\right) = 1.60 \text{ s}. \tag{23.2}$$

The reflection point is $z = 193$ m, so the shear SV wave produced travels to the surface at a steep angle and its material motion is mainly in the plane of the surface. The primary purpose of the figure is to demonstrate that these sorts of computations can be performed and various ray traces identified.

In addition to information about the elastic wave speeds based on arrival times, information about material strength of the asteroid material in the vicinity of the charge can be obtained by the magnitude of the material motion at the seismometer. Based on the velocity of the material and the distance from the charge, it is possible to estimate the stress level near the source. Given the stress level, computations can be performed of the blast loading of the surface that determine the strength of the material leading to that stress. In one-dimensional motion, the relation between stress and material velocity is given by the Hugoniot jump condition $\tilde{\sigma} = \rho c_L u$, where $\tilde{\sigma}$ is the stress in compression and u is the material velocity. In a three-dimensional expansion of the wave, the material velocity drops off as $1/r$, where r is the distance from the source (the energy in the wave drops off as $1/r^2$).

23.3 Full-body global seismology

Given that sources can be sensed over the whole body due to its small size, small bodies may be the ideal setting for global seismology. Global seismology occurs when an object is vibrating and it is possible to ascertain the normal modes by examining the motions at various detection points. The concept has been described earlier in this book for its successful applications to both the Earth and the Sun. What is unique about small bodies is their irregular shape. Highly sophisticated software has been developed for analyzing seismic data on the Earth and the Sun that takes advantage of those bodies' nearly spherical shapes (Dahlen and Tromp, 1998; Lay and Wallace, 1995). On small bodies we do not have a symmetric shape to work with, and so tools to analyze the data need to be fully three-dimensional and the computations can be quite involved and time consuming (Walker *et al.*, 2006).

The question we hope to answer about internal structure can be viewed in two respects: qualitatively what does the interior look like and quantitatively what does the interior look like? Since we expect a paucity of seismometers and sources, we need to gather as much information as possible about the asteroid and we wonder whether the spectrum may have unique features that qualitatively answer the question of whether the interior is a solid or made of comminuted material. Let us first look at this question in one dimension, to see, given lots of symmetry, what might occur.

For a one-dimensional uniaxial-strain elastic strip of length L, with longitudinal sound speed c_L, and with free ends, the natural frequencies are given by $\nu_n = \omega_n/2\pi = nc_L/2L$, where $n = 1, 2, 3, \ldots$ The frequency ν_n is in hertz (Hz or cycles per second) and the angular frequency ω_n is in radians per second. The period of an oscillation is given by $T = 1/\nu_n$ in seconds. Examining this expression we see that a way to non-dimensionalize the frequencies is to divide by $c_L/2L$, the reciprocal of the amount of time for a wave to transit from one end of the strip, reflect off the far end, and return. Thus, to simplify the discussions, we consider the normalized frequency

$$\bar{\nu}_n = \frac{\nu_n}{c_L/2L} = \frac{\omega_n/2\pi}{c_L/2L} = n. \tag{23.3}$$

To recover the actual frequency one multiplies $\bar{\nu}_n$ by $c_L/2L$.

To begin considering the role of a characteristic size, which in our context would be block sizes within the asteroid, it is simplest to consider a one-dimensional strip of material as a series of masses connected by a spring (this is essentially what the finite element method does). We assume the first and last masses have a free exterior edge (no spring attached on their outer side), and assume that there are N masses. If all the springs have the same spring constant and all the masses have the same mass, then the normal mode frequencies are (Crawford, 1968)

$$\bar{\nu}_n = \frac{\omega_n/2\pi}{c_L/2L} = \frac{2N}{\pi}\sin\left(\frac{n\pi}{2N}\right), \quad n = 1, 2, \ldots, N. \tag{23.4}$$

To be specific, if we compute with 512 masses (hence 511 springs), Figure 23.2 bottom row shows a dot for each frequency in log space. The lower frequencies are quite similar to the homogeneous continuum result of $\bar{\nu}_n = n$, while as the frequency increases and approaches the normal mode where each mass moves in the opposite direction to its neighbor, the normal modes bunch together. This bunching is a result of the frequency getting close to $\bar{\nu}_N = 2N/\pi$, which frequency is determined by the size of the L/N characteristic size of the parts of the system. Thus, a qualitative behavior that can be seen in the spectrum is bunching of the natural frequencies based on a characteristic size. The finite masses example has a maximum frequency the system can support due to the inability to distinguish higher frequency motion given the finite number of sampling points. The continuum system does not have an upper cut-off.

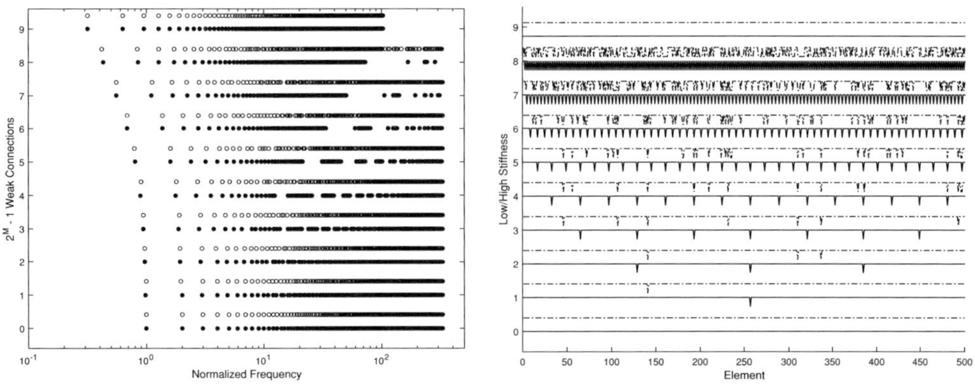

Figure 23.2 Left image shows the normal mode frequencies of a one-dimensional strip of 512 masses connected by springs as more weak springs are introduced. The darker points are when the weak springs are introduced in a symmetric fashion, while the open circles and dashed lines are when the weak springs are introduced in a random fashion. The right image shows the weak spring locations.

To explore what might happen if the asteroid in question was a collection of smaller rocks rather than one contiguous piece of material, we explore variations of the one-dimensional example. In the first, we break the length into equal size pieces, by changing the wave speed connecting the two masses (i.e., changing the spring constant), reducing it by a factor of $1/\sqrt{10}$. In Figure 23.2, the line labeled 1 represents 2 large parts made of the stiffer springs connected in the middle by a weak spring (the 1 means $2^1 - 1 = 1$ weak spring). The next line up labeled 2 has 4 large parts made of stiffer springs connected by 3 weak springs ($2^2 - 1 = 3$), and so on. The second line from the top labeled 8 has every other spring weak, while the top line labeled 9 has all weak springs (and so is simply Equation (23.3) with a lower sound speed). We note that there is structure to the natural frequencies – for example, large gaps appear in the spectrum as the size of the pieces becomes smaller, indicating substructure.

This looks exciting, but of course there is lots of symmetry in this example, in particular the size of the pieces. What happens if the size of the pieces is random? To address this question, we randomly distribute the weak springs uniformly across the length. When this is done, the resulting natural mode frequencies are displayed in Figure 23.2. Displayed is a specific example, not an average of many cases. We see that there is still structure in the distribution of the normal mode frequencies, but the gaps are not as large. Thus, a more subtle analysis is required, but there is still a structure to the normal mode frequencies that could be used to infer information about the (now randomly sized) internal structure.

Seeing the structure requires being able to measure modal frequencies over a large range of frequencies. In addition to the frequencies we expect for the lower modes, we want to make sure we can record frequencies in these regimes to look for any tell-tale signals of internal structure size.

The example of the one-dimensional strip of material is instructive, but the ultimate interest is in three-dimensional bodies. In three dimensions, we begin looking at the most

Table 23.3 *Spheroidal (S) and toroidal (T) mode normalized frequencies* $_n\bar{\nu}_l = (_n\omega_l/2\pi)/(c_L/2D) < 5$ *for a homogenous, isotropic elastic sphere with a Poisson's ratio of 0.25; multiplicity of each mode is* $2l + 1$*; there are 54 distinct modes and 530 modes counting multiplicities.*

Mode	$_n\bar{\nu}_l$	Mode	$_n\bar{\nu}_l$	Mode	$_n\bar{\nu}_l$	Mode	$_n\bar{\nu}_l$
$_0S_0$	1.632			$_0S_4$	1.841	$_0T_4$	1.873
$_1S_0$	3.857			$_1S_4$	2.963	$_1T_4$	3.570
				$_2S_4$	4.057	$_2T_4$	4.856
$_0S_1$	1.259	$_0T_1$	2.118	$_3S_4$	4.981		
$_1S_1$	2.489	$_1T_1$	3.343				
$_2S_1$	2.847	$_2T_1$	4.529	$_0S_5$	2.217	$_0T_5$	2.303
$_3S_1$	3.931			$_1S_5$	3.452	$_1T_5$	4.025
$_4S_1$	4.785			$_2S_5$	4.546		
$_0S_2$	0.970	$_0T_2$	0.919	$_0S_6$	2.581	$_0T_6$	2.721
$_1S_2$	1.788	$_1T_2$	2.623	$_1S_6$	4.101	$_1T_6$	4.472
$_2S_2$	3.061	$_2T_2$	3.865				
$_3S_2$	3.595			$_0S_7$	2.938	$_0T_7$	3.131
$_4S_2$	4.468			$_1S_7$	4.636	$_1T_7$	4.912
$_0S_3$	1.439	$_0T_3$	1.420	$_0S_8$	3.291	$_0T_8$	3.536
$_1S_3$	2.372	$_1T_3$	3.104	$_0S_9$	3.642	$_0T_9$	3.937
$_2S_3$	3.567	$_2T_3$	4.367	$_0S_{10}$	3.990	$_0T_{10}$	4.334
$_3S_3$	4.318			$_0S_{11}$	4.336	$_0T_{11}$	4.729
				$_0S_{12}$	4.682		

symmetric body, the sphere. The normalized natural frequencies of the sphere, assuming that Poisson's ratio is 0.25, which is typical for rock, are shown in Table 23.3, where the sphere has radius R and diameter $L = D = 2R$. Details of the solution technique can be found in Walker and Huebner (2004). The solutions are labeled according to their solution type based on spherical harmonics: essentially the radial behavior is indexed by n, where n indicates roughly quarter wavelengths along the radius, and l indicates wavelengths around the circumference. Spheroidal modes involve compressive material response, while toroidal modes are shear (in torsion) material response.

To explore the normal mode frequency dependence on the shape of the body, three computations were performed with the explicit finite element package LS-DYNA. These computations use as a common element the mass and density of Itokawa, namely a mass of 3.33×10^{10} kg (the actual mass of Itokawa is 3.51×10^{10} kg) and a density of 1.9 g/cm^3 (Fujiwara *et al.*, 2006). A sphere with this mass and density has a diameter of 322 m. We also consider an ellipsoid with the same mass, where the major axis is 664 m and the

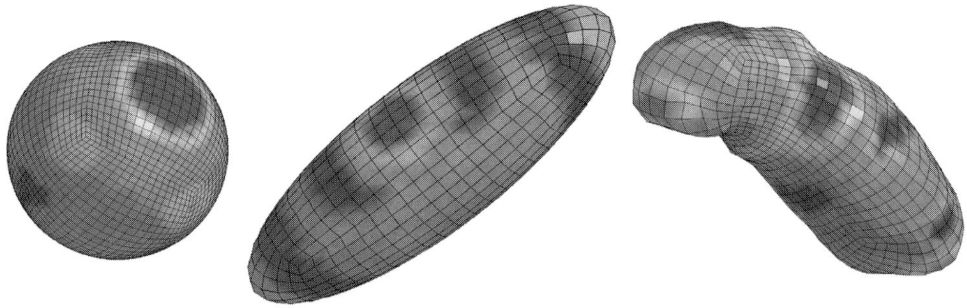

Figure 23.3 The three shapes explored in the normal mode computations: a sphere, an ellipsoid, and Itokawa. The shading is based on the surface acceleration magnitude at 10 s after the blast. A black and white version of this figure will appear in some formats. For the color version please refer to the plate section.

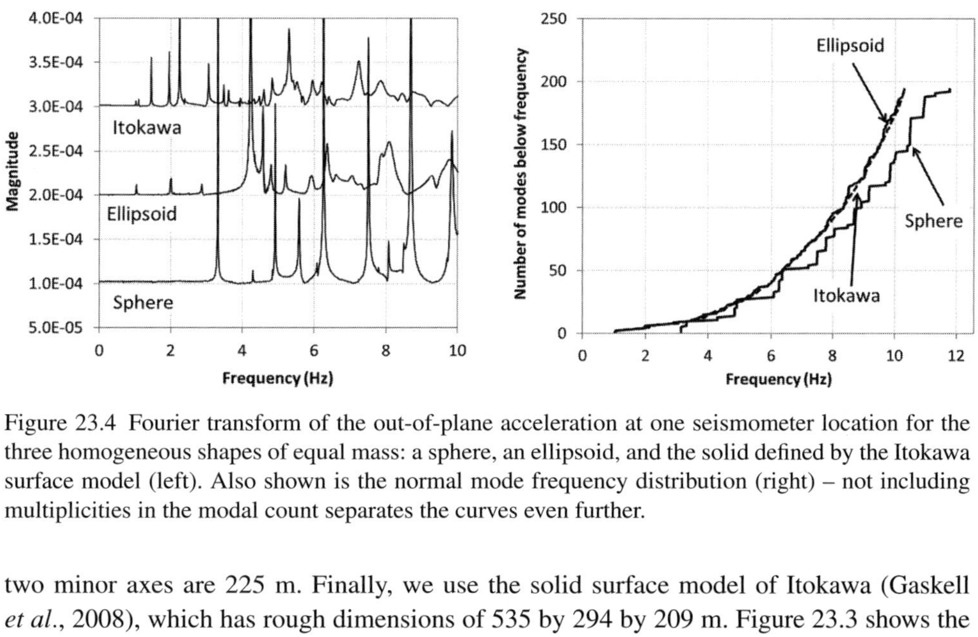

Figure 23.4 Fourier transform of the out-of-plane acceleration at one seismometer location for the three homogeneous shapes of equal mass: a sphere, an ellipsoid, and the solid defined by the Itokawa surface model (left). Also shown is the normal mode frequency distribution (right) – not including multiplicities in the modal count separates the curves even further.

two minor axes are 225 m. Finally, we use the solid surface model of Itokawa (Gaskell *et al.*, 2008), which has rough dimensions of 535 by 294 by 209 m. Figure 23.3 shows the shapes of the bodies. For the material, the longitudinal sound speed was $c_L = 2205$ m/s and Poisson's ratio was 0.25; these imply a shear wave speed $c_s = \sqrt{(1-2\nu)/(2-2\nu)}c_L = c_L/\sqrt{3} = 1273$ m/s. If a different longitudinal sound speed is of interest, the frequency in the plot should be multiplied by $c_L/2205$ m/s. If a different Poisson's ratio is of interest, the computations need to be redone for that Poisson's ratio. The seismic source is a 5 kg explosive sitting on the surface (see Appendix). A seismometer was placed away from the blast location, taking out-of-plane acceleration data at a rate of 2000 Hz ($\Delta t = 0.5$ ms). Data were collected for 100 s and are shown in Figure 23.4 (left). Though the peaks are identifiable with shorter data collection times, the peaks become narrower as more data are collected, helping to better identify the frequency of the normal mode. Longer data collection times determine how sharp a peak is going to be and how close a nearby peak can be and still be resolved.

For a qualitative assessment, we examine the distribution of normal mode frequencies, since that is a baseline from which deviations imply internal structure. LS-DYNA was used to compute the first 200 normal modes and their frequencies, and the results are shown in Figure 23.4 (right). For the sphere, there are 5 modal frequencies in the region from 0 to 5 Hz (${}_0T_2$, ${}_0S_2$, ${}_0S_1$, ${}_0T_3$, and ${}_0S_3$, making 27 counting multiplicities). The spheroidal modes lead to large peaks in the Fourier transform, but it is difficult to identify the toroidal modes ${}_0T_2$ and ${}_0T_3$, in part because the explosive loading condition is symmetric and does not directly excite toroidal modes (the latter can be seen in the rise of the ${}_0S_3$ mode, and presumably ${}_0T_2$ is in the rise of the ${}_0S_2$ peak). For the ellipsoid, there are 19 modal frequencies up to 5 Hz (26 counting multiplicities). For the homogeneous Itokawa model, there are 25 modal frequencies up to 5 Hz (there are no multiplicities).

It is interesting to note that for all three shapes there are similar numbers of modes, though outer surface symmetry leads to irregularities in the normal mode frequency distribution. In contrast to the sphere, by 7.5 Hz the Itokawa normal mode frequency distribution almost exactly matches a cubic: if $N(\nu)$ is the total number of normal modes less than or equal to frequency ν, then $N = C\nu^3$ for a constant $C\,(= 0.174\ \mathrm{s}^3$, in this case). This pattern agrees with the one-dimensional line model above, where $N = n = \bar{\nu}$: in both situations the number of modes is a constant times the frequency raised to the spatial dimension of the problem. Though it takes considerable effort and instrument resolution to count the number of modal frequencies in the Fourier transform, their distribution provides a clue to internal structure. The deviancy of the sphere from the ellipsoid and Itokawa indicate its special structure, though in this case the structure is all the symmetries of the sphere.

The specific normal mode frequencies can be used to help reconstruct the internal structure of the asteroid. In an asteroid mission, the spacecraft orbiting the asteroid would photograph it in detail and orbit it, which would lead to accurate measurements of mass and shape. Analysis would begin with the computation of the normal modes of a homogeneous, isotropic body using the initial data on wave speeds from arrival times in the ray-tracing phase of the analysis. Then various modifications of internal structure, both shape and material properties, would be tried to better match the actual measured modal frequencies.

Though we are mostly discussing techniques for asteroids of Eros' size or smaller (otherwise, larger charges need to be used), the last column of Table 23.1 shows a rough estimate of the lowest-in-frequency mode's period (so it is the maximum modal period). The estimate was roughly computed as the reciprocal of our scaling factor, $2L/c_L$, where L was taken as the largest extent of the asteroid and c_L was approximated by multiplying the density by 1 (km/s)/(g/cm^3). For comparison, Earth's longest period mode is ${}_0S_2$ at 53.9 minutes. We expect that global seismology techniques will be most effective on smaller bodies, which appears to correspond to lowest model modal frequencies of 0.01 to 1 Hz and corresponding periods of 1 to 100 s.

23.4 Gathering the data

Up to now, our discussion has focused on what the seismic data can tell us about the structure of the asteroid or comet nucleus. But how are we to gather the data? A number of seismometers have been developed that have the required sensitivity, but for our application we would like to reduce the mass. The Apollo seismometers, for example, detected 0.5 kg of explosive at 1.2 km distance and 2.7 kg of explosive at 2.7 km distance (Kovach *et al.*, 1973; Kovach and Watkins, 1973b). Kovach and Chyba (2001) show space-application seismometer sensitivities, and in particular state that in the region of 1 to 10 Hz there exist sensitivities on the order of 10^{-7} to 10^{-8} g (Earth g) acceleration. Their sensitivity chart drops off to about 10^{-6} g acceleration at 100 Hz.

A concern with small-body seismology is that dust, loose regolith, rubble-pile-like loose internal structure, and low gravity will make it difficult to attach the seismometer to the asteroid to obtain a good signal. Ideas for attaching seismometers range from using impactors to embed the seismometer into the surface to using glues or pads of sticky material to attach to the asteroid surface. Some of these ideas need to be explored with low-gravity flight tests.

A difficulty with digital sampling, when looking for normal modes, is that aliasing can occur, meaning that high frequencies are indistinguishable from lower frequencies. As an example, if the sampling rate is $\frac{\pi}{2}$ then the curves $\sin(t)$ and $-\sin(3t)$ are indistiguishable, since in the digital sample they are identical, namely $0, 1, 0, -1, 0, \ldots$ If both frequencies are present in the modal vibrations, it is not possible to separate them out. There are various approaches to handling this problem. One approach is to design a mechanical band-pass filter in the seismometer such that signals with a frequency above the highest value that is being digitally sampled are removed. Another approach is to determine the appropriate time window for taking a sample, since the higher frequency vibrations damp out more quickly. Finally, if the data-receiving antenna is outside Earth's atmosphere, perhaps analog measurements and transmissions could be considered: if the power is high enough, these could be used, allowing very fine data analysis to be performed on the receiving end.

There is ongoing research into micro-electromechanical-based seismometers (MEMs seismometers are seismometers on a chip) and whether they are sensitive enough to permit a significant reduction in weight of the seismometer package, potentially allowing a larger number of seismometers to be distributed on the surface of the body.

There is also the possibility of using imaging to determine ground motion. High-speed cameras are now available that have frame rates on the order of the Δts we have been discussing in this chapter. For example, we could measure wave arrivals of interest on a 100 m size object with framing rates on the order of 1 kHz. In a laboratory impact test on a 1 m diameter granite sphere, dust was observed to have launched on the side opposite the impact point (Walker *et al.*, 2013). With a high enough framing rate, such motion could potentially be seen. Since it is desirable to image as much of the asteroid as possible, successful implementation of imaging will probably entail, for example, distributing diffraction gratings or small prisms or reflectors on the surface, such that small motions lead to a frequency shift and/or amplitude shift in the reflected light that could be observed and measured with a (color) standoff telescope. This is how the modes of the Sun have been studied telescopically, observing the motion of features on the surface; a challenge with this approach is that huge amounts of data will be gathered, so analysis routines will need to be developed in advance and executed on the spacecraft.

23.5 Summary

We have shown that seismology on small bodies has considerable promise for determining interior structure and material properties. Hopefully this method will soon see application in missions to asteroids and comets, thus providing the internal structure, properties, and origins of these bodies as well as providing the information required to deflect or disperse them for engineering purposes or to prevent an impact with Earth.

23.6 Appendix

This appendix presents how loads were transferred from detailed explosive computations to the full-body finite element analysis model of the asteroid body. The blast loads themselves were computed in Walker (2003): a 5 kg charge of C4 explosive sitting on the surface leads to an in-plane seismic moment of $M_{\theta\theta} + M_{rr} = 6 \times 10^{14}$ dyne cm and a downward momentum of $p_z = 5 \times 10^8$ g cm/s; surface properties are $\rho = 2.7$ g/cm^3, $c_L = 4025$ m/s, Poisson's ratio $= 0.25$, a slope of 1 on the shock-speed/particle speed Hugoniot, and a Drucker–Prager linear strength–pressure relation with slope 1, zero pressure strength 50 MPa, and a maximum strength of 200 MPa. That paper describes in detail how these values are computed; we note that the moment and momentum scale with explosive charge size (and both increase when the charge is buried at shallow depth). For archival purposes, we state how the seismic moment and downward momentum, computed in an Eulerian computation, are transferred to the finite element Lagrangian computation.

Based on radial symmetry assumptions, an initial velocity is given to the nodes surrounding the node where the charge is positioned. We work in local coordinates at the charge location where the x-y plane is tangent to the surface and the z-direction, normal to the surface, points up out of the asteroid body. Assuming a computational mesh comprising "brick" elements, the velocity (v_x, v_y, v_z) at the nine nodes in the x-y plane nearest to and including the source location (the surface center node is the source location) is given by

$$\begin{array}{lll} (-v_{xb}, v_{xb}, -v_{zb}) & (0, v_{xa}, -v_{za}) & (v_{xb}, v_{xb}, -v_{zb}) \\ (-v_{xa}, 0, -v_{za}) & (0, 0, -v_{za}) & (v_{xa}, 0, -v_{za}) \\ (-v_{xb}, -v_{xb}, -v_{zb}) & (0, -v_{xa}, -v_{za}) & (v_{xb}, -v_{xb}, -v_{za}) \end{array}$$

The next plane down of nodes (vertices) repeats this same initial velocity assignment. All other nodes' velocities are initially zero. To approximately compute the moment, it is assumed that the z-axis is an axis of rotational symmetry. Δx is the element edge length in the x- and y-directions (assumed square) and Δz is the element edge length in the down direction. We assume the wavefront $v_r = v_{xa}$ is at location $r = \Delta x = c_L t$. Along the line connecting the element vertices in the x-direction, we have $(r, v_r) = (0, 0)$, $(\Delta x, v_{xa})$, and $(2\Delta x, 0)$ The velocity behavior in the z-direction is $(z, v_r) = (0, v_{xa})$, $(\Delta z, v_{xa})$, and $(2\Delta z, 0)$. The finite element basis functions specify that the velocity components of these points connect by a straight line. Integration with the radial symmetry assumption yields

$$M_{\theta\theta} + M_{rr} = \frac{1}{t}\int_V \rho v_r r\, dV = \frac{7}{2}\pi\rho c_L v_{xa}(\Delta x)^2 \Delta z. \tag{23.5}$$

The relation between v_{xa} and v_{xb} is determined by matching the 0 degree expression (along the x-axis) with the $0 = 45°$ expression ($x = y$-direction); expressions for v_{xa} and v_{xb} are

$$\begin{aligned} v_{xa} &= \frac{M_{\theta\theta} + M_{rr}}{(7/2)\pi\rho c_L(\Delta x)^2\Delta z}, \\ v_{xb} &= \left(\frac{3}{4\sqrt{2}} - \frac{1}{8}\right) v_{xa} \approx 0.40533 v_{xa}. \end{aligned} \tag{23.6}$$

Similarly, using the same symmetry assumptions,

$$\begin{aligned} p_z &= \frac{5}{2}\pi\rho v_{za}(\Delta x)^2 \Delta z, \\ v_{za} &= \frac{p_z}{(5/2)\pi\rho(\Delta x)^2\Delta z}, \\ v_{zb} &= \left(\frac{9}{4\sqrt{2}} - 1\right) v_{za} \approx 0.591 v_{za}. \end{aligned} \tag{23.7}$$

These expressions allow us to transfer the seismic impulse to the finite element mesh and then run our computations.

Part VI

Interdisciplinary research involving terrestrial seismology

24

The interpretation of seismic travel times

GUUST NOLET

24.1 Introduction

For a long time, (geo)physicists have considered the Earth as a spherically symmetric body, compositionally divided into an iron core, a silicate mantle, and a crust mostly built from melts rising to the surface. However, the surface of the Earth, divided into oceans and continents, betrays the existence of lateral differences that cannot be ignored. None of our oceans is older than 200 million years, whereas there are indications that the first continents started to form 4.3 billion years ago. The processes that impose such differences must operate at depth. The fact that heat cannot escape sufficiently fast by conduction, yet the mantle is not molten, leaves no doubt that some kind of convective engine operates to cool the mantle and to drive the observable motions of the much more rigid plates covering the surface (Jaupart and Mareschal, 2010).

To study the dynamic processes in the interior, we can inspect various physical fields for signs of lateral heterogeneity: the gravity field, the magnetic field, the response of the electromagnetic field to forcings in the atmosphere and beyond, and the seismic wavefield after a strong earthquake. We are often interested in *anomalies*, i.e., in deviations from spherical symmetry. Here we shall concentrate on the seismic field, for the simple reason that it has the most powerful information about variations in elastic properties in the interior that are themselves influenced by temperature variations and – probably to a lesser degree – by chemical differences. The analysis of seismic waves has given us detailed images of wave speed variations in the interior of the Earth and Sun. On Mars, Mercury, or one of the many moons in the Solar System, we shall at best be able to observe the seismic field with a handful of seismometers, and we wish to extract as much information from such seismic recordings as possible. Fortunately, recent developments in terrestrial seismic tomography have led to significant progress in analyzing the observed field.

When an earthquake releases potential energy, it emits both compressional (P) and transverse (S) waves. To first order, the waves visible on a seismogram can be explained with the spherically symmetric structure of the Earth. At high frequency ($\sim$1 Hz), P and S waves can be assumed to follow narrow trajectories (rays) and refract according to

Extraterrestrial Seismology, eds. Vincent C. H. Tong and Rafael A. García. Published by Cambridge University Press.

Snell's law, just as light waves in optics. Since the local speed of seismic waves generally increases with depth, the refraction is such that they return to the surface – or reflect from interior discontinuities, the most important of which are the core–mantle and the inner core boundary. The surface also acts as a reflector. Waves that depart from the source at very shallow angles may reflect multiple times at the surface, and interfere destructively at high frequency. At lower frequency ($\lesssim 0.1$ Hz), the sum of such multiple reflections shows up as a dispersive "surface wave" of high amplitude because its energy is trapped at shallow depth and spreads out in only two dimensions, as opposed to the "body" waves of P and S type.

Even a superficial inspection of seismograms, and their differences from station to station, shows that there must be deviations from spherical symmetry in the interior of our planet. Ideally, one would wish to find a three-dimensional model of variations in density ρ, compressibility κ, and shear modulus μ that allows us to reproduce the observed deviations as they appear in the seismic records. For several reasons, this goal so far has proved elusive, certainly at frequencies above about 0.03 Hz. If crystals in the rock line up, the elastic response of the rock depends on the direction and polarization of the wave, potentially increasing the number of unknowns from two (κ and μ) to 21 independent constants of a fourth-order elasticity tensor. Inverting for the distribution of such anisotropy is impossible without making drastic simplifications. Even the density and the two isotropic elastic parameters are usually conveniently reduced to just two intrinsic velocities, the compressional velocity $\alpha = \sqrt{(\kappa + 4\mu/3)/\rho}$ and the shear velocity $\beta = \sqrt{\mu/\rho}$, since the sensitivity of the seismogram to pure density variations is minimal.

In principle, the full waveform contains all available information in a seismogram. However, trying to model the waveforms themselves proves to be a very non-linear exercise, demanding very large computational resources. The time of the wave arrivals responds more linearly to changes in α and β than do the full waveforms (Mercerat and Nolet, 2013), and trying to fit arrival time (or phase) proves less complicated and often more robust. Historically, seismologists have picked "onset times" of seismic waves, and this is still the preferred method to quickly locate an earthquake. Given a location of the quake hypocenter, a (spherically symmetric) reference model is used to predict the onset times for P waves and possibly other phases. After comparison with observed onset times, one inverts the differences in a linearized inversion scheme to adjust the hypocenter until a satisfactory agreement is obtained. The agreement is never 100%, though, not only because of errors in the time picks, but also because the Earth deviates from spherical symmetry. In the 1970s this led to the first attempts to interpret such differences in terms of lateral variations in the Earth's mantle (Dziewonski *et al.*, 1977).

Onset times provide only one measure that is related to wave velocity. Since they denote the time of the earliest arrival, this wave follows a trajectory (the "ray path") that renders the travel time minimal. The stationarity of the travel time with respect to small deviations in the ray path is known as Fermat's principle, and one can show this to be equivalent to satisfying Snell's law of refraction, as in optics (e.g., Nolet, 2008). An example of an onset is shown in Figure 24.1. Since this is a signal from a nuclear explosion, the onset is sharper

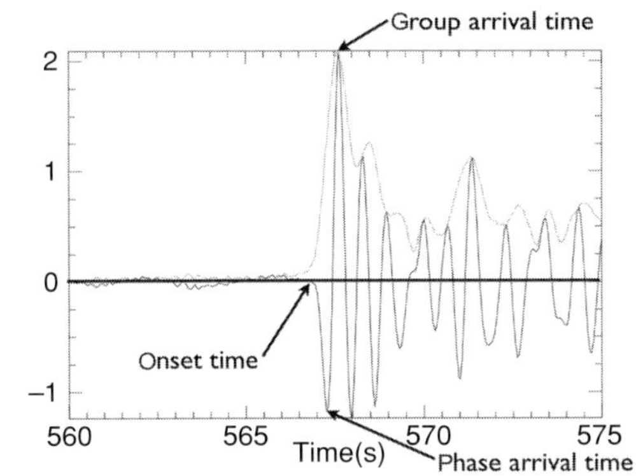

Figure 24.1 The P wave from a nuclear explosion near Lop Nor in NW China, on August 16, 1990, recorded in a seismic station in L'Aquila, Italy, at an epicentral distance of 54°. The arrival time of a P wave can be defined by its onset if this is clearly visible above the noise. The horizontal axis gives the time since the detonation of the bomb. The dotted line denotes the envelope of the recorded motion, and its maximum is identified with the "group arrival time." Any particular phase, such as the first zero crossing or (as here) the first minimum, can be used to define a "phase arrival time."

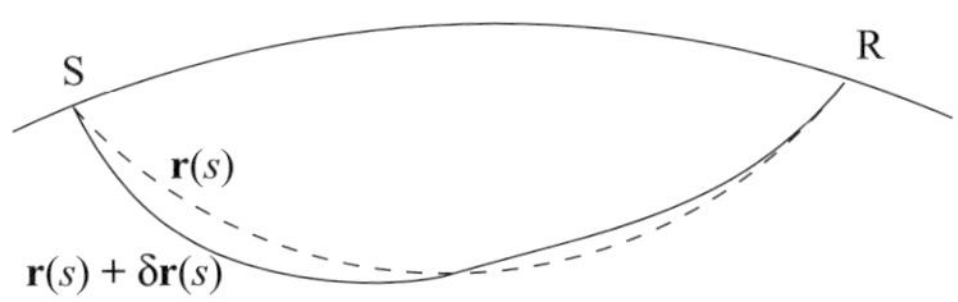

Figure 24.2 Fermat's principle states that the wave follows a trajectory that renders the travel time stationary – usually a minimum. We assume this is the case for the path $\mathbf{r}(s) + \delta\mathbf{r}(s)$ between source S and receiver R (solid line). The path for a simple, spherically symmetric planet bends away from this predicted path, but is more regular and easier to compute (broken line). However, since the travel time is stationary, the deviation of the true travel time along the path $\mathbf{r}(s) + \delta\mathbf{r}(s)$ with respect to that predicted for the one following $\mathbf{r}(s)$ is of second order.

than is usually the case for large earthquakes, where the rupture may have a more gradual build-up of energy release. The "onset" time T of the P wave from source to receiver along a path $\mathbf{r}(s)$ with a velocity locally given by $\alpha(\mathbf{r})$ is give by a line integral:

$$T_{\mathrm{rs}} = \int_{\mathbf{r}(s)} \frac{ds}{\alpha(\mathbf{r})} \,. \tag{24.1}$$

In general, we do not know the path $\mathbf{r}(s)$ precisely, since it is a function of the velocity distribution $\alpha(\mathbf{r})$. Instead we use an approximate path calculated for a symmetric reference "background" model such as the Preliminary Reference Earth Model, PREM (Dziewonski and Anderson, 1981), or AK135 (Kennett *et al.*, 1995). The fact that this path is computed for a slightly different velocity distribution (say $\alpha + \delta\alpha$) introduces an error in the predicted time. This "path error" is however of second order because of Fermat's principle (see Figure 24.2).

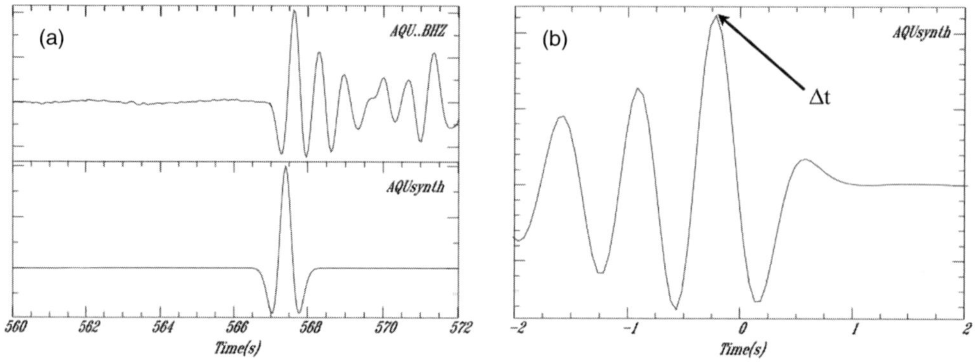

Figure 24.3 (a) The seismogram from Figure 24.1, compared with a "synthetic" arrival predicted from a reference Earth model. The station is AQU (L'Aquila, Italy) and the deflection is, over a broad frequency range, proportional to the ground velocity. (b) The cross-correlation of these two seismograms has a maximum that shows that the predicted P wave arrives about 0.25 s earlier than the observed one.

In practice, we wish to invert for the small deviations $\delta\alpha$ from the reference model. Fermat's principle and a Taylor expansion allows us to linearize (24.1):

$$\delta T_{\rm rs} = \int_{\mathbf{r}(s)+\delta\mathbf{r}} \delta\left[\frac{1}{\alpha(\mathbf{r})}\right] ds \approx -\int_{\mathbf{r}(s)} \frac{1}{\alpha(\mathbf{r})^2}\delta\alpha(\mathbf{r})ds. \tag{24.2}$$

The onset time defines what Brillouin (1960) calls the "signal velocity." For a purely acoustic or elastic wave, energy cannot travel faster than this. Onset times offer the advantage of being easy to interpret using (24.2). They are not always easy to observe, certainly not in the presence of noise. In refraction surveys with artificial sources it may even be impossible to pick the onset unambiguously, and many seismologists revert to picking the first maximum or minimum to define the arrival time. Since they pick a particular phase in the cyclical motion, this defines a "phase velocity." A third velocity definition is provided by the arrival of the maximum in the envelope that defines the sum of kinetic and potential energy (or rather the square root of this sum). Though not commonly in use for the timing of body waves, the "group" velocity defined in this way finds frequent application in the study of surface waves. In a homogenous, isotropic and non-attenuating medium, the signal velocity, group velocity, and phase velocity are identical, and equal to the intrinsic velocity of the homogeneous rock. Figure 24.1 shows the times representing the different velocities for a P wave.

The onset time, if it exists as an observable sharp onset, represents a non-dispersive velocity: it rises only instantaneously if all frequencies arrive at the same time. In practice, this means that the wave spectrum should contain sufficiently high frequencies. If not, the onset will be gradual, it will take time for the signal to arise above the noise, and a (positive) error in the arrival time results. This is one reason why delays measured by cross-correlation are often preferred. The procedure of cross-correlation is illustrated in Figure 24.3. In its most straightforward application, one computes a theoretical or "synthetic" seismogram $u_{\rm th}(t)$ for the body wave that arrives at the time predicted by a "background" model (Figure 24.3a). The cross-correlation of this wave with the observed seismogram $u_{\rm obs}(t)$ is defined as

$$\gamma(t) = \int_{\tau_1}^{\tau_2} u_{\rm obs}(\tau) u_{\rm th}(\tau - t) d\tau \,, \tag{24.3}$$

where $[\tau_1, \tau_2]$ is the interval containing the observed wave. The cross-correlation delay between the observed and predicted wave is defined as the time where $\gamma(t)$ reaches a maximum (Figure 24.3b).

This cross-correlation delay represents neither a group velocity nor a phase velocity. It is a new type of measurement, and requires its own theory to link it to the intrinsic velocity encountered by the wave on its way between source and receiver. Only if the medium, apart from a few discontinuities, is smooth enough that ray theory is applicable, and the wave is not dispersive, can we say that group = phase = onset = cross-correlation delay.

24.2 Interpreting the cross-correlation delay

How can we establish the correct relationship between such cross-correlation delay and the intrinsic velocity of the wave in the medium? For the onset time only an anomaly *on* the ray path $\mathbf{r}(s)$ in (24.2) influences the ray. The ray path has a width that is infinitesimal. But energy arriving in a finite time window $[\tau_1, \tau_2]$ can have traveled elsewhere, and will influence $\gamma(t)$ as long as it is not delayed beyond τ_2.

Mathematically, we can deal with the problem using first-order perturbation theory in which we first compute the zero-order wavefield in the "unperturbed" or "background" medium. To reduce the computational burden, this is usually a simple, spherically symmetric model. The exact solution for the wave excited by a point source in a non-attenuating homogeneous medium with intrinsic velocity α is given by

$$u(t) = \frac{1}{r} u_{\rm src}(t - r/\alpha) \,, \tag{24.4}$$

with $u_{\rm src}(t)$ the "source time function." In analogy with this, the ray theoretical solution for a wavefield from a point source s to a receiver r in a smoothly varying medium can be approximated by

$$u(t) = \frac{1}{\mathcal{R}_{\rm rs}} u_{\rm src}(t - T_{\rm rs}) \,, \tag{24.5}$$

where $\mathcal{R}_{\rm rs}$ is the geometrical spreading that describes the amplitude decrease, and $T_{\rm rs}$ is the travel time from (24.1). In a spherically symmetric medium, analytical expressions for $\mathcal{R}_{\rm rs}$ and $T_{\rm rs}$ exist. For a medium varying in two or three dimensions, one solves for the geometrical spreading using the theory of dynamic ray tracing (Červený, 1985). Attenuation can be dealt with by filtering the wave in the frequency (ω) domain with a low-pass filter of the form $e^{-\omega t^*/2}$, where t^* depends on the distance traveled but is of the order of 1 s for

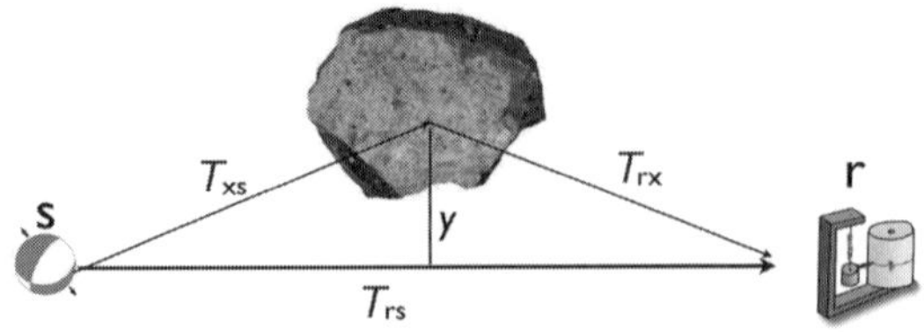

Figure 24.4 This figure illustrates the principle of first-order or Born scattering in a homogeneous background medium. The direct ray from the earthquake S arrives at the seismograph r after T_{rs} seconds. A velocity anomaly x is located at a distance y from the direct ray path, and the wave that scatters from this anomaly arrives after $T_{xs} + T_{rx}$ seconds. If it arrives within the cross-correlation window $[\tau_1, \tau_2]$ it may influence the maximum of the cross-correlation function $\gamma(t)$. Following seismological practice, the earthquake is represented by a beachball in which the dark areas represent the quadrants where the Earth is under tension, while white areas are under compression.

teleseismic P waves, 4 s for S waves, and by correcting α for the dispersion associated with attenuation (Nolet, 2008).

The zero-order wavefield is perturbed if there are deviations from the reference model. These heterogeneities act like secondary sources that emit scattered waves (Figure 24.4). Though the figure shows only one scatterer, the Earth may have many. The "first-order" or "Born" approximation implies that we neglect the possibility that a wave scatters first from one anomaly and then from another before being recorded by the seismograph. We also neglect the (hopefully small) energy loss suffered by the zero-order wavefield from scattering and assume the wave that is incident on each scatterer is the zero-order wave as computed for the background model, unaffected by earlier scattering events. To find the perturbed wavefield we sum over all scatterers.

Since the path length for the scattered wave $\delta u(t)$ is longer (Figure 24.4) it always arrives later than the direct wave, and will not influence the onset time. But if $T_{xs} + T_{rx} < \tau_2$ it may influence the cross-correlation. This is shown in Figure 24.5. For a short pulse that arrives with a delay in excess of the dominant period, the cross-correlation delay is not affected, as shown in the left part of the figure. But a low-frequency pulse as shown on the right does influence the cross-correlation delay. This introduces a dispersion associated with the presence of velocity anomalies. Exploiting the information inherent with this dispersion, by measuring cross-correlation delays in different frequency bands, is the goal of "multiple frequency" tomography. It is another reason to prefer cross-correlation delays over onsets.

One can show that, again to first order, the cross-correlation delay is proportional to the wavefield perturbation δu (Nolet, 2008):

$$\delta T = -\frac{\int_{-\infty}^{\infty} \dot{u}(t')\delta u(t')dt'}{\int_{-\infty}^{\infty} \ddot{u}(t')u(t')dt'}, \tag{24.6}$$

where the dot indicates differentiation with respect to time. From (24.6) it is clear that the delay depends on the shape of $u(t)$ and the scattered waveform $\delta u(t)$, or alternatively, their frequency content.

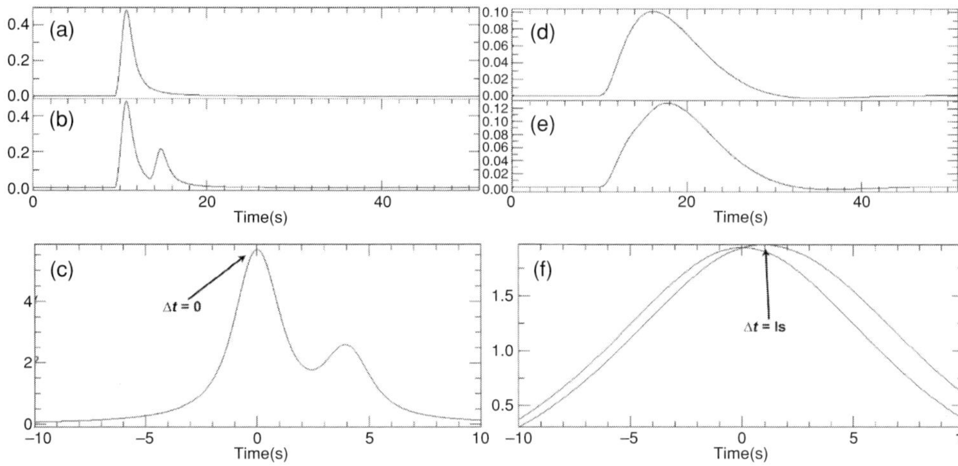

Figure 24.5 (Left) (a) Seismogram in the (homogeneous) background model of Figure 24.4, (b) with scattered wave added, (c) since the wave is high frequency, the scattered wave does not influence the location of the maximum in the cross-correlation. (Right) As before, but now for low frequency, (d) the unperturbed seismogram, (e) with scatterer superimposed, and (f) the resulting shift of 1 s in the cross-correlogram.

Equation (24.6) does not yet give us the desired relationship between perturbations $\delta\alpha(\mathbf{r})$ in the smooth background model and the observed cross-correlation delay δT_{xcor}. For that we use the first-order scattering theory that relates $\delta u(t)$ linearly to $\delta\alpha(\mathbf{r})$. Since δT_{xcor} depends linearly on $\delta u(t)$, which depends linearly on $\delta\alpha(\mathbf{r})$, the result of all these linearizations must also be linear. The delay is given by a volume integral:

$$\delta T_{\text{xcor}} = \int_V K(\mathbf{r})\delta\alpha(\mathbf{r})dV \,. \tag{24.7}$$

An example of the sensitivity kernel $K(\mathbf{r})$ is shown in Figure 24.6.

An analogy with optics may be illustrative at this point. If a light wave hits an obstacle, some of the energy may diffract around it. If this diffracted wave arrives at the sensor with a phase delay less than π, it is said to be in the first Fresnel zone. The width of the Fresnel zone depends on the wavelength of the wave, and therefore decreases with frequency. In our case, the volume occupied by the kernel $K(\mathbf{r})$ depends on the spectrum of $u(t)$. If the bandwidth of the pulse $u(t)$ is wide enough (at least one octave), the width of the kernel $K(\mathbf{r})$ is approximately that of the second Fresnel zone, where the phase delay is between π and 2π.

For very small scatterers, when we are in a regime known as "Rayleigh scattering," with scattered energy proportional to the fourth power of frequency, the perturbed field δu will be enriched in high frequencies. The single-scattering or Born approximation becomes therefore less accurate as we approach higher frequencies. However, for a scale length of 100 km – probably representative for many anomalies in the Earth's upper mantle – the

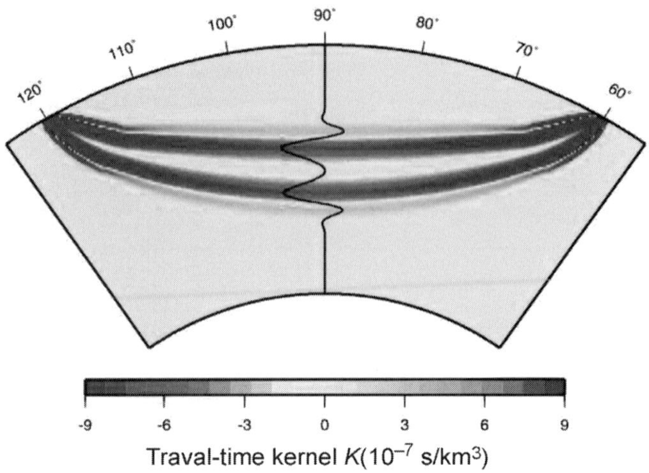

Figure 24.6 Example of a sensitivity kernel $K(\mathbf{r})$ for the cross-correlation delay of an S-wave with a dominant period of 20 s in the upper mantle of the Earth, as calculated for model PREM. The sensitivity at the outer edge of the kernel, which is in the second Fresnel zone, is reversed in sign.

linearity of (24.7) is accurate enough as long as the velocity anomaly remains below about 10% (Mercerat and Nolet, 2013). In fact, if we sum over many small scatterers, only low frequencies will interfere constructively. Large anomalies can be seen as the sum of many small ones, and the spectral content of the scattered wave becomes much more like that of the incoming wave.

A remarkable, and at first sight counterintuitive feature of the kernel in Figure 24.6 is that the sensitivity is zero where the ray path for the onset time is located. This can be understood if one imagines the anomaly in Figure 24.4 to be located on the direct ray path (i.e., $y = 0$). In that case δu arrives at the same time as the zero-order wave, and if its waveform is the same as that of $u(t)$, all it can do is change the amplitude of the direct wave.

24.3 Solving underdetermined inverse problems

We may have many observed delays δT_i, each one giving one constraint of the form $\delta T_i = \int K_i(\mathbf{r})\delta\alpha(\mathbf{r})dV$. To solve this, we parameterize the model by developing the seismic velocity in terms of basis functions h_j: $\delta\alpha(\mathbf{r}) = \sum_j m_j h_j(\mathbf{r})$. There is a large choice of basis functions, but the most commonly used are voxels ($h_j = 1$ in voxel j and 0 outside) and spherical harmonic functions. Wavelets are gaining increased attention, since they are localized in space like voxels but oscillate with a dominant wavelength much like spherical harmonics. Wavelets thus offer a compromise between a "local" resolution with voxels and a "spectral" resolution with spherical harmonics (Simons *et al.*, 2011). In extraterrestrial

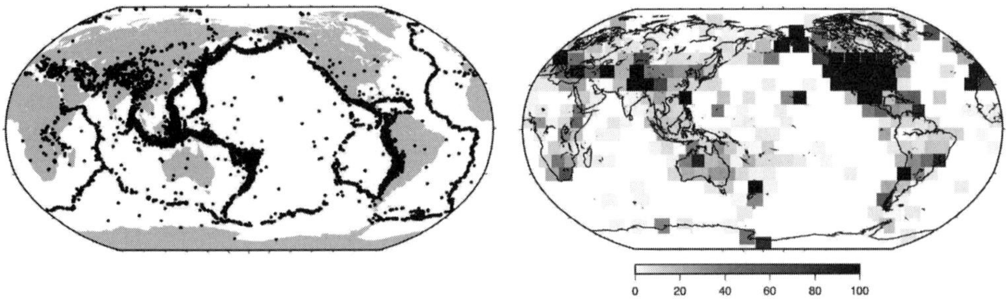

Figure 24.7 (Left) Global distribution of strong earthquakes (magnitude>5) from 2000 to 2014. (Right) Density of seismic stations present in the database of the IRIS Data Management Center. This database is somewhat incomplete, especially for Asia, but nevertheless clearly shows the lack of data in the oceanic domain when compared to the stations on the continents. The grayscale shows the number of stations per 10^6 km^2.

seismology we shall have only few data, and the $h_j(\mathbf{r})$ may represent homogeneous shells. But whatever the choice of basis, the result of parameterization is a linear system of the form

$$\boldsymbol{Am} = \boldsymbol{d} \tag{24.8}$$

with data $\boldsymbol{d} = (\delta T_1, \delta T_2, ..., \delta T_N)$, model $\boldsymbol{m} = (m_1, m_2, ..., m_M)$ and matrix elements

$$A_{ij} = \int K_i(\mathbf{r})h_j(\mathbf{r})dV. \tag{24.9}$$

The number of seismographs capable of permanently recording seismic events at large distances on Earth is not exactly known. It is probably of the order 10^4, almost all of them on land. Many of these are high-quality, digital, broad-band seismographs that are rapidly replacing more classical instrumentation. To this should be added the stations that are part of the deployment of temporary networks for geophysical exploration of the deep Earth. A recent (2012) count of the instruments listed in the database of the IRIS Data Management Center in Seattle, which distributes digital seismograms to the academic community, gave a total of 13,986 stations (Figure 24.7).

The distribution of seismic sources is governed by the configuration of rigid plates covering the Earth and this distribution is far from uniform – earthquakes occur preferentially at or near the plate boundaries (Figure 24.7). For S waves, the addition of surface waves may help significantly to resolve the structure in the lithosphere and upper mantle, down to about 600 km depth (surface waves are dominated by multiply reflecting S waves). A popular recent global model for shear wave velocity is S40RTS (Ritsema *et al.*, 2011). Figure 24.8 shows this model at a depth of 200 km. At this depth it is evident that the oceans are characterized by slow velocity, indicating higher temperatures, whereas the older parts of the continents are fast.

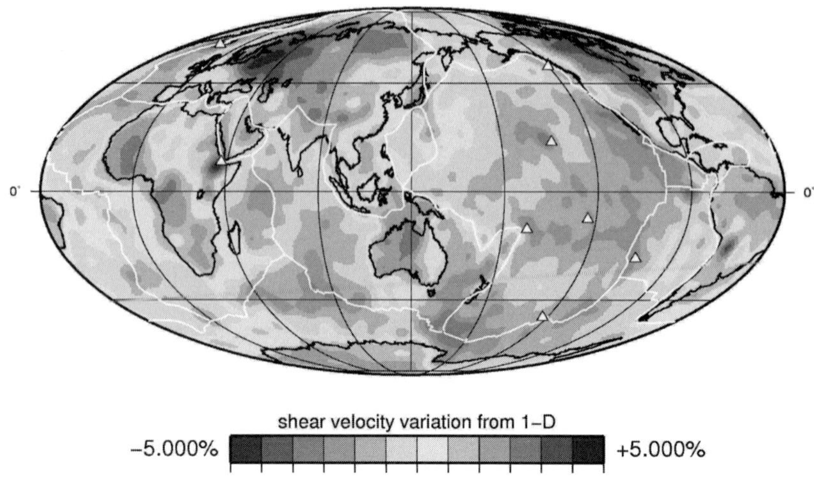

Figure 24.8 A map view of S velocity anomalies (with respect to the PREM model) at a depth of 200 km in global model S40RTS (Ritsema *et al.*, 2011). A black and white version of this figure will appear in some formats. For the color version please refer to the plate section.

Even with the addition of surface waves we face a very uneven coverage. Whereas we can reach a high resolution in some regions, we lack useful resolution in other areas, such as the mantle under the Pacific Ocean. One option is to limit the tomographic images to very long wavelengths only. However, the problem with that approach is that many features of interest in the Earth have small length scales that sometimes, though not always, can be resolved. The preferred strategy is therefore to allow for the linear system to be *under*determined: some small length scales are allowed by the parameterization but not constrained by the data. We are forced to constrain the unresolved parameters in some way, most commonly by looking for the smoothest model that satisfies the data. This is fundamentally a subjective constraint, because there is no geophysical evidence that would prefer smooth models over rapidly varying ones. We may also introduce objective constraints not present in the data set: We know for example the range of velocities permitted by silicates and can constrain the m_j to some range of values, a technique known as "damping," because it generally biases $\delta\alpha(\mathbf{r})$ towards zero. A special form of damping is to search for a model that can be represented by the minimum number of wavelets and still satisfy the data within their estimated errors ("compressed sensing") – this can potentially retain sharp boundaries while smoothing over ill-resolved regions (Simons *et al.*, 2011). Collectively, smoothing and damping efforts are known as "regularization" techniques.

A simple way to study the bias introduced by regularization is to compute a synthetic data vector for a known model, add random errors to it, invert for a regularized inversion of such data, and compare the resulting model with the one that generated the data. If such resolution tests are repeated with different realizations for the errors, one can also study the effect of data errors on the solutions.

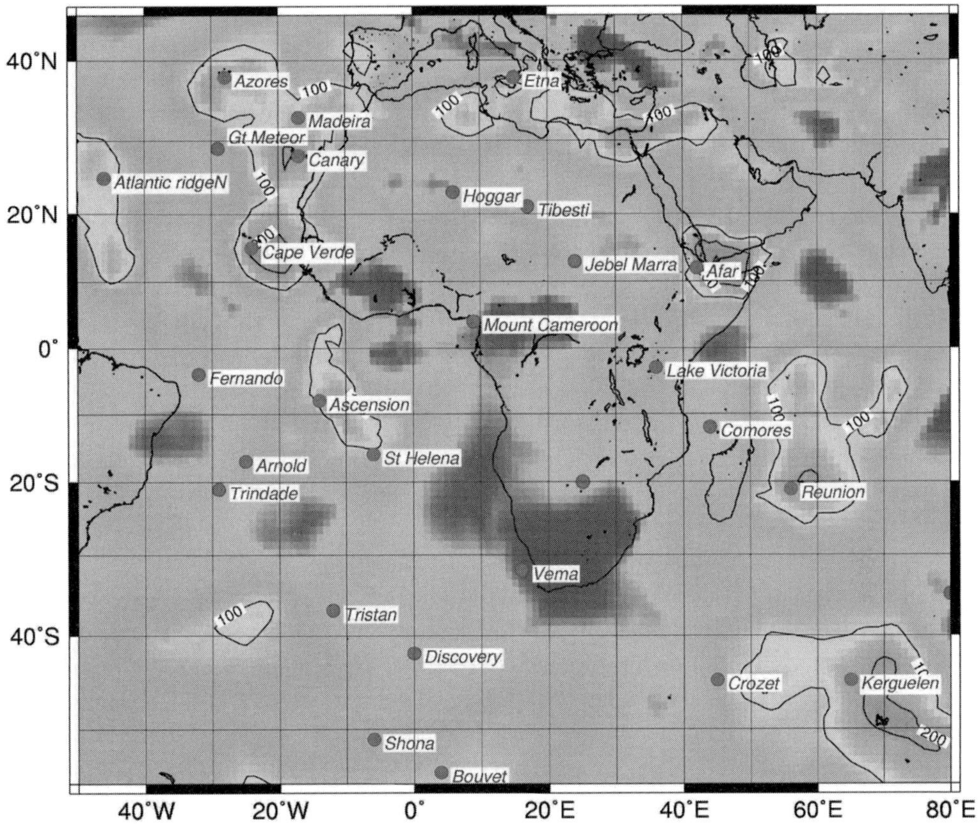

Figure 24.9 A map view of model PRI-P05 (Montelli *et al.*, 2006) converted into a temperature anomaly at a depth of 800 km, showing a correlation between high thermal anomalies of 100–200 K (yellow/orange) and the location of "hotspots" with high observed heat flux at the surface (red dots), indicating plumes rising from the lower mantle. A black and white version of this figure will appear in some formats. For the color version please refer to the plate section.

24.4 Epilogue

Even if future space missions succeed in putting seismometers on other planets or moons in the Solar System, the number of these will always be small, and we shall be happy if we can obtain the first-order radial structure of a planet like Mars. In the absence of plate tectonics, seismic activity is mainly due to tidal strains and volcanic activity. The magnitudes of such sources are limited and the probability that we may observe free oscillations is small. Thus, our spatial sampling of the interior of such bodies is limited to the path traveled by the waves between the sources and the few stations that are available. If all we can observe are onset times, our information on the radial velocity structure is limited to that given by a few ray paths. Our only means of optimizing the depth coverage is to estimate delays in several frequency bands, such that we have information over a wide range of

Fresnel volumes, averaging over different depth intervals, for each source-station pair. Such "finite-frequency" methods, recently developed in global seismic tomography for imaging in three dimensions, therefore also offer the tool of choice for a one-dimensional (depth) interpretation of extraterrestrial seismic data. Finite-frequency tomography proved its worth on Earth by providing the first images of thermal plumes rising in the lower mantle (Figure 24.9, Montelli *et al.*, 2006).

In helioseismology, finite-frequency methods are also being used (Birch *et al.*, 2001). Paradoxically, by observing Doppler shifts, helioseismologists can sample the wavefield at any place on the visible surface of the Sun with a "station" density that is far superior to that available to terrestrial seismologists, who are hampered by the limited number of stations on land, and the virtual absence of stations in the oceanic domain. The Helioseismic and Magnetic Imager (HMI) satellite, launched on February 11, 2010, is equipped with two 4096 pixel cameras that observe Doppler shifts in the Fe I absorption line (6173 Å), enabling us to measure the vibrations of the gas in the photosphere. This mission follows on from the Michelson Doppler Imager aboard the Solar and Heliospheric Observatory (SOHO).

It is not just the planets that pose a challenge for seismologists to obtain crucial data. The lack of seismic stations in the oceanic domain visible in Figure 24.7 poses a major obstacle to the imaging of mantle plumes. To fill this void, seismologists have begun to develop robots that float at 1–2 km depth (Simons *et al.*, 2009; Hello *et al.*, 2011). Equipped with a hydrophone, these "floats" are able to record the acoustic signals in the water column that are generated when an incoming P wave from an earthquake at large distance hits the bottom of the ocean. This concept can be extended beyond the Earth. In NASA's Titan Saturn System Mission, a hot air balloon or "aerobot" is planned to circumnavigate Titan. It could be equipped with a microphone to pick up seismic waves in Titan's atmosphere. Jupiter and Venus are also obvious candidates for aerobots – in fact, two Russian weather balloons floated briefly in the Venus atmosphere in 1985.

The theory of elastic wave propagation needed to interpret seismograms in terms of velocity models (and thus tomographic images) has developed rapidly and can be considered mature. As computing resources become more and more powerful, advances are to be expected in the interpretation of waveforms, which allow us to extract all of the information that is available through seismic measurements. This will also allow us to get as much information out of seismometers on other planets as we can. On Earth, we expect that significant progress will still be obtained by densifying station arrays on the continents and by moving into the hitherto almost inaccessible oceanic domain.

25

Interpreting cross-correlations in seismology

SHRAVAN HANASOGE

25.1 Introduction

Wavefields in the Sun and Earth, measured using various instruments, display small-amplitude fluctuations, essentially zero-mean random processes. More careful investigation reveals that these are seismic fluctuations generated by stochastic processes such as convection in the Sun and by anthropogenic activity and non-linear ocean wave interactions on Earth. The distinguishing feature of imaging with noise is that we do not know the dynamical details of the source (unlike in classical earthquake tomography) and we consequently seek a statistical description of the wavefield.

Here, we discuss the interpretation of the cross-correlation measurement in solar and terrestrial scenarios. The most common measurement in the context of noise is the ensemble-averaged cross-correlation. The physics of such measurements can be markedly different from raw wavefield measurements (as in classical tomography). The helioseismic theory of stochastic oscillations in conjunction with computational adjoint optimization, widely used in terrestrial seismic applications, provides a unified framework for addressing outstanding inverse problems in these areas. The seismic methodologies and challenges in each of these fields, despite the vastly different physics, are intimately related and we highlight these similarities in this article.

25.1.1 Seismology of the Sun

The Sun, our nearest star, serves as an astrophysical benchmark, contributing to our understanding of stellar evolution, stellar interiors and atmospheres, etc. Life on Earth is directly affected by the magnetic variability of the Sun. Our climate is sensitive to variations in the irradiance of the Sun caused by cyclic magnetic activity. Space instrumentation and telecommunications are susceptible to solar high-energy eruptive events (e.g., for reviews, see, Schrijver and Zwaan, 2000). Predictive models of solar magnetic activity (e.g., Pesnell, 2008) depend on our knowledge of internal global-scale fluid motions of the Sun, constraints that can only be obtained seismically.

Extraterrestrial Seismology, eds. Vincent C. H. Tong and Rafael A. García. Published by Cambridge University Press.

Subsequent to the discovery of oscillations on the surface of the Sun some 50 years ago (Leighton *et al.*, 1962), there has been significant progress in uncovering its properties. Helioseismology (see Christensen-Dalsgaard, 2002; Gizon and Birch, 2005; Gizon *et al.*, 2010, for reviews), like its terrestrial analog, is a collection of methods applied to infer the interior structure and dynamics of the Sun through the interpretation of surface observations of its oscillations. The earliest efforts in this field were focused on determining the background solar stratification and internal rotation. With the development of data analysis techniques such as ring analysis, acoustic holography, and time-distance helioseismology, it became possible in theory to study the 3D structure and dynamics of the Sun (for reviews, see, e.g., Gizon and Birch, 2005; Gizon *et al.*, 2010, and references therein).

Theory turned to reality with the launch of the Solar Oscillations Investigations/Michelson Doppler Imager (SOI/MDI) instrument (Scherrer *et al.*, 1995) onboard the Solar and Heliospheric Observatory (SOHO) satellite, leading to numerous discoveries of the dynamics of local inhomogeneities such as sunspots and convective flow systems. The successor to MDI, the Helioseismic and Magnetic Imager (HMI; Schou *et al.*, 2012) instrument onboard the Solar Dynamics Observatory (SDO), takes much higher quality, better resolved observations of the visible full disk of the Sun, with fewer sources of systematic errors. Simple models of solar structure where the stratification is only in the radial direction are in excellent agreement with those derived from stellar evolution codes, and the predicted resonant modes of oscillation closely match observed modal frequencies (e.g., Christensen-Dalsgaard, 2002). However the 3D Sun still remains a challenge, with a number of outstanding problems such as the structure and dynamics of sunspots, the spectrum of interior convection (e.g., Hanasoge *et al.*, 2012b; Gizon and Birch, 2012) and internal meridional circulation. With the increasing availability of computational resources, demand for greater accuracy in the interpretation of helioseismic measurements and the advent of billion-dollar instruments taking high-quality observations, computational methods of helioseismology that enable probing the 3D structure of the Sun have increasingly become relevant.

25.1.2 The helioseismic wavefield

Overturning supersonic turbulent convection, driven by radiative thermal losses at the photosphere (surface), is the cause of wave generation in the Sun. It is well modeled by a stationary and laterally homogeneous random process (e.g., Woodard, 1997; Gizon and Birch, 2002), where source location, amplitude, and excitation phase are random variables. Since the exact distribution of all relevant sources of wave excitation is unknown, raw time series of wavefield velocities contain no useful seismic information. Second-order correlations of the stochastic wavefield produce statistically significant measurements (Duvall *et al.*, 1993b). The interplay between wave damping, mode inertia, and the excitation process results in a helioseismic modal power spectrum that peaks around 3 mHz.

Woodard (1997) and Gizon and Birch (2002) were among the first to develop the theoretical framework to interpret cross-correlation measurements in the scenario where the sources are stochastic. The basic idea is to relate a seismic measurement, such as a travel time or amplitude, to the properties of the underlying medium, i.e., to pose an *inverse problem* that requires manipulating observations to deduce properties of the medium. The typical inverse problem in seismology may be described by

$$\int d\mathbf{x}\, K(\mathbf{x})\, \delta m = \delta\mu, \tag{25.1}$$

where the measurement $\delta\mu$, termed the misfit, denotes the deviation of a seismic measurement from some prediction, δm is the correction to our model of the medium m that results in a reduction of the misfit, $K(\mathbf{x})$ is the gradient of the misfit with respect to the model, also termed a sensitivity kernel, and the integration is over the relevant spatial coordinate $\mathbf{x}$. A principal challenge lies in computing the sensitivity kernel $K(\mathbf{x})$, since, in a 3D problem, it may represent the gradient of the misfit with respect to hundreds of thousands of parameters, e.g., the sound-speed, Alfvén vector, and flow velocities in the Sun (Hanasoge *et al.*, 2012a).

25.1.3 Seismic fluctuations of Earth

Terrestrial seismic noise is generated at a range of temporal frequencies, by human activity, storms, oceanic wave microseisms (Longuet-Higgins, 1950), and the ocean-excited low-frequency hum of Earth (e.g., Nawa *et al.*, 1998). Seismic noise is used as a compelling alternative to earthquake tomography to image the crust. Most importantly, it enables the study of temporal variations of the crust (e.g., Wegler and Sens-Schönfelder, 2007) and volcanoes (e.g., Brenguier *et al.*, 2007).

Under ideal conditions of azimuthally uniform illumination, representation theorems (e.g., Snieder *et al.*, 2010) state that the cross-correlation is essentially the Green function between the stations. In other words, an equally weighted sum of Green functions between every source (over all space; constant amplitude) and the stations is equivalent to the Green function between the stations. This approximation allows for the treatment of cross-correlations using earthquake-tomographic methods. However, Earth noise is typically anisotropic, and in such a scenario this correspondence does not hold. Consequently a more rigorous theory that fully addresses the physics of cross-correlations is necessary.

25.2 Cross-correlations

Correlation measurements, especially in the scenario of noise tomography, are complicated by the confluence of disparate concepts such as ensemble averaging, stochastic excitation, mode equipartition, multiple distributed sources, and the cross-correlation itself. This may lead to confusion in appreciating the physical flavor of the correlation measurement, different from the classical tomographic analog, i.e., wavefield displacement. An obvious difference is the time variable, which in classical tomography is the source–receiver propagation delay, as opposed to cross-correlation tomography, where it is connected to the relative propagation delay between the source(s) and the two receivers.

The mathematical theory of cross-correlations is much easier to follow in the temporal Fourier domain, and hence we work only with transformed quantities. The cross-correlation between two time series turns into a product in the temporal Fourier domain, rendering the analysis somewhat simpler (for a more detailed introduction, see, e.g., Nolet, 2008). Let the Green function for a given medium and for a given wave operator be denoted by $G(\mathbf{x}, \mathbf{x}'; t)$, where the source is at point $\mathbf{x}'$, receiver at $\mathbf{x}$, and time is t. The Fourier transform of the Green function is written as $G(\mathbf{x}, \mathbf{x}'; \omega)$, and to lighten the notational burden, we cease to explicitly state frequency ω unless required. The wavefield $\phi(\mathbf{x}, \omega)$ excited by some finite realization of sources $S(\mathbf{x}', \omega)$ is described by Green's theorem:

$$\phi(\mathbf{x}) = \int d\mathbf{x}'\, G(\mathbf{x}, \mathbf{x}')\, S(\mathbf{x}'), \tag{25.2}$$

where we have ceased altogether to explicitly state the dependence on frequency ω.

25.2.1 A deterministic source

The simplest case to start with would be the interpretation of the cross-correlation of the wavefield between a pair of stations due to a deterministic source. Consequently, this is not a statistical measurement and the theory is not hampered by the additional layers of complication. For further simplicity, let us choose a point source at $\mathbf{x}_s$, $S(\mathbf{x}') = \delta(\mathbf{x}' - \mathbf{x}_s)\, \mathcal{R}(\omega)$, where $\mathcal{R}(\omega)$ is the Fourier transform of the source-time function. The wave fields at two stations α, β and the cross-correlation $\mathcal{C}(\omega)$ are given by

$$\phi(\mathbf{x}_\alpha) = G(\mathbf{x}_\alpha, \mathbf{x}_s)\, \mathcal{R}, \qquad \phi(\mathbf{x}_\beta) = G(\mathbf{x}_\beta, \mathbf{x}_s)\, \mathcal{R}, \tag{25.3}$$

$$\mathcal{C}(\omega) = G^*(\mathbf{x}_\alpha, \mathbf{x}_s)\, G(\mathbf{x}_\beta, \mathbf{x}_s) |\mathcal{R}|^2. \tag{25.4}$$

It is seen that Equation (25.4) contains a product of Green functions. This is the basic difference between a classical wavefield measurement and the cross-correlation. We revisit this equation again at the end of this section.

25.2.2 Stochastic sources

For a given realization of sources, the finite-time estimate of the cross-correlation $\mathcal{C}(\omega)$ between the wavefields measured at two points α, β may be rewritten in terms of Green functions:

$$\mathcal{C} = \int d\mathbf{x}' \int d\mathbf{x}\, G^*(\mathbf{x}_\alpha, \mathbf{x})\, G(\mathbf{x}_\beta, \mathbf{x}')\, S^*(\mathbf{x})\, S(\mathbf{x}'). \tag{25.5}$$

Ensemble averaging over many realizations of sources moves us closer to the expected value of the cross-correlation (assuming it exists). The expected value of the cross-correlation, i.e., the infinite-time estimate of (25.5), denoted as $\langle \mathcal{C} \rangle$, is given by

$$\langle \mathcal{C} \rangle = \int d\mathbf{x}' \int d\mathbf{x}\, G^*(\mathbf{x}_\alpha, \mathbf{x})\, G(\mathbf{x}_\beta, \mathbf{x}')\, \langle S^*(\mathbf{x})\, S(\mathbf{x}') \rangle. \tag{25.6}$$

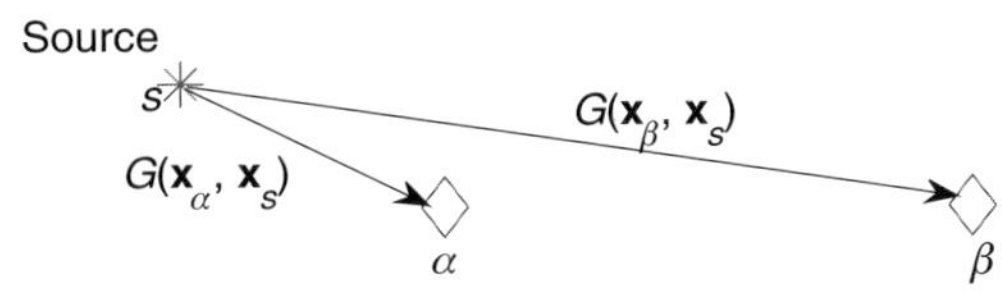

Figure 25.1 Pictorial depiction of cross correlations due to a source at $\mathbf{x}_s$, as described in Equations (25.4) and (25.7). A source excites waves which propagate to the two stations α, β. The cross-correlation will show energy at time lags corresponding to the relative travel delay between the sources and the two stations.

Source distributions are typically assumed to be spatially uncorrelated (not necessarily true but it renders the theory computable), i.e., $\langle S^*(\mathbf{x})\, S(\mathbf{x}')\rangle = \mathcal{P}(\omega)\,\delta(\mathbf{x}-\mathbf{x}')\,\sigma(\mathbf{x})$, where $\sigma(\mathbf{x})$ is the distribution of source amplitudes and $\mathcal{P}(\omega)$ is the temporal power spectrum of the oscillations. This reduces the cross-correlation to the following form:

$$\mathcal{C} \approx \langle \mathcal{C} \rangle = \int d\mathbf{x}\, G^*(\mathbf{x}_\alpha, \mathbf{x})\, G(\mathbf{x}_\beta, \mathbf{x})\, \sigma(\mathbf{x})\, \mathcal{P}(\omega), \tag{25.7}$$

where the assumption is that the finite-time cross-correlation approximates the expected value. A pictorial depiction of the meaning of this equation is shown in Figure 25.1.

We draw the attention of the reader to the similarity between Equations (25.4) and (25.7), which are identical when $\sigma(\mathbf{x}) = \delta(\mathbf{x} - \mathbf{x}_s)$ (a scenario where the stochastic source is a point source). Thus, the interpretation of cross-correlations in seismology is the same for deterministic and stochastic sources.

25.3 Adjoint optimization

Real-world inverse problems can be functions of large numbers of parameters and, as a consequence, ill-posed and computationally expensive. In problems of seismology, especially for travel-time tomography, prominent arrivals (such as the f and p_n acoustic waves in the Sun and the various P and S wave arrivals on Earth) are well understood. The implication is that we have an accurate starting model and the use of a gradient-based method to gradually improve the model so as to march towards a misfit minimum is justifiable. However, when adopting a multi-iteration inversion strategy, the presence of 3D inhomogeneities requires that the evaluation of kernels proceed by numerical means.

The adjoint method provides an algorithm that renders the evaluation of sensitivity kernels computationally tractable. The method of adjoint optimization, possibly introduced in its current form first by Lions (1971), is widely used in the control of fluid flow, airfoil design, meteorological prediction, global helioseismology, and terrestrial seismology (e.g., see Tromp *et al.*, 2010; Hanasoge *et al.*, 2011, for a full list of references).

Typical measurements in seismology are energy, waveform difference, and travel times, all of which may be written as linear functionals of the cross-correlation. In order to relate the cross-correlation to model parameters, we study variations of the cross-correlation. This

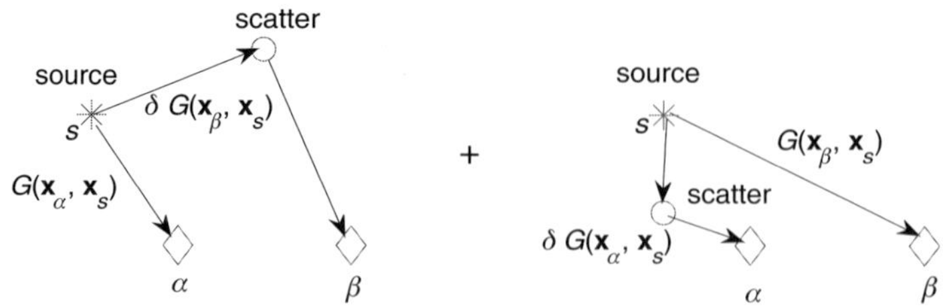

Figure 25.2 Single scattering as captured by cross-correlations, a pictorial depiction of Equation (25.9). The term "scatter" signifies our imperfect knowledge of the structure model, translating to an uncertainty in our expression for the Green function, i.e., $\delta G(\mathbf{x}_\alpha, \mathbf{x}_s)$ and $\delta G(\mathbf{x}_\beta, \mathbf{x}_s)$. This results in a deviation in the prediction from measurement. The first contribution (left of the plus sign) consists of a correlation between a wave generated at point $\mathbf{x}_s$ propagated to station α and a wave to location $\mathbf{x}_s$, where our knowledge of the structure model is inaccurate, which is thereby "scattered" and arrives at measurement point β. The second contribution (right of the plus sign) is essentially the same except with points α, β reversed (also see, e.g., Hanasoge *et al.*, 2012b). We only consider single scattering here; with higher-order scattering terms, the theory becomes tedious and very complicated.

variation attempts to address the difference between prediction and observation, and arises from our imperfect knowledge of the structure and sources:

$$\delta C = \delta C_{\text{structure}} + \delta C_{\text{source}}. \tag{25.8}$$

Using Equation (25.7), we derive expressions for these quantities:

$$\delta C_{\text{structure}} = \int d\mathbf{x}' \, [\delta G^*(\mathbf{x}_\alpha, \mathbf{x}')\, G(\mathbf{x}_\beta, \mathbf{x}') + G^*(\mathbf{x}_\alpha, \mathbf{x}')\, \delta G(\mathbf{x}_\beta, \mathbf{x}')]\, S(\mathbf{x}')\mathcal{P}, \tag{25.9}$$

$$\delta C_{\text{source}} = \int d\mathbf{x}' G^*(\mathbf{x}_\alpha, \mathbf{x}')\, G(\mathbf{x}_\beta, \mathbf{x}')\, \delta S(\mathbf{x}')\, \mathcal{P}, \tag{25.10}$$

where Equation (25.9) reflects our imperfect knowledge of the structure model and Equation (25.10) encapsulates the error in our source model. Structural deviations as measured by cross-correlations are complicated; the physical meaning of Equation (25.9) is depicted in Figure 25.2.

25.4 Imaging magnetoconvection in the Sun

Sunspots are prominent features, thought to be manifestations of magnetoconvection in the Sun. Imaging their subsurface structure and dynamics is an outstanding problem and is of broad physical import. Typically, observations of travel times of waves that propagate through sunspots are used as inputs in inversions. Despite the presence of strong anisotropies induced by magnetic fields, these measurements are interpreted in terms of changes to an effective isotropic wavespeed (i.e., sound speed) and plasma flows (e.g., Gizon and Birch, 2005).

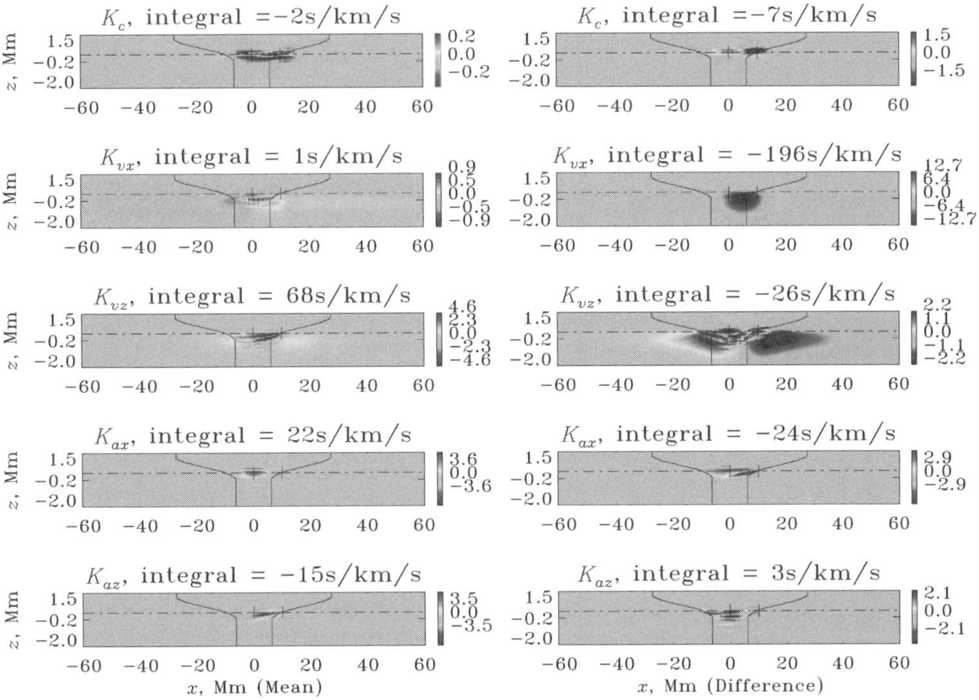

Figure 25.3 f-mode (surface) wavespeed kernels for a point-pair separated by a distance of 10 Mm. The rows display sound speed, flow, and anisotropic (Alfvén) wavespeeds, in that order. Kernels on the left column are for the mean travel time (the average of the negative and positive branch times) between the point pair, whereas the panels on the right column show kernels for the difference travel time (the difference between the negative and positive branch times). The horizontal dot-dash line denotes the height at which observations are made in the quiet Sun and the symbols mark the measurement points. The f-mode is seen to be significantly affected by the spot, as seen in the loss in symmetry of the kernels. Signatures of magnetoacoustic slow and fast modes and hints of conversion to acoustic p_1 may be discerned upon close examination. (Figure taken from Hanasoge *et al.*, 2012a, copyright (2012) by the American Physical Society.) A black and white version of this figure will appear in some formats. For the color version please refer to the plate section.

Adjoint optimization may be used to determine the appropriate parameterization of the structural properties of the solar magnetic interior. Hanasoge *et al.* (2012a) showed that seven different wavespeeds fully characterize the inverse problem in helioseismology: the isotropic sound speed, a flow-advection three-component vector velocity and a three-component vector anisotropic Alfvén velocity. The structure of magnetic media is sensed by compressible slow and fast magnetoacoustic modes and incompressive Alfvén waves, each of which can propagate at a different wavespeed. Hanasoge *et al.* (2012a) showed that even in the case of weak magnetic fields, significant errors may be incurred in the inferences if magnetic anisotropies are not carefully accounted for. In Figure 25.3, the sensitivity of the surface f-mode of the Sun to an embedded magnetic flux tube is depicted.

With kernels such as these, computed using the adjoint method, the accurate imaging of magnetoconvection in the Sun becomes plausible. A significant issue with magnetic regions is that translation invariance is lost, i.e., we do not have the luxury of translating kernels from one horizontal position to another. In other words, each pixel is privileged, placing an onerous computational price on performing the inversion, since one must in principle evaluate kernels around each observational pixel. This is an impossible feat in helioseismology owing to the vast numbers of observations. MDI alone records velocities at approximately 1 million points on the solar photosphere, and with the advent of HMI, each wavefield image comprises more than 16 million points. Computing kernels around every one of these points is neither computationally feasible nor is there sufficient independent information to justify such a massive calculation.

Adjoint methods address this issue by privileging some pixels over others, i.e., creating a collection of sources and receivers, such that the computational cost scales only with the number of sources. The procedure, a detailed account of which may be found in Hanasoge *et al.* (2011), allows for the assimilation of cross-correlations of sources with vast numbers of receivers. Indeed, this may prove to be extremely useful in the helioseismic scenario, especially when imaging the structure and dynamics of sunspots.

25.5 Imaging the Earth with surface waves

Terrestrial seismic cross-correlations, unlike their helioseismic cousins, consist almost solely of surface-wave signals. Inference of Earth's surface structure from these measurements focuses on measuring travel times and connecting these to the average wavespeed of the underlying medium (e.g., Shapiro *et al.*, 2005). Therefore, for the purposes of discussion here, we consider waves propagating on a 2D surface, governed by a simple 2D wave equation:

$$\rho\partial_t^2\phi + \rho\Gamma\partial_t\phi - \boldsymbol{\nabla}\cdot(\rho c^2\boldsymbol{\nabla}\phi) = S(\mathbf{x}, t), \tag{25.11}$$

where ρ is density, Γ attenuation, $\mathbf{x}$ a 2D space, t time, ϕ the wave displacement, $\boldsymbol{\nabla} = (\partial_x, \partial_y)$ the covariant spatial derivative, $S(\mathbf{x}, t)$ the source, and c wavespeed. For the simple case considered here, we assume a single mode with constant wavespeed c. The Green function $G(\mathbf{x}, \mathbf{x}'; t)$ for the displacement at $(\mathbf{x}, t)$ due to a spatio-temporal delta source at $(\mathbf{x}', 0)$ is the solution to

$$(\rho\partial_t^2 + \rho\Gamma\partial_t - \rho c^2\nabla^2)G(\mathbf{x}, \mathbf{x}'; t) = \delta(\mathbf{x} - \mathbf{x}')\delta(t). \tag{25.12}$$

This equation is explicitly solvable; the Green function in the temporal Fourier domain is given by

$$G(\mathbf{x}, \mathbf{x}'; \omega) = \frac{1}{\rho}H_0^{(1)}\left(\frac{\omega}{c}|\mathbf{x} - \mathbf{x}'|\sqrt{1 + i\frac{\Gamma}{\omega}}\right), \tag{25.13}$$

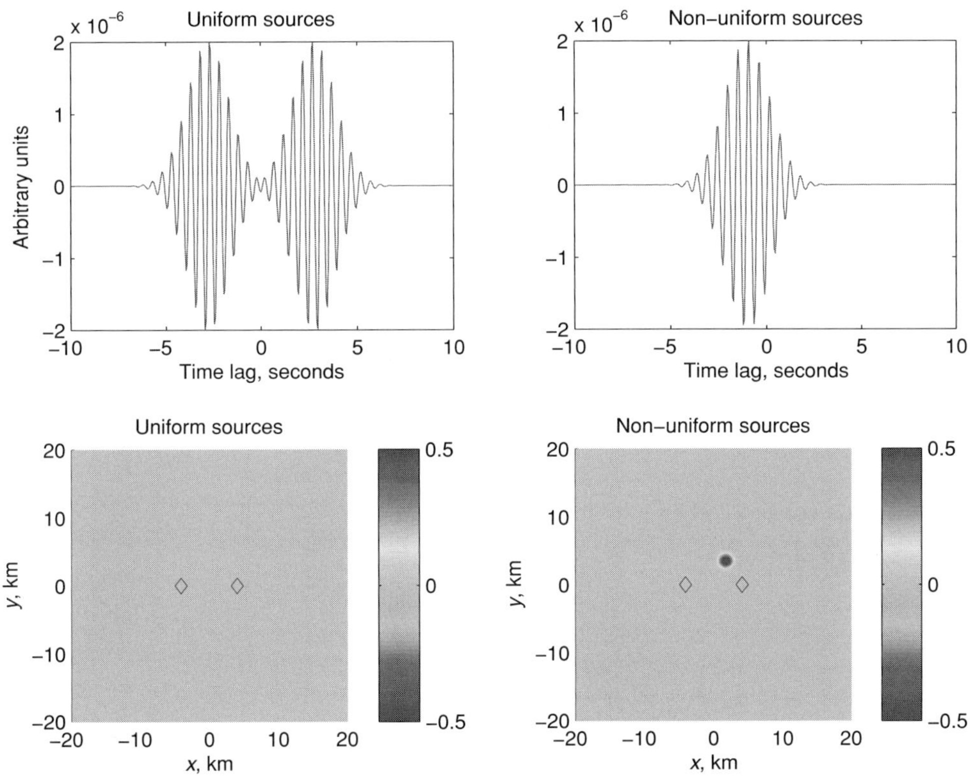

Figure 25.4 Cross-correlation (upper panels) for the source distributions in the corresponding bottom panels with stations marked by the diamonds. For the uniform model of sources, the cross-correlation has significant energy at both positive and negative time lags, peaking approximately at the time taken to travel between the two stations. However, the highly non-uniform distribution on the lower-right panel excites waves that arrive at the two stations nearly simultaneously, i.e., the difference in paths between the spot and the two stations is nearly zero. The cross-correlation is thus almost (but not quite) centered around zero time. These sorts of interpretations are very useful in terrestrial noise tomography where sources are indeed highly non-uniform. (Reproduction of Figure 7 of Hanasoge *et al.*, 2012b, Oxford University Press.) A black and white version of this figure will appear in some formats. For the color version please refer to the plate section.

where ω is temporal frequency and $H_0^{(1)}$ is the Hankel function of the first kind. A significant issue in Earth noise tomography is the anisotropy of wave excitation, typically not properly accounted for. Based on this simple formulation, we show examples of the sensitivity of the cross-correlation to the input source distribution in Figure 25.4.

25.5.1 A toy inverse problem

Cross-correlation measurements are commonly interpreted using theories that are meant for classical tomography. This can result in a systematic theoretical bias in the inversion

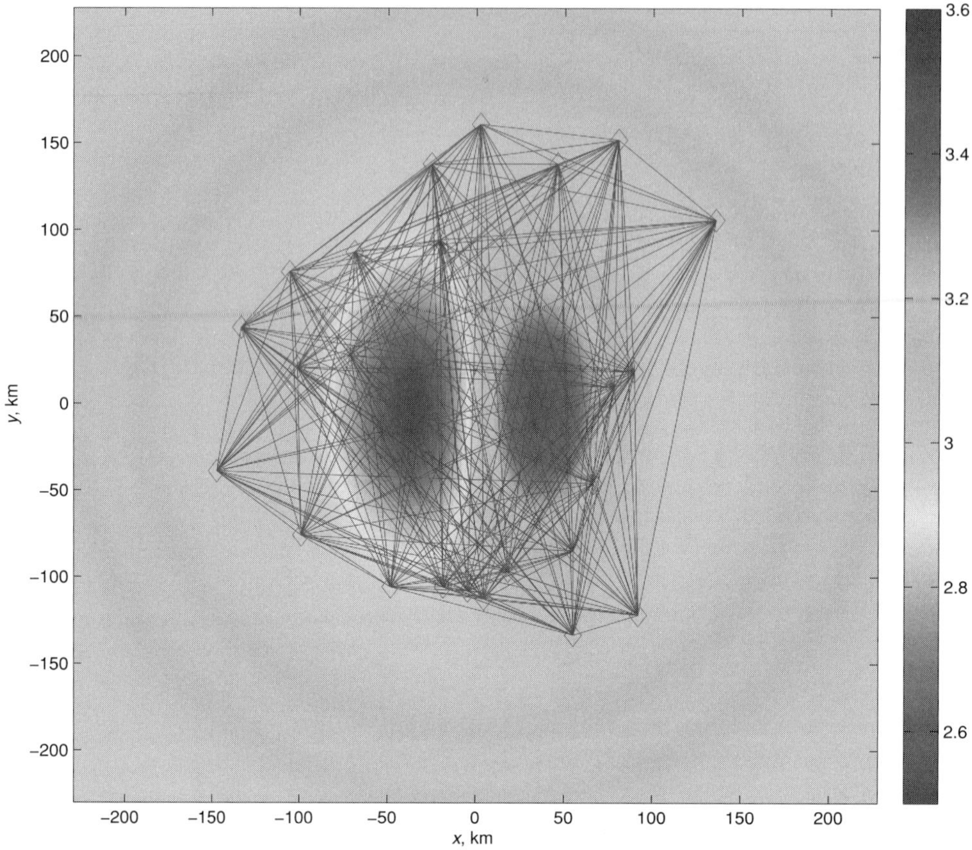

Figure 25.5 A toy problem to investigate the fidelity of cross-correlation tomography in a typical geophysical scenario. The diamonds indicate receivers surrounded by a ring of sources (the light-blue ring). The question is whether the $\pm 20\%$ wavespeed perturbation within the network can be successfully imaged. The ray-path coverage associated with the network is excellent, suggesting that the eventual inversion will likely be good. (Reproduction of Figure 5 of Hanasoge, 2014, Oxford University Press.). A black and white version of this figure will appear in some formats. For the color version please refer to the plate section.

(Hanasoge, 2014). Further, the anisotropy of illumination is an issue that can hamper inferences. With a theory of cross-correlations (e.g., Tromp *et al.*, 2010), we may avoid these pitfalls and fully account for finite-frequency effects. In a simple toy example illustrated in Figure 25.5, we consider a network of stations surrounding a $\pm 20\%$ wavespeed perturbation, illuminated by a ring of sources outside the network. In this ideal scenario, where the ray coverage of the anomaly and the illumination are excellent, we ask whether the central anomaly can be accurately inferred and investigate theoretical bias. With the adjoint-method, we can compute sensitivity kernels for every pair of stations, for each measurement

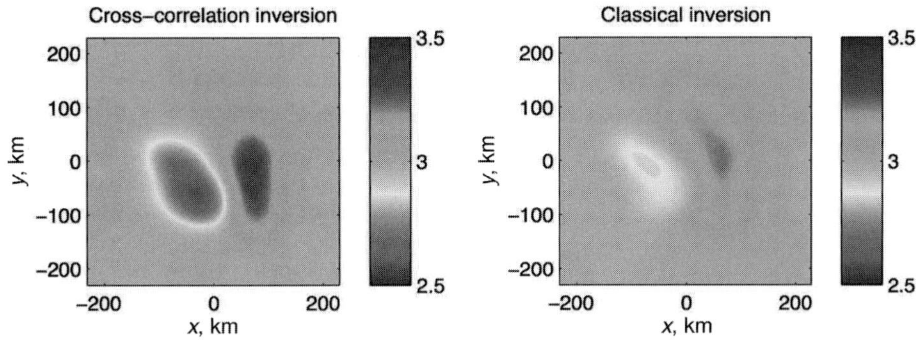

Figure 25.6 Inversions obtained using cross-correlation theory (left panel) and classical source–receiver description (right panel). The true model is shown in Figure 25.5 and "observations" for this model are estimated using the cross-correlation theory described by Hanasoge (2014). While neither inversion is perfect owing to the finite set of measurements, the classical inversion (based on the classical source–receiver description) appears to show larger errors in comparison to the cross-correlation inversion (based on the theory of cross-correlations). A black and white version of this figure will appear in some formats. For the color version please refer to the plate section.

and for each parameter (such as sound speed, attenuation, density, and source). The total gradient with respect to each parameter is given by a sum of all sensitivity kernels weighted by the travel-time difference between prediction and measurement associated with that pair of receivers. For instance, the total gradient of the misfit with respect to wave speed for travel-time measurements, denoted by G_c, is defined as

$$G_c(\mathbf{x}) = \sum_{\alpha,\beta} \Delta\tau_{\alpha\beta}\, K_c(\mathbf{x}; \mathbf{x}_\alpha, \mathbf{x}_\beta), \tag{25.14}$$

where $K_c(\mathbf{x}; \mathbf{x}_\alpha, \mathbf{x}_\beta)$ is the sensitivity kernel for wavespeed for travel-time measurements between stations α and β and $\Delta\tau_{\alpha\beta}$ is the difference between predicted and observed travel times for that station pair. For the toy problem of Figure 25.5, the "measurements" comprise correlations estimated using the theory described in Hanasoge (2014) and the starting model has a uniform wavespeed. Using the Gauss–Newton method, we perform a one-step inversion for the wavespeed, interpreting the "measurements" using cross-correlation theory and alternatively a classical source–receiver description. In Figure 25.6, we show the systematic theoretical bias that can be introduced when the latter approximation is invoked.

25.6 Outlook

In this review, we focused on the theory of cross-correlations applied to infer interior properties of the Sun and Earth. A number of important results in seismology rest on interpreting measurements of noise cross-correlations. Contemporary seismology is a precision science that relies on a combination of accurate theory, computational methods, and high-quality observations. The development of such methods is thus critical to future progress in seismology. The adjoint-method theory of cross-correlation tomography (Tromp *et al.*, 2010; Hanasoge *et al.*, 2011) discussed here was formulated using methods prevalent in helio- and terrestrial seismology.

Noise tomography is an exciting field and a broad range of problems in both the Sun and Earth remain to be addressed. Together with large repositories of data and novel theoretical developments, the accurate imaging of the interiors of the Sun and Earth is a plausible reality.

26

Waveform tomography in geophysics and helioseismology

LAURA J. COBDEN, ANDREAS FICHTNER, AND VINCENT C. H. TONG

26.1 What is waveform tomography?

Seismic tomography – in which we construct images of a body's interior using seismic waves – is an inverse problem; that is, our goal is to find a model that fits a set of existing data observations. This is much less straightforward than the reverse, forward problem (i.e., generating synthetic data from an existing model) due to the fact that multiple models can fit the same data or, in other words, the solution is non-unique. Furthermore, there may be parts of the model to which the data have no sensitivity, and small errors in the data can propagate into significant errors in the model (e.g., Trampert, 1998). In order to transform data space into model space, a seismic modeling algorithm is required that can generate synthetic data from an initial model, and then update the initial model to minimize the misfit between synthetic and observed data. There is therefore always a trade-off between the computational efficiency of the modeling algorithm and the accuracy or resolution with which it can represent the real seismic structure. Which kind of modeling algorithm is employed in a given situation depends very much on the nature of the structure being imaged, the quality of the data, and the available computational resources.

Traditionally, seismic tomography has used the travel times of wave phases between a seismic source and receiver to infer the sound-speed structure along the path between them. This so-called travel-time tomography, more typically referred to as "time–distance helioseismology" when applied to the Sun, is based upon ray theory, which assumes that waves travel with an infinitely high frequency, in much the same way as light rays propagating through a medium with smoothly varying refractive index, occasionally encountering a sharp interface. Under this approximation the seismic energy propagates along infinitesimally narrow geometric "ray paths." The travel time along a given ray path is then, according to Fermat's principle, simply the line integral of the wave speed along that path:

$$T = \int_l \frac{ds}{c},$$

Extraterrestrial Seismology, eds. Vincent C. H. Tong and Rafael A. García. Published by Cambridge University Press.

where l is the ray path, c is the wave speed, T is the travel time, and ds is an increment along the ray path.

Obtaining models of the solar or terrestrial wave-speed structure using travel-time tomography is mathematically simple and computationally efficient. Furthermore, as long as the wavelength of the waves is very much less than their path length, the ray approximation gives a reasonable representation of the region being studied. However, the ray approximation becomes inadequate for complex structures that vary on length scales that are smaller than or comparable with the Fresnel zone of the propagating waves (e.g., Williamson and Worthington, 1993). In reality, seismic waves with a finite frequency do not travel along ray paths between source and receiver, but will sample extensive regions outside the geometric ray path due to scattering. This modifies the structure of the propagating waveform and causes travel times to deviate from their ray-theoretical values.

The idea of waveform tomography is that, during the inversion process, we aim to fit the entire signal waveform recorded on a seismogram, rather than just the travel times of certain peaks within that waveform. Forward modeling of seismic waveforms can proceed by solving the full wave equation, rather than relying on simple approximations such as ray theory to simulate wave propagation. In this case, complex phenomena such as diffractions are accounted for implicitly, rather than imposing explicit conditions to divide the ray into transmitted and reflected components. Modeling entire seismic waveforms requires vastly more computational resources than modeling only the travel times, but it allows us to exploit additional information contained in the amplitudes and/or phases of incoming waves. This extra information may, for example, provide constraints on the density and attenuation structure in a region, which cannot be extracted from travel-time analysis alone. Furthermore, by taking into account the low-frequency components of the seismic signal, the spatial resolution is only limited by the source and receiver distributions. It is thus possible to image structures at sub-wavelength scales, significantly better than the limit of the first Fresnel zone in travel-time tomography (e.g., Pratt *et al.*, 1996).

An additional advantage of waveform tomography relates to data processing. In travel-time tomography, travel times are not a directly recorded parameter, but must be picked somehow from the seismic record. This picking can be highly subjective, depending on human judgement, and it may be difficult to extract the timing of a specific peak from triplicated waveforms (i.e., where multiple waveforms arrive at the same receiver within a short and possibly overlapping time interval). Since waveform tomography makes use of the entire wave train recorded at a receiver, these subjective data-picking errors are not introduced, although errors due to noise will still affect the accuracy of the inversion.

With the increasing availability of cheap random access memory and parallelized computer systems (supercomputers) in the past two decades, the application of waveform tomography has become increasingly feasible. In situations where the resolution of small-scale 3D structures is paramount, and sufficiently high quality data are available, waveform tomography should be the imaging method of choice.

26.2 Seismic wave propagation in complex media

The first step in performing waveform tomography is to be able to simulate, accurately, wave propagation through the medium of interest (i.e., the solar or Earth's interior), via numerical solution of the wave equation. For the Earth, the goal is to solve the seismic wave equation, the theoretical derivations of which can be found, for instance, in Aki and Richards (2002). On the Sun, we are ultimately required to solve a highly complex magnetoacoustic wave equation (discussed in Section 26.5).

The earliest fully numerical models of seismic wave propagation through heterogeneous Earth models were mostly based on finite difference methods (e.g., Alterman and Karal, 1968). Finite differencing is a popular method for solving differential equations computationally. It requires construction of a grid. At each node of the grid, values of continuous functions are calculated approximately, by replacing the differentials of the function with discrete difference equations, e.g., $\partial u/\partial t \approx [u(t+\Delta t) - u(t)]/\Delta t$. The advantage of this method is that it can produce accurate solutions and allows modeling of complex media. Finite difference (FD) methods for wave propagation rapidly gained popularity in the 1980s, thanks to both increasing computational resources and the implementation of the staggered grid scheme (Virieux, 1984) originally introduced in fluid dynamics (Harlow and Welch, 1965). In the staggered grid, material properties and dynamic field variables are defined on different sets of grid points, which leads to a substantial reduction of numerical dispersion. Owing to their robustness and simplicity, FD methods continue to be widely used, especially in industrial applications. A comprehensive and modern treatment of FD methods can be found in Moczo *et al.* (2002).

A major drawback of FD methods is the inability to adapt the numerical grid to complex subsurface structure and topography. In exploration geophysics, where most surveying is done in small ($<10^2$ km) regions of relatively flat topography, this is usually not a major concern, but FD methods become increasingly inappropriate at longer length scales on the Earth, when wave propagation effects related to surface topography and undulating seismic discontinuities cannot be ignored. This has led to the development of finite element (FE) methods for the solution of the wave equation. Among the family of FE methods, the spectral element method (SEM) is currently considered to provide the optimal balance between flexibility, accuracy, and computational requirements (e.g., Komatitsch and Vilotte, 1998). In the SEM, the model volume is divided into a number of non-overlapping, hexahedral elements. Within each element, the wavefield and material properties are represented by continuous functions (basis functions), typically Lagrange polynomials of degree 4 to 8. The shape of each element is also defined by low-degree polynomials, and continuity of displacement along the boundaries is enforced. Thus, complex structures and topography are better accounted for than with the discrete FD procedure, and free-surface boundary conditions are implicitly included, unlike in FD modeling where one must explicitly define whether a boundary is reflecting or absorbing wave energy. In contrast to other, more standard, FE methods, the SEM leads to a diagonal mass matrix, which significantly reduces computational costs, since the inversion of a large matrix is not required.

26.3 Solution of the full waveform inverse problem

Finite difference methods and the SEM allow us to generate high-fidelity synthetic seismograms for a given Earth or solar model. Since our goal in tomography is to derive a model that best represents a given dataset, the second component to performing waveform tomography is an algorithm that will update an initial-guess model, in order to minimize the misfit between synthetic and observed seismograms. This requires that we (1) define a misfit functional and (2) choose an optimization scheme to locate the global minimum.

Designing suitable misfit functionals remains an active area of research, and good solutions tend to be strongly application-dependent. Many applications are based on the L_2 norm: Find the model m that minimizes the L_2 distance χ between the observed seismograms $u_0^r(t)$ and numerically computed (synthetic) seismograms $u^r(t;m)$:

$$\chi(m) = \frac{1}{2}\sum_r \int [u^r(t;m) - u_0^r(t)]^2 dt, \tag{26.1}$$

where r denotes the index of a seismic receiver at position x^r. In Equation (26.1), additional data-specific error weights may be introduced. However, the first applications of waveform tomography based on this algorithm (e.g., Bamberger *et al.*, 1982; Gauthier *et al.*, 1986) revealed two problems: (1) The dependence of χ on the model m can be highly nonlinear, such that gradient-based optimization schemes tend to lock into local minima; and (2) the L_2 waveform difference is dominated by large-amplitude waves. Valuable information contained, for instance, in small-amplitude body waves is almost entirely disregarded. Thus, careful choice of a suitable misfit functional is essential for the successful solution of a waveform tomographic inverse problem.

Alternative suggestions for misfit functionals in terrestrial waveform tomography include variants of the L_1 norm (e.g., Brossier *et al.*, 2009), as well as instantaneous phase and amplitude misfits (e.g., Bozdag *et al.*, 2011). Fichtner *et al.* (2008, 2013) proposed the use of time- and frequency-dependent phase misfits. In a first step, observed and synthetic seismograms are transformed to the time–frequency domain using, for instance, a moving-window Fourier transform. This results in complex-valued time–frequency representations of observations and synthetics. The L_2 distance between the observed phase $\phi_0^r(t,\omega)$ and the synthetic phase $\phi^r(t,\omega;m)$ is then used to define the misfit:

$$\chi(m) = \frac{1}{2}\sum_r \int\int [\phi^r(t,\omega;m) - \phi_0^r(t,\omega)]^2 dt d\omega, \tag{26.2}$$

where again additional weights may be incorporated. In Figure 26.1, we illustrate the measurement of the time–frequency phase misfit for the case of a north–south (N–S) component seismogram recorded at station ARGB in eastern Turkey, at a distance of 9.52° from the earthquake epicentre. Those parts of the recordings where observations and synthetics are not approximately in phase must be eliminated in order to produce meaningful measurements of phase difference. This particular misfit functional does not

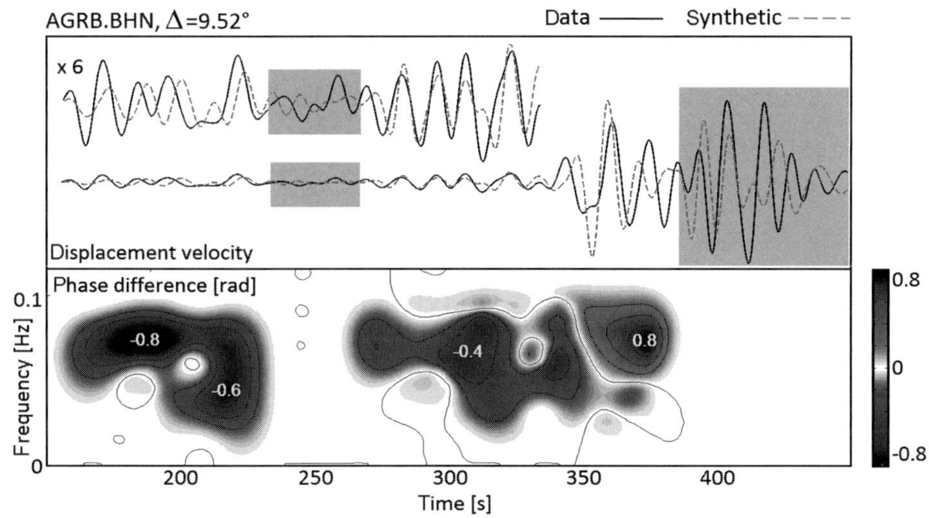

Figure 26.1 Measurement of a time–frequency phase misfit. (Top) Observed (solid) and synthetic (dashed) seismograms for station AGRB in western Turkey (distance to earthquake $\Delta = 9.52°$; see Figure 26.2 for source–receiver geometry). Oscillations prior to 340 s correspond to low-amplitude body waves amplified by a factor of 6 for better visibility. Large-amplitude surface waves appear from 340 to 450 s. Gray shading marks intervals that are excluded from the misfit calculation because waveform differences are too large for phase difference measurements. (Bottom) Phase difference, $\varphi - \varphi_0$, between observations and synthetics as a function of time and frequency. The phase differences between the small-amplitude body waves are as significant as those between large-amplitude surface waves. A black and white version of this figure will appear in some formats. For the color version please refer to the plate section.

require the presence of well-defined body or surface waves; it is comparatively insensitive to the choice of initial model, and it is not biased by the amplitudes of the waveforms.

Owing to the large number of free parameters in a waveform tomographic inversion (thousands to millions), the optimization problem is usually solved using gradient-based techniques, including conjugate gradients or Newton-like methods. Gradient-based optimization algorithms rely on an initial model that is sufficiently close to the global minimum. Since good initial models are often not available, multi-scale strategies (e.g., Virieux and Operto, 2009) have been developed in order to prevent locking into local minima. Multi-scale strategies are based on the observation that misfit functionals are better behaved, i.e., closer to being convex, when low-frequency waves are considered. The iterative optimization therefore starts with the lowest available frequencies, which are used to improve a potentially crude model. After several iterations at low frequencies, the refined model is used as the initial model for the next few iterations at higher frequencies. This cycle is repeated until the model explains the highest-frequency waveforms.

Gradient-based optimizations can be computed with two different approaches: adjoint methods or scattering-integral methods. Adjoint methods (e.g., Fichtner *et al.*, 2006)

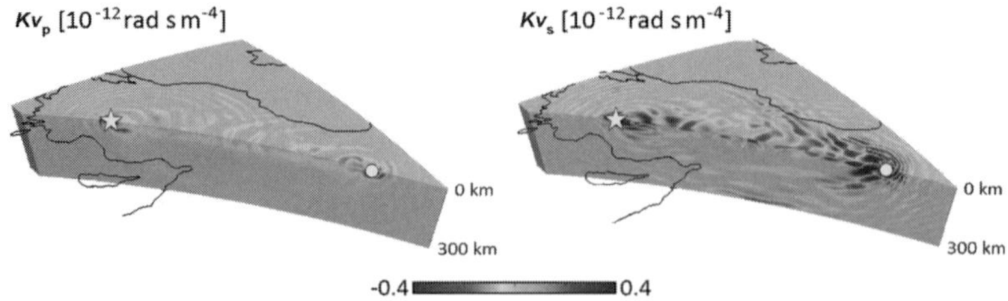

Figure 26.2 Sensitivity kernels $K_{\nu p}$ and $K_{\nu s}$ for the phase difference measurement in Figure 26.1. The positions of the earthquake source and receiver (station ARGB) are indicated by a star and filled circle respectively. Regions that are brighter or darker than average are those where perturbations in υ_p (for the case of $K_{\nu p}$) and υ_s (for the case of $K_{\nu s}$) affect the phase difference measurement. Thus, the kernels indicate where the current model should be modified to reduce the phase misfit. A black and white version of this figure will appear in some formats. For the color version please refer to the plate section.

are closely related to time-reversal techniques in seismic exploration, and to acoustic holography in helioseismology. The difference between observations and synthetics recorded at an array of receivers is time-reversed and re-injected at the receiver positions as sources in a time-reversed or *adjoint* simulation. The resulting wavefield focuses where the synthetic model differs from the true model, and the quality of the focusing depends on the source–receiver coverage. The time correlation of the adjoint wavefield and the regular wavefield yields sensitivity kernels, $K(x)$, which relate infinitesimal variations in the misfit functional, $\delta\chi$, to infinitesimal variations of the model, δm. The sensitivity kernels for the phase difference measurements shown in Figure 26.1 with respect P-velocity and S-velocity ($K_{\nu p}$ and $K_{\nu s}$) are shown in Figure 26.2.

The computation of sensitivity kernels via adjoint methods requires one forward and one time-reversed adjoint wavefield simulation per seismic source, irrespective of the number of receivers. It follows that adjoint methods are particularly efficient when the number of sources is small, and the number of receivers is large. Scattering-integral methods (e.g., Chen *et al.*, 2007) are most efficient for exactly the opposite scenario: many sources and few receivers. Making use of source–receiver reciprocity (Aki and Richards, 2002), scattering-integral methods compute sensitivity kernels using receiver-side Green functions i.e., wavefields excited by sources located at receiver positions.

26.4 Application of waveform tomography in geophysics

The first practical applications of waveform tomography based on solving the wave equation were in exploration geophysics, using controlled (i.e., man-made) seismic sources (e.g., Zhou *et al.*, 1995). Tests on borehole seismic data demonstrated that for surveys spanning a few tens of meters, waveform tomography could image structures as small as 2 m, compared with travel-time tomography, which could only resolve down to about 10 m (e.g., Pratt *et al.*, 1996). Following this success, waveform tomography was then applied to wide-angle seismic data (survey dimensions: tens to hundreds of kilometers) to image deep crustal structures (e.g., Pratt *et al.*, 1996; Shipp and Singh, 2002). Again, it was shown that waveform tomography could resolve structures up to an order of magnitude smaller than those imaged with travel-time tomography. More recently, it has been possible to apply waveform inversion in non-commercial settings, such as the imaging of buried impact craters (Morgan *et al.*, 2011), and on continental length scales (10^4 km) (Fichtner *et al.*, 2009; Tape *et al.*, 2010), with the latter relying on natural sources (i.e., earthquakes). Advances towards global models of the Earth's mantle from SEM-based waveform inversions are also being made (e.g., French *et al.*, 2013).

Although waveform tomography is now being used on a wide range of length scales on the Earth, each application is characterized by its own set of challenges and limitations. For example, a key feature in exploration seismics is the use of dense arrays involving large numbers of seismic sources, easily exceeding several thousand. Specific work-arounds are needed to ensure computational efficiency in the wavefield modeling of such large numbers of sources. On the other hand, while fewer sources are typically present in regional or global seismology, a major issue is the heterogeneous distribution of both the earthquake sources (dependent on plate tectonics) and seismic receivers (constrained by financial, geographical, and political factors), leading to large null spaces in the Earth model. Additionally, since the sources are earthquakes, which occur many kilometers below the surface, large uncertainties exist on the source–time function of each event, as well as their spatial coordinates.

The strong scale and source dependence of waveform tomography largely prevents the development of blackbox-like algorithms for waveform tomography that produce excellent Earth models without human interaction. We summarize below key strategies that have been or are being applied on different length scales in the Earth. In Figure 26.3 we show examples of waveform tomography on different length scales.

26.4.1 Exploration seismics

Exploration seismics is mostly oriented towards the discovery and characterization of natural resources, and seismic surveys typically operate on length scales from a few hundred meters to a few tens of kilometers. The biggest problem to overcome in waveform inversion is the need to model hundreds to thousands of sources.

Where significant numbers of sources are present, waveform modeling with FD or FE methods is best handled in the frequency domain (Marfurt, 1984). In particular, only a limited number of frequencies may need to be included to retrieve a satisfactory inversion of the structure (e.g., Pratt *et al.*, 1996), and each frequency can be represented by a small discrete system of equations. Solving the wave-propagation problem for a new source then reduces to a simple vector-matrix multiplication, which is relatively fast to compute.

Wave propagation in exploration seismics is usually modeled with an acoustic wave equation, as opposed to the more correct elastic wave equation. This means that only the early-arriving compressional waves can be exploited in the tomography; shear waves,

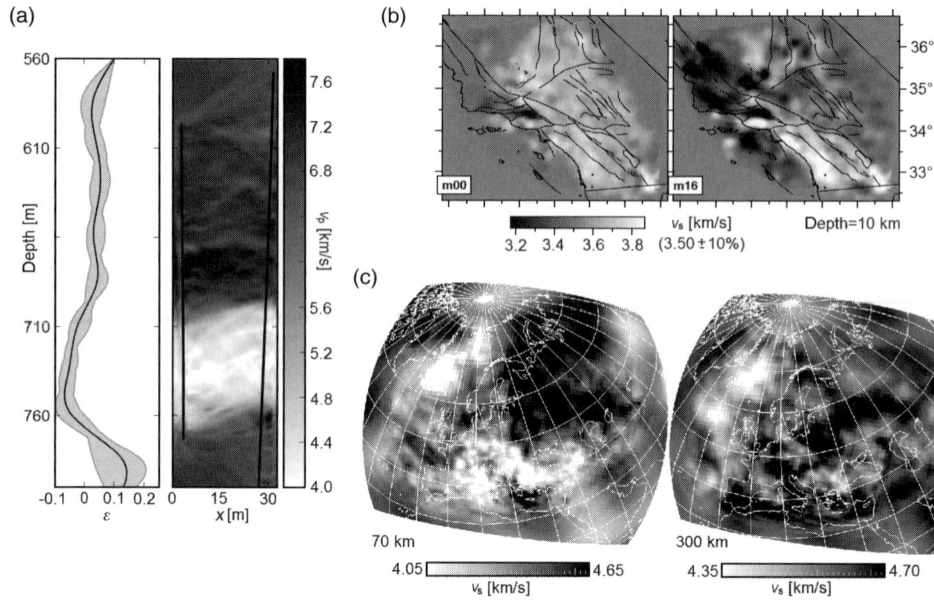

Figure 26.3 Examples of waveform tomographies on various scales in the Earth. (a) Crosshole seismic tomography. The 1D profile of elliptical anisotropy ε (left panel) was found using simulated annealing to avoid trapping in local minima. The P-velocity model (right panel) was obtained subsequently via 2D frequency-domain waveform tomography. Boreholes are indicated by black lines. (Modified from Afanasiev *et al.*, 2014.) (b) Waveform tomography of the Southern California crust (Tape *et al.*, 2010) obtained with 3D spectral element simulations in the time domain. Shown are the initial model (left) and the final model (right) obtained after 15 conjugate-gradient iterations. (c) Waveform tomography of Europe and Western Asia, also based on spectral-element time-domain modeling. (Modified after Fichtner *et al.*, 2013.)

which do not exist in acoustic media, must be disregarded in the data analysis. Furthermore, simulations are usually limited to 2D or 2.5D, for which wave amplitudes are not correctly modeled. In response to these drawbacks, exploration applications are currently moving towards 3D simulations (e.g., Warner *et al.*, 2013), and simulations based on the elastic wave equation (e.g., Guasch *et al.*, 2012). These are performed in the time domain, because the explicit solution of the large frequency-domain system of equations is currently not feasible for the 3D elastic case.

On a kilometer scale, the Earth is characterized by strong heterogeneities, which complicate the construction of an initial model sufficiently close to the global minimum of the misfit functional. Today, initial models are mostly constructed using classical travel-time tomography, which constrains the long-wavelength properties of the subsurface. Alternatively, Monte-Carlo-type searches may be used to estimate the broad patterns of an Earth model described by a small number of parameters. This approach was taken, for instance, by Afanasiev *et al.* (2014) (Figure 26.3a).

26.4.2 Regional and global seismology

Finding suitable initial models is a lesser problem on regional (10^3–10^4 km) to global scales, where radially symmetric reference models (e.g., Dziewonski and Anderson, 1981) explain the travel times of the major seismic wave arrivals to within a few seconds. The challenge lies in the inherent three-dimensionality of the problem and the need to account fully for elastic effects, anisotropy, and visco-elastic dissipation. Consequently, the computational costs of solving the forward problem are orders of magnitude larger than in 2D acoustic simulations that can be justified in certain exploration-scale applications. The seismic sources for waveform tomography on length scales exceeding several kilometers are mostly naturally occurring earthquakes, although the use of ambient seismic noise is an area of active research (e.g., Tromp *et al.*, 2010). Owing to the enormous computational requirements, regional- to global-scale full waveform tomographies have only become feasible in the past few years. Tape *et al.* (2010) presented the first full-waveform tomographic model of the Southern California crust, revealing a close correlation between seismic velocity structure and major fault systems (Figure 26.3b). On continental scales, full waveform tomography has established clear relations between seismic anisotropy and mantle flow. (Fichtner *et al.*, 2009). An example for Europe and Western Asia is presented in Figure 26.3c.

The highly irregular distribution of both earthquakes and seismic stations over the globe generally results in a heterogeneous resolution. This differs from exploration applications, where more regular source–receiver geometries are available. The quantification of resolution is thus a challenging aspect of regional and global waveform tomography which has so far received little attention. Most resolution analysis is limited to synthetic inversions of chequerboard-like input patterns. Information about the curvature of the misfit functional near its minimum, obtained via second-order adjoints, provides resolution proxies in the vicinity of the optimal model (e.g., Fichtner and Trampert, 2011). However, practicable methods for global uncertainty analysis in large-scale full waveform tomography are not currently available.

26.5 Application of waveform tomography in helioseismology

Seismic waves on the Sun are generated in the convection zone, a turbulent layer making up the outer 30% of the Sun. Above the convection zone lies the photosphere, a layer often taken to be the "solar surface." It is believed that seismic waves are excited continuously and stochastically by numerous sources in a narrow layer just beneath ($\sim$75 $\pm$ 50 km below) the photosphere, as a result of supersonic downflow plumes formed at integranular lanes, which shock the immediate sub-surface layers, channeling fluid kinetic energy into acoustic radiation (e.g., Nordlund *et al.*, 2009). The rapidly downwelling material forms thin "intergranular lanes," which surround broad, upwelling regions known as granules (Figure 26.4). The waves are acoustic in nature, and when they reach the photosphere, provided they are below a certain frequency known as the acoustic cut-off frequency of approximately 5.55 mHz, they are prevented from penetrating it due to its sharp density

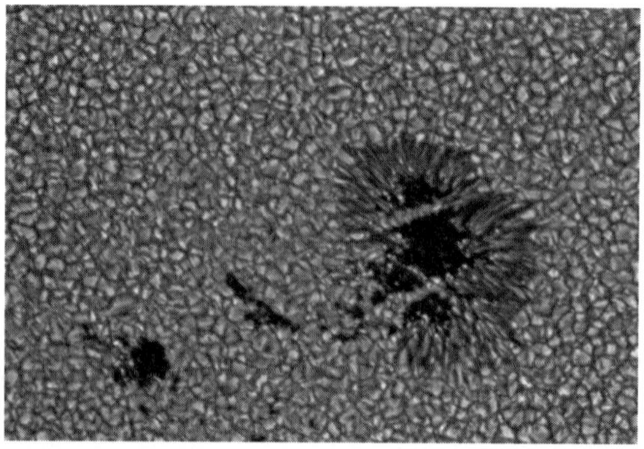

Figure 26.4 Photograph of the photosphere, containing a sunspot (right-hand side). It has a central dark region (umbra) surrounded by radial filaments (penumbra). The smaller polygons are convection cells, known as granules, typically 1–1.5 Mm across. Bright regions are upwelling, dark lines (intergranular lanes) are downwelling. The surface is highly dynamic, with each granule typically lasting about 10 minutes before being replaced by new granules. (Image courtesy of NASA.)

gradient. Each time a wave bounces off the photosphere it causes vertical particle motion, which can be measured via the Doppler shifting of absorption lines in the solar spectrum.

In reality the oscillations recorded at any particular point on the photosphere represent the combined effect of many waves arriving at that point. This produces "noisy" seismograms. In order to extract a coherent signal from the noise, most solar tomography is based on the cross-correlation of pairs of seismic traces to extract the Green function, i.e., the signal that would be obtained if there were a single source beneath the location of one of the data recorders and a receiver at the other.

Full waveform tomography using the approach outlined in Sections 26.1–26.3 has not yet been applied to real solar data. However, waveform tomography has great potential for enhancing the imaging of localized 3D structures near the surface of the Sun, in particular sunspots, and the development of suitable inversion algorithms is already underway (e.g., Hanasoge *et al.*, 2011). Sunspots are dark regions of intense magnetism, typically between ten to a few tens of megameters across and, at 4000 K, about 2000 degrees cooler than their surroundings. They have lifetimes lasting anything between a few hours and several months. Although they are well-resolved photographically at the visible surface of the Sun (Figure 26.4), their structure below the photosphere cannot be observed directly and remains somewhat of a mystery. Tomographic imaging has been applied to probe their subsurface structure, and time–distance helioseismology has provided tantalizing first images of the structures and flows underneath sunspots (e.g., Zhao *et al.*, 2001). However, Gizon *et al.* (2009) raised concerns over the effect of data filtering procedures on the validity of tomographic models. The presence of magnetic fields causes anisotropy, which needs to be carefully modeled (Hanasoge *et al.*, 2012). In addition, the resolution that can be achieved by time–distance methods is limited by the fact that in the shallowest layers of the Sun the

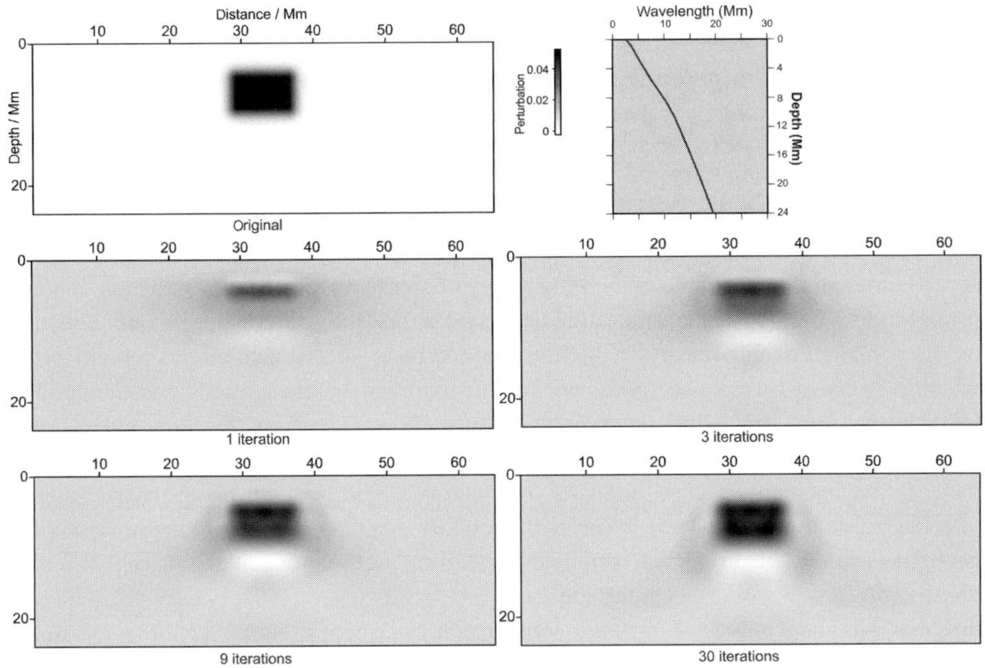

Figure 26.5 Two-dimensional synthetic test showing recovery of a small-sunspot-like structure from full waveform inversion, after Cobden *et al.* (2011). Top row: input seismic structure (left) and seismic wavelength as a function of depth (right) in the Sun's outermost convection zone. Other rows: sound-speed structure recovered from waveform inversion after various iteration numbers of the inversion algorithm. The wavefield is modeled using an acoustic wave equation without convection or magnetic field effects, and assuming a uniform distribution of acoustic sources near the surface. The inversion can retrieve sharp vertical and horizontal boundaries, which are notoriously difficult to obtain in travel-time inversions, as well as accurate imaging of the spatial dimensions and anomaly magnitude of a structure with a size comparable to the signal wavelength.

ratio of the seismic wavelength to path length is high. In particular, structures underneath sunspots have typically been imaged to depths of a little more than 10 Mm, but below this depth, the diameter of sunspots is only two or three times greater than the seismic wavelength at most, and thus their interior structure is potentially beyond the resolution of the ray approximation.

In response to this limitation, several finite-frequency approximations, such as the Fresnel zone, Born, and Rytov approximations, which allow for a single scattering event between the source and receiver, have been applied to solar acoustic data, and such approximations have allowed solar subsurface structure to be imaged to a greater depth and higher accuracy than with the ray approximation (e.g., Couvidat *et al.*, 2004). Exploratory numerical simulations of solar wave propagation in 2D, based on the acoustic wave equation, have indicated that the application of full waveform tomography to solar data could potentially allow an even finer spatial resolution (Tong *et al.*, 2003a,b; Cobden *et al.*, 2011, Figure 26.5), as well as the inference of additional parameters such as density and attenuation (Tong *et al.*, 2003c; Tong, 2005). Nonetheless, the application of waveform tomography to helioseismic data presents several major challenges that are not encountered in terrestrial seismology.

In terrestrial seismology, seismic waveforms are modified by the elastic properties of the effectively static media through which they are transmitted. The transmission speed of the waves (km/s) is much shorter than the time scales over which the material is convecting (mm–cm/year), and tomography gives us a "snapshot" image of structures that are evolving on time scales of millions of years. However, on the Sun, the wave speeds (km/s) are comparable to the time scales of some convective motions, such that the propagation velocities of acoustic waves between two surface points are dependent not only on the temperature and density of the intermediate material, but also on the velocity with which it is convecting. Wave speeds are further modified by the spatially and temporally variable magnetic fields that are characteristic of plasmas such as the Sun, and seismic perturbations inside sunspots (e.g., Braun and Birch, 2008) may be produced by both thermal and magnetic effects. Inferring the magnetic field strength using measurements of propagation times of seismic waves is particularly challenging because heliomagnetism generates seismic anisotropy between field-parallel and field-perpendicular wave paths, and thus, wave paths that intersect at 90° are required to study it, which likely have encountered different inhomogeneities along their mostly non-intersecting paths. At the same time, numerical modeling tells us that magnetic effects are non-trivial and sensitivity kernels that do not realistically account for them will lead to the simulation of incorrect wave travel times (Crouch *et al.*, 2011; Braun *et al.*, 2012). Thus, the need to include magnetic and convection effects makes complete numerical modeling of wave propagation much more challenging, both theoretically and computationally. Inclusion of magnetic effects in numerical models is a very recent development (Hanasoge *et al.*, 2011; Crouch *et al.*, 2011), but convective mass transfers have not yet been realistically incorporated.

A second problem in helioseismology is the stochastic and continuous nature of the acoustic sources, leading to the need to cross-correlate pairs of traces to extract a coherent signal. In order to converge at the desired Green function, and to enhance the signal-to-noise ratio, a large amount of data must be averaged. On the Sun, the rapid temporal evolution (hours to weeks) of the target structures prohibits extensive temporal averaging. To compensate, data are typically averaged over a patch or annulus of neighboring pixels. This spatial data-averaging would likely cancel out any improvements in the resolution of images that could be obtained in waveform tomography. Additionally, waveform amplitude analysis is hampered by the fact that the amplitudes of cross-correlated waveforms are influenced by the distribution of the acoustic sources (Cobden *et al.*, 2010), which in turn is unknown. A promising solution to these problems may lie in the adjoint method of Tromp *et al.* (2010) for computing sensitivity kernels for cross-correlation observations. In this case the misfit functional can be any feature of the cross-correlated signal (travel times, waveforms, etc.). Applying this method to the Sun, a small number of "master pixels" are selected to be cross-correlated with any number of other pixels; computation of sensitivity kernels requires only three numerical simulations per master pixel. This method relies on temporal averaging of cross-correlation time series rather than spatial averaging, thus avoiding the loss of resolution associated with spatial averaging, and it can accommodate non-uniform source distributions. It remains to be tested how successfully it can be applied to time-limited solar structures.

26.6 Future directions

In geophysics, waveform tomography is a relatively young, but fully established, method for obtaining high-resolution images of the Earth's interior. Active research is targeted at applying the method to increasingly larger areas and greater depths in the Earth, at shorter wavelengths with higher spatial resolution, as well as elucidation of physical properties beyond isotropic wavespeeds, such as density and attenuation. On the Sun, the application of full waveform tomography to real data remains a challenging but realistic prospect within our grasp. It will require the combination of a code that can realistically simulate both wave generation by stochastic sources and wave propagation through a fully magnetohydrodynamic medium, with an efficient inversion algorithm based on fitting cross-correlated time series. These three components have all been developed separately and future endeavors to combine them may allow us an unparalleled view into the internal structure of sunspots.

References

Abramowicz, M. A., and Fragile, P. C. 2013. Foundations of black hole accretion disk theory. *Living Rev. Relativity*, **16**(Jan.), 1.

Acevedo-Arreguin, L. A., Garaud, P., and Wood, T. S. 2013. Dynamics of the solar tachocline: III. Numerical solutions of the Gough and McIntyre model. *MNRAS*, **434**(Sept.), 720–741.

Aerts, C. 2013. Asteroseismology of binary stars and a compilation of core overshoot and rotational frequency values of OB stars. Pages 323–330 of Pavlovski, K., Tkachenko, A., and Torres, G. (eds), *Setting a New Standard in the Analysis of Binary Stars*. EAS Publications Series, vol. 64.

Aerts, C., Thoul, A., Daszyńska, J., *et al.* 2003. Asteroseismology of HD 129929: Core overshooting and nonrigid rotation. *Science*, **300**(June), 1926–1928.

Aerts, C., Christensen-Dalsgaard, J., and Kurtz, D. W. 2010. *Asteroseismology*. Springer.

Afanasiev, M. V., Pratt, G., Kamei, R., and McDowell, G. 2014. Waveform-based simulated annealing of crosshole transmission data: A semi-global method for estimating seismic anisotropy. *Geophys. J. Int.*, **193**(3), 1586–1607.

Ahmad, Q. R., Allen, R. C., Andersen, T. C., *et al.* 2002. Direct evidence for neutrino flavor transformation from neutral-current interactions in the Sudbury Neutrino Observatory. *Phys. Rev. Lett.*, **89**(1), 011301.

Aizenman, M., Smeyers, P., and Weigert, A. 1977. Avoided crossing of modes of non-radial stellar oscillations. *A&A*, **58**(June), 41.

Aki, K., and Chouet, B. 1975. Origin of coda waves: Source, attenuation, and scattering effects. *J. Geophys. Res.*, **80**, 3322–3342.

Aki, K., and Richards, P. G. 2002. *Quantitative Seismology*, 2nd edn, Sausalito, CA: University Science Books.

Allain, S. 1998. Modelling the angular momentum evolution of low-mass stars with core-envelope decoupling. *A&A*, **333**(May), 629–643.

Allred, J. C., Hawley, S. L., Abbett, W. P., and Carlsson, M. 2005. Radiative hydrodynamic models of the optical and ultraviolet emission from solar flares. *ApJ*, **630**(Sept.), 573–586.

Alterman, Z., and Karal, F. 1968. Propagation of elastic waves in layered media by finite difference methods. *Bull. Seism. Soc. Am.*, **58**, 367–398.

Alvan, L., Mathis, S., and Decressin, T. 2013. Coupling between internal waves and shear-induced turbulence in stellar radiation zones: The critical layers. *A&A*, **553**(May), A86.

Alvan, L., Brun, A. S., and Mathis, S. 2014. Theoretical seismology in 3D: Nonlinear simulations of internal gravity waves in solar-like stars. *A&A*, **565**(May), A42.

Andersen, J. 1991. Accurate masses and radii of normal stars. *A&A Rev.*, **3**, 91–126.

Anderson, D. L. 1972. Internal constitution of Mars. *J. Geophys. Res.*, **77**, 789–795.

Anderson, D. L., Miller, W. F., Latham, G. V., *et al.* 1977. Seismology on Mars. *J. Geophys. Res.*, **82**(Sept.), 4524–4546.

Anderson, E. R., Duvall, Jr., T. L., and Jefferies, S. M. 1990. Modeling of solar oscillation power spectra. *ApJ*, **364**(Dec.), 699–705.

Ando, H., and Osaki, Y. 1975. Nonadiabatic nonradial oscillations: An application to the five-minute oscillation of the sun. *PASJ*, **27**, 581–603.

Ando, H., Watanabe, E., Yutani, M., Shimizu, Y., and Nishimura, S. 1988. Search for small-amplitude oscillations in stars by a Fabry–Perot interferometer. *PASJ*, **40**, 249–262.

Antia, H. M., and Basu, S. 1994. Nonasymptotic helioseismic inversion for solar structure. *A&AS*, **107**(Nov.), 421–444.

Antia, H. M., and Basu, S. 2000. Temporal variations of the rotation rate in the solar interior. *ApJ*, **541**(Sept.), 442–448.

Antia, H. M., and Basu, S. 2005. The discrepancy between solar abundances and helioseismology. *ApJ*, **620**(Feb.), L129–L132.

Antia, H. M., and Chitre, S. M. 1999. Limits on the proton-proton reaction cross-section from helioseismology. *A&A*, **347**(July), 1000–1004.

Antia, H. M., Basu, S., and Chitre, S. M. 1998. Solar internal rotation rate and the latitudinal variation of the tachocline. *MNRAS*, **298**(Aug.), 543–556.

Antia, H. M., Basu, S., Hill, F., *et al.* 2001. Solar-cycle variation of the sound-speed asphericity from GONG and MDI data 1995–2000. *MNRAS*, **327**(Nov.), 1029–1040.

Antia, H. M., Chitre, S. M., and Thompson, M. J. 2003. On variation of the latitudinal structure of the solar convection zone. *A&A*, **399**(Feb.), 329–336.

Apollo 11 Mission Report 1969. NASA Manned Spacecraft Center MSC-00171, Houston, Texas.

Apollo 16 Mission Report 1972. NASA Manned Spacecraft Center MSC-07230, Houston, Texas.

Appourchaux, T. 2014. A crash course on data analysis in asteroseismology. In Pallé, P. (ed), *Asteroseismology*. Canary Islands Winter School of Astrophysics, vol. 22. Cambridge: Cambridge University Press.

Appourchaux, T., Catala, C., Catalano, S., *et al.* 1991. PRISMA: Probing Rotation and Interior of Stars. Microvariability and activity, Assessment study report. In Appourchaux, T. (ed), *SCI(91)5*.

Appourchaux, T., Catala, C., Cornelisse, J., *et al.* 1993. PRISMA: Probing Rotation and Interior of Stars. Microvariability and activity. In Appourchaux, T. (ed), *Phase-A study report, SCI(93)3*, European Space Agency.

Appourchaux, T., Toutain, T., Telljohann, U., *et al.* 1995. Frequencies and splittings of low-degree solar p modes: results of the Luminosity Oscillations Imager. *A&A*, **294**, L13–L16.

Appourchaux, T., Andersen, B. N., Frohlich, C., *et al.* 1997. In-flight performance of the Virgo Luminosity Oscillations Imager aboard SOHO. *Sol. Phys.*, **170**, 27–41.

Appourchaux, T., Michel, E., Auvergne, M., *et al.* 2008. CoRoT sounds the stars: P-mode parameters of Sun-like oscillations on HD 49933. *A&A*, **488**(Sept.), 705–714.

Appourchaux, T., Belkacem, K., Broomhall, A.-M., *et al.* 2010. The quest for the solar g modes. *A&A Rev.*, **18**(Feb.), 197–277.

Appourchaux, T., Chaplin, W. J., García, R. A., *et al.* 2012a. Oscillation mode frequencies of 61 main-sequence and subgiant stars observed by Kepler. *A&A*, **543**(July), A54.

Appourchaux, T., Benomar, O., Gruberbauer, M., *et al.* 2012b. Oscillation mode linewidths of main-sequence and subgiant stars observed by Kepler. *A&A*, **537**(Jan.), A134.

Araki, H. 2001. Focal processes of deep moonquakes. *J. Geod. Soc. Jpn.*, **47**(1), 508–513.

Araki, H., Tazawa, S., Node, H., *et al.* 2009. Lunar global shape and polar topography derived from Kaguya-LALT laser altimetry. *Science*, **323**, 897–900.

Arentoft, T., Kjeldsen, H., Bedding, T. R., *et al.* 2008. A multisite campaign to measure solar-like oscillations in Procyon. I. Observations, data reduction, and slow variations. *ApJ*, **687**(Nov.), 1180–1190.

Artru, J., Farges, T., and Lognonné, P. 2004. Acoustic waves generated from seismic surface waves: Propagation properties determined from Doppler sounding observation and normal-modes modeling. *Geophys. J. Int.*, **158**, 1067–1077.

Asplund, M., Grevesse, N., and Sauval, A. J. 2005 (Sept.). The solar chemical composition. Page 25 of Barnes, III, T. G., and Bash, F. N. (eds), *Cosmic Abundances as Records of Stellar Evolution and Nucleosynthesis*. Astronomical Society of the Pacific Conference Series, vol. 336.

Asplund, M., Grevesse, N., Sauval, A. J., and Scott, P. 2009. The chemical composition of the Sun. *ARA&A*, **47**(Sept.), 481–522.

Bachmann, K. T., and Brown, T. M. 1993. P-mode frequency variation in relation to global solar activity. *ApJ*, **411**(July), L45–L48.

Bachmann, K. T., Duvall, Jr., T. L., Harvey, J. W., and Hill, F. 1995. Frequencies of high degree solar oscillations. Page 156 of Ulrich, R. K., Rhodes, Jr., E. J., and Dappen, W. (eds), *GONG 1994. Helio- and Astro-Seismology from the Earth and Space*. Astronomical Society of the Pacific Conference Series, vol. 76.

Baer, J. J. 2010. *Development of an observational error model, and astrometric masses of 28 asteroids*. Ph.D. thesis, James Cook University, Australia.

Baglin, A., Weiss, W. W., and Bisnovatyi-Kogan, G. 1993 (Jan.). E. V. R. I. S on board MARS 94: the first space experiment devoted to stellar seismology. Page 758 of Weiss, W. W., and Baglin, A. (eds), *IAU Colloqium 137: Inside the Stars*. Astronomical Society of the Pacific Conference Series, vol. 40.

Baglin, A., Auvergne, M., Barge, P., *et al.* 2009 (Feb.). CoRoT: Description of the mission and early results. Pages 71–81 of Pont, F., Sasselov, D., and Holman, M. J. (eds), *Transiting Planets*. IAU Symposium, S253.

Baglin, A., Michel, E., and CoRoT Team. 2012 (Sept.). CoRoT: A few highlights and their impacts on understanding internal stellar structure. Page 492 of Shibahashi, H., Takata, M., and Lynas-Gray, A. E. (eds), *Progress in Solar/Stellar Physics with Helio- and Asteroseismology*. Astronomical Society of the Pacific Conference Series, vol. 462.

Bahcall, J. N., Pinsonneault, M. H., Basu, S., and Christensen-Dalsgaard, J. 1997. Are standard solar models reliable? *Phys. Rev. Lett.*, **78**(Jan.), 171–174.

Bahcall, J. N., Basu, S., and Pinsonneault, M. H. 1998. How uncertain are solar neutrino predictions? *Phys. Lett. B*, **433**, 1–8.

Bahcall, J. N., Basu, S., Pinsonneault, M., and Serenelli, A. M. 2005a. Helioseismological implications of recent solar abundance determinations. *ApJ*, **618**(Jan.), 1049–1056.

Bahcall, J. N., Basu, S., and Serenelli, A. M. 2005b. What is the neon abundance of the Sun? *ApJ*, **631**(Oct.), 1281–1285.

Bahcall, J. N., Serenelli, A. M., and Basu, S. 2006. 10,000 standard solar models: A Monte Carlo simulation. *ApJS*, **165**(July), 400–431.

Baldner, C. S., and Basu, S. 2008. Solar cycle related changes at the base of the convection zone. *ApJ*, **686**(Oct.), 1349–1361.

Baldner, C. S., and Schou, J. 2012. Effects of asymmetric flows in solar convection on oscillation modes. *ApJ*, **760**(Nov.), L1.

Ballot, J., Turck-Chièze, S., and García, R. A. 2004. Seismic extraction of the convective extent in solar-like stars: The observational point of view. *A&A*, **423**(Sept.), 1051–1061.

Ballot, J., García, R. A., and Lambert, P. 2006. Rotation speed and stellar axis inclination from p modes: How CoRoT would see other suns. *MNRAS*, **369**(July), 1281–1286.

Ballot, J., Lignières, F., Reese, D. R., and Rieutord, M. 2010. Gravity modes in rapidly rotating stars: Limits of perturbative methods. *A&A*, **518**(July), A30.

Ballot, J., Lignières, F., Prat, V., Reese, D. R., and Rieutord, M. 2012 (Sept.). 2D computations of g-modes in fast rotating stars. Page 389 of Shibahashi, H., Takata, M., and Lynas-Gray, A. E. (eds), *Progress in Solar/Stellar Physics with Helio- and Asteroseismology*. Astronomical Society of the Pacific Conference Series, vol. 462.

Ballot, J., Lignières, F., and Reese, D. R. 2013. Numerical exploration of oscillation modes in rapidly rotating stars. Page 91 of Goupil, M., Belkacem, K., Neiner, C., Lignières, F., and Green, J. J. (eds), *Studying Stellar Rotation and Convection*. Lecture Notes in Physics, Berlin: Springer Verlag, vol. 865.

Balmforth, N. J. 1992a. Solar pulsational stability. Part Three: Acoustical excitation by turbulent convection. *MNRAS*, **255**(Apr.), 639–649.

Balmforth, N. J. 1992b. Solar pulsational stability. I: Pulsation-mode thermodynamics. *MNRAS*, **255**(Apr.), 603–631.

Bamberger, A., Chavent, G., Hemon, Ch., and Lailly, P. 1982. Inversion of normal incidence seismograms. *Geophysics*, **47**, 757–770.

Banerdt, W. B, Chui, T., Griggs, C. E., *et al.* 2007. Using the Moon as a low-noise seismic detector for strange quark nuggets. *Nucl. Phys. B Proc. Suppl.*, **166**, 203–208.

Banerdt, W. B., Smrekar, S., Alkalai, L., *et al.* 2012. *InSight: An Integrated Exploration of the Interior of Mars*, 43rd Lunar and Planetary Science Conference, March 19–23, 2012 at The Woodlands, Texas. LPI Contribution No. 1659.

Barban, C., De Ridder, J., Mazumdar, A., *et al.* 2004a. Detection of solar-like oscillations in two red giant stars. Page 113 of Danesy, D. (ed), *SOHO 14 Helio- and Asteroseismology: Towards a Golden Future*. ESA Special Publication, vol. 559.

Barban, C., Hill, F., and Kras, S. 2004b. Simultaneous velocity-intensity spectral and cross-spectral fitting of helioseismic data. *ApJ*, **602**(Feb.), 516–527.

Barban, C., Matthews, J. M., De Ridder, J., *et al.* 2007. Detection of solar-like oscillations in the red giant star ϵ Ophiuchi by MOST spacebased photometry. *A&A*, **468**(June), 1033–1038.

Barnes, S. A. 2003. On the rotational evolution of solar- and late-type stars, its magnetic origins, and the possibility of stellar gyrochronology. *ApJ*, **586**(Mar.), 464–479.

Bass, H. E., and Chambers, J. P. 2001. Absorption of sound in the Martian atmosphere. *J. Acoust. Soc. Am.*, **109**, 3069–3071.

Bates, J. R., Lauderdale, W. W., and Kernaghan, H. 1979. *ALSEP Termination Report*, NASA reference Publication 1036, NASA Scientific and Technical Information Office.

Bastien, F. A., Stassun, K. G., Basri, G., and Pepper, J. 2013. An observational correlation between stellar brightness variations and surface gravity. *Nature*, **500**(Aug.), 427–430.

Basu, S. 2003. Stellar inversions. *Ap&SS*, **284**, 153–164.

Basu, S., and Antia, H. M. 1995. Helium abundance in the solar envelope. *MNRAS*, **276**(Oct.), 1402–1408.

Basu, S., and Antia, H. M. 1997. Seismic measurement of the depth of the solar convection zone. *MNRAS*, **287**(May), 189–198.

Basu, S., and Antia, H. M. 1999. Large-scale flows in the solar interior: Effect of asymmetry in peak profiles. *ApJ*, **525**(Nov.), 517–523.

Basu, S., and Antia, H. M. 2000. Effect of asymmetry in peak profiles on solar oscillation frequencies. *ApJ*, **531**(Mar.), 1088–1093.

Basu, S., and Antia, H. M. 2004. Constraining solar abundances using helioseismology. *ApJ*, **606**(May), L85–L88.

Basu, S., and Antia, H. M. 2008. Helioseismology and solar abundances. *Phys. Rep.*, **457**(Mar.), 217–283.

Basu, S., and Antia, H. M. 2010. Characteristics of solar meridional flows during solar cycle 23. *ApJ*, **717**(July), 488–495.

Basu, S., and Christensen-Dalsgaard, J. 1997. Equation of state and helioseismic inversions. *A&A*, **322**(June), L5–L8.

Basu, S., Antia, H. M., and Narasimha, D. 1994. Helioseismic measurement of the extent of overshoot below the solar convection zone. *MNRAS*, **267**(Mar.), 209.

Basu, S., Christensen-Dalsgaard, J., Chaplin, W. J., *et al.* 1997. Solar internal sound speed as inferred from combined BiSON and LOWL oscillation frequencies. *MNRAS*, **292**(Dec.), 243.

Basu, S., Däppen, W., and Nayfonov, A. 1999a. Helioseismic analysis of the hydrogen partition function in the solar interior. *ApJ*, **518**(June), 985–993.

Basu, S., Antia, H. M., and Tripathy, S. C. 1999b. Ring diagram analysis of near-surface flows in the Sun. *ApJ*, **512**(Feb.), 458–470.

Basu, S., Pinsonneault, M. H., and Bahcall, J. N. 2000. How much do helioseismological inferences depend on the assumed reference model? *ApJ*, **529**(Feb.), 1084–1100.

Basu, S., Christensen-Dalsgaard, J., and Thompson, M. J. 2002 (Jan.). SOLA inversions for the core structure of solar-type stars. Pages 249–252 of Battrick, B., Favata, F., Roxburgh, I. W., and Galadi, D. (eds), *Stellar Structure and Habitable Planet Finding*. ESA Special Publication, vol. 485.

Basu, S., Antia, H. M., and Bogart, R. S. 2004. Ring-diagram analysis of the structure of solar active regions. *ApJ*, **610**(Aug.), 1157–1168.

Basu, S., Chaplin, W. J., Elsworth, Y., New, R., and Serenelli, A. M. 2009. Fresh insights on the structure of the solar core. *ApJ*, **699**(July), 1403–1417.

Basu, S., Chaplin, W. J., and Elsworth, Y. 2010. Determination of stellar radii from asteroseismic data. *ApJ*, **710**(Feb.), 1596–1609.

Basu, S., Grundahl, F., Stello, D., *et al.* 2011. Sounding open clusters: Asteroseismic constraints from Kepler on the properties of NGC 6791 and NGC 6819. *ApJ*, **729**(Mar.), L10.

Basu, S., Broomhall, A.-M., Chaplin, W. J., and Elsworth, Y. 2012. Thinning of the Sun's magnetic layer: The peculiar solar minimum could have been predicted. *ApJ*, **758**(Oct.), 43.

Batalha, N. M., Borucki, W. J., Bryson, S. T., *et al.* 2011. Kepler's first rocky planet: Kepler-10b. *ApJ*, **729**(Mar.), 27–48.

Baturin, V. A., Däppen, W., Gough, D. O., and Vorontsov, S. V. 2000. Seismology of the solar envelope: Sound-speed gradient in the convection zone and its diagnosis of the equation of state. *MNRAS*, **316**(July), 71–83.

Baudin, F., Barban, C., Belkacem, K., *et al.* 2011. Amplitudes and lifetimes of solar-like oscillations observed by CoRoT. Red-giant versus main-sequence stars. *A&A*, **529**(May), A84.

Bazot, M., Bourguignon, S., and Christensen-Dalsgaard, J. 2012. A Bayesian approach to the modelling of α Cen A. *MNRAS*, **427**(Dec.), 1847–1866.

Beauduin, R., Lognonné, P., Montagner, J. P., Cacho, S., Karczewski, J. F. and Morand, M. 1996. The effect of the atmospheric pressure changes on seismic signals or how to improve the quality of a station. *Bull. Seismol. Soc. Am.*, **86**, 1760–1769.

Beck, C., Fabbian, D., Moreno-Insertis, F., Puschmann, K. G., and Rezaei, R. 2013. Thermodynamic fluctuations in solar photospheric three-dimensional convection simulations and observations. *A&A*, **557**(Sept.), A109.

Beck, P. G., Bedding, T. R., Mosser, B., *et al.* 2011. Kepler detected gravity-mode period spacings in a red giant star. *Science*, **332**(Apr.), 205.

Beck, P. G., Montalban, J., Kallinger, T., *et al.* 2012. Fast core rotation in red-giant stars as revealed by gravity-dominated mixed modes. *Nature*, **481**(Jan.), 55–57.

Bedding, T. R. 2014. Solar-like oscillations: An observational perspective. Pages 60-86 in Pallé, P. and Esteban, C. (eds), *Asteroseismology*. Canary Islands Winter School of Astrophysics, vol. 22. Cambridge: Cambridge University Press.

Bedding, T. R., Kjeldsen, H., Bouchy, F., *et al.* 2005. The non-detection of oscillations in Procyon by MOST: Is it really a surprise? *A&A*, **432**(Mar.), L43–L48.

Bedding, T. R., Kjeldsen, H., Arentoft, T., *et al.* 2007. Solar-like oscillations in the G2 subgiant β Hydri from dual-site observations. *ApJ*, **663**(July), 1315–1324.

Bedding, T. R., Kjeldsen, H., Campante, T. L., *et al.* 2010a. A multi-site campaign to measure solar-like oscillations in Procyon. II. Mode frequencies. *ApJ*, **713**(Apr.), 935–949.

Bedding, T. R., Huber, D., Stello, D., *et al.* 2010b. Solar-like oscillations in low-luminosity red giants: First results from Kepler. *ApJ*, **713**(Apr.), L176–L181.

Bedding, T. R., Mosser, B., Huber, D., *et al.* 2011. Gravity modes as a way to distinguish between hydrogen- and helium-burning red giant stars. *Nature*, **471**(Mar.), 608–611.

Behrend, R., Bernasconi, L., and Roy, R., *et al.* 2006. Four new binary minor planets: (854) Frostia, (1089) Tama, (1313) Berna, (4492) Debussy. *A&A*, **446**(Feb.), 1177–1184.

Belkacem, K., Goupil, M. J., Dupret, M. A., *et al.* 2011. The underlying physical meaning of the $\nu_{\rm max}$–$\nu_{\rm c}$ relation. *A&A*, **530**(June), A142.

Belkacem, K., Dupret, M. A., Baudin, F., *et al.* 2012. Damping rates of solar-like oscillations across the HR diagram: Theoretical calculations confronted to CoRoT and Kepler observations. *A&A*, **540**(Apr.), L7.

Belkacem, K., Samadi, R., Mosser, B., Goupil, M.-J., and Ludwig, H.-G. 2013. On the seismic scaling relations $\Delta\nu$–ρ and $\nu_{\rm max}$–$\nu_{\rm c}$. Page 61 of Shibahashi, H., and Lynas-Gray, A. E. (eds), *Progress in Physics of the Sun and Stars: A New Era in Helio- and Asteroseismology*. Astronomical Society of the Pacific Conference Series, vol. 479.

Belloche, A., Schuller, F., Parise, B., *et al.* 2011. The end of star formation in Chamaeleon I? A LABOCA census of starless and protostellar cores. *A&A*, **527**(Mar.), A145.

Bellot Rubio, L. R. 2010. The Evershed flow and the brightness of the penumbra. Page 193 of Hasan, S. S., and Rutten, R. J. (eds), *Magnetic Coupling between the Interior and Atmosphere of the Sun*. Springer Astrophysics and Space Science Proceedings.

Belmonte, J. A., Jones, A. R., Palle, P. L., and Roca Cortés, T. 1990a. Acoustic oscillations in the K2 III star Arcturus. *Ap&SS*, **169**(July), 77–84.

Belmonte, J. A., Pérez Hernández, F., and Roca Cortés, T. 1990b. Detection of acoustic oscillations in the F2 V star HD 155543. *A&A*, **231**(May), 383–390.

Belmonte, J. A., Jones, A. R., Palle, P. L., and Roca Cortés, T. 1990c. Global acoustic oscillations on Alpha Bootis. *ApJ*, **358**(Aug.), 595–609.

Ben-Menahem, A. 1975. Source parameters of the Siberian explosion of June 30, 1908, from analysis and synthesis of seismic signal at four stations. *Phys. Earth Planet. Inter.*, **11**, 1–35.

Benomar, O., Appourchaux, T., and Baudin, F. 2009. The solar-like oscillations of HD 49933: A Bayesian approach. *A&A*, **506**(Oct.), 15–32.

Bercovici, D., and Schubert, G. 1987. Jovian seismology. *Icarus*, **69**(Mar.), 557–565.

Berthomieu, G., Gonczi, G., Graff, P., Provost, J., and Rocca, A. 1978. Low-frequency gravity modes of a rotating star. *A&A*, **70**(Nov.), 597.

Bills, B. G., and Ferrari, A. J. 1977. A lunar density model consistent with topographic, gravitational, librational, and seismic data. *J. Geophys. Res.*, **82**, 1306–1314.

Birch, A. C., and Kosovichev, A. G. 2000. Travel time sensitivity kernels. *Sol. Phys.*, **192**(Mar.), 193–201.

Birch, A. C., Kosovichev, A. G., Price, G. H., and Schlottmann, R. B. 2001. The accuracy of the Born and ray approximations in time-distance helioseismology. *ApJ*, **561**(Nov.), L229–L232.

Birch, A. C., Kosovichev, A. G., and Duvall, Jr., T. L. 2004. Sensitivity of acoustic wave travel times to sound-speed perturbations in the solar interior. *ApJ*, **608**(June), 580–600.

Birch, A. C., Gizon, L., Hindman, B. W., and Haber, D. A. 2007. The linear sensitivity of helioseismic ring diagrams to local flows. *ApJ*, **662**(June), 730–737.

Birch, A. C., Braun, D. C., and Fan, Y. 2010. An estimate of the detectability of rising flux tubes. *ApJ*, **723**(Nov.), L190–L194.

Bogdan, T. J. 1997. A comment on the relationship between the modal and time-distance formulations of local helioseismology. *ApJ*, **477**(Mar.), 475–484.

Bogdan, T. J., and Cally, P. S. 1995. Jacket modes: Solar acoustic oscillations confined to regions surrounding sunspots and plage. *ApJ*, **453**(Nov.), 919.

Böhm-Vitense, E. 1958. Über die Wasserstoffkonvektionszone in Sternen verschiedener Effektivtemperaturen und Leuchtkräfte. Mit 5 Textabbildungen. *ZAp*, **46**, 108.

Böhm-Vitense, E. 2007. Chromospheric activity in G and K main-sequence stars, and what it tells us about stellar dynamos. *ApJ*, **657**(Mar.), 486–493.

Bonet, J. A., Márquez, I., Sánchez Almeida, J., Cabello, I., and Domingo, V. 2008. Convectively driven vortex flows in the Sun. *ApJ*, **687**(Nov.), L131–L134.

Bonet, J. A., Márquez, I., Sánchez Almeida, J., *et al.* 2010. SUNRISE/IMaX observations of convectively driven vortex flows in the Sun. *ApJ*, **723**(Nov.), L139–L143.

Borucki, W. J., Koch, D., Basri, G., *et al.* 2010. Kepler planet-detection mission: Introduction and first results. *Science*, **327**(Feb.), 977–980.

Bouchy, F., and Carrier, F. 2001. P-mode observations on α Cen A. *A&A*, **374**(July), L5–L8.

Bouchy, F., and Carrier, F. 2002. The acoustic spectrum of alpha Cen A. *A&A*, **390**(July), 205–212.

Bouchy, F., Pepe, F., and Queloz, D. 2001. Fundamental photon noise limit to radial velocity measurements. *A&A*, **374**(Aug.), 733–739.

Bouvier, J. 2013 (Sept.). Observational studies of stellar rotation. Pages 143–168 of *Role and Mechanisms of Angular Momentum Transport During the Formation and Early Evolution of Stars*. EAS Publications Series, vol. 62.

Bouvier, J., Forestini, M., and Allain, S. 1997. The angular momentum evolution of low-mass stars. *A&A*, **326**(Oct.), 1023–1043.

Bozdag, E., Trampert, J., and Tromp, J. 2011. Misfit functions for full waveform inversion based on instantaneous phase and envelope measurements. *Geophys. J. Int.*, **185**, 845–870.

Braithwaite, J. 2006. A differential rotation driven dynamo in a stably stratified star. *A&A*, **449**(Apr.), 451–460.

Braithwaite, J. 2008. On non-axisymmetric magnetic equilibria in stars. *MNRAS*, **386**(June), 1947–1958.

Braithwaite, J., and Spruit, H. C. 2004. A fossil origin for the magnetic field in A stars and white dwarfs. *Nature*, **431**(Oct.), 819–821.

Brandão, I. M., Cunha, M. S., and Christensen-Dalsgaard, J. 2014. On the inference of stellar ages and convective-core properties in main-sequence solar-like pulsators. *MNRAS*, **438**(Feb.), 1751–1761.

Brandenburg, A., Jennings, R. L., Nordlund, Å., *et al.* 1996. Magnetic structures in a dynamo simulation. *J. Fluid Mech.*, **306**(Jan.), 325–352.

Brandt, P. N., Scharmer, G. B., Ferguson, S., Shine, R. A., and Tarbell, T. D. 1988. Vortex flow in the solar photosphere. *Nature*, **335**(Sept.), 238–240.

Braun, D. C. 1995. Scattering of p-modes by sunspots: I. Observations. *ApJ*, **451**(Oct.), 859.

Braun, D. C., and Birch, A. C. 2008. Surface-focused seismic holography of sunspots: I. Observations. *Sol. Phys.*, **251**, 267–289.

Braun, D. C., Duvall, Jr., T. L., and Labonte, B. J. 1987. Acoustic absorption by sunspots. *ApJ*, **319**(Aug.), L27–L31.

Braun, D. C., Duvall, Jr., T. L., and Labonte, B. J. 1988. The absorption of high-degree p-mode oscillations in and around sunspots. *ApJ*, **335**(Dec.), 1015–1025.

Braun, D. C., Lindsey, C., Fan, Y., and Jefferies, S. M. 1992. Local acoustic diagnostics of the solar interior. *ApJ*, **392**(June), 739–745.

Braun, D. C., Birch, A. C., Rempel, M., and Duvall, T. L. 2012. Helioseismology of a realistic magnetoconvective sunspot simulation. *ApJ*, **744**(Jan.), 77.

Breger, M. 1993. Accuracy in variable-star work: The three-star single channel technique. Page 106 of Butler, C. J., and Elliott, I. (eds), *Stellar Photometry: Current Techniques and Future Developments*. IAU Colloquium 136.

Breger, M., Stich, J., Garrido, R., *et al.* 1993. Nonradial pulsation of the delta-Scuti star Bu-Cancri in the Praesepe cluster. *A&A*, **271**(Apr.), 482.

Brenguier, F., Shapiro, N. M., Campillo, M., Nercessian, A., and Ferrazzini, V. 2007. 3-D surface wave tomography of the Piton de la Fournaise volcano using seismic noise correlations. *Geophys. Res. Lett.*, **34**(Jan.), 2305.

Brillouin, L. 1960. *Wave Propagation and Group Velocity*. New York: Academic Press.

Britt, D. T., Yeomans, D., Housen, K., and Consolmagno, G. 2002. Asteroid density, porosity, and structure. *Asteroids III*, 485–500.

Brogi, M., Snellen, I. A. G., de Kok, R. J., *et al.* 2012. The signature of orbital motion from the dayside of the planet τ Boötis b. *Nature*, **486**(June), 502–504.

Brookes, J. R., Isaak, G. R., and van der Raay, H. B. 1978. A resonant-scattering solar spectrometer. *MNRAS*, **185**(Oct.), 1–18.

Broomhall, A.-M., Salabert, D., Chaplin, W. J., *et al.* 2012a. Misleading variations in estimated rotational frequency splittings of solar p modes: Consequences for helioseismology and asteroseismology. *MNRAS*, **422**(June), 3564–3573.

Broomhall, A.-M., Chaplin, W. J., Elsworth, Y., and Simoniello, R. 2012b. Quasi-biennial variations in helioseismic frequencies: Can the source of the variation be localized? *MNRAS*, **420**(Feb.), 1405–1414.

Brossier, R., Operto, S., and Virieux, J. 2009. Robust elastic frequency-domain full-waveform inversion using the L-1 norm. *Geophys. Res. Lett.,* **36**, L20310.

Brown, P., Spalding, R. E., ReVelle, D. O., Tagliaferri, E., and Worden, S. P. 2002. The flux of small near-Earth objects colliding with the Earth. *Nature*, **420**, 294–296.

Brown, T. 1981. The Fourier tachometer: Principles of operation and current status. Page 150 of Dunn, R. B. (ed), *Solar Instrumentation: What's Next?* Sunspot, NM: Sacramento Peak National Observatory.

Brown, T. M. 1985. Solar rotation as a function of depth and latitude. *Nature*, **317**(Oct.), 591–594.

Brown, T. M., and Christensen-Dalsgaard, J. 1990. A technique for estimating complicated power spectra from time series with gaps. *ApJ*, **349**(Feb.), 667–674.

Brown, T. M., Gilliland, R. L., Noyes, R. W., and Ramsey, L. W. 1991. Detection of possible p-mode oscillations on Procyon. *ApJ*, **368**(Feb.), 599–609.

Brown, T. M., Kennelly, E. J., Korzennik, S. G., *et al.* 1997. A radial velocity search for p-mode pulsations in eta Bootis. *ApJ*, **475**(Jan.), 322.

Brown, T. M., Latham, D. W., Everett, M. E., and Esquerdo, G. A. 2011. Kepler Input Catalog: Photometric calibration and stellar classification. *AJ*, **142**(Oct.), 112.

Browning, M. K., Miesch, M. S., Brun, A. S., and Toomre, J. 2006. Dynamo action in the solar convection zone and tachocline: Pumping and organization of toroidal fields. *ApJ*, **648**(Sept.), L157–L160.

Brummell, N. H., Hurlburt, N. E., and Toomre, J. 1998. Turbulent compressible convection with rotation: II. Mean flows and differential rotation. *ApJ*, **493**(Jan.), 955.

Brun, A. S. 2004. On the interaction between differential rotation and magnetic fields in the Sun. *Sol. Phys.*, **220**(Apr.), 333–345.

Brun, A. S., and Toomre, J. 2002. Turbulent convection under the influence of rotation: Sustaining a strong differential rotation. *ApJ*, **570**(May), 865–885.

Brun, A. S., and Zahn, J.-P. 2006. Magnetic confinement of the solar tachocline. *A&A*, **457**(Oct.), 665–674.

Brun, A. S., Turck-Chièze, S., and Zahn, J. P. 1999. Standard solar models in the light of new helioseismic constraints: II. Mixing below the convective zone. *ApJ*, **525**(Nov.), 1032–1041.

Brun, A. S., Miesch, M. S., and Toomre, J. 2004. Global-scale turbulent convection and magnetic dynamo action in the solar envelope. *ApJ*, **614**(Oct.), 1073–1098.

Brun, A. S., Miesch, M. S., and Toomre, J. 2011. Modeling the dynamical coupling of solar convection with the radiative interior. *ApJ*, **742**(Dec.), 79.

Brun, A. S., Browning, M. K., Dikpati, M., Hotta, H., and Strugarek, A. 2013. Recent advances on solar global magnetism and variability. *Space Sci. Rev.*, Nov, doi:10.1007/s11214-013-0028-0.

Bruntt, H., Kjeldsen, H., Buzasi, D. L., and Bedding, T. R. 2005. Evidence for granulation and oscillations in Procyon from photometry with the WIRE satellite. *ApJ*, **633**(Nov.), 440–446.

Bulow, R. C., Johnson, C. L., and Shearer, P. M. 2005. New events discovered in the Apollo lunar seismic data. *J. Geophys. Res.*, **110**, E10003, doi:10.1029/2005JE002414.

Bulow, R. C., Johnson, C. L., Bills, B. G., and Shearer, P. M. 2007. Temporal and spatial properties of some deep moonquake clusters. *J. Geophys. Res.*, **112**, E09003, doi:10.1029/2006JE002847.

Buzasi, D. L. 2000 (Apr.). Experiment design and data reduction for seismology: Lessons learned from WIRE. Page 9 of Teixeira, T., and Bedding, T. (eds), *The Third MONS Workshop: Science Preparation and Target Selection.*

Buzasi, D., Catanzarite, J., Laher, R., *et al.* 2000. The detection of multimodal oscillations on α Ursae Majoris. *ApJ*, **532**(Apr.), L133–L136.

Buzasi, D. L., Bruntt, H., Preston, H. L., and Clausen, J. V. 2004 (Dec.). High precision photometry with WIRE: Converting spectroscopic binaries to eclipsing systems. *Bull. AAS*, **36**, 1369.

Cacciani, A., and Fofi, M. 1978. The magneto-optical filter: II. Velocity field measurements. *Sol. Phys.*, **59**(Sept.), 179–189.

Cacciani, A., Dolci, M., Moretti, P. F., *et al.* 2001. Search for global oscillations on Jupiter with a double-cell sodium magneto-optical filter. *A&A*, **372**(June), 317–325.

Cacciani, A., Jefferies, S. M., Finsterle, W., *et al.* 2003 (Feb.). A new instrument for sounding the solar atmosphere. Pages 243–245 of Sawaya-Lacoste, H. (ed), *GONG+ 2002. Local and Global Helioseismology: the Present and Future*. ESA Special Publication, vol. 517.

Caffau, E., Ludwig, H.-G., Steffen, M., Freytag, B., and Bonifacio, P. 2011. Solar chemical abundances determined with a CO5BOLD 3D model atmosphere. *Sol. Phys.*, **268**(Feb.), 255–269.

Callen, H. B., and Welton, T. A. 1951. Irreversibility and generalized noise. *Phys. Rev.*, **83**, 34–40.

Cally, P. S., and Bogdan, T. J. 1993. Solar p-modes in a vertical magnetic field: Trapped and damped pi-modes. *ApJ*, **402**(Jan.), 721–732.

Cally, P. S., Crouch, A. D., and Braun, D. C. 2003. Probing sunspot magnetic fields with p-mode absorption and phase shift data. *MNRAS*, **346**(Dec.), 381–389.

Cameron, R., Schüssler, M., Vögler, A., and Zakharov, V. 2007. Radiative magnetohydrodynamic simulations of solar pores. *A&A*, **474**(Oct.), 261–272.

Cammarano, F., Lekic, V., Manga, M., Panning, M., and Romanowicz, B. 2006. Long-period seismology on Europa: 1. Physically consistent interior models. *J. Geophys. Res. (Planets)*, **111**(10), 12009.

Campante, T. L., Handberg, R., Mathur, S., *et al.* 2011. Asteroseismology from multi-month Kepler photometry: The evolved Sun-like stars KIC 10273246 and KIC 10920273. *A&A*, **534**(Oct.), A6.

Canuto, V. M., and Mazzitelli, I. 1991. Stellar turbulent convection: A new model and applications. *ApJ*, **370**(Mar.), 295–311.

Carlsson, M. 1998. Radiative transfer and radiation hydrodynamics. Page 163 of Vial, J. C., Bocchialini, K., and Boumier, P. (eds), *Space Solar Physics: Theoretical and Observational Issues in the Context of the SOHO Mission*. Lecture Notes in Physics, Berlin: Springer Verlag, vol. 507.

Carpenter, K. G., Schrijver, C. J., and Karovska, M. 2009. The Stellar Imager (SI) project: A deep space UV/optical interferometer (UVOI) to observe the Universe at 0.1 milli-arcsec angular resolution. *Ap&SS*, **320**(Apr.), 217–223.

Carpenter, K. G., Schrijver, C. J., Karovska, M., *et al.* 2010. Stellar Imager (SI): Developing and testing a predictive dynamo model for the Sun by imaging other stars. ArXiv:1011.5214.

Carrier, F., and Bourban, G. 2003. The acoustic spectrum of alpha Cen B. *A&A*, **406**(July), L23.

Carrier, F., Eggenberger, P., and Bouchy, F. 2005. New seismological results on the G0 IV η Bootis. *A&A*, **434**(May), 1085–1095.

Catala, C. 2009. PLATO: PLAnetary Transits and Oscillations of stars. *Exp. Astron.*, **23**(Mar.), 329–356.

Catala, C., Mangeney, A., Gautier, D., *et al.* 1995. COROT: A proposal to study stellar convection and internal rotation. Page 426 of Ulrich, R. K., Rhodes, Jr., E. J., and Dappen, W. (eds), *GONG 1994: Helio- and Astro-Seismology from the Earth and Space*. Astronomical Society of the Pacific Conference Series, vol. 76.

Ceillier, T., Eggenberger, P., García, R. A., and Mathis, S. 2013. Understanding angular momentum transport in red giants: The case of KIC 7341231. *A&A*, **555**(July), A54.

Červený, V. 1985. The application of ray tracing to the numerical modeling of seismic wavefields in complex structures. Pages 1–124 of Dohr, G. P. (ed), *Handbook of Geophysical Exploration*, vol. 15A. London: Geophysical Press.

Chaboyer, B., Demarque, P., and Pinsonneault, M. H. 1995. Stellar models with microscopic diffusion and rotational mixing: 1. Application to the Sun. *ApJ*, **441**(Mar.), 865–875.

Chakrabarti, S. K., and Manickam, S. G. 2000. Correlation among quasi-periodic oscillation frequencies and quiescent-state duration in black hole candidate GRS 1915+105. *ApJ*, **531**(Mar.), L41–L44.

Chandrasekhar, S. 1964. A general variational principle governing the radial and the non-radial oscillations of gaseous masses. *ApJ*, **139**(Feb.), 664.

Chaplin, W. J. 2006. *Music of the Sun: Story of Helioseismology*. London: Oneworld Publications.

Chaplin, W. J., and Miglio, A. 2013. Asteroseismology of solar-type and red-giant stars. *ARA&A*, **51**, 353–392.

Chaplin, W. J., Elsworth, Y., Howe, R., *et al.* 1996. BiSON performance. *Sol. Phys.*, **168**(Sept.), 1–18.

Chaplin, W. J., Elsworth, Y., Isaak, G. R., *et al.* 1997a. Solar p-mode linewidths from recent BiSON helioseismological data. *MNRAS*, **288**(July), 623–626.

Chaplin, W. J., Elsworth, Y., Howe, R., *et al.* 1997b. Techniques used in the analysis of data collected by the Birmingham Solar-Oscillations Network (BiSON): II. Frequency domain analysis and data merging. *A&AS*, **125**(Oct.), 195–205.

Chaplin, W. J., Christensen-Dalsgaard, J., Elsworth, Y., *et al.* 1999. Rotation of the solar core from BiSON and LOWL frequency observations. *MNRAS*, **308**(Sept.), 405–414.

Chaplin, W. J., Elsworth, Y., Isaak, G. R., Miller, B. A., and New, R. 2000. Variations in the excitation and damping of low-l solar p modes over the solar activity cycle. *MNRAS*, **313**(Mar.), 32–42.

Chaplin, W. J., Elsworth, Y., Isaak, G. R., *et al.* 2001. Changes to low-l solar p-mode frequencies over the solar cycle: Correlations on different time-scales. *MNRAS*, **322**(Mar.), 22–30.

Chaplin, W. J., Sekii, T., Elsworth, Y., and Gough, D. O. 2004. On the detectability of a rotation-rate gradient in the solar core. *MNRAS*, **355**(Dec.), 535–542.

Chaplin, W. J., Appourchaux, T., Baudin, F., *et al.* 2006. Solar FLAG hare and hounds: On the extraction of rotational p-mode splittings from seismic, Sun-as-a-star data. *MNRAS*, **369**(June), 985–996.

Chaplin, W. J., Appourchaux, T., Arentoft, T., *et al.* 2008. AsteroFLAG: First results from hare-and-hounds Exercise #1. *Astron. Nachr.*, **329**, 549.

Chaplin, W. J., Houdek, G., Karoff, C., Elsworth, Y., and New, R. 2009. Mode lifetimes of stellar oscillations: Implications for asteroseismology. *A&A*, **500**(June), L21–L24.

Chaplin, W. J., Kjeldsen, H., Christensen-Dalsgaard, J., *et al.* 2011a. Ensemble asteroseismology of solar-type stars with the NASA Kepler mission. *Science*, **332**(Apr.), 213–216.

Chaplin, W. J., Kjeldsen, H., Bedding, T. R., *et al.* 2011b. Predicting the detectability of oscillations in solar-type stars observed by Kepler. *ApJ*, **732**(May), 54.

Chaplin, W. J., Sanchis-Ojeda, R., Campante, T. L., *et al.* 2013a. Asteroseismic determination of obliquities of the exoplanet systems Kepler-50 and Kepler-65. *ApJ*, **766**(Apr.), 101.

Chaplin, W. J, Kjeldsen, H., Christensen-Dalsgaard, J., *et al.* 2013b. Kepler white paper: Asteroseismology of solar-like oscillators in a 2-wheel mission. ArXiv:1309.0702.

Chaplin, W. J., Basu, S., Huber, D., *et al.* 2014. Asteroseismic fundamental properties of solar-type stars observed by the NASA Kepler mission. *ApJS*, **210**(Jan.), 1.

Charbonneau, P. 2010. Dynamo models of the solar cycle. *Living Rev. Solar Phys.*, **7**(Sept.), 3.

Charbonneau, P., and MacGregor, K. B. 1993. Angular momentum transport in magnetized stellar radiative zones. II. The solar spin-down. *ApJ*, **417**(Nov.), 762.

Charbonneau, P., Christensen-Dalsgaard, J., Henning, R., *et al.* 1999. Helioseismic constraints on the structure of the solar tachocline. *ApJ*, **527**(Dec.), 445–460.

Charbonnel, C., and Talon, S. 2005. Influence of gravity waves on the internal rotation and Li abundance of solar-type stars. *Science*, **309**(Sept.), 2189–2191.

Chaty, S. 2007. The role of microquasars in astroparticle physics. Page 329 of Giovannelli, F., and Mannocchi, G. (eds), *Vulcano Workshop: Frontier Objects in Astrophysics and Particle Physics*.

Chaty, S. 2013. Optical/infrared observations unveiling the formation, nature and evolution of high-mass X-ray binaries. *Adv. Space Res.*, **52**(Dec.), 2132–2142.

Chen, P., Jordan, T. H., and Zhao, L. 2007. Full three-dimensional tomography: A comparison between the scattering-integral and adjoint-wavefield methods. *Geophys. J. Int.*, **170**, 175–181.

Chenet, H., Lognonné, Ph., Wieczorek, M., and Mizutani, H. 2006. Lateral variations of lunar crustal thickness from the Apollo seismic data set. *Earth Planet. Sci. Lett.*, **243**, 1–14.

Cheng, C. H., and Toksoz, M. N. 1978. Tidal stresses in the Moon. *J. Geophys. Res.*, **83**(Feb.), 845–853.

Cheung, M. C. M., Schüssler, M., Tarbell, T. D., and Title, A. M. 2008. Solar surface emerging flux regions: A comparative study of radiative MHD modeling and Hinode SOT observations. *ApJ*, **687**(Nov.), 1373–1387.

Chou, D.-Y., and Dai, D.-C. 2001. Solar cycle variations of subsurface meridional flows in the Sun. *ApJ*, **559**(Oct.), L175–L178.

Chou, D.-Y., Sun, M.-T., Huang, T.-Y., *et al.* 1995. Taiwan Oscillation Network. *Sol. Phys.*, **160**(Sept.), 237–243.

Chou, D.-Y., Yang, M.-H., Liang, Z.-C., and Sun, M.-T. 2009a. Measurements of absorption of acoustic waves in sunspots with direction filters and cross correlation. *ApJ*, **690**(Jan.), L23–L26.

Chou, D.-Y., Yang, M.-H., Zhao, H., Liang, Z.-C., and Sun, M.-T. 2009b. Spatial distributions of absorption, local suppression, and emissivity reduction of solar acoustic waves in magnetic regions. *ApJ*, **706**(Nov.), 909–916.

Chicarro, A. F. 2013. INSPIRE: A future European Mars Network Science Mission. 44th Lunar and Planetary Science Conference, March 18–22. 2013 in The Woodlands, Texas. LPI Contribution No. 1719.

Christensen, E. M., Batterson, S. A., Benson, H. E., *et al.* 1968. Lunar Surface Mechanical Properties. Pages 44–108 of *Surveyor VI Mission Report, Part II: Science Results*, Jet Propulsion Laboratory, Technical Report 32–1262, Pasadena.

Christensen-Dalsgaard, J. 1984. What will asteroseismology teach us? Page 11 of Mangeney, A., and Praderie, F. (eds), *Space Research in Stellar Activity and Variability*. Observatoire de Paris-Meudon.

Christensen-Dalsgaard, J. 1989. The effect of rotation on whole-disc Doppler observations of solar oscillations. *MNRAS*, **239**(Aug.), 977–994.

Christensen-Dalsgaard, J. 1993. On the asteroseismic HR diagram. Page 347 of Brown, T. M. (ed), *GONG 1992: Seismic Investigation of the Sun and Stars*. Astronomical Society of the Pacific Conference Series, vol. 42.

Christensen-Dalsgaard, J. 2002. Helioseismology. *Rev. Mod. Phys.*, **74**(Nov.), 1073–1129.

Christensen-Dalsgaard, J. 2004. Physics of solar-like oscillations. *Sol. Phys.*, **220**(Apr.), 137–168.

Christensen-Dalsgaard, J. 2012. Stellar model fits and inversions. *Astron. Nachr.*, **333**(Dec.), 914.

Christensen-Dalsgaard, J., and Di Mauro, M. P. 2007. Diffusion and helioseismology. Pages 3–16 of Straka, C. W., Lebreton, Y., and Monteiro, M. J. P. F. G. (eds), *Stellar Evolution and Seismic Tools for Asteroseismology*. EAS Publications Series, vol. 26.

Christensen-Dalsgaard, J., and Frandsen, S. 1983. Stellar 5 min oscillations. *Sol. Phys.*, **82**(Jan.), 469–486.

Christensen-Dalsgaard, J., and Gough, D. O. 1976. Towards a heliological inverse problem. *Nature*, **259**(Jan.), 89–92.

Christensen-Dalsgaard, J., and Gough, D. O. 1982. On the interpretation of five-minute oscillations in solar spectrum line shifts. *MNRAS*, **198**(Jan.), 141–171.

Christensen-Dalsgaard, J., and Schou, J. 1988 (Dec.). Differential rotation in the solar interior. Pages 149–153 of Rolfe, E. J. (ed), *Seismology of the Sun and Sun-Like Stars*. ESA Special Publication, vol. 286.

Christensen-Dalsgaard, J., and Thompson, M. J. 1997. On solar p-mode frequency shifts caused by near-surface model changes. *MNRAS*, **284**(Jan.), 527–540.

Christensen-Dalsgaard, J., Duvall, Jr., T. L., Gough, D. O., Harvey, J. W., and Rhodes, Jr., E. J. 1985. Speed of sound in the solar interior. *Nature*, **315**(May), 378–382.

Christensen-Dalsgaard, J., Thompson, M. J., and Gough, D. O. 1989. Differential asymptotic sound-speed inversions. *MNRAS*, **238**(May), 481–502.

Christensen-Dalsgaard, J., Gough, D. O., and Thompson, M. J. 1991. The depth of the solar convection zone. *ApJ*, **378**(Sept.), 413–437.

Christensen-Dalsgaard, J., Dappen, W., Ajukov, S. V., *et al.* 1996. The current state of solar modeling. *Science*, **272**(May), 1286–1292.

Christensen-Dalsgaard, J., Kjeldsen, H., and Mattei, J. A. 2001. Solar-like oscillations of semiregular variables. *ApJ*, **562**(Dec.), L141–L144.

Christensen-Dalsgaard, J., di Mauro, M. P., Houdek, G., and Pijpers, F. 2009. On the opacity change required to compensate for the revised solar composition. *A&A*, **494**(Jan.), 205–208.

Christensen-Dalsgaard, J., Monteiro, M. J. P. F. G., Rempel, M., and Thompson, M. J. 2011. A more realistic representation of overshoot at the base of the solar convective envelope as seen by helioseismology. *MNRAS*, **414**(June), 1158–1174.

Claret, A., and Giménez, A. 2010. The apsidal-motion test of stellar structure and evolution: An update. *A&A*, **519**(Sept.), A57.

Claverie, A., Isaak, G. R., McLeod, C. P., van der Raay, H. B., and Cortés, T. R. 1979. Solar structure from global studies of the 5-minute oscillation. *Nature*, **282**(Dec.), 591–594.

Clement, M. J. 1984. Normal modes of oscillation for rotating stars: II. Variational solutions. *ApJ*, **276**(Jan.), 724–736.

Cobden, L. J., Tong, C. H., and Warner, M. R. 2010. Influence of acoustic source density on cross-correlated signals: Implications for amplitude-based tomography in helioseismology. *ApJ*, **725**, 313–318.

Cobden, L. J., Tong, C. H., and Warner, M. R. 2011. Inversion of full acoustic wavefield in local helioseismology: A study with synthetic data. *ApJ*, **727**, 68.

Coleiro, A., and Chaty, S. 2013. Distribution of high-mass X-ray binaries in the Milky Way. *ApJ*, **764**(Feb.), 185.

Connes, P. 1985. Absolute astronomical accelerometry. *Ap&SS*, **110**(Mar.), 211–255.

Cooley, J. W., and Tukey, J. W. 1965. An algorithm for the machine calculation of complex Fourier series. *Math. Comput.*, **19**, 297–301.

Cooper, M. R., Kovach, R. L., and Watkins, J. S. 1974. Lunar near-surface structure. *Rev. Geophys. Space Phys.*, **12**, 291–308.

Cooper, M. R., and Kovach, R. L. 1975. Energy, frequency, and distance of moonquakes at the Apollo 17 site. *Proceedings of the 6th Lunar Science Conference*, 2863–2879.

Corsaro, E., Stello, D., Huber, D., *et al.* 2012. Asteroseismology of the open clusters NGC 6791, NGC 6811, and NGC 6819 from 19 months of Kepler photometry. *ApJ*, **757**(Oct.), 190.

Costa, J. E. S., and Kepler, S. O. 2008. The temporal changes of the pulsational periods of the pre-white dwarf PG 1159-035. *A&A*, **489**(Oct.), 1225–1232.

Couvidat, S., and Birch, A. C. 2006. Optimal Gaussian phase-speed filters in time-distance helioseismology. *Sol. Phys.*, **237**(Sept.), 229–243.

Couvidat, S., García, R. A., Turck-Chièze, S., *et al.* 2003. The rotation of the deep solar layers. *ApJ*, **597**(Nov.), L77–L79.

Couvidat, S., Birch, A. C., Kosovichev, A. G., and Zhao, J. 2004. Three-dimensional inversion of time-distance helioseismology data: Ray-path and Fresnel-zone approximations. *ApJ*, **607**(May), 554–563.

Couvidat, S., Zhao, J., Birch, A. C., *et al.* 2012a. Implementation and comparison of acoustic travel-time measurement procedures for the Solar Dynamics Observatory/Helioseismic and Magnetic Imager time–distance helioseismology pipeline. *Sol. Phys.*, **275**(Jan.), 357–374.

Couvidat, S., Schou, J., Shine, R. A., *et al.* 2012b. Wavelength dependence of the Helioseismic and Magnetic Imager (HMI) instrument onboard the Solar Dynamics Observatory (SDO). *Sol. Phys.*, **275**(Jan.), 285–325.

Covas, E., Moss, D., and Tavakol, R. 2005. Dynamo models and differential rotation in late-type rapidly rotating stars. *A&A*, **429**(Jan.), 657–665.

Cowling, T. G., and Newing, R. A. 1949. The oscillations of a rotating star. *ApJ*, **109**(Jan.), 149.

Cranmer, S. R., Bastien, F. A., Stassun, K. G., and Saar, S. H. 2014. Stellar granulation as the source of high-frequency flicker in Kepler light curves. *ApJ*, **781**(Feb.), 124.

Crawford, Jr., F. S. 1968. *Waves*. New York: McGraw-Hill Book Company.

Creevey, O. L., Monteiro, M. J. P. F. G., Metcalfe, T. S., *et al.* 2007. The complementary roles of interferometry and asteroseismology in determining the mass of solar-type stars. *ApJ*, **659**(Apr.), 616–625.

Criswell, D. R., and Lindsay, J. F. 1974. Thermal moonquakes and booming dunes (abstract). *Lunar Science V*, 151–153.

Crouch, A. D., Cally, P. S., Charbonneau, P., Braun, D. C., and Desjardins, M. 2005. Genetic magnetohelioseismology with Hankel analysis data. *MNRAS*, **363**(Nov.), 1188–1204.

Crouch, A. D., Birch, A. C., Braun, D. C., and Clack, C. T. M. 2011. Helioseismic probing of the subsurface structure of sunspots. Pages 384–388 of Choudhary, D. P., and Strassmeier, K. G. (eds), *Physics of Sun and Star Spots*. IAU Symposium 273.

Cunha, M. S., and Metcalfe, T. S. 2007. Asteroseismic signatures of small convective cores. *ApJ*, **666**(Sept.), 413–422.

Cunha, M. S., Aerts, C., Christensen-Dalsgaard, J., *et al.* 2007. Asteroseismology and interferometry. *A&A Rev.*, **14**(Nov.), 217–360.

Dahlen, F. A., and Tromp, J. 1998. *Theoretical Global Seismology*. Princeton: Princeton University Press.

Dainty, A. M., and M. N. Toksöz 1981. Seismic codas on the Earth and the Moon: A comparison. *Phys. Earth Planet. Inter.*, **26**, 250–260.

Dainty, A. M., Toksöz, M. N., Anderson, K. R., *et al.* 1974. Seismic scattering and shallow structure of the Moon in Oceanus Procellarum. *The Moon*, **9**, 11–29.

Daubar, I. J., McEwen, A. S., Byrne, S., Kennedy, M. R., and Ivanov, B. 2013. The current martian cratering rate. *Icarus*, **225**(July), 506–516.

Davis, P. M. 1993. Meteoroid impacts as seismic sources on Mars. *Icarus*, **105**, 469–478.

De Oliveira Fialho, F. 2007. *Définition et implémentation des corrections instrumentales de la mission spatiale CoRoT*. Ph.D. thesis, Université Pierre et Marie Curie, Paris, France.

De Ridder, J., Barban, C., Carrier, F., *et al.* 2006. Discovery of solar-like oscillations in the red giant ε Ophiuchi. *A&A*, **448**(Mar.), 689–695.

De Ridder, J., Barban, C., Baudin, F., *et al.* 2009. Non-radial oscillation modes with long lifetimes in giant stars. *Nature*, **459**(May), 398–400.

Decressin, T., Mathis, S., Palacios, A., *et al.* 2009. Diagnoses to unravel secular hydrodynamical processes in rotating main sequence stars. *A&A*, **495**(Feb.), 271–286.

Degroote, P., Briquet, M., Auvergne, M., *et al.* 2010. Detection of frequency spacings in the young O-type binary HD 46149 from CoRoT photometry. *A&A*, **519**(Sept.), A38.

Deheuvels, S., and Michel, E. 2011. Constraints on the structure of the core of subgiants via mixed modes: The case of HD 49385. *A&A*, **535**(Nov.), A91.

Deheuvels, S., Bruntt, H., Michel, E., *et al.* 2010. Seismic and spectroscopic characterization of the solar-like pulsating CoRoT target HD 49385. *A&A*, **515**(June), A87.

Deheuvels, S., García, R. A., Chaplin, W. J., *et al.* 2012. Seismic evidence for a rapidly rotating core in a lower-giant-branch star observed with Kepler. *ApJ*, **756**(Sept.), 19.

Deheuvels, S., Doğan, G., Goupil, M. J., *et al.* 2014. Seismic constraints on the radial dependence of the internal rotation profiles of six Kepler subgiants and young red giants. *A&A*, **564**(Apr.), A27.

Delahunty, A. K., and Pike, W. T. 2013. Metal-armouring for shock protection of MEMS. *Sensors and Actuators*, **215**, 36–43.

Deming, D., Mumma, M. J., Espenak, F., *et al.* 1989. A search for p-mode oscillations of Jupiter: Serendipitous observations of nonacoustic thermal wave structure. *ApJ*, **343**(Aug.), 456–467.

Denissenkov, P. A., Pinsonneault, M., Terndrup, D. M., and Newsham, G. 2010. Angular momentum transport in solar-type stars: Testing the timescale for core-envelope coupling. *ApJ*, **716**(June), 1269–1287.

Derr, J. S. 1969. Free oscillations of new lunar models. *Phys. Earth Planet. Inter.*, **2**, 61–68.

Deubner, F.-L. 1975. Observations of low wavenumber nonradial eigenmodes of the Sun. *A&A*, **44**(Nov.), 371–375.

Deupree, R. G. 2011. Structure of uniformly rotating stars. *ApJ*, **735**(July), 69.

Dikpati, M., and Charbonneau, P. 1999. A Babcock–Leighton flux transport dynamo with solar-like differential rotation. *ApJ*, **518**(June), 508–520.

Dikpati, M., and Gilman, P. A. 2006. Simulating and predicting solar cycles using a flux-transport dynamo. *ApJ*, **649**(Sept.), 498–514.

Dintrans, B., and Rieutord, M. 2000. Oscillations of a rotating star: A non-perturbative theory. *A&A*, **354**(Feb.), 86–98.

Dintrans, B., Brandenburg, A., Nordlund, Å., and Stein, R. F. 2005. Spectrum and amplitudes of internal gravity waves excited by penetrative convection in solar-type stars. *A&A*, **438**(July), 365–376.

Domingo, V., Fleck, B., and Poland, A. I. 1995. The SOHO mission: An overview. *Sol. Phys.*, **162**, 1–37.

Donati, J.-F., and Landstreet, J. D. 2009. Magnetic fields of nondegenerate stars. *ARA&A*, **47**(Sept.), 333–370.

Donea, A. 2011. Seismic transients from flares in solar cycle 23. *Space Sci. Rev.*, **158**(July), 451–469.

Donea, A.-C., and Lindsey, C. 2005. Seismic emission from the solar flares of 2003 October 28 and 29. *ApJ*, **630**(Sept.), 1168–1183.

Donea, A. C., Lindsey, C., and Braun, D. 1999. Helioseismic holography: A technique for understanding solar flares. *Romanian Astron. J.*, **9**, 71.

Dorman, J., Evans, S., Nakamura, Y., and Latham, G. 1978. On the time-varying properties of the lunar seismic meteoroid population. *Proceedings of the 9th Lunar and Planetary Science Conference*, 3615–3626.

Doschek, G. A., Warren, H. P., and Young, P. R. 2013. Chromospheric evaporation in an M1.8 flare observed by the Extreme-ultraviolet Imaging Spectrometer on Hinode. *ApJ*, **767**(Apr.), 55.

Drummond, R., Vandenbussche, B., Aerts, C., De Oliveira Fialho, F., and Auvergne, M. 2006. Jitter correction algorithms for the COROT satellite mission. *PASP*, **118**(June), 874.

Dubus, G., Hameury, J.-M., and Lasota, J.-P. 2001. The disc instability model for X-ray transients: Evidence for truncation and irradiation. *A&A*, **373**(July), 251–271.

Duennebier, F. 1976. Thermal movement of the regolith. *Proceedings of the 7th Lunar Science Conference*, 1073–1086.

Duennebier, F., and Sutton, G.H. 1974. Thermal moonquakes. *J. Geophys. Res.*, **79**(29), 4351–4363.

Duennebier, F., Dorman, J., Lammlein, D., Latham, G., and Nakamura, Y. 1975. Meteoroid flux from passive seismic experiment data. *Proceedings of the 6th Lunar Science Conference*, 2417–2426.

Duennebier, F. K., Nakamura, Y, Latham, G. V., and Dorman, H. J. 1976. Meteoroid storms detected on the Moon. *Science*, **192**, 1000–1002.

Duez, V., and Mathis, S. 2010. Relaxed equilibrium configurations to model fossil fields: I. A first family. *A&A*, **517**(July), A58.

Duez, V., Braithwaite, J., and Mathis, S. 2010. On the stability of non-force-free magnetic equilibria in stars. *ApJ*, **724**(Nov.), L34–L38.

Dupret, M.-A., Thoul, A., Scuflaire, R., *et al.* 2004. Asteroseismology of the β Cep star HD 129929: II. Seismic constraints on core overshooting, internal rotation and stellar parameters. *A&A*, **415**(Feb.), 251–257.

Dupret, M.-A., Belkacem, K., Samadi, R., *et al.* 2009. Theoretical amplitudes and lifetimes of non-radial solar-like oscillations in red giants. *A&A*, **506**(Oct.), 57–67.

Duvall, Jr., T. L. 1982. A dispersion law for solar oscillations. *Nature*, **300**(Nov.), 242.

Duvall, Jr., T. L., and Birch, A. C. 2010. The vertical component of the supergranular motion. *ApJ*, **725**(Dec.), L47–L51.

Duvall, T. L., and Hanasoge, S. M. 2013. Subsurface supergranular vertical flows as measured using large distance separations in time-distance helioseismology. *Sol. Phys.*, **287**(Oct.), 71–83.

Duvall, Jr., T. L., and Harvey, J. W. 1983. Observations of solar oscillations of low and intermediate degree. *Nature*, **302**(Mar.), 24–27.

Duvall, Jr., T. L., Dziembowski, W. A., Goode, P. R., *et al.* 1984. Internal rotation of the sun. *Nature*, **310**(July), 22–25.

Duvall, Jr., T. L., Harvey, J. W., and Pomerantz, M. A. 1986. Latitude and depth variation of solar rotation. *Nature*, **321**(May), 500.

Duvall, Jr., T. L., Jefferies, S. M., Harvey, J. W., Osaki, Y., and Pomerantz, M. A. 1993a. Asymmetries of solar oscillation line profiles. *ApJ*, **410**(June), 829–836.

Duvall, Jr., T. L., Jefferies, S. M., Harvey, J. W., and Pomerantz, M. A. 1993b. Time-distance helioseismology. *Nature*, **362**(Apr.), 430–432.

Duvall, T. L., D'Silva, S., Jefferies, S. M., Harvey, J. W., and Schou, J. 1996. Downflows under sunspots detected by helioseismic tomography. *Nature*, **379**(Jan.), 235–237.

Duvall, Jr., T. L., Kosovichev, A. G., Scherrer, P. H., *et al.* 1997. Time-distance helioseismology with the MDI instrument: Initial results. *Sol. Phys.*, **170**, 63–73.

Dziembowski, W., and Kosovichev, A. 1987. Low frequency oscillations in slowly rotating stars: I. General properties. *Acta Astronomica*, **37**, 313.

Dziembowski, W. A. 1977. Light and radial velocity variations in a nonradially oscillating star. *Acta Astron.*, **27**, 203–211.

Dziembowski, W. A., and Goode, P. R. 1992. Effects of differential rotation on stellar oscillations: A second-order theory. *ApJ*, **394**(Aug.), 670–687.

Dziembowski, W. A., and Pamyatnykh, A. A. 1991. A potential asteroseismological test for convective overshooting theories. *A&A*, **248**(Aug.), L11–L14.

Dziembowski, W. A., and Pamyatnykh, A. A. 2008. The two hybrid B-type pulsators: ν Eridani and 12 Lacertae. *MNRAS*, **385**(Apr.), 2061–2068.

Dziembowski, W. A., Goode, P. R., and Libbrecht, K. G. 1989. The radial gradient in the Sun's rotation. *ApJ*, **337**(Feb.), L53–L57.

Dziembowski, W. A., Pamyatnykh, A. A., and Sienkiewicz, R. 1990. Solar model from helioseismology and the neutrino flux problem. *MNRAS*, **244**(June), 542–550.

Dziewonski, A. M., and Anderson, D. L. 1981. Preliminary reference Earth model. *Phys. Earth Planet. Inter.*, **25**(June), 297–356.

Dziewonski, A. M., Hager, B. H., and O'Connell, R. J. 1977. Large-scale heterogeneities in the lower mantle. *J. Geophys. Res.*, **82**(Jan.), 239–255.

Eckart, C. 1960. *Hydrodynamics of Oceans and Atmospheres*. New York: Pergamon.

Eddington, A. S. 1925. Circulating currents in rotating stars. *The Observatory*, **48**(Mar.), 73–75.

Edmonds, P. D., and Cram, L. E. 1995. A search for global acoustic oscillations on α^1 Cen and β Hyi. *MNRAS*, **276**(Oct.), 1295.

Edwards, W. N. 2008. Meteor generated infrasound: Theory and observation. Pages 355–408 of Le Pichon, A. (ed), *Infrasound Monitoring for Atmospheric Studies*. New York: Springer.

Edwards, W. N., Eaton, D. W., and Brown, P. G. 2008. Seismic observations of meteors: Coupling theory and observations, *Rev. Geophys.* **46**, RG4007.

Eff-Darwich, A., and Korzennik, S. G. 2013. The dynamics of the solar radiative zone. *Sol. Phys.*, **287**(Oct.), 43–56.

Eggenberger, P., Miglio, A., Montalban, J., *et al.* 2010. Effects of rotation on the evolution and asteroseismic properties of red giants. *A&A*, **509**(Jan.), A72.

Eggenberger, P., Montalbán, J., and Miglio, A. 2012a. Angular momentum transport in stellar interiors constrained by rotational splittings of mixed modes in red giants. *A&A*, **544**(Aug.), L4.

Eggenberger, P., Haemmerlé, L., Meynet, G., and Maeder, A. 2012b. Impact of rotation and disc lifetime on pre-main sequence lithium depletion of solar-type stars. *A&A*, **539**(Mar.), A70.

Elliott, J. R., and Kosovichev, A. G. 1998. The adiabatic exponent in the solar core. *ApJ*, **500**(June), L199–L202.

Elsworth, Y., Howe, R., Isaak, G. R., McLeod, C. P., and New, R. 1990. Evidence from solar seismology against non-standard solar-core models. *Nature*, **347**(Oct.), 536–539.

Endal, A. S., and Sofia, S. 1978. The evolution of rotating stars: II. Calculations with time-dependent redistribution of angular momentum for 7- and 10-solar-mass stars. *ApJ*, **220**(Feb.), 279–290.

Epstein, C. R., and Pinsonneault, M. H. 2014. How good a clock is rotation? The stellar rotation-mass-age relationship for old field stars. *ApJ*, **780**(Jan.), 159.

Espinosa Lara, F., and Rieutord, M. 2013. Self-consistent 2D models of fast-rotating early-type stars. *A&A*, **552**(Apr.), A35.

Ewing, M., Latham, G., Press, F., *et al.* 1971. Seismology of the Moon and implications on internal structure, origin and evolution. Pages 155–172 of de Jager, C. (ed), *Highlights of Astronomy*. Dordrecht: D. Reidel.

Eyer, L., and Mowlavi, N. 2008. Variable stars across the observational HR diagram. *J. Phys. Conf. Ser.*, **118**(1), 012010.

Eyer, L., Holl, B., Pourbaix, D., *et al.* 2013. The Gaia Mission. *Central Eur. Astrophys. Bull.*, **37**, 115–126.

Fan, Y., Braun, D. C., and Chou, D.-Y. 1995. Scattering of p-modes by sunspots: II. Calculations of phase shifts from a phenomenological model. *ApJ*, **451**(Oct.), 877.

Fares, R., Donati, J.-F., Moutou, C., *et al.* 2009. Magnetic cycles of the planet-hosting star τ Bootis: II. A second magnetic polarity reversal. *MNRAS*, **398**(Sept.), 1383–1391.

Favata, F. 2004 (Jan.). The Eddington baseline mission. Pages 3–11 of Favata, F., Aigrain, S., and Wilson, A. (eds), *Stellar Structure and Habitable Planet Finding*. ESA Special Publication, vol. 538.

Favre, A. 1969. In Statistical equations of turbulent gases, *Problems of Hydrodynamics and Continuum Mechanics*, Society for Industrial and Applied Mathematics.

Featherstone, N. A., Hindman, B. W., and Thompson, M. J. 2011. Ring-analysis flow measurements of sunspot outflows. *J. Phys. Conf. Ser.*, **271**(1), 012002.

Feigelson, E. D., and Montmerle, T. 1999. High-energy processes in young stellar objects. *ARA&A*, **37**, 363–408.

Felipe, T., Braun, D., Crouch, A., and Birch, A. 2012. Scattering of the f-mode by small magnetic flux elements from observations and numerical simulations. *ApJ*, **757**(Oct.), 148.

Fernandes, J., and Monteiro, M. J. P. F. G. 2003. HR diagram and asteroseismic analysis of models for beta Hydri. *A&A*, **399**(Feb.), 243–251.

Fichtner, A., and Trampert, J. 2011. Resolution analysis in full waveform inversion. *Geophys. J. Int.*, **187**, 1604–1624.

Fichtner, A., Bunge, H.-P., and Igel, H. 2006. The adjoint method in seismology. *Phys. Earth Planet. Inter.*, **157**, 86–104.

Fichtner, A., Kennett, B. L. N., Igel, H., and Bunge, H.-P. 2008. Theoretical background for continental- and global-scale full-waveform inversion in the time-frequency domain. *Geophys. J. Int.*, **175**, 665–685.

Fichtner, A., Kennett, B. L. N., Igel, H., and Bunge, H.-P. 2009. Full seismic waveform tomography for upper-mantle structure in the Australasian region using adjoint methods. *Geophys. J. Int.*, **179**, 1703–1725.

Fichtner, A., Trampert, J., Cupillard, P., *et al.* 2013. Multiscale full waveform inversion. *Geophys. J. Int.*, **194**, 534–556.

Fisher, G. H., Canfield, R. C., and McClymont, A. N. 1985. Flare loop radiative hydrodynamics: VII. Dynamics of the thick target heated chromosphere. *ApJ*, **289**(Feb.), 434.

Fleming, W. P. 1895. Stars having peculiar spectra: Eleven new variable stars. *ApJ*, **1**(May), 411–415.

Fletcher, S. T., Chaplin, W. J., Elsworth, Y., and New, R. 2009. Efficient pseudo-global fitting for helioseismic data. *ApJ*, **694**(Mar.), 144–150.

Foglizzo, T., García, R. A., Boumier, P., *et al.* 1998. Are solar acoustic modes correlated? *A&A*, **330**(Feb.), 341–350.

Fogtmann-Schulz, A., Hinrup, B., Van Eylen, V., *et al.* 2014. Accurate parameters of the oldest known rocky-exoplanet hosting system: Kepler-10 revisited. *ApJ*, **781**(Feb.), 67.

Forbriger, T., Widmer-Schnidrig, R., Wielandt, E., Hayman, M., and Ackerley, N. 2010. Magnetic field background variations can limit the resolution of seismic broad-band sensors. *Geophys. J. Int.*, **183**, 303–312.

Forgács-Dajka, E., and Petrovay, K. 2002. Dynamics of the fast solar tachocline: I. Dipolar field. *A&A*, **389**(July), 629–640.

Fortney, J. J., and Nettelmann, N. 2010. The interior structure, composition, and evolution of giant planets. *Space Sci. Rev.*, **152**(May), 423–447.

Fossat, E. 1991. The IRIS network for full disk helioseismology: Present status of the programme. *Sol. Phys.*, **133**(May), 1–12.

Frandsen, S. 1987. An upper limit on p-mode amplitudes in Beta Hyi. *A&A*, **181**(July), 289–292.

Frandsen, S., Carrier, F., Aerts, C., *et al.* 2002. Detection of solar-like oscillations in the G7 giant star xi Hya. *A&A*, **394**(Oct.), L5–L8.

Frankel, A., and Clayton, R. W. 1986. Finite difference simulations of seismic scattering: Implications for the propagation of short-period seismic waves in the crust and models of crustal heterogeneity. *J. Geophys. Res.*, **91**(May), 6465–6489.

French, S. W., Lekic, V., and Romanowicz, B. 2013. Waveform tomography reveals channeled flow at the base of the oceanic asthenosphere. *Science*, **334**, 783–787.

Fridlund, M., Gough, D. O., Jones, A., *et al.* 1995. STARS: A proposal for a dedicated space mission to study stellar structure and evolution. Page 416 of Ulrich, R. K., Rhodes, Jr., E. J., and Dappen, W. (eds), *GONG 1994: Helio- and Astro-Seismology from the Earth and Space*. Astronomical Society of the Pacific Conference Series, vol. 76.

Frohlich, C., and Nakamura, Y. 2006. Possible extra-solar-system cause for certain lunar seismic events. *Icarus*, **185**, 21–28.

Frohlich, C., and Nakamura, Y. 2009. The physical mechnisms of deep moonquakes and intermediate-depth earthquakes: How similar and how different? *Phys. Earth Planet. Inter.*, **173**, 365–374.

Fröhlich, C., Romero, J., Roth, H., *et al.* 1995. VIRGO: Experiment for helioseismology and solar irradiance monitoring. *Sol. Phys.*, **162**, 101–128.

Fröhlich, C., Crommelynck, D. A., Wehrli, C., *et al.* 1997. In-flight performance of the Virgo solar irradiance instruments on SOHO. *Sol. Phys.*, **175**(Oct.), 267–286.

Fröhlich, H.-E., Frasca, A., Catanzaro, G., *et al.* 2012. Magnetic activity and differential rotation in the young Sun-like stars KIC 7985370 and KIC 7765135. *A&A*, **543**(July), A146.

Fujiwara, A., Kawaguchi, J., D. K. Yeomans, *et al.* 2006. The rubble-pile asteroid Itokawa as observed by Hayabusa. *Science*, **312**, 1330–1334.

Fuller, J., Lai, D., and Storch, N. I. 2014. Non-radial oscillations in rotating giant planets with solid cores: Application to Saturn and its rings. *Icarus*, **231**(Mar.), 34–50.

Gabriel, A. H., Grec, G., Charra, J., *et al.* 1995. Global oscillations at low frequency from the SOHO mission (GOLF). *Sol. Phys.*, **162**, 61–99.

Gabriel, A. H., Charra, J., Grec, G., *et al.* 1997. Performance and early results from the GOLF instrument flown on the SOHO mission. *Sol. Phys.*, **175**(Oct.), 207–226.

Gagnepain-Beyneix, J., Lognonné, P., Chenet, H., Lombardi, D., and Spohn, T. 2006. A seismic model of the lunar mantle and constraints on temperature and mineralogy. *Phys. Earth Planet. Inter.*, **159**, 140–166.

Gangi, A. F., and Yen, T. E. 1979. Velocity structure of the shallow lunar crust. *The Moon and The Planets*, **20**, 439–468.

Gallet, F., and Bouvier, J. 2013. Improved angular momentum evolution model for solar-like stars. *A&A*, **556**(Aug.), A36.

Gandorfer, A., Solanki, S. K., Woch, J., *et al.* 2011. The Solar Orbiter mission and its Polarimetric and Helioseismic Imager (SO/PHI). *J. Phys. Conf. Ser.*, **271**(1), 012086.

García, R. A., Pallé, P. L., Turck-Chièze, S., *et al.* 1998. High-frequency peaks in the power spectrum of solar velocity observations from the GOLF experiment. *ApJ*, **504**(Sept.), L51.

García, R. A., Corbard, T., Chaplin, W. J., *et al.* 2004. About the rotation of the solar radiative interior. *Sol. Phys.*, **220**(Apr.), 269–285.

Garcia, R., Lognonné, P., and Bonnin, X. 2005. Detecting atmospheric perturbations produced by Venus quakes. *Geophys. Res. Lett.*, **32**, L16205.

García, R. A., Turck-Chièze, S., Boumier, P., *et al.* 2005. Global solar Doppler velocity determination with the GOLF/SOHO instrument. *A&A*, **442**(Oct.), 385–395.

García, R. A., Turck-Chièze, S., Jiménez-Reyes, S. J., *et al.* 2007. Tracking solar gravity modes: The dynamics of the solar core. *Science*, **316**(June), 1591–1593.

García, R. A., Mathur, S., Ballot, J., *et al.* 2008a. Influence of low-degree high-order p-mode splittings on the solar rotation profile. *Sol. Phys.*, **251**(Sept.), 119–133.

García, R. A., Jiménez, A., Mathur, S., *et al.* 2008b. Update on g-mode research. *Astron. Nachr.*, **329**, 476.

García, R. A., Mathur, S., Salabert, D., *et al.* 2010. CoRoT reveals a magnetic activity cycle in a Sun-like star. *Science*, **329**(Aug.), 1032.

García, R. A., Salabert, D., Ballot, J., *et al.* 2011. New insights on the solar core. *J. Phys. Conf. Ser.*, **271**(1), 012046.

Garcia, R. F., Gagnepain-Beyneix, J., Chevrot, S., and Lognonné, P. 2011. Very preliminary reference Moon model. *Phys. Earth Planet. Inter.*, **188**, 96–113.

Garcia, R. F., Gagnepain-Beyneix, J., Chevrot, S., and Lognonné, P. 2012. Erratum to "Very preliminary reference Moon model" by R. F. Garcia, J. Gagnepain-Beyneix, S. Chevrot, P. Lognonné [*Phys. Earth Planet. Inter.* 188 (2011) 96–113], *Phys. Earth Planet. Inter.*, **202–203**, 89–91.

García, R. A., Ceillier, T., Mathur, S., and Salabert, D. 2013 (Dec.). Measuring reliable surface rotation rates from Kepler photometric observations. Page 129 of Shibahashi, H., and Lynas-Gray, A. E. (eds), *Progress in Physics of the Sun and Stars: A New Era in Helio- and Asteroseismology*. Astronomical Society of the Pacific Conference Series, vol. 479.

García, R. A., Pérez Hernández, F., Benomar, O., *et al.* 2014a. Study of KIC 8561221 observed by Kepler: An early red giant showing depressed dipolar modes. *A&A*, **563**(Mar.), A84.

García, R. A., Ceillier, T., Salabert, D., *et al.* 2014b. Towards asteroseismically calibrated age-rotation-activity relations for Kepler solar-like stars. ArXiv:1403.7155.

Garcia, R. F., Gagnepain-Beyneix, J., Chevrot, S., and Lognonné, P. 2011. Very preliminary reference Moon model. *Phys. Earth Planet. Inter.*, **188**(Sept.), 96–113.

Gaskell, R., Saito, J., and Ishiguro, M., *et al.* 2008. *Itokawa Shape Model V1.0. HAY-A-AMICA-5-ITOKAWASHAPE-V1.0*. Technical Report NASA Planetary Data System.

Gaulme, P., and Mosser, B. 2005. Coupling of acoustic waves to clouds in the jovian troposphere. *Icarus*, **178**(Nov.), 84–96.

Gaulme, P., Schmider, F. X., Gay, J., *et al.* 2008. SYMPA, a dedicated instrument for Jovian seismology: II. Real performance and first results. *A&A*, **490**(Nov.), 859–871.

Gaulme, P., Schmider, F.-X., Gay, J., Guillot, T., and Jacob, C. 2011. Detection of Jovian seismic waves: A new probe of its interior structure. *A&A*, **531**(July), A104.

Gaulme, P., McKeever, J., Rawls, M. L., *et al.* 2013. Red giants in eclipsing binary and multiple-star systems: Modeling and asteroseismic analysis of 70 candidates from Kepler data. *ApJ*, **767**(Apr.), 82.

Gauthier, O., Virieux, J., and Tarantola, A. 1986. Two-dimensional nonlinear inversion of seismic wave-forms: Numerical results. *Geophysics*, **51**, 1387–1403.

Gelly, B., Gree, G., and Fossat, E. 1986. Evidence for global pressure oscillations in Procyon and Alpha Centauri. *A&A*, **164**(Aug.), 383–394.

Georgobiani, D., Kosovichev, A. G., Nigam, R., Nordlund, Å., and Stein, R. F. 2000. Numerical simulations of oscillation modes of the solar convection zone. *ApJ*, **530**(Feb.), L139–L142.

Georgobiani, D., Stein, R. F., and Nordlund, Å. 2003. Asymmetry reversal in solar acoustic modes. Pages 279–282 of Sawaya-Lacoste, H. (ed), *GONG+ 2002. Local and Global Helioseismology: The Present and Future*. ESA Special Publication, vol. 517.

Georgobiani, D., Zhao, J., Kosovichev, A. G., *et al.* 2007. Local helioseismology and correlation tracking analysis of surface structures in realistic simulations of solar convection. *ApJ*, **657**(Mar.), 1157–1161.

Germano, M., Piomelli, U., Moin, P., and Cabot, W. H. 1991. A dynamic subgrid-scale eddy viscosity model. *Phys. Fluids*, **3**(July), 1760–1765.

Giacconi, R., Gursky, H., Paolini, F. R., and Rossi, B. B. 1962. Evidence for X-rays from sources outside the Solar System. *Phys. Rev. Lett.*, **9**(Dec.), 439–443.

Giles, P. M., Duvall, T. L., Scherrer, P. H., and Bogart, R. S. 1997. A subsurface flow of material from the Sun's equator to its poles. *Nature*, **390**(Nov.), 52–54.

Gilliland, R. L. 2008. Photometric oscillations of low-luminosity red giant stars. *AJ*, **136**(Aug.), 566–579.

Gilliland, R. L., and Brown, T. M. 1988. Time-resolved CCD photometry of an ensemble of stars. *PASP*, **100**(June), 754–765.

Gilliland, R. L., and Brown, T. M. 1992. Limits to CCD ensemble photometry precision and prospects for asteroseismology. *PASP*, **104**, 582–591.

Gilliland, R. L., Brown, T. M., Duncan, D. K., *et al.* 1991. Time-resolved CCD photometry of an ensemble of stars in the open cluster M67. *AJ*, **101**(Feb.), 541–561.

Gilliland, R. L., Brown, T. M., Kjeldsen, H., *et al.* 1993. A search for solar-like oscillations in the stars of M67 with CCD ensemble photometry on a network of 4 M telescopes. *AJ*, **106**(Dec.), 2441–2476.

Gilliland, R. L., Brown, T. M., Christensen-Dalsgaard, J., *et al.* 2010. Kepler asteroseismology program: Introduction and first results. *PASP*, **122**(Feb.), 131–143.

Gilman, P. A., and Howe, R. 2003 (Feb.). Meridional motion and the slope of isorotation contours. Pages 283–285 of Sawaya-Lacoste, H. (ed), *GONG+ 2002. Local and Global Helioseismology: The Present and Future*. ESA Special Publication, vol. 517.

Gilman, P. A., Morrow, C. A., and Deluca, E. E. 1989. Angular momentum transport and dynamo action in the Sun: Implications of recent oscillation measurements. *ApJ*, **338**(Mar.), 528–537.

Gizon, L., and Birch, A. C. 2002. Time-distance helioseismology: The forward problem for random distributed sources. *ApJ*, **571**(June), 966–986.

Gizon, L., and Birch, A. C. 2004. Time-distance helioseismology: Noise estimation. *ApJ*, **614**(Oct.), 472–489.

Gizon, L., and Birch, A. C. 2005. Local helioseismology. *Living Rev. Sol. Phys.*, **2**(Nov.), 6.

Gizon, L., and Birch, A. C. 2012. Helioseismology challenges models of solar convection. *Proc. Nat. Acad. Sci.*, **109**(July), 11896–11897.

Gizon, L., and Solanki, S. K. 2003. Determining the inclination of the rotation axis of a Sun-like star. *ApJ*, **589**(June), 1009–1019.

Gizon, L., Duvall, Jr., T. L., and Larsen, R. M. 2000. Seismic tomography of the near solar surface. *J. Astrophys. Astron.*, **21**(June), 339.

Gizon, L., Duvall, T. L., and Schou, J. 2003. Wave-like properties of solar supergranulation. *Nature*, **421**(Jan.), 43–44.

Gizon, L., Schunker, H., Baldner, C. S., *et al.* 2009. Helioseismology of sunspots: A case study of NOAA Region 9787. *Space Sci. Rev.*, **144**(Apr.), 249–273.

Gizon, L., Birch, A. C., and Spruit, H. C. 2010. Local helioseismology: Three-dimensional imaging of the solar interior. *ARA&A*, **48**(Sept.), 289–338.

Gizon, L., Ballot, J., Michel, E., *et al.* 2013. Seismic constraints on rotation of Sun-like star and mass of exoplanet. *Proc. Nat. Acad. Sci.*, **110**(Aug.), 13267–13271.

Goins, N. R., and Lazarewicz, A. R. 1979. Martian seismicity. *Geophys. Res. Lett.*, **6**, 368–370.

Goins, N. R., Dainty, A. M., and Toksöz, M. N. 1977. The deep seismic structure of the Moon, *Proceedings of the 8th Lunar Science Conference*, 471–486.

Goins, N. R., Dainty, A. M., and Toksöz, M. N. 1981a. Seismic energy release of the Moon. *J. Geophys. Res.*, **86**, 378–388.

Goins, N. R., Dainty, A. M., and Toksöz, M. N. 1981b. Structure of the lunar crust at highland site Apollo Station 16. *Geophys. Res. Lett.*, **8**, 29–32.

Goins, N. R., Dainty, A. M., and Toksöz, M. N. 1981c. Lunar seismology: The internal structure of the Moon. *J. Geophys. Res.*, **86**, 5061–5074.

Gold, T., and Soter, S. 1970. Apollo 12 seismic signal: Indication of a deep layer of powder. *Science*, **169**, 1071–1075.

Goldreich, P., and Keeley, D. A. 1977. Solar seismology: II. The stochastic excitation of the solar p-modes by turbulent convection. *ApJ*, **212**(Feb.), 243–251.

Goldreich, P., and Nicholson, P. D. 1989. Tides in rotating fluids. *ApJ*, **342**(July), 1075–1078.

Goldreich, P., Murray, N., Willette, G., and Kumar, P. 1991. Implications of solar p-mode frequency shifts. *ApJ*, **370**(Apr.), 752–762.

Goldreich, P., Murray, N., and Kumar, P. 1994. Excitation of solar p-modes. *ApJ*, **424**(Mar.), 466–479.

Golombek, M. P. 2002. A revision of Mars seismicity from surface faulting. *Lunar and Planetary Science Conference*, XXXIII, Abstract #1244.

Golombek, M. P., Banerdt, W. B., Tanaka, K. L., and Tralli, D. M. 1992. A prediction of Mars seismicity from surface faulting. *Science*, **258**(Nov.), 979–981.

Goode, P. R., and Thompson, M. J. 1992. The effect of an inclined magnetic field on solar oscillation frequencies. *ApJ*, **395**(Aug.), 307–315.

Goode, P. R., Gough, D., and Kosovichev, A. G. 1992. Localized excitation of solar oscillations. *ApJ*, **387**(Mar.), 707–711.

Goodricke, J. 1783. A series of observations on, and a discovery of, the period of the variation of the light of the bright star in the Head of Medusa, called Algol. In a letter from John Goodricke, Esq. to the Rev. Anthony Shepherd, D. D. F. R. S. and Plumian Professor at Cambridge. *Phil. Trans. R. Soc. London Ser. I*, **73**, 474–482.

Goodricke, J., and Englefield, H. C. 1785. Observations of a new variable star. By John Goodricke, Esq.; Communicated by Sir H. C. Englefield, Bart. F. R. S. and A. S. *Phil. Trans. R. Soc. London Ser. I*, **75**, 153–164.

Gough, D. 2013. What have we learned from helioseismology, what have we really learned, and what do we aspire to learn? *Sol. Phys.*, **287**(Oct.), 9–41.

Gough, D. O. 1984. On the rotation of the Sun. *Phil. Trans. R. Soc. London Ser. A*, **313**(Nov.), 27–38.

Gough, D. O. 1990. Comments on helioseismic inference. Page 283 of Osaki, Y., and Shibahashi, H. (eds), *Progress of Seismology of the Sun and Stars*. Lecture Notes in Physics, Berlin: Springer Verlag, vol. 367.

Gough, D. O., and Kosovichev, A. G. 1993. Seismic analysis of stellar P-mode spectra. Page 351 of Brown, T. M. (ed), *GONG 1992. Seismic Investigation of the Sun and Stars*. Astronomical Society of the Pacific Conference Series, vol. 42.

Gough, D. O., and McIntyre, M. E. 1998. Inevitability of a magnetic field in the Sun's radiative interior. *Nature*, **394**(Aug.), 755–757.

Gough, D. O., and Thompson, M. J. 1990. The effect of rotation and a buried magnetic field on stellar oscillations. *MNRAS*, **242**(Jan.), 25–55.

Gough, D. O., Kosovichev, A. G., Toomre, J., *et al.* 1996. The seismic structure of the Sun. *Science*, **272**(May), 1296–1300.

Goulty, N. R. 1979. Tidal triggering of deep moonquakes. *Phys. Earth Planet. Int.*, **19**(May), 52–58.

Goupil, M. J., Dziembowski, W. A., and Fontaine, G. 1998. On some observational consequences of nonlinearities in stellar pulsations. *Baltic Astron.*, **7**(Mar.), 21–41.

Goupil, M. J., Mosser, B., Marques, J. P., *et al.* 2013. Seismic diagnostics for transport of angular momentum in stars: II. Interpreting observed rotational splittings of slowly rotating red giant stars. *A&A*, **549**(Jan.), A75.

Gowen, R. A., Smith, A., Fortes, A. D., *et al.* 2011. Penetrators for in situ subsurface investigations of Europa. *Adv. Space Res.*, **48**, 725–742.

Grec, G., Fossat, E., and Vernin, J. 1976. A spectrophotometer for the study of long period solar photospheric oscillations. *A&A*, **50**(July), 221–225.

Grec, G., Fossat, E., and Pomerantz, M. 1980. Solar oscillations: Full disk observations from the geographic South Pole. *Nature*, **288**(Dec.), 541–544.

Grec, G., Fossat, E., and Pomerantz, M. A. 1983. Full-disk observations of solar oscillations from the geographic South Pole: Latest results. *Sol. Phys.*, **82**(Jan.), 55–66.

Greenberg, J. M., and Huebner, W. F. 2002. Summary of the Workshop on Geophysical and Geological Properties of NEOs: 'Know Your Enemy'. Pages 419–432 of *International Seminars on Nuclear War and Planetary Emergencies*, vol. 26.

Grevesse, N., and Sauval, A. J. 1998. Standard solar composition. *Space Sci. Rev.*, **85**(May), 161–174.

Grimm, H.-J., Gilfanov, M., and Sunyaev, R. 2002. The Milky Way in X-rays for an outside observer: Log(N)–Log(S) and luminosity function of X-ray binaries from RXTE/ASM data. *A&A*, **391**(Sept.), 923–944.

Gruberbauer, M., Kallinger, T., Weiss, W. W., and Guenther, D. B. 2009. On the detection of Lorentzian profiles in a power spectrum: A Bayesian approach using ignorance priors. *A&A*, **506**(Nov.), 1043–1053.

Gruberbauer, M., Guenther, D. B., and Kallinger, T. 2012. Toward a new kind of asteroseismic grid fitting. *ApJ*, **749**(Apr.), 109.

Grundahl, F., Kjeldsen, H., Christensen-Dalsgaard, J., Arentoft, T., and Frandsen, S. 2007. Stellar oscillations network group. *Commun. Asteroseismol.*, **150**(June), 300.

Grundahl, F., Christensen-Dalsgaard, J., Gråe Jørgensen, U., *et al.* 2011. SONG: Getting ready for the prototype. *J. Phys. Conf. Ser.*, **271**(1), 012083.

Gruneisen, R. 1998. High precision photometry from space with COROT: Effect of spatial image jitter on the photometric performances of CCDs. Page 217 of Beletic, J., and Amico, P. (eds), *Optical Detectors for Astronomy*. Astrophysics and Space Science Library, vol. 228.

Guasch, L., Warner, M., Nangoo, T., *et al.* 2012. Elastic 3D full-waveform inversion, 82nd Annual International Meeting, SEG, Expanded Abstracts, SI6.8.

Gudkova, T., Mosser, B., Provost, J., *et al.* 1995. Seismological comparison of giant planet interior models. *A&A*, **303**(Nov.), 594.

Gudkova, T., Lognonné, P., and Gagnepain-Beyneix, J. 2011. Large impacts detected by the Apollo seismometers: Impactor mass and source cutoff frequency estimations. *Icarus*, **211**, 1049–1065.

Gudkova, T., Lognonné, P., Miljković, K., and Gagnepain-Beyneix, J. 2014. Impact cutoff frequency–momentum scaling law inverted from Apollo seismic data. *Earth Planet. Sci. Lett.*

Guillot, T. 1999. Interior of giant planets inside and outside the Solar System. *Science*, **286**(Oct.), 72–77.

Guillot, T. 2005. The interiors of giant planets: Models and outstanding questions. *Annu. Rev. Earth Planet. Sci.*, **33**(Jan.), 493–530.

Guillot, T., Stevenson, D. J., Hubbard, W. B., and Saumon, D. 2004. The interior of Jupiter. Pages 35–57 of Bagenal, F., Dowling, T. E., and McKinnon, W. B. (eds), *Jupiter: The Planet, Satellites and Magnetosphere*. Cambridge: Cambridge University Press.

Haber, D. A., and Hindman, B. W. 2005. Solar meridional flows: Recent findings. *Highlights of Astronomy*, **13**(Jan.), 431.

Haber, D. A., Hindman, B. W., Toomre, J., *et al.* 2002. Evolving submerged meridional circulation cells within the upper convection zone revealed by ring-diagram analysis. *ApJ*, **570**(May), 855–864.

Haber, D. A., Hindman, B. W., Toomre, J., and Thompson, M. J. 2004. Organized subsurface flows near active regions. *Sol. Phys.*, **220**(Apr.), 371–380.

Hall, R. C. 1977. *Lunar Impact: A History of Project Ranger,* NASA History Series, NASA SP-4210, Washington, D.C.

Hammel, H. B., Beebe, R. F., Ingersoll, A. P., *et al.* 1995. HST imaging of atmospheric phenomena created by the impact of comet Shoemaker-Levy 9. *Science*, **267**, 1288–1296.

Hammel, H. B., Wong, M. H., Clarke, J. T., *et al.* 2010. Jupiter after the 2009 impact: Hubble Space Telescope imaging of the impact-generated debris and its temporal evolution. *Astron. J. Lett.*, **715**, L150.

Hanasoge, S. M. 2013. The influence of noise sources on cross-correlation amplitudes. *Geophys. J. Int.*, **192**(Jan.), 295–309.

Hanasoge, S. M. 2014. Measurements and kernels for source-structure inversions in noise tomography. *Geophys. J. Int.*, **196**(Feb.), 971–985.

Hanasoge, S. M., Duvall, Jr., T. L., and Couvidat, S. 2007. Validation of helioseismology through forward modeling: Realization noise subtraction and kernels. *ApJ*, **664**(Aug.), 1234–1243.

Hanasoge, S. M., Duvall, Jr., T. L., and DeRosa, M. L. 2010. Seismic constraints on interior solar convection. *ApJ*, **712**(Mar.), L98–L102.

Hanasoge, S. M., Birch, A., Gizon, L., and Tromp, J. 2011. The adjoint method applied to time-distance helioseismology. *ApJ*, **738**(Sept.), 100.

Hanasoge, S., Birch, A., Gizon, L., and Tromp, J. 2012a. Seismic probes of solar interior magnetic structure. *Phys. Rev. Lett.*, **109**(10), 101101.

Hanasoge, S. M., Duvall, T. L., and Sreenivasan, K. R. 2012b. Anomalously weak solar convection. *Proc. Nat. Acad. Sci.*, **109**(July), 11928–11932.

Handberg, R., and Campante, T. L. 2011. Bayesian peak-bagging of solar-like oscillators using MCMC: A comprehensive guide. *A&A*, **527**(Mar.), A56.

Hansen, C. J., Kawaler, S. D., and Trimble, V. 2004. *Stellar Interiors: Physical Principles, Structure, and Evolution*. New York: Springer.

Harlow, F. H., and Welch, J. E. 1965. Numerical calculation of time-dependent viscous incompressible flow of fluid with free surface. *Phys. Fluids*, **8**, 2182–2189.

Hartlep, T., Kosovichev, A. G., Zhao, J., and Mansour, N. N. 2011. Signatures of emerging subsurface structures in acoustic power maps of the Sun. *Sol. Phys.*, **268**(Feb.), 321–327.

Hartlep, T., Zhao, J., Kosovichev, A. G., and Mansour, N. N. 2013. Solar wave-field simulation for testing prospects of helioseismic measurements of deep meridional flows. *ApJ*, **762**(Jan.), 132.

Harvey, J. 1985. High-resolution helioseismology. Page 199 of Rolfe, E., and Battrick, B. (eds), *Future Missions in Solar, Heliospheric & Space Plasma Physics*. ESA Special Publication, vol. 235.

Harvey, J., Tucker, R., and Britanik, L. 1998. High resolution upgrade of the GONG instruments. Page 209 of Korzennik, S. (ed), *Structure and Dynamics of the Interior of the Sun and Sun-like Stars*. ESA Special Publication, vol. 418.

Harvey, J. W. 1988. Techniques for observing stellar oscillations. Page 497 of Christensen-Dalsgaard, J., and Frandsen, S. (eds), *Advances in Helio- and Asteroseismology*. IAU Symposium, vol. 123.

Harvey, J. W., Hill, F., Hubbard, R., *et al.* 1996. The Global Oscillation Network Group (GONG) Project. *Science*, **272**(May), 1284.

Hathaway, D. H. 2012. Supergranules as probes of the Sun's meridional circulation. *ApJ*, **760**(Nov.), 84.

Hedman, M. M., and Nicholson, P. D. 2013. Kronoseismology: Using density waves in Saturn's C ring to probe the planet's interior. *AJ*, **146**(July), 12.

Hekker, S., Barban, C., Baudin, F., *et al.* 2010. Oscillation mode lifetimes of red giants observed during the initial and first anticentre long run of CoRoT. *A&A*, **520**(Sept.), A60.

Hekker, S., Gilliland, R. L., Elsworth, Y., *et al.* 2011. Characterization of red giant stars in the public Kepler data. *MNRAS*, **414**(July), 2594–2601.

Helled, R., and Guillot, T. 2013. Interior models of Saturn: Including the uncertainties in shape and rotation. *ApJ*, **767**(Apr.), 113.

Hello, Y., Ogé, A., Sukhovich, A., and Nolet, G. 2011. Modern mermaids: New floats image the deep Earth. *EOS Trans.*, **92**(4 Oct.), 337–338.

Hempel, S., Knapmeyer, M., Jonkers, A. R. T., and Oberst, J. 2012. Uncertainty of Apollo deep moonquake locations and implications for future network designs. *Icarus*, **220**(Aug.), 971–980.

Hill, F. 1984. The effects of a nearly 100% duty cycle in observations of solar oscillations. Pages 271–277 of Ulrich, R. K., Harvey, J., Rhodes, Jr., E. J., and Toomre, J. (eds), *Solar Seismology from Space*. NASA JPL.

Hill, F. 1988. Rings and trumpets: Three-dimensional power spectra of solar oscillations. *ApJ*, **333**(Oct.), 996–1013.

Hill, F., and Newkirk, Jr., G. 1985. On the expected performance of a solar oscillation network. *Sol. Phys.*, **95**(Feb.), 201–219.

Hill, F., Stark, P. B., and Stebbins, R. T., *et al.* 1996. The solar acoustic spectrum and eigenmode parameters. *Science*, **272**(May), 1292–1295.

Hill, F., Anderson, E., Howe, R., *et al.* 1998. Estimated mode parameters from the fitting of GONG spectra. Page 231 of Korzennik, S. (ed.), *Structure and Dynamics of the Interior of the Sun and Sun-like Stars*. ESA Special Publication, vol. 418.

Hill, F., Baldner, C. S., García, R. A., Roth, M., and Schunker, H. 2013. Where to go from here: The future of helio- and astero-seismology. Pages 401–408 of Jain, K., Tripathy, S. C., Hill, F., Leibacher, J. W., and Pevtsov, A. A. (eds), *Fifty Years of Seismology of the Sun and Stars*. Astronomical Society of the Pacific Conference Series, vol. 478.

Hill, H. A., Stebbins, R. T., and Brown, T. M. 1976. Recent oblateness observations: Data, interpretation and significance for earlier work. Page 622 of Sanders, J. H., and Wapstra, A. H. (eds), *Atomic Masses and Fundamental Constants*, vol. 5. New York: Plenum.

Hilton, J. L. 2002. Asteroid masses and densities. *Asteroids III*, 103–112.

Hopf, T., Kumar, S., Karl, W. J., and Pike, W. T. 2010. Shock protection of penetrator-based instrumentation via a sublimation approach. *Adv. Space Res.*, **45**, 460–467.

Horvath, P., Latham, G., Nakamura, Y., and Dorman, H. J. 1980. Lunar near-surface shear wave velocities at the Apollo landing sites as inferred from spectral amplitude ratios. *J. Geophys. Res.*, **85**, 6572–6578.

Houdek, G., and Gough, D. O. 2007. An asteroseismic signature of helium ionization. *MNRAS*, **375**(Mar.), 861–880.

Houdek, G., Balmforth, N. J., Christensen-Dalsgaard, J., and Gough, D. O. 1999. Amplitudes of stochastically excited oscillations in main-sequence stars. *A&A*, **351**(Nov.), 582–596.

Howe, R. 2009. Solar interior rotation and its variation. *Living Rev. Sol. Phys.*, **6**(Feb.), 1.

Howe, R., and Hill, F. 1998. Estimating low-degree mode parameters from GONG data using the leakage matrix. Page 237 of Korzennik, S. (ed), *Structure and Dynamics of the Interior of the Sun and Sun-like Stars*. ESA Special Publication, vol. 418.

Howe, R., and Thompson, M. J. 1998. A strategy for fitting partially blended ridges in GONG solar p-mode spectra. *A&AS*, **131**(Sept.), 539–548.

Howe, R., Christensen-Dalsgaard, J., Hill, F., *et al.* 2000. Dynamic variations at the base of the solar convection zone. *Science*, **287**(Mar.), 2456–2460.

Howe, R., Christensen-Dalsgaard, J., and Hill, F., *et al.* 2005. Solar convection-zone dynamics, 1995–2004. *ApJ*, **634**(Dec.), 1405–1415.

Howe, R., Christensen-Dalsgaard, J., Hill, F., *et al.* 2009. A note on the torsional oscillation at solar minimum. *ApJ*, **701**(Aug.), L87–L90.

Howe, R., Christensen-Dalsgaard, J., Hill, F., *et al.* 2013. The high-latitude branch of the solar torsional oscillation in the rising phase of cycle 24. *ApJ*, **767**(Apr.), L20.

Howell, S. B., Sobeck, C., Haas, M., *et al.* 2014. The K2 mission: Characterization and early results. *PASP*, **126**(Apr.), 398–408.

Hu, H., Tout, C. A., Glebbeek, E., and Dupret, M.-A. 2011. Slowing down atomic diffusion in subdwarf B stars: Mass loss or turbulence? *MNRAS*, **418**(Nov.), 195–205.

Huber, D., Bedding, T. R., Stello, D., *et al.* 2011. Testing scaling relations for solar-like oscillations from the main sequence to red giants using Kepler data. *ApJ*, **743**(Dec.), 143.

Huber, D., Chaplin, W. J., Christensen-Dalsgaard, J., *et al.* 2013a. Fundamental properties of Kepler planet-candidate host stars using asteroseismology. *ApJ*, **767**(Apr.), 127.

Huber, D., Carter, J. A., Barbieri, M., *et al.* 2013b. Stellar spin-orbit misalignment in a multiplanet system. *Science*, **342**(Oct.), 331–334.

Huber, D., Silva Aguirre, V., Matthews, J. M., *et al.* 2014. Revised stellar properties of Kepler targets for the quarter 1-16 transit detection run. *ApJS*, **211**(Mar.), 2.

Hudson, H. S., Fisher, G. H., and Welsch, B. T. 2008. Flare energy and magnetic field variations. Page 221 of Howe, R., Komm, R. W., Balasubramaniam, K. S., and Petrie, G. J. D. (eds), *Subsurface and Atmospheric Influences on Solar Activity*. Astronomical Society of the Pacific Conference Series, vol. 383.

Huebner, W. F., Cellino, A., Cheng, A. F., and Greenberg, J. M. 2001. NEOs: Physical properties. Pages 309–34 of *International Seminars on Nuclear War and Planetary Emergencies*, vol. 25.

Hueso, R., Sánchez-Lavega, A., and Guillot, T. 2002. A model for large-scale convective storms in Jupiter. *J. Geophys. Res. (Planets)*, **107**(Oct.), 5075.

Hunt, M. L., and Vriend, N. M. 2010. Booming sand dunes. *Annu. Rev. Earth Planet. Sci.*, 2010.38:281–301, doi:10.1146/annurev-earth-040809-152336.

Iben, Jr., I. 1971. Globular cluster stars: Results of theoretical evolution and pulsation studies compared with the observations. *PASP*, **83**(Dec.), 697.

Ibrahim, M. I. 2012. *The elastic properties of carbonaceous chondrites*. Ph.D. thesis, University of Calgary, Alberta, Canada.

Ilonidis, S., Zhao, J., and Kosovichev, A. G. 2011. Detection of emerging sunspot regions in the solar interior by helioseismology. *AGU Fall Meeting Abstracts*, Dec., A1.

Irwin, P. G. J. 2009. *Giant Planets of Our Solar System*. Springer-Praxis.

Jackiewicz, J., Gizon, L., and Birch, A. C. 2008. High-resolution mapping of flows in the solar interior: Fully consistent OLA inversion of helioseismic travel times. *Sol. Phys.*, **251**(Sept.), 381–415.

Jackiewicz, J., Nettelmann, N., Marley, M., and Fortney, J. 2012a. Forward and inverse modeling for jovian seismology. *Icarus*, **220**(Aug.), 844–854.

Jackiewicz, J., Birch, A. C., Gizon, L., *et al.* 2012b. Multichannel three-dimensional SOLA inversion for local helioseismology. *Sol. Phys.*, **276**(Feb.), 19–33.

Jackson, J. D., and Fox, R. F. 1999. *Classical Electrodynamics*, 3rd edn. New York: John Wiley & Sons.

Jacoutot, L., Kosovichev, A. G., Wray, A. A., and Mansour, N. N. 2008a. Numerical simulation of excitation of solar oscillation modes for different turbulent models. *ApJ*, **682**(Aug.), 1386–1391.

Jacoutot, L., Kosovichev, A. G., Wray, A., and Mansour, N. N. 2008b. Realistic numerical simulations of solar convection and oscillations in magnetic regions. *ApJ*, **684**(Sept.), L51–L54.

Jain, R., and Haber, D. 2002. Solar p-modes and surface magnetic fields: Is there an acoustic emission? MDI/SOHO observations. *A&A*, **387**(June), 1092–1099.

Jaupart, C., and Mareschal, J.-C. 2010. *Heat Generation and Transport in the Earth*. Cambridge: Cambridge University Press.

Jensen, J. M., and Pijpers, F. P. 2003. Sensitivity kernels for time-distance inversion based on the Rytov approximation. *A&A*, **412**(Dec.), 257–265.

Jensen, J. M., Duvall, Jr., T. L., Jacobsen, B. H., and Christensen-Dalsgaard, J. 2001. Imaging an emerging active region with helioseismic tomography. *ApJ*, **553**(June), L193–L196.

Jiménez-Reyes, S. J., Chaplin, W. J., García, R. A., *et al.* 2008. SolarFLAG hare and hounds: Estimation of p-mode frequencies from Sun-as-star helioseismology data. *MNRAS*, **389**(Oct.), 1780–1790.

Johnston, D. H., and Toksöz, M. N. 1977. Internal structure and properties of Mars. *Icarus*, **32**, 73–84.

Jones, S. F. 2009. *Elastic wave velocity, porosity, and pore geometry of ordinary chondrites and artificially shocked samples*. M.Phil. thesis, University of Calgary, Alberta, Canada.

Kallinger, T., Mosser, B., Hekker, S., *et al.* 2010a. Asteroseismology of red giants from the first four months of Kepler data: Fundamental stellar parameters. *A&A*, **522**(Nov.), A1.

Kallinger, T., Weiss, W. W., Barban, C., *et al.* 2010b. Oscillating red giants in the CoRoT exofield: Asteroseismic mass and radius determination. *A&A*, **509**(Jan.), A77.

Kallinger, T., Hekker, S., Mosser, B., *et al.* 2012. Evolutionary influences on the structure of red-giant acoustic oscillation spectra from 600d of Kepler observations. *A&A*, **541**(May), A51.

Kanamori, H. 1977. The energy release in great earthquakes. *J. Geophys. Res.*, **82**, 2981–2987.

Kanamori, H. 1994. Mechanics of earthquakes. *Annu. Rev. Earth Planet. Sci.*, **22**, 207–237.

Kanamori, H. 2004. Some fluid-mechanical problems in geophysics: waves in the atmosphere and fault lubrication. *Fluid Dyn. Res.*, **34**, 1–19.

Kanamori, H., and Anderson, D. L. 1975a. Theoretical basis of some empirical relations in seismology. *Bull. Seismol. Soc. Am.*, **65**, 1073–1095.

Kanamori, H., and Anderson, D. L. 1975b. Amplitude of the Earth's free oscillations and long-period characteristics of the earthquake source. *J. Geophys. Res.*, **80**(Mar.), 1075–1078.

Karato, S. 2013. Geophysical constraints on the water content of the lunar mantle and its implications for the origin of the Moon. *Earth Planet. Sci. Lett.*, **384**, 144–153.

Karoff, C., and Kjeldsen, H. 2008. Evidence that solar flares drive global oscillations in the Sun. *ApJ*, **678**(May), L73–L76.

Karoff, C., Metcalfe, T. S., Chaplin, W. J., *et al.* 2013. Sounding stellar cycles with Kepler: II. Ground-based observations. *MNRAS*, **433**(Aug.), 3227–3238.

Kato, S., and Fukue, J. 1980. Trapped radial oscillations of gaseous disks around a black hole. *PASJ*, **32**, 377.

Kato, S., and Nakamura, K. E. 1998. Transition radius from cooling-dominated to advection-dominated regimes in two temperature disks. *PASJ*, **50**(Dec.), 559–566.

Kawaler, S. D. 1988. Angular momentum loss in low-mass stars. *ApJ*, **333**(Oct.), 236–247.

Kawaler, S. D., Sekii, T., and Gough, D. 1999. Prospects for measuring differential rotation in white dwarfs through asteroseismology. *ApJ*, **516**(May), 349–365.

Kawaler, S. D., Reed, M. D., Quint, A. C., *et al.* 2010. First Kepler results on compact pulsators: II. KIC 010139564, a new pulsating subdwarf B (V361 Hya) star with an additional low-frequency mode. *MNRAS*, **409**(Dec.), 1487–1495.

Kawamura, T., Tanaka, S., Kobayashi, N., and Lognonné, P. 2010. The Lunar Surface Gravimeter as a lunar seismometer. In *Ground-Based Geophysics on the Moon*, January 21–22, 2010, Tempe, Arizona. LPI Contribution No. 1530.

Kawamura, T., Morota, T., Kobayashi, N., and Tanaka, S. 2011. Cratering asymmetry on the Moon: New insight from the Apollo Passive Seismic Experiment. *J. Geophys. Res.*, **38**, L15201, doi: 10.1029/2011GL048047.

Keihm S. J., and Langseth, M. G., Jr. 1973. Surface brightness temperatures at the Apollo 17 heat flow site: Thermal conductivity of the upper 15 cm of regolith. *Proceedings of the 4th Lunar Science Conference, Geochim. Cosmochim. Acta, Suppl.* **4**, 2503–2513.

Kennett, B. L. N., Engdahl, E. R., and Buland, R. 1995. Constraints on seismic velocities in the Earth from traveltimes. *Geophys. J. Int.*, **122**(July), 108–124.

Kenyon, S. L., Lawrence, J. S., Ashley, M. C. B., *et al.* 2006. Atmospheric scintillation at Dome C, Antarctica: Implications for photometry and astrometry. *PASP*, **118**(June), 924–932.

Kepler, S. O., Mukadam, A., Winget, D. E., *et al.* 2000. Evolutionary timescale of the pulsating white dwarf G117-B15A: The most stable optical clock known. *ApJ*, **534**(May), L185–L188.

Khan, A., and Mosegaard, K. 2001. New information on the deep lunar interior from an inversion of lunar free oscillation periods. *Geophys. Res. Lett.*, **28**, 1791–1794.

Khan, A., and Mosegaard, K. 2002. An inquiry into the Lunar interior: A non-linear inversion of the Apollo lunar seismic data. *J. Geophys. Res.*, **107**, E6, doi:10.1029/2001JE001658.

Khan, A., Mosegaard, K., and Rasmussen, K. L. 2000. A new seismic velocity model for the Moon from a Monte Carlo inversion of the Apollo lunar seismic data. *Geophys. Res. Lett.*, **27**, 1591–1594.

Kholikov, S. 2013. A search for helioseismic signature of emerging active regions. *Sol. Phys.*, **287**(Oct.), 229–237.

Kholikov, S., Serebryanskiy, A., and Jackiewicz, J. 2014. Meridional flow in the solar convection zone: I. Measurements from GONG data. *ApJ*, **784**(Apr.), 145.

Khomenko, E., and Collados, M. 2008. Magnetohydrostatic sunspot models from deep subphotospheric to chromospheric layers. *ApJ*, **689**(Dec.), 1379–1387.

Khomenko, E., Kosovichev, A., Collados, M., Parchevsky, K., and Olshevsky, V. 2009. Theoretical modeling of propagation of magnetoacoustic waves in magnetic regions below sunspots. *ApJ*, **694**(Mar.), 411–424.

Kippenhahn, R., and Weigert, A. 1990. *Stellar Structure and Evolution*. Berlin: Springer-Verlag.

Kippenhahn, R., Weigert, A., and Weiss, A. 2013. *Stellar Structure and Evolution*, 2nd edn. Berlin, Heidelberg: Springer.

Kitiashvili, I. N., Kosovichev, A. G., Wray, A. A., and Mansour, N. N. 2009. Traveling waves of magnetoconvection and the origin of the Evershed effect in sunspots. *ApJ*, **700**(Aug.), L178–L181.

Kitiashvili, I. N., Bellot Rubio, L. R., Kosovichev, A. G., *et al.* 2010a. Explanation of the sea-serpent magnetic structure of sunspot penumbrae. *ApJ*, **716**(June), L181–L184.

Kitiashvili, I. N., Kosovichev, A. G., Wray, A. A., and Mansour, N. N. 2010b. Mechanism of spontaneous formation of stable magnetic structures on the Sun. *ApJ*, **719**(Aug.), 307–312.

Kitiashvili, I. N., Kosovichev, A. G., Mansour, N. N., and Wray, A. A. 2011a. Excitation of acoustic waves by vortices in the quiet Sun. *ApJ*, **727**(Feb.), L50.

Kitiashvili, I. N., Kosovichev, A. G., Mansour, N. N., and Wray, A. A. 2011b. Numerical MHD simulations of solar magnetoconvection and oscillations in inclined magnetic field regions. *Sol. Phys.*, **268**(Feb.), 283–291.

Kitiashvili, I. N., Kosovichev, A. G., Mansour, N. N., Lele, S. K., and Wray, A. A. 2012. Vortex tubes of turbulent solar convection. *Phys. Scr.*, **86**(1), 018403.

Kitiashvili, I. N., Kosovichev, A. G., Lele, S. K., Mansour, N. N., and Wray, A. A. 2013. Ubiquitous solar eruptions driven by magnetized vortex tubes. *ApJ*, **770**(June), 37.

Kivelson, M. G., Khurana, K. K., Russell, C. T., *et al.* 2000. Galileo magnetometer measurements: A stronger case for a subsurface ocean at Europa. *Science*, **289**(Aug.), 1340–1343.

Kjeldsen, H., and Bedding, T. R. 1995. Amplitudes of stellar oscillations: The implications for asteroseismology. *A&A*, **293**(Jan.), 87–106.

Kjeldsen, H., Bedding, T. R., Viskum, M., and Frandsen, S. 1995. Solarlike oscillations in eta Boo. *AJ*, **109**(Mar.), 1313–1319.

Kjeldsen, H., Bedding, T. R., Baldry, I. K., *et al.* 2003. Confirmation of solar-like oscillations in η Bootis. *AJ*, **126**(Sept.), 1483–1488.

Kjeldsen, H., Bedding, T. R., Butler, R. P., *et al.* 2005. Solar-like oscillations in α Centauri B. *ApJ*, **635**(Dec.), 1281–1290.

Kjeldsen, H., Bedding, T. R., and Christensen-Dalsgaard, J. 2008a. Correcting stellar oscillation frequencies for near-surface effects. *ApJ*, **683**(Aug.), L175–L178.

Kjeldsen, H., Bedding, T. R., Arentoft, T., *et al.* 2008b. The amplitude of solar oscillations using stellar techniques. *ApJ*, **682**(Aug.), 1370–1375.

Kjeldsen, H., Christensen-Dalsgaard, J., Handberg, R., *et al.* 2010. The Kepler asteroseismic investigation: Scientific goals and first results. *Astron. Nachr.*, **331**(Dec.), 966.

Knapmeyer, M. 2008. Location of seismic events using inaccurate data from very sparse networks. *Geophys. J. Int.*, **175**(Dec.), 975–991.

Knapmeyer, M., Oberst, J., Hauber, E., *et al.* 2006. Working models for spatial distribution and level of Mars' seismicity. *J. Geophys. Res. (Planets)*, **111**(10), 11006.

Kochan, H., Feibig, W., Konopka, U., *et al.* 2000. CASSE – The ROSETTA lander comet acoustic surface sounding experiment: Status of some aspects, the technical realisation and laboratory simulations. *Planet. Space Sci.*, **48**, 385–399.

Komatitsch, D., and Vilotte, J. P. 1998. The spectral element method: An efficient tool to simulate the seismic response of 2D and 3D geological structures. *Bull. Seismol. Soc. Am.,* **88**, 368–392.

Komatitsch, D., and Tromp, J. 2002. Spectral-element simulations of global seismic wave propagation: I. Validation. *Geophys. J. Int.*, **149**(May), 390–412.

Komm, R. W., Gu, Y., Hill, F., Stark, P. B., and Fodor, I. K. 1999. Multitaper spectral analysis and wavelet denoising applied to helioseismic data. *ApJ*, **519**(July), 407–421.

Komm, R. W., Howe, R., and Hill, F. 2000. Solar-cycle changes in GONG p-mode widths and amplitudes 1995–1998. *ApJ*, **531**(Mar.), 1094–1108.

Komm, R. W., González Hernández, I., and Hill, F., *et al.* 2013. Subsurface meridional flow from HMI using the ring-diagram pipeline. *Sol. Phys.*, **287**(Oct.), 85–106.

Kormendy, J., and Ho, L. C. 2013. Coevolution (or not) of supermassive black holes and host galaxies. *ARA&A*, **51**(Aug.), 511–653.

Kornilov, V. 2012. Stellar scintillation on large and extremely large telescopes. *MNRAS*, **426**(Oct.), 647–655.

Kornilov, V., Sarazin, M., Tokovinin, A., Travouillon, T., and Voziakova, O. 2012. Comparison of the scintillation noise above different observatories measured with MASS instruments. *A&A*, **546**(Oct.), A41.

Korzennik, S. G. 2008. YAOPBM–II: Extension to higher degrees and to shorter time series. *J. Phys. Conf. Ser.*, **118**(1), 012082.

Korzennik, S. G., Rabello-Soares, M. C., Schou, J., and Larson, T. P. 2013. Accurate characterization of high-degree modes using MDI observations. *ApJ*, **772**, 87.

Kosovichev, A. G. 1986. Simulating thermal and gasdynamic processes in solar-flare pulse phases. *Bull. Crimean Astrophys. Observatory*, **75**, 6.

Kosovichev, A. G. 1996. Tomographic imaging of the Sun's interior. *ApJ*, **461**(Apr.), L55.

Kosovichev, A. G. 2006a. Direct observations of acoustic waves excited by solar flares and their propagation in sunspot regions. Page 154 of Leibacher, J., Stein, R. F., and Uitenbroek, H. (eds), *Solar MHD Theory and Observations: A High Spatial Resolution Perspective*. Astronomical Society of the Pacific Conference Series, vol. 354.

Kosovichev, A. G. 2006b. Properties of flares-generated seismic waves on the Sun. *Sol. Phys.*, **238**(Oct.), 1–11.

Kosovichev, A. G. 2006c. Sunquake sources and wave propagation. In *Proceedings of SOHO 18/GONG 2006/HELAS I, Beyond the Spherical Sun*. ESA Special Publication, vol. 624.

Kosovichev, A. G. 2007. The cause of photospheric and helioseismic responses to solar flares: high-energy electrons or protons? *ApJ*, **670**(Nov.), L65–L68.

Kosovichev, A. G. 2009. Solar oscillations. Pages 547–559 of Guzik, J. A. and Bradley, P. A. (eds), *Stellar Pulsation*. American Institute of Physics Conference Series, vol. 1170.

Kosovichev, A. G. 2011. Advances in global and local helioseismology: An introductory review. Pages 3–642 of Rozelot, J.-P., and Neiner, C. (eds), *The Pulsations of the Sun and the Stars*. Lecture Notes in Physics, Berlin: Springer Verlag, vol. 832.

Kosovichev, A. G. 2012a. Local helioseismology of sunspots: Current status and perspectives. *Sol. Phys.*, **279**(Aug.), 323–348.

Kosovichev, A. G. 2012b (May). Physics of sunquakes events observed with SDO. Page 109.03 of *American Astronomical Society Meeting Abstracts 220*, vol. 220.

Kosovichev, A. G. 2014 (Feb.). Sunquakes and starquakes. Pages 349–352 of Guzik, J. A., Chaplin, W. J., Handler, G., and Pigulski, A. (eds), *Precision Asteroseismology*. IAU Symposium 301.

Kosovichev, A. G., and Duvall, Jr., T. L. 1997 (Dec.). Acoustic tomography of solar convective flows and structures. Pages 241–260 of Pijpers, F. P., Christensen-Dalsgaard, J., and Rosenthal, C. S. (eds), *SCORe'96: Solar Convection and Oscillations and their Relationship*. Astrophysics and Space Science Library, Springer, vol. 225.

Kosovichev, A. G., and Zhao, J., in preparation

Kosovichev, A. G., and Zharkova, V. V. 1995. Seismic response to solar flares: Theoretical predictions. Page 341 of *Helioseismology*. ESA Special Publication, vol. 376.

Kosovichev, A. G., and Zharkova, V. V. 1998. X-ray flare sparks quake inside the Sun. *Nature*, **393**(May), 317.

Kosovichev, A. G., and Zharkova, V. V. 2001. Magnetic energy release and transients in the solar flare of 2000 July 14. *ApJ*, **550**(Mar.), L105–L108.

Kosovichev, A. G., Schou, J., Scherrer, P. H., *et al.* 1997. Structure and rotation of the solar interior: Initial results from the MDI Medium-L Program. *Sol. Phys.*, **170**, 43–61.

Kosovichev, A. G., Duvall, Jr., T. L., and Scherrer, P. H. 2000. Time-distance inversion methods and results (Invited Review). *Sol. Phys.*, **192**(Mar.), 159–176.

Kostiuk, N. D., and Pikelner, S. B. 1975. Gasdynamics of a flare region heated by a stream of high-velocity electrons. *Soviet Ast.*, **18**(Apr.), 590–599.

Kovach, R. L., and Chyba, C. F. 2001. Seismic detectability of a subsurface ocean on Europa. *Icarus*, **150**, 279–287.

Kovach, R. L., and Watkins, J. S. 1973a. The velocity structure of the lunar crust. *The Moon*, **7**, 63–75.

Kovach, R. L., and Watkins, J. S. 1973b. Apollo 17 seismic profiling: Probing the lunar crust. *Science*, **180**, 1063–1064.

Kovach, R. L., Watkins, J. S., and Landers, T. 1971. Active Seismic Experiment. In *Apollo 14 Preliminary Science Report (prepared by NASA Manned Spacecraft Center)*. Washington, D.C.: U.S. Government Printing Office.

Kovach, R. L., Watkins, J. S., and Talwani, P. 1972. Active Seismic Experiment. In *Apollo 16 Preliminary Science Report (prepared by NASA Manned Spacecraft Center)*. Washington, D.C.: U.S. Government Printing Office.

Kovach, R. L., Watkins, J. S., and Talwani, P. 1973. Lunar Seismic Profiling Experiment. In *Apollo 17 Preliminary Science Report NASA-SP-330 (prepared by Johnson Space Center, NASA)*. Washington, D.C.: U.S. Government Printing Office 1-10–1-12.

Koyama, J., and Nakamura, Y. 1979. Re-examination of the lunar seismic velocity structure, abstract, *Proceedings 10th Lunar and Planetary Science Conference*, 685–686.

Koyama, J., and Nakamura,Y. 1980. Focal mechanism of deep moonquakes. *Proceedings 11th Lunar Planetary Science Conference*, 1835–1865.

Kraft, R. P. 1967. Studies of stellar rotation: V. The dependence of rotation on age among solar-type stars, *ApJ*, **150**, 551.

Krasinsky, G. A., Pitjeva, E. V., Vasilyev, M. V., and Yagudina, E. I. 2002. Hidden mass in the asteroid belt. *Icarus*, **158**(Jul), 98–105.

Krishnamurthi, A., Pinsonneault, M. H., Barnes, S., and Sofia, S. 1997. Theoretical models of the angular momentum evolution of solar-type stars. *ApJ*, **480**(May), 303.

Ksanfomaliti, L. V., Zubkova, V. M., Morozov, N. A., and Petrova, E. V. 1982. Microseisms at the Venera 13 and Venera 14 landing sites. *Sov. Astron. Lett.*, **8**, 241–242.

Kumar, P., and Lu, E. 1991. The location of the source of high-frequency solar acoustic oscillations. *ApJ*, **375**(July), L35–L39.

Kurtz, D. W. 1978. 12.15 minute light variations in Przybylski's star, HD 101065. *Information Bulletin on Variable Stars*, **1436**(June), 1.

Kurtz, D. W., and Martinez, P. 2000. Observing roAp Stars with WET: A primer. *Baltic Astron.*, **9**, 253–353.

Kurtz, D. W., and Shibahashi, H. 1986. An analysis of the L = 1 dipole oscillation in HR 3831 (HD 83368). *MNRAS*, **223**(Dec.), 557–579.

Labeyrie, A., Le Coroller, H., and Dejonghe, J. 2008. Steps toward hypertelescopes on Earth and in space. In *Optical and Infrared Interferometry*. Society of Photo-Optical Instrumentation Engineers (SPIE) Conference Series, vol. 7013.

Labeyrie, A., Le Coroller, H., Dejonghe, J., *et al.* 2009. Luciola hypertelescope space observatory: Versatile, upgradable high-resolution imaging, from stars to deep-field cosmology. *Exp. Astron.*, **23**(Mar.), 463–490.

Lammlein, D. R. 1977. Lunar seismicity and tectonics. *Phys. Earth Planet. Inter.*, **14**(June), 224–273.

Lammlein, D. R., Latham, G. V., Dorman, J., Nakamura, Y., and Ewing, M. 1974. Lunar seismicity, structure, and tectonics. *Rev. Geophys. Space Phys.*, **12**, 1–21.

Langseth, M. G., Keihm, S. J., and Peters, K. 1976. Revised lunar heat-flow values. *Proceedings 7th Lunar Science Conference*, 3143–3171.

Lanza, A. F. 2013. Current state of the modelling of photospheric activity. Page 215 of Suárez, J. C., Garrido, R., Balona, L. A., and Christensen-Dalsgaard, J. (eds), *Advances in Solid State Physics*. Berlin: Springer, vol. 31.

Larose, E., Khan, A., Nakamura, Y., and Campillo, M. 2005. Lunar subsurface investigated from correlation of seismic noise. *Geophys. Res. Lett.*, **32**, L16201, doi:10.1029/2005GL023518.

Larson, T., and Schou, J. 2011. HMI global helioseismology data analysis pipeline. *J. Phys. Conf. Ser.*, **271**(1), 012062.

Laster, S. J., and Press, F. 1968. A new estimate of lunar seismicity due to meteorite impact. *Phys. Earth Planet. Int.*, **1**, 151–154.

Lasota, J.-P. 2001. The disc instability model of dwarf novae and low-mass X-ray binary transients. *New Astron. Rev.*, **45**(June), 449–508.

Latham, G. V., Ewing, M., Press, F., and Sutton, G. 1969a. Passive seismic experiment. *Science*, **165**, 241–250.

Latham, G. V., Ewing, M., Press, F., *et al.* 1969b. Passive Seismic Experiment. In *Apollo 11 Preliminary Science Report*, NASA SP-214.

Latham, G. V., Ewing, M., Dorman, J., *et al.* 1970. Seismic data from man-made impacts on the Moon. *Science*, **170**, 620–626.

Latham, G., Ewing, M., Dorman, J., *et al.* 1971. Moonquakes. *Science*, **174**, 687–692.

Latham, G. V., Ewing, M., Press, F., *et al.* 1972. Passive Seismic Experiment. In *Apollo 15 Preliminary Science Report*, NASA Scientific and Technical Information Office, Washington, D.C., 8-1–8-25.

Latham, G. V., Ewing, M., Press, F., *et al.* 1973. Passive Seismic Experiment. In *Apollo 17 Preliminary Science Report*, NASA Scientific and Technical Information Office, NASA Sp-330, Washington, D.C., 11-1–11-9.

Lay, T., and Wallace, T. C. 1995. *Modern Global Seismology*. Orlando, Florida: Academic Press.

Lazarewicz, A. R., Anderson, D. L., Anderson, K., *et al.* (eds). 1981. *The Viking Seismometry Final Report*. NASA Contract Report, vol. 3408.

Lazrek, M., Baudin, F., Bertello, L., *et al.* 1997. First results on p modes from GOLF experiment. *Sol. Phys.*, **175**(Oct.), 227–246.

Le Bihan, B., and Burrows, A. 2013. Pulsation frequencies and modes of giant exoplanets. *ApJ*, **764**(Feb.), 18.

Leahy, D. A., Elsner, R. F., and Weisskopf, M. C. 1983. On searches for periodic pulsed emission: The Rayleigh test compared to epoch folding. *ApJ*, **272**(Sept.), 256–258.

Leconte, J., and Chabrier, G. 2012. A new vision of giant planet interiors: Impact of double diffusive convection. *A&A*, **540**(Apr.), A20.

Ledoux, P. 1951. The nonradial oscillations of gaseous stars and the problem of Beta Canis Majoris. *ApJ*, **114**(Nov.), 373.

Ledoux, P., and Walraven, T. 1958. Variable stars. *Handbuch der Physik*, **51**, 353–604.

Lee, U., and Baraffe, I. 1995. Pulsational stability of rotating main sequence stars: The second order effects of rotation on the nonadiabatic oscillations. *A&A*, **301**(Sept.), 419.

LeFeuvre, M., and Wieczorek, M. A. 2008. Nonuniform cratering of the terrestrial planets. *Icarus*, **197**, 291–306.

LeFeuvre, M., and Wieczorek, M. A. 2011. Nonuniform cratering of the Moon and a revised crater chronology of the inner Solar System. *Icarus*, **214**, 1–20.

Lee, U., Jeffery, C. S., and Saio, H. 1992. Line profile variations caused by low-frequency non-radial pulsations of rapidly rotating stars: II. *MNRAS*, **254**(Jan.), 185–191.

Lehner, F. E., Witt, E. O., Miller, W. F., and Gurney, R. D. 1962. A seismometer for lunar experiments. *J. Geophys. Res.*, **67**, 4779–4786.

Leibacher, J. W., and Stein, R. F. 1971. A new description of the solar five-minute oscillation. *Astrophys. Lett.*, **7**, 191–192.

Leighton, R. B., Noyes, R. W., and Simon, G. W. 1962. Velocity fields in the solar atmosphere: I. Preliminary report. *ApJ*, **135**(Mar.), 474.

Lenz, P., and Breger, M. 2005. Period04 User Guide. *Commun. Asteroseismol.*, **146**(June), 53–136.

Lesieur, M., Métais, O., and Comte, P. 2005. *Large-Eddy Simulations of Turbulence*. Cambridge: Cambridge University Press.

Libbrecht, K. G. 1988. Solar p-mode phenomenology. *ApJ*, **334**(Nov.), 510–516.

Libbrecht, K. G. 1989. Solar p-mode frequency splittings. *ApJ*, **336**(Jan.), 1092–1097.

Libbrecht, K. G. 1992. On the ultimate accuracy of solar oscillation frequency measurements. *ApJ*, **387**(Mar.), 712.

Libbrecht, K. G., and Kaufman, J. M. 1988. Frequencies of high-degree solar oscillations. *ApJ*, **324**(Jan.), 1172–1183.

Libbrecht, K. G., and Woodard, M. F. 1990. Solar-cycle effects on solar oscillation frequencies. *Nature*, **345**(June), 779–782.

Lignières, F., and Georgeot, B. 2008. Wave chaos in rapidly rotating stars. *Phys. Rev. E*, **78**(1), 016215.

Lignières, F., and Georgeot, B. 2009. Asymptotic analysis of high-frequency acoustic modes in rapidly rotating stars. *A&A*, **500**(June), 1173–1192.

Lignières, F., Rieutord, M., and Reese, D. 2006. Acoustic oscillations of rapidly rotating polytropic stars: I. Effects of the centrifugal distortion. *A&A*, **455**(Aug.), 607–620.

Lindal, G. F. 1992. The atmosphere of Neptune: An analysis of radio occultation data acquired with Voyager 2. *AJ*, **103**(Mar.), 967–982.

Lindsey, C., and Braun, D. C. 1990. Helioseismic imaging of sunspots at their antipodes. *Sol. Phys.*, **126**(Mar.), 101–115.

Lindsey, C., and Braun, D. C. 2000a. Basic principles of solar acoustic holography (Invited Review). *Sol. Phys.*, **192**(Mar.), 261–284.

Lindsey, C., and Braun, D. C. 2000b. Seismic images of the far side of the Sun. *Science*, **287**(Mar.), 1799–1801.

Lindsey, C., and Braun, D. C. 2005a. The acoustic showerglass: I. Seismic diagnostics of photospheric magnetic fields. *ApJ*, **620**(Feb.), 1107–1117.

Lindsey, C., and Braun, D. C. 2005b. The acoustic showerglass: II. Imaging active region subphotospheres. *ApJ*, **620**(Feb.), 1118–1131.

Lindsey, C., Braun, D. C., González Hernández, I., and Donea, A. 2011. Computational seismic holography of acoustic waves in the solar interior. Pages 81–106 of Ramirez, F. A. M. (ed), *Holography: Different Fields of Application*. Intech.

Lingenfelter, R. E., and Schubert, G. 1973. Evidence for convection in planetary interiors from first-order topography. *The Moon*, **7**, 172–180.

Linkin, V., Harri, A. M., Lipatov, A., *et al.* 1998. A sophisticated lander for scientific exploration of Mars: Scientific objectives and implementation of the Mars96 Small Station. *Planet. Space Sci.*, **46**, 717–737.

Lions, J. L. 1971. *Optimal Control of Systems Governed by Partial Differential Equations*. Berlin: Springer-Verlag.

Lisle, J. P., Rast, M. P., and Toomre, J. 2004. Persistent north-south alignment of the solar supergranulation. *ApJ*, **608**(June), 1167–1174.

Liu, Q. Z., van Paradijs, J., and van den Heuvel, E. P. J. 2006. Catalogue of high-mass X-ray binaries in the Galaxy (4th edition). *A&A*, **455**(Sept.), 1165–1168.

Liu, Q. Z., van Paradijs, J., and van den Heuvel, E. P. J. 2007. A catalogue of low-mass X-ray binaries in the Galaxy, LMC, and SMC (Fourth edition). *A&A*, **469**(July), 807–810.

Livshits, M. A., Badalian, O. G., Kosovichev, A. G., and Katsova, M. M. 1981. The optical continuum of solar and stellar flares. *Sol. Phys.*, **73**(Oct.), 269–288.

Lognonné, P. 2005. Planetary seismology. *Annu. Rev. Earth Planet. Sci.*, **33**, 571–604.

Lognonné, P., and Johnson, C. L. 2007. Planetary seismology. Pages 69–122 of Shubert, G. (ed), *Treatise on Geophysics, 10, Planets and Moons*. Elsevier.

Lognonné, P., and Mosser, B. 1993. Planetary seismology. *Surv. Geophysics*, **14**, 239–302.

Lognonné, P., Mosser, B., and Dahlen, F. A. 1994. Excitation of Jovian seismic waves by the Shoemaker-Levy 9 cometary impact. *Icarus*, **110**(Aug.), 180–195.

Lognonné, P., Beyneix, J. G., Banerdt, W. B., *et al.* 1996. Ultra broad band seismology on InterMarsNet. *Planet. Space Sci.*, **44**(Nov.), 1237–1249.

Lognonné, P., Zharkov, V. N., Karczewski, J. K., *et al.* 1998. The Seismic Optimism Experiment. *Planet. Space Sci.*, **46**, 739–747.

Lognonné, P., Gagnepain-Beyneix, J., and Chenet, H. 2003. A new seismic model of the Moon: Implications for structure, thermal evolution and formation of the Moon. *Earth Planet. Sci. Lett.*, **211**, 27–44.

Lognonné, P., LeFeuvre, M., Johnson, C. L., and Weber, R. C. 2009. Moon meteoritic seismic hum: Steady state prediction. *J. Geophys. Res.*, **114**, E12003.

Lognonné, P., Spiga, A., Hurst, K., *et al.* 2012. Martian atmospheric induced micro-seismic noise generation: Large eddy simulations. *43rd Lunar and Planetary Science Conference*, March 19–23, 2012 at The Woodlands, Texas, LPI Contribution No. 1659.

Longuet-Higgins, M. S. 1950. A theory of the origin of microseisms. *Phil. Trans. R. Soc. London Ser. A*, **243**(Sept.), 1–35.

Lorenz, R. D., and Nakamura, Y. 2013. Viking seismometer record: Data restoration and dust devil search. *44th Lunar and Planetary Science Conference*, abstract 1178.

Loudin, M. G. 1979. Structural-compositional models of the lunar interior and interpretation of observed estimates of the Moon's fundamental spheroidal free oscillations. M.S. thesis, Pennsylvania State University.

Lynden-Bell, D., and Ostriker, J. P. 1967. On the stability of differentially rotating bodies. *MNRAS*, **136**, 293.

MacDonald, G. J. F. 1960. Stress history of the Moon. *Planet. Space Sci.*, **2**, 249–255.

MacGregor, K. B., and Brenner, M. 1991. Rotational evolution of solar-type stars: I. Main-sequence evolution. *ApJ*, **376**(July), 204–213.

MacGregor, K. B., Jackson, S., Skumanich, A., and Metcalfe, T. S. 2007. On the structure and properties of differentially rotating, main-sequence stars in the 1–2 M_{solar} range. *ApJ*, **663**(July), 560–572.

Maeder, A. 2009. *Physics, Formation and Evolution of Rotating Stars*. Berlin, Heidelberg: Springer.

Maeder, A., and Zahn, J.-P. 1998. Stellar evolution with rotation: III. Meridional circulation with μ-gradients and non-stationarity. *A&A*, **334**(June), 1000–1006.

Magic, Z., Serenelli, A., Weiss, A., and Chaboyer, B. 2010. On using the color-magnitude diagram morphology of M67 to test solar abundances. *ApJ*, **718**(Aug.), 1378–1387.

Maillard, J. P., and Michel, G. 1982. A high resolution Fourier transform spectrometer for the Cassegrain focus at the CFH telescope. Pages 213–222 of *ASSL Vol. 92: IAU Colloquium 67: Instrumentation for Astronomy with Large Optical Telescopes*.

Marchis, F., Hestroffer, D., and Descamps, P., *et al.* 2006. A low density of 0.8 g cm^{-3} for the Trojan binary asteroid 617 Patroclus. *Nature*, **439**, 565–567.

Marfurt, K. J. 1984. Accuracy of finite-difference and finite-element modeling of the scalar and elastic wave-equations. *Geophysics,* **49**, 533–549.

Markey, P., and Tayler, R. J. 1973. The adiabatic stability of stars containing magnetic fields: II. Poloidal fields. *MNRAS*, **163**, 77.

Marley, M. S., and Porco, C. C. 1993. Planetary acoustic mode seismology: Saturn's rings. *Icarus*, **106**(Dec.), 508.

Marques, J. P., Goupil, M. J., Lebreton, Y., *et al.* 2013. Seismic diagnostics for transport of angular momentum in stars: I. Rotational splittings from the pre-main sequence to the red-giant branch. *A&A*, **549**(Jan.), A74.

Marsden, S. C., Petit, P., Jeffers, S. V., *et al.* 2013. A Bcool magnetic snapshot survey of solar-type stars. ArXiv:1311.3374.

Martić, M., Lebrun, J.-C., Appourchaux, T., and Korzennik, S. G. 2004. P-mode frequencies in solar-like stars: I. Procyon A. *A&A*, **418**(Apr.), 295–303.

Mathis, S. 2009. Transport by gravito-inertial waves in differentially rotating stellar radiation zones: I. Theoretical formulation. *A&A*, **506**(Nov.), 811–828.

Mathis, S., and de Brye, N. 2011. Low-frequency internal waves in magnetized rotating stellar radiation zones: I. Wave structure modification by a toroidal field. *A&A*, **526**(Feb.), A65.

Mathis, S., and de Brye, N. 2012. Low-frequency internal waves in magnetized rotating stellar radiation zones: II. Angular momentum transport with a toroidal field. *A&A*, **540**(Apr.), A37.

Mathis, S., and Zahn, J.-P. 2004. Transport and mixing in the radiation zones of rotating stars: I. Hydrodynamical processes. *A&A*, **425**(Oct.), 229–242.

Mathis, S., and Zahn, J.-P. 2005. Transport and mixing in the radiation zones of rotating stars: II. Axisymmetric magnetic field. *A&A*, **440**(Sept.), 653–666.

Mathis, S., Talon, S., Pantillon, F.-P., and Zahn, J.-P. 2008. Angular momentum transport in the Sun's radiative zone by gravito-inertial waves. *Sol. Phys.*, **251**(Sept.), 101–118.

Mathur, S., Turck-Chièze, S., Couvidat, S., and García, R. A. 2007. On the characteristics of the solar gravity mode frequencies. *ApJ*, **668**(Oct.), 594–602.

Mathur, S., Eff-Darwich, A., García, R. A., and Turck-Chièze, S. 2008. Sensitivity of helioseismic gravity modes to the dynamics of the solar core. *A&A*, **484**(June), 517–522.

Mathur, S., García, R. A., Catala, C., *et al.* 2010. The solar-like CoRoT target HD 170987: Spectroscopic and seismic observations. *A&A*, **518**(July), A53.

Mathur, S., Hekker, S., Trampedach, R., *et al.* 2011a. Granulation in red giants: Observations by the Kepler mission and three-dimensional convection simulations. *ApJ*, **741**(Nov.), 119.

Mathur, S., Handberg, R., Campante, T. L., *et al.* 2011b. Solar-like oscillations in KIC 11395018 and KIC 11234888 from 8 months of Kepler data. *ApJ*, **733**(June), 95.

Mathur, S., Metcalfe, T. S., Woitaszek, M., *et al.* 2012. A uniform asteroseismic analysis of 22 solar-type stars observed by Kepler. *ApJ*, **749**(Apr.), 152.

Mathur, S., Bruntt, H., Catala, C., *et al.* 2013. Study of HD 169392A observed by CoRoT and HARPS. *A&A*, **549**(Jan.), A12.

Mathur, S., García, R. A., Ballot, J., *et al.* 2014. Magnetic activity of F stars observed by Kepler. *A&A*, **562**(Feb.), A124.

Matt, S. P., MacGregor, K. B., Pinsonneault, M. H., and Greene, T. P. 2012. Magnetic braking formulation for Sun-like stars: Dependence on dipole field strength and rotation rate. *ApJ*, **754**(Aug.), L26.

Matthews, J. M., Kuschnig, R., Walker, G. A. H., *et al.* 2000. Ultraprecise photometry from space: The MOST Microsat mission. Pages 74–75 of Szabados, L. and Kurtz, D. (eds), *IAU Colloquium 176: The Impact of Large-Scale Surveys on Pulsating Star Research*. Astronomical Society of the Pacific Conference Series, vol. 203.

Matthews, J. M., Kusching, R., Guenther, D. B., *et al.* 2004. No stellar p-mode oscillations in space-based photometry of Procyon. *Nature*, **430**(July), 51–53.

Mazumdar, A. 2005. Asteroseismic diagrams for solar-type stars. *A&A*, **441**(Oct.), 1079–1086.

Mazumdar, A., Michel, E., Antia, H. M., and Deheuvels, S. 2012. Seismic detection of acoustic sharp features in the CoRoT target HD 49933. *A&A*, **540**(Apr.), A31.

Mazumdar, A., Monteiro, M. J. P. F. G., Ballot, J., *et al.* 2014. Measurement of acoustic glitches in solar-type stars from oscillation frequencies observed by Kepler. *ApJ*, **782**(Feb.), 18.

McClintock, J. E., and Remillard, R. A. 2006. Black hole binaries. Pages 157–213 of Lewin, W. and van der Klis, M. (eds), *Compact Stellar X-ray Sources*. Cambridge: Cambridge University Press.

McGarr, A. G., Latham, V., and Gault, D. E. 1969. Meteoroid impacts as sources of seismicity on the Moon. *J. Geophys. Res.*, **74**, 5981–5994.

McGlaun, J. M., Thompson, S. L., and Elrick, M. G. 1990. CTH: A three-dimensional shock wave physics code. *Int. J. Impact Engng.*, **10**, 351–360.

McQuillan, A., Mazeh, T., and Aigrain, S. 2013. Stellar rotation periods of the Kepler objects of interest: A dearth of close-in planets around fast rotators. *ApJ*, **775**(Sept.), L11.

Mercerat, E. D., and Nolet, G. 2013. On the linearity of cross-correlation delay times in finite-frequency tomography. *Geophys. J. Int.*, **192**(Feb.), 681–687.

Mestel, L. 1965. "Fossil" magnetic fields. Page 420 of Lust, R. (ed), *Stellar and Solar Magnetic Fields*. IAU Symposium 22.

Mestel, L., Tayler, R. J., and Moss, D. L. 1988. The mutual interaction of magnetism, rotation and meridian circulation in stellar radiative zones. *MNRAS*, **231**(Apr.), 873–885.

Mészáros, P. 2006. Gamma-ray bursts. *Rep. Prog. Phys.*, **69**(Aug.), 2259–2321.

Mészáros, S., Holtzman, J., García Pérez, A. E., *et al.* 2013. Calibrations of atmospheric parameters obtained from the first year of SDSS-III APOGEE observations. *AJ*, **146**(Nov.), 133.

Metcalfe, T. S., Dziembowski, W. A., Judge, P. G., and Snow, M. 2007. Asteroseismic signatures of stellar magnetic activity cycles. *MNRAS*, **379**(July), L16–L20.

Metcalfe, T. S., Creevey, O. L., and Christensen-Dalsgaard, J. 2009. A stellar model-fitting pipeline for asteroseismic data from the Kepler mission. *ApJ*, **699**(July), 373–382.

Metcalfe, T. S., Monteiro, M. J. P. F. G., Thompson, M. J., *et al.* 2010a. A precise asteroseismic age and radius for the evolved Sun-like star KIC 11026764. *ApJ*, **723**(Nov.), 1583–1598.

Metcalfe, T. S., Basu, S., Henry, T. J., *et al.* 2010b. Discovery of a 1.6 year magnetic activity cycle in the exoplanet host star ι Horologii. *ApJ*, **723**(Nov.), L213–L217.

Metcalfe, T. S., Chaplin, W. J., Appourchaux, T., *et al.* 2012. Asteroseismology of the solar analogs 16 Cyg A and B from Kepler observations. *ApJ*, **748**(Mar.), L10.

Metcalfe, T. S., Buccino, A. P., Brown, B. P., *et al.* 2013. Magnetic activity cycles in the exoplanet host star epsilon Eridani. *ApJ*, **763**(Feb.), L26.

Michel, E., Baglin, A., Auvergne, M., *et al.* 2008. CoRoT measures solar-like oscillations and granulation in stars hotter than the Sun. *Science*, **322**, 558–560.

Miesch, M. S. 2005. Large-scale dynamics of the convection zone and tachocline. *Living Rev. Sol. Phys.*, **2**(Apr.), 1.

Miesch, M. S., Elliott, J. R., Toomre, J., *et al.* 2000. Three-dimensional spherical simulations of solar convection: I. Differential rotation and pattern evolution achieved with laminar and turbulent states. *ApJ*, **532**(Mar.), 593–615.

Miesch, M. S., Brun, A. S., De Rosa, M. L., and Toomre, J. 2008. Structure and evolution of giant cells in global models of solar convection. *ApJ*, **673**(Jan.), 557–575.

Miglio, A., Montalbán, J., Baudin, F., *et al.* 2009. Probing populations of red giants in the galactic disk with CoRoT. *A&A*, **503**(Sept.), L21–L24.

Miglio, A., Montalbán, J., Carrier, F., *et al.* 2010. Evidence for a sharp structure variation inside a red-giant star. *A&A*, **520**(Sept.), L6.

Miglio, A., Brogaard, K., Stello, D., *et al.* 2012a. Asteroseismology of old open clusters with Kepler: Direct estimate of the integrated red giant branch mass-loss in NGC 6791 and 6819. *MNRAS*, **419**(Jan.), 2077–2088.

Miglio, A., Morel, T., Barbieri, M., *et al.* 2012b. Solar-like pulsating stars as distance indicators: G-K giants in the CoRoT and Kepler fields. In Reylé, C., Robin, A., and Schultheis, M. (eds), *Assembling the Puzzle of the Milky Way*, Le Grand-Bornand, France, EPJ Web of Conferences, vol. 19(Feb.), 5012.

Miglio, A., Chiappini, C., Morel, T., *et al.* 2013a (Mar.). Differential population studies using asteroseismology: Solar-like oscillating giants in CoRoT fields LRc01 and LRa01. Page 3004 of 40th Liège International Astrophysical Colloquium Ageing Low Mass Stars. EPJ Web of Conferences, vol. 43.

Miglio, A., Chiappini, C., Morel, T., *et al.* 2013b. Galactic archaeology: Mapping and dating stellar populations with asteroseismology of red-giant stars. *MNRAS*, **429**(Feb.), 423–428.

Miglio, A., Chaplin, W. J., Farmer, R., *et al.* 2014. Prospects for detecting asteroseismic binaries in Kepler data. *ApJ*, **784**(Mar.), L3.

Militzer, B., Hubbard, W. B., Vorberger, J., Tamblyn, I., and Bonev, S. A. 2008. A massive core in Jupiter predicted from first-principles simulations. *ApJ*, **688**(Nov.), L45–L48.

Miller, M. C., Lamb, F. K., and Psaltis, D. 1998. Sonic-point model of kilohertz quasi-periodic brightness oscillations in low-mass X-ray binaries. *ApJ*, **508**(Dec.), 791–830.

Mimoun, D., Wieczorek, M., Alkalai, L., *et al.* 2012. Farside explorer: Unique science from a mission to the farside of the Moon. *Exp. Astron.*, **33**, 529–585.

Minshull, T. A., and Goulty, N. R. 1988. The influence of tidal stresses on deep moonquake activity. *Phys. Earth Planet. Int.*, **52**(Oct.), 41–55.

Mitra-Kraev, U., and Thompson, M. J. 2007. Meridional flow profile measurements with SOHO/MDI. *Astron. Nachr.*, **328**(Dec.), 1009–1012.

Moczo, P., Kristek, J., Vavryčuk, V., Archuleta, R. J., and Halada, L. 2002. 3D heterogeneous staggered-grid finite-difference modeling of seismic motion with volume harmonic and arithmetic averaging of elastic moduli and densities. *Bull. Seismol. Soc. Am.*, **92**, 3042–3066.

Moin, P., Squires, K., Cabot, W., and Lee, S. 1991. A dynamic subgrid-scale model for compressible turbulence and scalar transport. *Phys. Fluids*, **3**(Nov.), 2746–2757.

Molenda-Żakowicz, J., Bruntt, H., Sousa, S., *et al.* 2010. Asteroseismology of solar-type stars with Kepler: III. Ground-based data. *Astron. Nachr.*, **331**(Dec.), 981.

Montalbán, J., Miglio, A., Noels, A., *et al.* 2013. Testing convective-core overshooting using period spacings of dipole modes in red giants. *ApJ*, **766**(Apr.), 118.

Monteiro, M. J. P. F. G. 2009. *Evolution and Seismic Tools for Stellar Astrophysics*. Springer.

Monteiro, M. J. P. F. G., and Thompson, M. J. 1998. Looking for variations with latitude of the base of the solar convection zone. Page 819 of Korzennik, S. (ed), *Structure and Dynamics of the Interior of the Sun and Sun-like Stars*. ESA Special Publication, vol. 418.

Monteiro, M. J. P. F. G., Christensen-Dalsgaard, J., and Thompson, M. J. 1994. Seismic study of overshoot at the base of the solar convective envelope. *A&A*, **283**(Mar.), 247–262.

Montelli, R., Nolet, G., Dahlen, F. A., and Masters, G. 2006. A catalogue of deep mantle plumes: New results from finite-frequency tomography. *Geochem., Geophys., Geosyst.*, **7**(Nov.), 11007.

Moradi, H., Baldner, C., Birch, A. C., *et al.* 2010. Modeling the subsurface structure of sunspots. *Sol. Phys.*, **267**(Nov.), 1–62.

Morgan, J. V., Warner, M. R., Collins, G. S., *et al.* 2011. Full waveform tomographic images of the peak ring at the Chicxulub impact crater, *J. Geophys. Res.*, **116**, B06303.

Moss, D. 1992. Magnetic fields and differential rotation in stars. *MNRAS*, **257**(Aug.), 593–601.

Mosser, B. 1990. The pressure mode oscillation spectrum of a rotating gaseous sphere: Application to Jupiter. *Icarus*, **87**(Sept.), 198–209.

Mosser, B. 1995. Propagation and trapping of global oscillations in the Jovian troposphere and stratosphere. *A&A*, **293**(Jan.), 586–593.

Mosser, B., Mekarnia, D., Maillard, J. P., *et al.* 1993. Seismological observations with a Fourier transform spectrometer: Detection of Jovian oscillations. *A&A*, **267**(Jan.), 604–622.

Mosser, B., Galdemard, P., Lagage, P., *et al.* 1996. Impact seismology: A search for primary pressure waves following impacts A and H. *Icarus*, **121**(June), 331–340.

Mosser, B., Maillard, J. P., Mekarnia, D., and Gay, J. 1998. New limit on the p-mode oscillations of Procyon obtained by Fourier transform seismometry. *A&A*, **340**(Dec.), 457–462.

Mosser, B., Maillard, J. P., and Mékarnia, D. 2000. New attempt at detecting the jovian oscillations. *Icarus*, **144**(Mar.), 104–113.

Mosser, B., Baudin, F., Lanza, A. F., *et al.* 2009. Short-lived spots in solar-like stars as observed by CoRoT. *A&A*, **506**(Oct.), 245–254.

Mosser, B., Belkacem, K., Goupil, M.-J., *et al.* 2010. Red-giant seismic properties analyzed with CoRoT. *A&A*, **517**(July), A22.

Mosser, B., Barban, C., Montalbán, J., *et al.* 2011. Mixed modes in red-giant stars observed with CoRoT. *A&A*, **532**(Aug.), A86.

Mosser, B., Elsworth, Y., Hekker, S., *et al.* 2012a. Characterization of the power excess of solar-like oscillations in red giants with Kepler. *A&A*, **537**(Jan.), A30.

Mosser, B., Goupil, M. J., Belkacem, K., *et al.* 2012b. Spin down of the core rotation in red giants. *A&A*, **548**(Dec.), A10.

Mosser, B., Dziembowski, W. A., Belkacem, K., *et al.* 2013. Period-luminosity relations in evolved red giants explained by solar-like oscillations. *A&A*, **559**(Nov.), A137.

Motta, S. E., Belloni, T. M., Stella, L., Muñoz-Darias, T., and Fender, R. 2014. Precise mass and spin measurements for a stellar-mass black hole through X-ray timing: The case of GRO J1655-40. *MNRAS*, **437**(Jan.), 2554–2565.

Muller, P. M., and Sjogren, W. L. 1968. Mascons: Lunar mass concentrations. *Science*, **161**, 680–684.

Nagashima, K. 2010 (May). Local helioseismology with solar optical telescope onboard Hinode. Page 319.04 of *American Astronomical Society Meeting Abstracts 216*. Bulletin of the American Astronomical Society, vol. 41.

Nagashima, K., Zhao, J., Kosovichev, A. G., and Sekii, T. 2011. Detection of supergranulation alignment in polar regions of the Sun by helioseismology. *ApJ*, **726**(Jan.), L17.

Nakamura, T., Noguchi, T., and Tanaka, M., *et al.* 2011. Itokawa dust particles: A direct link between S-type asteroids and ordinary chondrites. *Science*, **333**(6046), 1113–1116.

Nakamura, Y. 1976. Seismic energy transmission in the lunar surface zone determined from signals generated by movement of lunar rovers. *Bull. Seismol. Soc. Am.*, **66**, 593–606.

Nakamura, Y. 1977. Seismic energy transmission in an intensively scattering environment. *J. Geophys.*, **43**, 389–399.

Nakamura, Y. 1978. A_1 moonquakes: Source distribution and mechanism. *Proceedings of the 9th Lunar and Planetary Science Conference*, 3589–3607.

Nakamura, Y. 1983. Seismic velocity structure of the lunar mantle. *J. Geophys. Res.*, **88**(Jan.), 677–686.

Nakamura, Y. 2003. New identification of deep moonquakes in the Apollo lunar seismic data. *Phys. Earth Planet. Int.*, **139**, 197–205.

Nakamura, Y. 2005. Farside deep moonquakes and deep interior of the Moon. *J. Geophys. Res.*, **110**(Jan.), E01001.

Nakamura, Y. 2010. Lunar seismology: Current status and future challenges. *DGG-Mitteilungen,* **3/2010**, 4–13.

Nakamura, Y. 2011. Timing problem with the Lunar Module impact data as recorded by the LSPE and corrected near-surface structure at the Apollo 17 landing site. *J. Geophys. Res.*, **116**, E12005, doi:10.1029/2011JE003972.

Nakamura, Y., and Anderson, D. L. 1979. Martian wind activity detected by a seismometer at Viking Lander 2 site. *Geophys. Res. Lett.*, **6**, 499–502.

Nakamura, Y., Lammlein, D., Latham, G., *et al.* 1973. New seismic data on the state of the deep lunar interior. *Science*, **181**, 49–51.

Nakamura, Y., Latham, G., Lammlein, D., *et al.* 1974a. Deep lunar interior inferred from recent seismic data. *Geophys. Res. Lett.*, **1**, 137–140.

Nakamura, Y., Dorman, J., Duennebier, F., *et al.* 1974b. High-frequency lunar teleseismic events, *Proceedings of the 5th Lunar Science Conference* (Suppl. 5, *Geochim Cosmochim Acta*), **3**, 2883–2890.

Nakamura, Y., Latham, G. V., Dorman, H. J., and Duennebier, F. K. 1976. Seismic structure of the moon: A summary of current status. *Proceedings of the 7th Lunar Science Conference*, 3113–3121.

Nakamura, Y., Latham, G. V., Dorman, H. J., *et al.* 1979. Shallow moonquakes: Depth, distribution and implications as to the present state of the lunar interior. *Proceedings of the 10th Lunar and Planetary Science Conference*, 2299–2309.

Nakamura, Y., Latham, G. V., and Dorman, H. J. 1980. How we processed Apollo lunar seismic data. *Phys. Earth Planet. Int.*, **21**, 218–224.

Nakamura, Y., Latham, G. V., Dorman, H. J., and Harris, J. E. 1981. *Passive Seismic Experiment Long-Period Event Catalog, Final version, 1969 day 202 – 1977 day 273*. Galveston Geophysics Laboratory, Galveston, UTIG Technical Report No. 18, available online: ftp://utig.ig.utexas.edu/pub/PSE/catsrepts/levent.1008.

Nakamura, Y., Latham, G. V., and Dorman, H. J. 1982. Apollo Lunar Seismic Experiment: Final summary. Proceedings 13th Lunar and Planetary Science Conference, pt. 1. *J. Geophys. Res.*, **87** suppl., A117–A123.

Nather, R. E., Winget, D. E., Clemens, J. C., Hansen, C. J., and Hine, B. P. 1990. The Whole Earth Telescope: A new astronomical instrument. *ApJ*, **361**(Sept.), 309–317.

Nawa, K., Suda, N., Fukao, Y., *et al.* 1998. Incessant excitation of the Earth's free oscillations. *Earth, Planets, and Space*, **50**(Jan.), 3–8.

Neal, C. R., Banerdt, W. B., Alkalai, L., *et al.* 2011. Lunette: A two-lander discovery-class geophysics mission to the Moon. *42nd Lunar and Planetary Science Conference*, March 7–11, 2011 at The Woodlands, Texas, LPI Contribution No. 1608.

Neiner, C., Floquet, M., Samadi, R., *et al.* 2012. Stochastic gravito-inertial modes discovered by CoRoT in the hot Be star HD 51452. *A&A*, **546**(Oct.), A47.

Nettelmann, N., Helled, R., Fortney, J. J., and Redmer, R. 2013a. New indication for a dichotomy in the interior structure of Uranus and Neptune from the application of modified shape and rotation data. *Planet. Space Sci.*, **77**(Mar.), 143–151.

Nettelmann, N., Püstow, R., and Redmer, R. 2013b. Saturn layered structure and homogeneous evolution models with different EOSs. *Icarus*, **225**(July), 548–557.

Nielsen, M. B., Gizon, L., Schunker, H., and Karoff, C. 2013. Rotation periods of 12 000 main-sequence Kepler stars: Dependence on stellar spectral type and comparison with $v \sin i$ observations. *A&A*, **557**(Sept.), L10.

Nigam, R., and Kosovichev, A. G. 1998. Measuring the Sun's eigenfrequencies from velocity and intensity helioseismic spectra: Asymmetrical line profile-fitting formula. *ApJ*, **505**(Sept.), L51.

Nigam, R., Kosovichev, A. G., Scherrer, P. H., and Schou, J. 1998. Asymmetry in velocity and intensity helioseismic spectra: A solution to a long-standing puzzle. *ApJ*, **495**(Mar.), L115.

Nobili, L., Turolla, R., Zampieri, L., and Belloni, T. 2000. A Comptonization model for phase-lag variability in GRS 1915+105. *ApJ*, **538**(Aug.), L137–L140.

Nolet, G. 2008. *A Breviary of Seismic Tomography*. Cambridge: Cambridge University Press.

Nordlund, Å. 1985. Solar convection. *Sol. Phys.*, **100**(Oct.), 209–235.

Nordlund, Å., and Stein, R. F. 2001. Solar oscillations and convection. I. Formalism for radial oscillations. *ApJ*, **546**(Jan.), 576–584.

Nordlund, Å., Stein, R. F., and Asplund, M. 2009. Solar surface convection. *Living Rev. Sol. Phys.*, **6**(Apr.), 2.

Nowak, M. A., and Wagoner, R. V. 1991. Diskoseismology: Probing accretion disks. I. Trapped adiabatic oscillations. *ApJ*, **378**(Sept.), 656–664.

Nowak, M. A., and Wagoner, R. V. 1992. Diskoseismology: Probing accretion disks. II. G-modes, gravitational radiation reaction, and viscosity. *ApJ*, **393**(July), 697–707.

Oberst, J. 1987. Unusually high stress drops associated with shallow moonquakes. *J. Geophys. Res.*, **92**, 1397–1405.

Oberst, J. 1989. *Meteoroids near the Earth-Moon system as inferred from temporal and spatial distribution of impacts detected by the Lunar Seismic Network*. Ph.D. thesis, University of Texas, Austin, TX.

Oberst, J., and Nakamura, Y. 1987. Distinct meteoroid families identified on the lunar seismograms. *J. Geophys. Res.*, **92**, 769–773.

Oberst, J., and Nakamura, Y. 1991. A search for clustering among the meteoroid impacts detected by the Apollo lunar seismic network. *Icarus*, **91**, 315–325.

Olbers, W. 1850. Materialien zu einer Lebensbeschreibung der beiden Astronomen David und Johannes Fabricius, von W. Olbers. *Astron. Nachr.*, **31**(Sept.), 129.

Olver, F. W., Lozier, D. W., Boisvert, R. F., and Clark, C. W. 2010. *NIST Handbook of Mathematical Functions*. Cambridge: Cambridge University Press.

Ortiz, J. L., Aceituno, F. J., Quesada, J. A., *et al.* 2006. Detection of sporadic impact flashes on the Moon: Implications for the luminous efficiency of hypervelocity impacts and derived terrestrial impact rates. *Icarus*, **184**, 319–326.

Ortiz, A., Bellot Rubio, L. R., and Rouppe van der Voort, L. 2010. Downflows in sunspot umbral dots. *ApJ*, **713**(Apr.), 1282–1291.

Osaki, J. 1975. Nonradial oscillations of a 10 solar mass star in the main-sequence stage. *PASJ*, **27**, 237–258.

Ostro, S. J., Benner, L. A. M., and Nolan, M. C., *et al.* 2004. Radar observations of asteroid 25143 Itokawa (1998 SF36). *Meteor. Planet Sci.*, **39**, 407–424.

Otí Floranes, H., Christensen-Dalsgaard, J., and Thompson, M. J. 2005. The use of frequency-separation ratios for asteroseismology. *MNRAS*, **356**(Jan.), 671–679.

Ouazzani, R. 2011. *La rotation et son interaction avec les oscillations dans les étoiles*. Ph.D. thesis, Université Pierre et Marie Curie (Paris 6), France.

Paik, H. J., and Venkateswara, K. Y. 2009. Gravitational wave detection on the Moon and the moons of Mars. *Adv. Space Res.*, **43**, 167–170.

Pallavicini, R., Serio, S., and Vaiana, G. S. 1977. A survey of soft X-ray limb flare images: The relation between their structure in the corona and other physical parameters. *ApJ*, **216**(Aug.), 108–122.

Pamyatnykh, A. A., Handler, G., and Dziembowski, W. A. 2004. Asteroseismology of the β Cephei star ν Eridani: Interpretation and applications of the oscillation spectrum. *MNRAS*, **350**(May), 1022–1028.

Panning, M., Lekic, V., Manga, M., Cammarano, F., and Romanowicz, B. 2006. Long-period seismology on Europa: 2. Predicted seismic response. *J. Geophys. Res. (Planets)*, **111**(10), 12008.

Panning, M. P., Beucler, E., Drilleau, M., Mocquet, A., Lognonné, P., and Banerdt, W. B. 2015. Verifying single-station seismic approaches using Earth-based data: Preparation for data return from the InSight mission to Mars. *Icarus*, **248**, 230–242.

Pápics, P. I., Briquet, M., Baglin, A., *et al.* 2012. Gravito-inertial and pressure modes detected in the B3 IV CoRoT target HD 43317. *A&A*, **542**(June), A55.

Parchevsky, K. V., and Kosovichev, A. G. 2007. Three-dimensional numerical simulations of the acoustic wave field in the upper convection zone of the Sun. *ApJ*, **666**(Sept.), 547–558.

Parchevsky, K. V., and Kosovichev, A. G. 2009a. Excitation, propagation and conversion of helioseismic MHD waves in strong field regions. Page 61 of Dikpati, M., Arentoft, T., González Hernández, I., Lindsey, C., and Hill, F. (eds), *Solar-Stellar Dynamos as Revealed by Helio- and Asteroseismology: GONG 2008/SOHO 21*. Astronomical Society of the Pacific Conference Series, vol. 416.

Parchevsky, K. V., and Kosovichev, A. G. 2009b. Numerical simulation of excitation and propagation of helioseismic MHD waves: Effects of inclined magnetic field. *ApJ*, **694**(Mar.), 573–581.

Parchevsky, K., Kosovichev, A., Khomenko, E., Olshevsky, V., and Collados, M. 2010. Numerical simulation of excitation and propagation of helioseismic MHD waves in magnetostatic models of sunspots. ArXiv:1002.1117.

Parker, E. N. 1979. Sunspots and the physics of magnetic flux tubes: I. The general nature of the sunspot; II. Aerodynamic drag. *ApJ*, **230**(June), 905–923.

Parker, E. N. 1993. A solar dynamo surface wave at the interface between convection and nonuniform rotation. *ApJ*, **408**(May), 707–719.

Pasek, M., Georgeot, B., Lignières, F., and Reese, D. R. 2011. Regular modes in rotating stars. *Phys. Rev. Lett.*, **107**(12), 121101.

Pasek, M., Lignières, F., Georgeot, B., and Reese, D. R. 2012. Regular oscillation sub-spectrum of rapidly rotating stars. *A&A*, **546**(Oct.), A11.

Patrón, J., González Hernández, I., Chou, D.-Y., *et al.* 1997. Comparison of two fitting methods for ring diagram analysis of very high l solar oscillations. *ApJ*, **485**(Aug.), 869–874.

Pätzold, M., Andert, T. P., Asmar, S. W., *et al.* 2011. Asteroid 21 Lutetia: Low mass, high density. *Science*, **334**(Oct.), 491–492.

Paxton, B., Bildsten, L., Dotter, A., *et al.* 2011. Modules for Experiments in Stellar Astrophysics (MESA). *ApJS*, **192**(Jan.), 3.

Paxton, B., Cantiello, M., Arras, P., *et al.* 2013. Modules for Experiments in Stellar Astrophysics (MESA): Planets, oscillations, rotation, and massive stars. *ApJS*, **208**(Sept.), 4.

Pepe, F., Mayor, M., Delabre, B., *et al.* 2000 (Aug.). HARPS: A new high-resolution spectrograph for the search of extrasolar planets. Pages 582–592 of Iye, M., and Moorwood, A. F. (eds), *Optical and IR Telescope Instrumentation and Detectors*. Society of Photo-Optical Instrumentation Engineers (SPIE) Conference Series, vol. 4008.

Perdang, J. 1988. Stellar acoustic ray patterns. Page 125 of Christensen-Dalsgaard, J., and Frandsen, S. (eds), *Advances in Helio- and Asteroseismology*. IAU Symposium 123.

Pérez Hernández, F., and Christensen-Dalsgaard, J. 1994. The phase function for stellar acoustic oscillations: III. The solar case. *MNRAS*, **269**(July), 475.

Pérez Hernández, F., and Christensen-Dalsgaard, J. 1998. The phase function for stellar acoustic oscillations: IV. Solar-like stars. *MNRAS*, **295**(Apr.), 344–352.

Pesnell, W. D. 2008. Predictions of solar cycle 24. *Sol. Phys.*, **252**(Oct.), 209–220.

Pesnell, W. D., Thompson, B. J., and Chamberlin, P. C. 2012. The Solar Dynamics Observatory (SDO). *Sol. Phys.*, **275**(Jan.), 3–15.

Peterson, J. 1993. *Observations and Modeling of Seismic Background Noise*. US Geological Survey Open-file Report, 93–322.

Petersson, N. A., and Sjogreen, B. 2010. *Reference guide to WPP version 2.0*. Lawrence Livermore National Laboratory Technical Report LLNL-TR-422928.

Petrovic, J. J. 2001. Review: Mechanical properties of meteorites and their constituents. *J. Mater. Sci.*, **36**, 1579–1583.

Phillips, R. J. 1991. Expected rates of marsquakes. Pages 35–38 of *Scientific Rationale and Requirements for a Global Seismic Network on Mars*. Lunar and Planetary Institute Technical Report.

Pijpers, F. P., and Thompson, M. J. 1992. Faster formulations of the optimally localized averages method for helioseismic inversions. *A&A*, **262**(Sept.), L33–L36.

Pinsonneault, M. H., Kawaler, S. D., Sofia, S., and Demarque, P. 1989. Evolutionary models of the rotating sun. *ApJ*, **338**(Mar.), 424–452.

Pipin, V. V., and Kosovichev, A. G. 2013. The mean-field solar dynamo with a double cell meridional circulation pattern. *ApJ*, **776**(Oct.), 36.

Podolak, M., Podolak, J. I., and Marley, M. S. 2000. Further investigations of random models of Uranus and Neptune. *Planet. Space Sci.*, **48**(Feb.), 143–151.

Pottasch, E. M., Butcher, H. R., and van Hoesel, F. H. J. 1992. Solar-like oscillations on Alpha Centauri A. *A&A*, **264**(Oct.), 138–146.

Pötzi, W., and Brandt, P. N. 2005. Is solar plasma sinking down in vortices. *Hvar Observatory Bulletin*, **29**, 61–70.

Pratt, R., Song, Z.-M., Williamson, P., and Warner, M. 1996. Two-dimensional velocity models from wide-angle seismic data by wavefield inversion. *Geophys. J. Int.*, **124**, 323–340.

Press, F., and Ewing, M. 1951. Propagation of elastic waves in a floating ice sheet. *Trans. Am. Geophys. Union*, **32**, 673–678.

Press, F., Buwalda, P., and Neugebauer, M. 1960. A lunar seismic experiment. *J. Geophys. Res.*, **65**, 3097–3105.

Press, W. H., and Rybicki, G. B. 1989. Fast algorithm for spectral analysis of unevenly sampled data. *ApJ*, **338**(Mar.), 277–280.

Press, W. H., Teukolsky, S. A., Vetterling, W. T., and Flannery, B. P. 1992. *Numerical Recipes in FORTRAN: The Art of Scientific Computing* (2nd edn). Cambridge: Cambridge University Press.

Priest, E. R. 2003. Solar magnetohydrodynamics. Pages 217–237 of Dwivedi, B. N. (ed), *Dynamic Sun*. Cambridge: Cambridge University Press.

Provost, J., Mosser, B., and Berthomieu, G. 1993. A new asymptotic formalism for jovian seismology. *A&A*, **274**(July), 595–611.

Qin, C., Muirhead, A. C., and Zhong, S. 2012. Correlation of deep moonquakes and mare basalts: Implications for lunar mantle structure and evolution. *Icarus*, **220**, 100–105.

Queloz, D., Mayor, M., Weber, L., *et al.* 2000. The CORALIE survey for southern extra-solar planets: I. A planet orbiting the star Gliese 86. *A&A*, **354**(Feb.), 99–102.

Quirion, P.-O., Christensen-Dalsgaard, J., and Arentoft, T. 2010. Automatic determination of stellar parameters via asteroseismology of stochastically oscillating stars: Comparison with direct measurements. *ApJ*, **725**(Dec.), 2176–2189.

Rabello-Soares, M. C., Christensen-Dalsgaard, J., Rosenthal, C. S., and Thompson, M. J. 1999. Effects of line asymmetries on the determination of solar internal structure. *A&A*, **350**(Oct.), 672–679.

Rabello-Soares, M. C., Basu, S., Christensen-Dalsgaard, J., and Di Mauro, M. P. 2000. The potential of solar high-degree modes for structure inversion. *Sol. Phys.*, **193**(Apr.), 345–356.

Rahoui, F., Chaty, S., Rodriguez, J., *et al.* 2010. Long-term multi-wavelength studies of GRS 1915+105. I. A high-energy and mid-infrared focus with RXTE/INTEGRAL and Spitzer. *ApJ*, **715**(June), 1191–1202.

Rajaguru, S. P., Basu, S., and Antia, H. M. 2001. Ring diagram analysis of the characteristics of solar oscillation modes in active regions. *ApJ*, **563**(Dec.), 410–418.

Rajaguru, S. P., Wachter, R., Sankarasubramanian, K., and Couvidat, S. 2010. Local helioseismic and spectroscopic analyses of interactions between acoustic waves and a sunspot. *ApJ*, **721**(Oct.), L86–L91.

Rashba, T. I., Semikoz, V. B., Turck-Chièze, S., and Valle, J. W. F. 2007. Probing the internal solar magnetic field through g modes. *MNRAS*, **377**(May), 453–458.

Rauer, H., Catala, C., Aerts, C., *et al.* 2013. The PLATO 2.0 Mission. ArXiv:1310.0696.

Reed, W. M. 1893. Stars that have been found variable in the photograph. *AJ*, **13**(May), 62–63.

Reese, D., Lignières, F., and Rieutord, M. 2006. Acoustic oscillations of rapidly rotating polytropic stars: II. Effects of the Coriolis and centrifugal accelerations. *A&A*, **455**(Aug.), 621–637.

Reese, D., Lignières, F., and Rieutord, M. 2008. Regular patterns in the acoustic spectrum of rapidly rotating stars. *A&A*, **481**(Apr.), 449–452.

Reese, D. R., MacGregor, K. B., Jackson, S., Skumanich, A., and Metcalfe, T. S. 2009. Pulsation modes in rapidly rotating stellar models based on the self-consistent field method. *A&A*, **509**(Oct.), 183–188.

Reiners, A. 2006. Rotation- and temperature-dependence of stellar latitudinal differential rotation. *A&A*, **446**(Jan.), 267–277.

Reiners, A., and Mohanty, S. 2012. Radius-dependent angular momentum evolution in low-mass stars. I. *ApJ*, **746**(Feb.), 43.

Reinhold, T., Reiners, A., and Basri, G. 2013. Rotation and differential rotation of active Kepler stars. *A&A*, **560**(Dec.), A4.

Remillard, R. A., and McClintock, J. E. 2006. X-ray properties of black-hole binaries. *ARA&A*, **44**(Sept.), 49–92.

Rempel, M. 2007. Origin of solar torsional oscillations. *ApJ*, **655**(Jan.), 651–659.

Rempel, M. 2011. Subsurface magnetic field and flow structure of simulated sunspots. *ApJ*, **740**(Oct.), 15.

Rempel, M., Schüssler, M., and Knölker, M. 2009. Radiative magnetohydrodynamic simulation of sunspot structure. *ApJ*, **691**(Jan.), 640–649.

Retter, A., Bedding, T. R., Buzasi, D. L., Kjeldsen, H., and Kiss, L. L. 2003. Oscillations in Arcturus from WIRE photometry. *ApJ*, **591**(July), L151–L154.

ReVelle, D. 1976. On meteor generated infrasound. *J. Geophys. Res.*, **81**, 1217–1229.

Reville, V., Brun, A. S., Matt, S. P., Strugarek, A., and Pinto, P. 2014. The effect of magnetic topology on thermally-driven wind: Towards a general formulation of the braking law. *ApJ*, in press.

Rhodes, Jr., E. J., Didkovsky, L., Chumak, O. V., and Scherrer, P. H. 1995. Status of the Mt. Wilson–Crimean–Kazakhstan High-Degree Helioseismology Network. Page 398 of Ulrich, R. K., Rhodes, Jr., E. J., and Dappen, W. (eds), *GONG 1994. Helio- and Astro-Seismology from the Earth and Space*. Astronomical Society of the Pacific Conference Series, vol. 76.

Richard, O., Vauclair, S., Charbonnel, C., and Dziembowski, W. A. 1996. New solar models including helioseismological constraints and light-element depletion. *A&A*, **312**(Aug.), 1000–1011.

Richardson, J. E., Melosh, H. J., and Greenberg, R. 2004. Impact-induced seismic activity on Asteroid 433 Eros: A surface modification process. *Science*, **306**, 1526–1529.

Ricker, G. R., Latham, D. W., Vanderspek, R. K., *et al.* 2010 (Jan.). Transiting Exoplanet Survey Satellite (TESS). Page 450.06 of *American Astronomical Society Meeting Abstracts 215*. Bulletin of the American Astronomical Society, vol. 42.

Rickett, J. E., and Claerbout, J. F. 2000. Calculation of the Sun's acoustic impulse response by multi-dimensional spectral factorization. *Sol. Phys.*, **192**(Mar.), 203–210.

Rieutord, M., and Espinosa Lara, F. 2013. Ab initio modelling of steady rotating stars. Page 49 of Goupil, M., Belkacem, K., Neiner, C., Lignières, F., and Green, J. J. (eds), *Studying Stellar Rotation and Convection: Theoretical Background and Seismic Diagnostics*. Lecture Notes in Physics, Berlin: Springer Verlag, vol. 865.

Ritsema, J., Deuss, A., van Heijst, H. J., and Woodhouse, J. H. 2011. S40RTS: A degree-40 shear-velocity model for the mantle from new Rayleigh wave dispersion, teleseismic traveltime and normal-mode splitting function measurements. *Geophys. J. Int.*, **184**(Mar.), 1223–1236.

Roberson, F. I., and Kaula, W. M. 1972. Orbital Science Investigations Part D: Apollo 15 Laser Altimeter. In *Apollo 15 Preliminary Science Report*, NASA Scientific and Information Office SP-289, Washington D.C., 25-48–25-50.

Roberts, I. 1889. On a photographic method for determining variability in stars. [Abstract]. *R. Soc. London Proc. Ser. I*, **47**, 137–138.

Robinson, M. S., Brylow, S. M., Tschimmel, M., *et al.* 2010. Lunar Reconnaissance Orbiter Camera (LROC) instrument overview. *Space Sci. Rev.*, **150**, 81–124.

Roca Cortés, T., and Pallé, P. L. 2014. The Mark I helioseismic experiment: I. Measurements of the solar gravitational redshift (1976–2013). ArXiv:1406.5944.

Rogers, F. J., Swenson, F. J., and Iglesias, C. A. 1996. OPAL equation-of-state tables for astrophysical applications. *ApJ*, **456**(Jan.), 902.

Rogers, T. M., and Glatzmaier, G. A. 2005. Gravity waves in the Sun. *MNRAS*, **364**(Dec.), 1135–1146.

Ronan, R. S., and Labonte, B. J. 1994. Intermediate degree p-mode frequency splittings near solar maximum. *Sol. Phys.*, **149**(Jan.), 1–21.

Roulston, M. S., and Ahrens, T. J. 1997. Impact mechanics and frequency of SL9-type events on Jupiter. *Icarus*, **126**, 138–147.

Roxburgh, I. W. 2002. Background to the Eddington mission. Pages 11–15 of Battrick, B., Favata, F., Roxburgh, I. W., and Galadi, D. (eds), *Stellar Structure and Habitable Planet Finding*. ESA Special Publication, vol. 485.

Roxburgh, I. W. 2006. 2-dimensional models of rapidly rotating stars: II. Hydrostatic and acoustic models with $\Omega = \Omega(r, \theta)$. *A&A*, **454**(Aug.), 883–888.

Roxburgh, I. W. 2010. Asteroseismology of solar and stellar models. *Ap&SS*, **328**(July), 3–11.

Roxburgh, I. W. 2015. Surface layer independent model fitting by phase matching: Theory and application to HD49933 and HD177153 (aka Perky). *A&A*, **574**(Feb.), A45.

Roxburgh, I. W., and Vorontsov, S. V. 1994. Seismology of the solar envelope: The base of the convective zone as seen in the phase shift of acoustic waves. *MNRAS*, **268**(June), 880.

Roxburgh, I. W., and Vorontsov, S. V. 2003. The ratio of small to large separations of acoustic oscillations as a diagnostic of the interior of solar-like stars. *A&A*, **411**(Nov.), 215–220.

Roxburgh, I. W., and Vorontsov, S. V. 2013. On the use of the ratio of small to large separations in asteroseismic model fitting. *A&A*, **560**(Dec.), A2.

Royer, F., Zorec, J., and Gómez, A. E. 2007. Rotational velocities of A-type stars: III. Velocity distributions. *A&A*, **463**(Feb.), 671–682.

Rubio da Costa, F., Zuccarello, F., Labrosse, N., *et al.* 2011 (June). Solar flares: Observations vs simulations. Pages 182–184 of Bonanno, A., de Gouveia Dal Pino, E., and Kosovichev, A. G. (eds), *Advances in Plasma Astrophysics*. IAU Symposium 274.

Rudiger, G., and Kitchatinov, L. L. 1997. The slender solar tachocline: A magnetic model. *Astron. Nachr.*, **318**(Aug.), 273.

Rüdiger, G., Kitchatinov, L. L., and Elstner, D. 2012. Helicity and dynamo action in magnetized stellar radiation zones. *MNRAS*, **425**(Sept.), 2267–2276.

Russell, C. T., Raymond, C. A., and Coradini, A., *et al.* 2012. Dawn at Vesta: Testing the protoplanetary paradigm. *Science*, **336**(6082), 684–686.

Saar, S. H., and Brandenburg, A. 2002. A new look at dynamo cycle amplitudes. *Astron. Nachr.*, **323**(July), 357–360.

Saio, H. 1981. Rotational and tidal perturbations of nonradial oscillations in a polytropic star. *ApJ*, **244**(Feb.), 299–315.

Salabert, D., Leibacher, J., Appourchaux, T., and Hill, F. 2009a. Measurement of low signal-to-noise ratio solar p-modes in spatially resolved helioseismic data. *ApJ*, **696**(May), 653–667.

Salabert, D., García, R. A., Pallé, P. L., and Jiménez-Reyes, S. J. 2009b. The onset of solar cycle 24. What global acoustic modes are telling us. *A&A*, **504**(Sept.), L1–L4.

Salabert, D., Régulo, C., Ballot, J., García, R. A., and Mathur, S. 2011. About the p-mode frequency shifts in HD 49933. *A&A*, **530**(June), A127.

Samadi, R., Belkacem, K., Ludwig, H.-G., *et al.* 2013. Stellar granulation as seen in disk-integrated intensity: II. Theoretical scaling relations compared with observations. *A&A*, **559**(Nov.), A40.

Sana, H., de Mink, S. E., de Koter, A., *et al.* 2012. Binary interaction dominates the evolution of massive stars. *Science*, **337**(July), 444–446.

Sanz-Forcada, J., Stelzer, B., and Metcalfe, T. S. 2013. ιHorologi, the first coronal activity cycle in a young solar-like star. *A&A*, **553**(May), L6.

Saumon, D., and Guillot, T. 2004. Shock compression of deuterium and the interiors of Jupiter and Saturn. *ApJ*, **609**(July), 1170–1180.

Savonije, G. J. 2005. Unstable quasi g-modes in rotating main-sequence stars. *A&A*, **443**(Nov.), 557–570.

Savonije, G. J., Papaloizou, J. C. B., and Alberts, F. 1995. Nonadiabatic tidal forcing of a massive uniformly rotating star. *MNRAS*, **277**(Nov.), 471.

Scargle, J. D. 1982. Studies in astronomical time series analysis: II. Statistical aspects of spectral analysis of unevenly spaced data. *ApJ*, **263**(Dec.), 835–853.

Schad, A., Timmer, J., and Roth, M. 2012. Measuring the solar meridional flow from perturbations of eigenfunctions of global oscillations. *Astron. Nachr.*, **333**(Dec.), 991.

Schad, A., Timmer, J., and Roth, M. 2013. Global helioseismic evidence for a deeply penetrating solar meridional flow consisting of multiple flow cells. *ApJ*, **778**(Dec.), L38.

Schatzman, E. 1993. Transport of angular momentum and diffusion by the action of internal waves. *A&A*, **279**(Nov.), 431–446.

Scherrer, P. H., Bogart, R. S., Bush, R. I., *et al.* 1995. The Solar Oscillations Investigation: Michelson Doppler Imager. *Sol. Phys.*, **162**(Dec.), 129–188.

Scherrer, P. H., Schou, J., Bush, R. I., *et al.* 2012. The Helioseismic and Magnetic Imager (HMI) investigation for the Solar Dynamics Observatory (SDO). *Sol. Phys.*, **275**(Jan.), 207–227.

Schmider, F.-X., Fossat, E., and Mosser, B. 1991. Possible detection of Jovian global oscillations. *A&A*, **248**(Aug.), 281–291.

Schmider, F. X., Gay, J., Gaulme, P., *et al.* 2007. SYMPA, a dedicated instrument for Jovian seismology: I. Principle and performance. *A&A*, **474**(Nov.), 1073–1080.

Schmidt, B. E., Thomas, P. C., and Bauer, J. M., *et al.* 2008 (March). Hubble takes a look at Pallas: Shape, size and surface. Page 2502 of *Lunar and Planetary Institute Science Conference Abstracts*. Lunar and Planetary Institute Technical Report, vol. 39.

Schmitt, J., Bouchy, F., and Bestaux, J. L. 2000. First results of the EMILIE spectrograph with the AAA system on extrasolar planets. Page 607 of Garzón, G., Eiroa, C., de Winter, D., and Mahoney, T. J. (eds), *Disks, Planetesimals, and Planets*. Astronomical Society of the Pacific Conference Series, vol. 219.

Schou, J. 1992. *On the analysis of helioseismic data*. Ph.D. thesis, Aarhus University, Aarhus, Denmark.

Schou, J., and Bogart, R. S. 1998. Flows and horizontal displacements from ring diagrams. *ApJ*, **504**(Sept.), L131–L134.

Schou, J., and Brown, T. M. 1994. Generation of artificial helioseismic time-series. *A&A Suppl. Ser.*, **107**(Nov.), 541–550.

Schou, J., and Buzasi, D. L. 2001 (Jan.). Observations of p-modes in α Cen. Pages 391–394 of Wilson, A. and Pallé, P. L. (eds), *SOHO 10/GONG 2000 Workshop: Helio- and Asteroseismology at the Dawn of the Millennium*. ESA Special Publication, vol. 464.

Schou, J., Howe, R., Basu, S., *et al.* 2002. A comparison of solar p-mode parameters from the Michelson Doppler Imager and the Global Oscillation Network Group: Splitting coefficients and rotation inversions. *ApJ*, **567**(Mar.), 1234–1249.

Schou, J., Scherrer, P. H., Bush, R. I., *et al.* 2012. Design and ground calibration of the Helioseismic and Magnetic Imager (HMI) instrument on the Solar Dynamics Observatory (SDO). *Sol. Phys.*, **275**(Jan.), 229–259.

Schrijver, C. J., and Zwaan, C. 2000. *Solar and Stellar Magnetic Activity*. New York: Cambridge University Press.

Schunker, H., Cameron, R. H., Gizon, L., and Moradi, H. 2011. Constructing and characterising solar structure models for computational helioseismology. *Sol. Phys.*, **271**(July), 1–26.

Schunker, H., Gizon, L., Cameron, R. H., and Birch, A. C. 2013. Helioseismology of sunspots: How sensitive are travel times to the Wilson depression and to the subsurface magnetic field? *A&A*, **558**(Oct.), A130.

Schüssler, M., and Vögler, A. 2006. Magnetoconvection in a sunspot umbra. *ApJ*, **641**(Apr.), L73–L76.

Schuster, G. 2009. *Seismic Interferometry*, Cambridge: Cambridge University Press.

Sekii, T., Kosovichev, A. G., Zhao, J., *et al.* 2007. Initial helioseismic observations by Hinode/SOT. *PASJ*, **59**(Nov.), 637.

Sens-Schönfelder, C., and Larose, E. 2010. Lunar noise correlation, imaging and monitoring. *Earthq. Sci.*, **23**, 519–530.

Shapiro, N. M., Campillo, M., Stehly, L., and Ritzwoller, M. H. 2005. High-resolution surface-wave tomography from ambient seismic noise. *Science*, **307**(Mar.), 1615–1618.

Shipp, R. M., and Singh, S. C. 2002. Two-dimensional full wavefield inversion of wide-aperture marine seismic streamer data. *Geophys. J. Int.*, **151**, 325–344.

Shiraishi, H., Tanaka, S., Fujimura, A., and Hayakawa, H. 2008. The present status of the Japanese Penetrator Mission: LUNAR-A. *Adv. Space Res.*, **42**, 386–393.

Shirley, J. H. 1985. Shallow moonquakes and large shallow earthquakes: A temporal correlation. *Earth Planet. Sci. Lett.*, **76**, 241–253.

Showman, A. P., and Guillot, T. 2002. Atmospheric circulation and tides of "51 Pegasus b-like" planets. *A&A*, **385**(Apr.), 166–180.

Sierks, H., Lamy, P., Barbieri, C., *et al.* 2011. Images of Asteroid 21 Lutetia: A remnant planetesimal from the early Solar System. *Science*, **334**(6055), 487–490.

Sills, A., and Pinsonneault, M. H. 2000. Rotation of horizontal-branch stars in globular clusters. *ApJ*, **540**(Sept.), 489–503.

Silva Aguirre, V., Chaplin, W. J., Ballot, J., *et al.* 2011. Constructing a one-solar-mass evolutionary sequence using asteroseismic data from Kepler. *ApJ*, **740**(Oct.), L2.

Silva Aguirre, V., Casagrande, L., Basu, S., *et al.* 2012. Verifying asteroseismically determined parameters of Kepler stars using Hipparcos parallaxes: Self-consistent stellar properties and distances. *ApJ*, **757**(Sept.), 99.

Silva Aguirre, V., Basu, S., Brandão, I. M., *et al.* 2013. Stellar ages and convective cores in field main-sequence stars: First asteroseismic application to two Kepler targets. *ApJ*, **769**(June), 141.

Silvotti, R., Schuh, S., Janulis, R., *et al.* 2007. A giant planet orbiting the 'extreme horizontal branch' star V391 Pegasi. *Nature*, **449**(Sept.), 189–191.

Simons, F. J., Nolet, G., Georgief, P., *et al.* 2009. On the potential of recording earthquakes for global seismic tomography by low-cost autonomous instruments in the oceans. *J. Geophys. Res. (Solid Earth)*, **114**(May), 5307.

Simons, F. J., Loris, I., Nolet, G., *et al.* 2011. Solving or resolving global tomographic models with spherical wavelets, and the scale and sparsity of seismic heterogeneity. *Geophys. J. Int.*, **187**(Nov.), 969–988.

Skartlien, R., Stein, R. F., and Nordlund, Å. 2000. Excitation of chromospheric wave transients by collapsing granules. *ApJ*, **541**(Sept.), 468–488.

Skumanich, A. 1972. Time scales for Ca II emission decay, rotational braking, and lithium depletion. *ApJ*, **171**(Feb.), 565.

Smagorinsky, J. 1963. General circulation experiments with the primitive equations. *Mon. Wea. Rev.*, **91**, 99.

Smith, A., Crawford, I. A., Gowen, R. A., *et al.* 2009. Lunar Net: A proposal in response to an ESA M3 call in 2010 for a medium sized mission. *Exp. Astron.*, **33**, 587–644.

Smith, D., Zuber, M. T., Neumann, G. A., *et al.* 2010. Initial observations from the Lunar Orbiter Laser Altimeter (LOLA). *Geophys. Res. Lett.*, **37**, L18204, doi:10.1029/2010GL043751.

Snieder, R., Fan, Y., Slob, E., and Wapenaar, K. 2010. Equipartitioning is not sufficient for Green's function extraction. *Earthquake Sci.*, **23**(Oct.), 403–415.

Sohl, F., Wagner, F. W., Hussmann, H., and Grott, M. 2009. Planetary interiors. Pages 200–224 of Trümper, J. E. (ed.), *Landolt-Börnstein Numerical Data and Functional*

Relationships in Science and Technology, New Series, Group VI: Astronomy and Astrophysics, vol. 4, subvol. B.

Soszyński, I., Udalski, A., Szymański, M. K., *et al.* 2009. The Optical Gravitational Lensing Experiment. The OGLE-III Catalog of Variable Stars. IV. Long-period variables in the Large Magellanic Cloud. *Acta Astron.*, **59**(Sept.), 239–253.

Soulat, L., Schmider, F.-X., Robbe-Dubois, S., *et al.* 2012 (Sept.). Echoes: A new instrumental concept of spectro-imaging for Jovian seismology. In *Space Telescopes and Instrumentation 2012: Optical, Infrared, and Millimeter Wave*. Society of Photo-Optical Instrumentation Engineers (SPIE) Conference Series, vol. 8442.

Spada, F., Lanzafame, A. C., and Lanza, A. F. 2010. A semi-analytic approach to angular momentum transport in stellar radiative interiors. *MNRAS*, **404**(May), 641–660.

Speake, C. C. 1996. Forces and force gradients due to patch fields and contact-potential differences. *Class. Quantum Grav.*, **13**, A291–A297.

Spiegel, E. A., and Zahn, J.-P. 1992a. The solar tachocline. *A&A*, **265**(Nov.), 106–114.

Spiegel, E. A., and Zahn, J.-P. 1992b. The turbulent tachycline. Page 188 of Giampapa, M. S., and Bookbinder, J. A. (eds), *Cool Stars, Stellar Systems, and the Sun*. Astronomical Society of the Pacific Conference Series, vol. 26.

Spruit, H. C. 1999. Differential rotation and magnetic fields in stellar interiors. *A&A*, **349**(Sept.), 189–202.

Spruit, H. C. 2002. Dynamo action by differential rotation in a stably stratified stellar interior. *A&A*, **381**(Jan.), 923–932.

Spruit, H. C. 2003. Origin of the torsional oscillation pattern of solar rotation. *Sol. Phys.*, **213**(Mar.), 1–21.

Spruit, H. C., and Bogdan, T. J. 1992. The conversion of p-modes to slow modes and the absorption of acoustic waves by sunspots. *ApJ*, **391**(June), L109–L112.

Stebbins, J. 1911. The measurement of the light of stars with a selenium photometer, with an application to the variations of Algol. *Pop. Astron.*, **19**(Feb.), 65–80.

Stebbins, J. 1915. The electrical photometry of stars. *Science*, **41**(June), 809–813.

Stebbins, J., and Brown, F. C. 1907. A determination of the Moon's light with a selenium photometer. *ApJ*, **26**(Dec.), 326.

Stein, R. F., and Nordlund, Å. 2001. Solar oscillations and convection: II. Excitation of radial oscillations. *ApJ*, **546**(Jan.), 585–603.

Stein, R. F., and Nordlund, Å. 2012. On the formation of active regions. *ApJ*, **753**(July), L13.

Steiner, O., Franz, M., Bello González, N., *et al.* 2010. Detection of vortex tubes in solar granulation from observations with SUNRISE. *ApJ*, **723**(Nov.), L180–L184.

Stello, D., Kjeldsen, H., Bedding, T. R., and Buzasi, D. 2006. Oscillation mode lifetimes in ξ Hydrae: Will strong mode damping limit asteroseismology of red giant stars? *A&A*, **448**(Mar.), 709–715.

Stello, D., Bruntt, H., Kjeldsen, H., *et al.* 2007. Multisite campaign on the open cluster M67: II. Evidence for solar-like oscillations in red giant stars. *MNRAS*, **377**(May), 584–594.

Stello, D., Bruntt, H., Preston, H., and Buzasi, D. 2008. Oscillating K giants with the WIRE satellite: Determination of their asteroseismic masses. *ApJ*, **674**(Feb.), L53–L56.

Stello, D., Chaplin, W. J., Bruntt, H., *et al.* 2009a. Radius determination of solar-type stars using asteroseismology: What to expect from the Kepler mission. *ApJ*, **700**(Aug.), 1589–1602.

Stello, D., Chaplin, W. J., Basu, S., Elsworth, Y., and Bedding, T. R. 2009b. The relation between $\Delta\nu$ and $\nu_{\rm max}$ for solar-like oscillations. *MNRAS*, **400**(Nov.), L80–L84.

Stello, D., Meibom, S., Gilliland, R. L., *et al.* 2011. An asteroseismic membership study of the red giants in three open clusters observed by Kepler: NGC 6791, NGC 6819, and NGC 6811. *ApJ*, **739**(Sept.), 13.

Stello, D., Huber, D., Bedding, T. R., *et al.* 2013. Asteroseismic classification of stellar populations among 13,000 red giants observed by Kepler. *ApJ*, **765**(Mar.), L41.

Stello, D., Compton, D. L., Bedding, T. R., *et al.* 2014. Non-radial oscillations in M-giant semi-regular variables: Stellar models and *Kepler* observations. *ApJ*, **788**, 10.

Strugarek, A., Brun, A. S., and Zahn, J.-P. 2011a. Magnetic confinement of the solar tachocline: The oblique dipole. *Astron. Nachr.*, **332**(Dec.), 891.

Strugarek, A., Brun, A. S., and Zahn, J.-P. 2011b. Magnetic confinement of the solar tachocline: II. Coupling to a convection zone. *A&A*, **532**(Aug.), A34.

Sturrock, P. A. 1966. Model of the high-energy phase of solar flares. *Nature*, **211**(Aug.), 695–697.

Suárez, J. C., Goupil, M. J., and Morel, P. 2006. Effects of moderately fast shellular rotation on adiabatic oscillations. *A&A*, **449**(Apr.), 673–685.

Sutton, G. H., and Latham, G. V. 1964. Analysis of a feedback controlled seismometer. *J. Geophys. Res.*, **69**, 3865–3882.

Svestka, Z. 1986. On the varieties of solar flares. Pages 332–355 of Neidig, D. F. (ed), *The Lower Atmosphere of Solar Flares*. Proceedings of the Solar Maximum Mission Symposium, Sunspot, NM, Aug. 20–24, 1985 (A87-26201 10-92).

Tagger, M., and Pellat, R. 1999. An accretion-ejection instability in magnetized disks. *A&A*, **349**(Sept.), 1003–1016.

Takata, M., and Saio, H. 2013. Rosette modes of oscillations of rotating stars caused by close degeneracies: I. Axisymmetric modes. *PASJ*, **65**(June), 68.

Talon, S., and Charbonnel, C. 2005. Hydrodynamical stellar models including rotation, internal gravity waves, and atomic diffusion: I. Formalism and tests on Pop I dwarfs. *A&A*, **440**(Sept.), 981–994.

Talon, S., and Charbonnel, C. 2010 (Apr.). Light elements as diagnostics on the structure and evolution of low-mass stars. Pages 365–374 of Charbonnel, C., Tosi, M., Primas, F., and Chiappini, C. (eds), *Light Elements in the Universe*. IAU Symposium 268.

Tanaka, S., Mitani, T., Otake, H., *et al.* 2013. Present status of the lunar lander project SELENE-2. *44th Lunar and Planetary Science Conference*, March 18–22. 2013 in The Woodlands, Texas, LPI Contribution No. 1719.

Tanimoto, T., Eitzel, M., and Yano, T. 2008. The noise cross-correlation approach for Apollo 17 LSPE data: Diurnal change in seismic parameters in shallow lunar crust. *J. Geophys. Res.*, **113**, E08011, doi:10.1029/2007JE3016.

Tape, C., Liu, Q., Maggi, A., and Tromp, J. 2010. Seismic tomography of the southern California crust based on spectral-element and adjoint methods. *Geophys. J. Int.*, **180**, 433–462.

Tassoul, M. 1980. Asymptotic approximations for stellar nonradial pulsations. *ApJS*, **43**(Aug.), 469–490.

Tauris, T. M., and van den Heuvel, E. P. J. 2006. Formation and evolution of compact stellar X-ray sources. Pages 623–665 of Lewin, W., and van der Klis, M. (eds), *Compact Stellar X-ray Sources*. Cambridge: Cambridge University Press.

Tauzin, B., Debayle, E., Quantin, C., and Coltice, N. 2013. Seismoacoustic coupling induced by the breakup of the 15 February 2013 Chelyabinsk meteor. *Geophys. Res. Lett.*, **40**, 3522–3526.

Tayler, R. J. 1973. The adiabatic stability of stars containing magnetic fields: I. Toroidal fields. *MNRAS*, **161**, 365.

Teanby, N., and Wookey, J. 2011. Seismic detection of meteorite impacts on Mars. *Phys. Earth Planet. Inter.*, **186**, 70–80.

Telting, J. H., Østensen, R. H., Baran, A. S., *et al.* 2012. Three ways to solve the orbit of KIC 11 558 725: A 10-day beaming sdB+WD binary with a pulsating subdwarf. *A&A*, **544**(Aug.), A1.

Theobald, M. L., Fox, P. A., and Sofia, S. 1994. A subgrid-scale resistivity for magnetohydrodynamics. *Phys. Plasmas*, **1**(Sept.), 3016–3032.

Thomas, J. H., and Weiss, N. O. 2012. *Sunspots and Starspots*. Cambridge: Cambridge University Press.

Thomas, J. H., Cram, L. E., and Nye, A. H. 1982. Five-minute oscillations as a subsurface probe of sunspot structure. *Nature*, **297**(June), 485–487.

Thompson, M. J., and Christensen-Dalsgaard, J. 2002. On inverting asteroseismic data. Pages 95–101 of Battrick, B., Favata, F., Roxburgh, I. W., and Galadi, D. (eds), *Stellar Structure and Habitable Planet Finding*. ESA Special Publication, vol. 485.

Thompson, M. J., Christensen-Dalsgaard, J., Miesch, M. S., and Toomre, J. 2003. The internal rotation of the Sun. *ARA&A*, **41**, 599–643.

Titarchuk, L., and Osherovich, V. 2000. The global normal disk oscillations and the persistent low-frequency quasi-periodic oscillations in X-ray binaries. *ApJ*, **542**(Oct.), L111–L114.

Toksöz, M. N., Press, F., Anderson, K., *et al.* 1972a. Velocity structure and properties of the lunar crust. *The Moon*, **4**, 490–504.

Toksöz, M. N., Solomon, S. C., Minear, J. W., and Johnston, D. H. 1972b. Thermal evolution of the Moon. *The Moon*, **4**, 190–213.

Toksöz, M. N., Dainty, A. M., Solomon, S. C., and Anderson, K. R. 1974. Structure of the Moon. *Rev. Geophys. Space Phys.*, **12**, No. 4, 539–567.

Toksoz, M. N., Goins, N. R., and Cheng, C. H. 1977. Moonquakes: Mechanisms and relation to tidal stresses. *Science*, **196**(May), 979–981.

Tomczyk, S., Streander, K., Card, G., *et al.* 1995. An instrument to observe low-degree solar oscillations. *Sol. Phys.*, **159**(June), 1–21.

Tomczyk, S., Jiménez Reyes, S. J., Jiménez, A., and Pallé, P. L. 2000 (May). The ECHO (Experiment for Coordinated Helioseismic Observations) Network. Page 804 of *AAS/Solar Physics Division Meeting #31*. Bulletin of the American Astronomical Society, vol. 32.

Tong, C. H., 2005. Imaging sunspots using seismic methods. *Phil. Trans. R. Soc. Math., Phys. Eng. Sci.*, **363**, 2761–2775.

Tong, C. H., Thompson, M. J., Warner, M. R., Rajaguru, S. P., and Pain, C. C. 2003a. Acoustic wave propagation in the Sun: Implications for wave field and time-distance helioseismology. *ApJ*, **582,** L121–L124.

Tong, C. H., Thompson, M. J., Warner, M. R., and Pain, C. C. 2003b. Helioseismic signals and wave field helioseismology. *ApJ*, **593,** 1242–1248.

Tong, C. H., Thompson, M. J., Warner, M. R., and Pain, C. C. 2003c. The significance of density and attenuation in the local helioseismology. *ApJ,* **596**, L263–L266.

Toomre, J., Augustson, K. C., Brown, B. P., *et al.* 2012 (Sept.). New era in 3-D modeling of convection and magnetic dynamos in stellar envelopes and cores. Page 331 of Shibahashi, H., Takata, M., and Lynas-Gray, A. E. (eds), *Progress in Solar/Stellar Physics with Helio- and Asteroseismology*. Astronomical Society of the Pacific Conference Series, vol. 462.

Toriumi, S., and Yokoyama, T. 2010. Two-step emergence of the magnetic flux sheet from the solar convection zone. *ApJ*, **714**(May), 505–516.

Toriumi, S., Ilonidis, S., Sekii, T., and Yokoyama, T. 2013. Probing the shallow convection zone: Rising motion of subsurface magnetic fields in the solar active region. *ApJ*, **770**(June), L11.

Toutain, T., and Gouttebroze, P. 1993. Visibility of solar p-modes. *A&A*, **268**(Feb.), 309–318.

Toutain, T., Appourchaux, T., Baudin, F., *et al.* 1997. Tri-phonic helioseismology: Comparison of solar p modes observed by the helioseismology instruments aboard SOHO. *Sol. Phys.*, **175**(Oct.), 311–328.

Townsend, R. H. D. 1997. Spectroscopic modelling of non-radial pulsation in rotating early-type stars. *MNRAS*, **284**(Feb.), 839–858.

Trampert, J. 1998. Global seismic tomography: The inverse problem and beyond. *Inverse Probl.*, **14**, 371–385.

Tromp, J., Luo, Y., Hanasoge, S., and Peter, D. 2010. Noise cross-correlation sensitivity kernels. *Geophys. J. Int.*, **183**(Nov.), 791–819.

Tsai, V. C., Nettles, M., Ekström, G., and Dziewonski, A. M. 2005. Multiple CMT source analysis of the 2004 Sumatra earthquake. *Geophys. Res. Lett.*, **32**, L17304, doi:10.1029/2005GL023813.

Turck-Chièze, S., and Couvidat, S. 2011. Solar neutrinos, helioseismology and the solar internal dynamics. *Rep. Prog. Phys.*, **74**(8), 086901.

Turck-Chièze, S., Nghiem, P., Couvidat, S., and Turcotte, S. 2001. Solar internal composition and nuclear reaction rates in the light of helioseismology. *Sol. Phys.*, **200**(May), 323–342.

Turck-Chièze, S., Palacios, A., Marques, J. P., and Nghiem, P. A. P. 2010. Seismic and dynamical solar models: I. The impact of the solar rotation history on neutrinos and seismic indicators. *ApJ*, **715**(June), 1539–1555.

Turck-Chièze, S., Carton, P.-H., Barrière, J.-C., *et al.* 2012 (Sept.). Solar global oscillations of low-degree modes (GOLD): The status of the multi-channel resonance spectrometer GOLF-NG. Page 240 of Shibahashi, H., Takata, M., and Lynas-Gray, A. E. (eds), *Progress in Solar/Stellar Physics with Helio- and Asteroseismology*. Astronomical Society of the Pacific Conference Series, vol. 462.

Ulrich, R. K. 1970. The five-minute oscillations on the solar surface. *ApJ*, **162**(Dec.), 993.

Ulrich, R. K. 1986. Determination of stellar ages from asteroseismology. *ApJ*, **306**(July), L37–L40.

Unno, W., Osaki, Y., Ando, H., Saio, H., and Shibahashi, H. 1989. *Nonradial Oscillations of Stars*. Berlin: Springer-Verlag.

Utsu, T. 2002. Relationships between magnitude scales. In Lee, W. H. K., *et al.* (ed.), *International Handbook of Erthquake and Engineering Seismology, Part A*. Amsterdam, Boston: Academic Press.

Uytterhoeven, K., Briquet, M., Bruntt, H., *et al.* 2010. Ground-based follow-up in relation to Kepler asteroseismic investigation. *Astron. Nachr.*, **331**(Dec.), 993.

van der Klis, M. 2005 (Oct.). Timing neutron stars. Pages 345–358 of Burderi, L., Antonelli, L. A., D'Antona, F., *et al.* (eds), *Interacting Binaries: Accretion, Evolution, and Outcomes*. American Institute of Physics Conference Series, vol. 797.

van der Klis, M., Wijnands, R. A. D., Horne, K., and Chen, W. 1997. Kilohertz quasi-periodic oscillation peak separation is not constant in Scorpius X-1. *ApJ*, **481**(June), L97–L100.

van Leeuwen, F. 2007. *Hipparcos, the New Reduction of the Raw Data*. Astrophysics and Space Science Library, vol. 350. Springer.

van Saders, J. L., and Pinsonneault, M. H. 2013. Fast star, slow star; old star, young star: Subgiant rotation as a population and stellar physics diagnostic. *ApJ*, **776**(Oct.), 67.

VandenBerg, D. A., Gustafsson, B., Edvardsson, B., Eriksson, K., and Ferguson, J. 2007. A constraint on Z_{solar} from fits of isochrones to the color-magnitude diagram of M67. *ApJ*, **666**(Sept.), L105–L108.

Veverka, J., Thomas, P., Harch, A., *et al.*, 1999. NEAR encounter with asteroid 253 Mathilde: Overview. *Icarus*, **140**(1), 3–16.

Veverka, J., Robinson, M., and Thomas, P., *et al.* 2000. NEAR at Eros: Imaging and spectral results. *Science*, **289**(5487), 2088–2097.

Vinnik, L., Chenet, H., Gagnepain-Beyneix, J., and Lognonné, P. 2001. First seismic receiver functions on the Moon. *Geophys. Res. Lett.*, **28**, 3031–3034.

Virieux, J., 1984. Sh-wave propagation in heterogeneous media: Velocity-stress finite-difference method. *Geophysics*, **49**, 1933–1942.

Virieux, J., and Operto, S. 2009. An overview of full-waveform inversion in exploration geophysics. *Geophysics*, **74**, WCC1–WCC26.

Vögler, A., Shelyag, S., Schüssler, M., *et al.* 2005. Simulations of magneto-convection in the solar photosphere: Equations, methods, and results of the MURaM code. *A&A*, **429**(Jan.), 335–351.

Vorontsov, S. V. 1989. The inverse problem of helioseismology: The speed of sound in the solar interior. *Pisma Astron. Zh.*, **15**(Jan.), 48–60.

Vorontsov, S. V., and Jefferies, S. M. 2013. Modeling solar oscillation power spectra. II. Parametric model of spectral lines observed in Doppler-velocity measurements. *ApJ*, **778**(Nov.), 75.

Vorontsov, S. V., and Zharkov, V. N. 1981. The natural oscillations of giant planets: Effects of rotation and ellipticity. *AZh*, **58**(Oct.), 1101–1114.

Vorontsov, S. V., Zharkov, V. N., and Lubimov, V. M. 1976. The free oscillations of Jupiter and Saturn. *Icarus*, **27**(Jan.), 109–118.

Vorontsov, S. V., Gudkova, T. V., and Zharkov, V. N. 1989. Jovian seismology. *Soviet Astron. Lett.*, **15**(July), 278.

Vorontsov, S. V., Baturin, V. A., and Pamiatnykh, A. A. 1991. Seismological measurement of solar helium abundance. *Nature*, **349**(Jan.), 49–51.

Vorontsov, S. V., Baturin, V. A., Ayukov, S. V., and Gryaznov, V. K. 2013. Helioseismic calibration of the equation of state and chemical composition in the solar convective envelope. *MNRAS*, **430**(Apr.), 1636–1652.

Vostreys, R. W. 1980. *Data User's Note: Apollo Seismological Investigations*. National Space Science Data Center/World Data Center A for Rockets and Satellites, NASA Goddard Space Flight Center, Greenbelt, Maryland.

Wagoner, R. V. 1999. Relativistic diskoseismology. *Phys. Rep.*, **311**(Apr.), 259–269.

Wagoner, R. V. 2012. Diskoseismology and QPOs confront black hole spin. *ApJ*, **752**(June), L18.

Wagoner, R. V., Silbergleit, A. S., and Ortega-Rodríguez, M. 2001. "Stable" quasi-periodic oscillations and black hole properties from diskoseismology. *ApJ*, **559**(Sept.), L25–L28.

Walker, J. D. 2003. Loading sources for seismological investigation of asteroids and comets. *Int. J. Impact Engng*, **29**, 757–769.

Walker, J. D., and Huebner, W. F. 2004. Seismological investigation of asteroid and comet interiors. Pages 234–265 of Belton, M. J. S., Morgan, T. H., Samarasinha, N. H., and Yeomans, D. K. (eds), *Mitigation of Hazardous Comets and Asteroids*. Cambridge: Cambridge University Press.

Walker, J. D., Sagebiel, E. J., and Huebner, W. F. 2006. A preliminary analysis of seismological techniques to study Eros and other asteroids. *Adv. Space Res.*, **37**, 142–152.

Walker, J. D., Chocron, S., Durda, D. D., *et al.* 2013. Scale size effect in momentum enhancement. *Proc. Eng.*, **58**(0), 240–250. Proceedings of the 12th Hypervelocity Impact Symposium.

Walter, C. M., Marley, M. S., Hunten, D. M., *et al.* 1996. A search for seismic waves from the impact of the SL/9 R fragment. *Icarus*, **121**(June), 341–350.

Walterscheid, R. L., Brinkman, D. G., and Schubert, G. 2000. Wave disturbances from the comet SL-9 impacts into Jupiter's atmosphere. *Icarus,* **145**, 140–146.

Wang, Y., Noyes, R. W., Tarbell, T. D., and Title, A. M. 1995. Vorticity and divergence in the solar photosphere. *ApJ*, **447**(July), 419.

Warner, M. R., Ratcliffe, A., Nangoo, T., *et al.* 2013. Anisotropic 3-D full-waveform inversion. *Geophysics,* **78**, R59–R80.

Watkins, J. S., and Kovach, R. L. 1973. Seismic investigation of the lunar regolith, *Proceedings of the 4th Lunar Science Conference* (Supplement 4, *Geochim. Cosmochim. Acta*), **3**, 2561–2564.

Weber, R. C., and Knapmeyer, M. 2012. Deep moonquake focal mechanisms: Recovery and implications. Page 1466 of *Lunar and Planetary Science Conference*, vol. 43.

Weber, R. C., Bills, B. G., and Johnson, C. L. 2009. Constraints on deep moonquake focal mechanisms through analyses of tidal stress. *J. Geophys. Res. (Planets)*, **114**(May), 5001.

Weber, R. C., Garcia, R., Johnson, C. L., *et al.* 2010a. The use of deep moonquakes for constraining the internal structure of the Moon. *AGU Fall Meeting Abstracts*, U51B–0037.

Weber, R. C., Bills, B. G., and Johnson, C. L. 2010b. A simple physical model for deep moonquake occurrence times. *Phys. Earth Planet. Inter.*, **182**, 152–160.

Weber, R. C., Lin, P.-Y., Garnero, E. J., Williams, Q., and Lognonné, P. 2011. Seismic detection of the lunar core. *Science*, **331**(Jan.), 309–312.

Wedemeyer, S., Freytag, B., Steffen, M., Ludwig, H.-G., and Holweger, H. 2004. Numerical simulation of the three-dimensional structure and dynamics of the non-magnetic solar chromosphere. *A&A*, **414**(Feb.), 1121–1137.

Wegler, U., and Sens-Schönfelder, C. 2007. Fault zone monitoring with passive image interferometry. *Geophys. J. Int.*, **168**(Mar.), 1029–1033.

Whitaker, E. A. 1972. Artificial lunar impact craters: Four new identifications. In *Apollo 16 Preliminary Science Report*, NASA Scientific and Technical Information Office, NASA SP-315, Washington, D.C., 29-39–29-45.

White, T. R., Bedding, T. R., Stello, D., *et al.* 2011. Calculating asteroseismic diagrams for solar-like oscillations. *ApJ*, **743**(Dec.), 161.

White, T. R., Bedding, T. R., Gruberbauer, M., *et al.* 2012. Solving the mode identification problem in asteroseismology of F stars observed with Kepler. *ApJ*, **751**(June), L36.

White, T. R., Huber, D., Maestro, V., *et al.* 2013. Interferometric radii of bright Kepler stars with the CHARA Array: theta Cygni and 16 Cygni A and B. *MNRAS*, **433**(Aug.), 1262–1270.

Wieczorek, M. A., Jolliff, B. L., Khan, A., *et al.* 2006. The constitution and structure of the lunar interior. *Rev. Mineral. Geochem.*, **60**, 221–364.

Wieczorek, M. A., Neumann, G. A., Nimmo, F., *et al.* 2013. The crust of the Moon as seen by GRAIL. *Science*, **339**, 671–675.

Williams, J. P. 2001. Acoustic environment of the martian surface. *J. Geophys. Res.*, **106**, 5033–5041.

Williamson, P., and Worthington, M. 1993. Resolution limits in ray tomography due to wave behavior: Numerical experiments. *Geophysics*, **58**, 727–735.

Willson, R. C. 1984. Measurements of solar total irradiance and its variability. *Space Sci. Rev.*, **38**(Aug.), 203–242.

Wilson, O. C. 1978. Chromospheric variations in main-sequence stars. *ApJ*, **226**(Dec.), 379–396.

Winget, D. E., Nather, R. E., Clemens, J. C., *et al.* 1991. Asteroseismology of the DOV star PG 1159-035 with the Whole Earth Telescope. *ApJ*, **378**(Sept.), 326–346.

Winget, D. E., Nather, R. E., Clemens, J. C., *et al.* 1994. Whole Earth Telescope observations of the DBV white dwarf GD 358. *ApJ*, **430**(Aug.), 839–849.

Wolff, C. L. 1972. Free oscillations of the Sun and their possible stimulation by solar flares. *ApJ*, **176**(Sept.), 833.

Woodard, M., and Hudson, H. 1983. Solar oscillations observed in the total irradiance. *Sol. Phys.*, **82**(Jan.), 67–73.

Woodard, M. F. 1997. Implications of localized, acoustic absorption for heliotomographic analysis of sunspots. *ApJ*, **485**(Aug.), 890–894.

Woodard, M. F. 2002. Solar subsurface flow inferred directly from frequency-wavenumber correlations in the seismic velocity field. *ApJ*, **565**(Jan.), 634–639.

Woodard, M. F. 2009. Helioseismic measurement of large-scale solar flows. Page 15 of Dikpati, M., Arentoft, T., González Hernández, I., Lindsey, C., and Hill, F. (eds), *Solar-Stellar Dynamos as Revealed by Helio- and Asteroseismology: GONG 2008/SOHO 21*. Astronomical Society of the Pacific Conference Series, vol. 416.

Woodard, M. F., and Noyes, R. W. 1985. Change of solar oscillation eigenfrequencies with the solar cycle. *Nature*, **318**(Dec.), 449–450.

Yamada, R., Garcia, R. F., Lognonné, P., *et al.* 2011. Optimisation of seismic network design: Application to a geophysical international lunar network. *Planet. Space Sci.*, **59**, 343–354.

Yamada, R., Garcia, R. F., Lognonné, P., *et al.* 2013. On the possibility of lunar core phase detection using new seismometers for soft-landers in future lunar missions. *Planet. Space Sci.*, **81**, 18–31.

Yates, M. T. 1967. *The Use of Spent Stages as Seismic Sources for Lunar Seismological Experiments*. Bellcom, Inc., technical memorandum TM-67-1012-11, 21.

Yeomans, D. K., Antreasian, P. G., and Barriot, J.-P., *et al.* 2000. Radio science results during the NEAR-Shoemaker spacecraft rendezvous with Eros. *Science*, **289**(5487), 2085–2088.

Young, A. T. 1967. Photometric error analysis: VI. Confirmation of Reiger's theory of scintillation. *AJ*, **72**(Aug.), 747.

Zaatri, A., Corbard, T., Roth, M., González Hernández, I., and von der Lühe, O. 2008. Comparison of geometrical mapping for ring diagram analysis. *J. Phys. Confer. Ser.*, **118**(1), 012090.

Zahn, J.-P. 1992. Circulation and turbulence in rotating stars. *A&A*, **265**(Nov.), 115–132.

Zahn, J.-P., Talon, S., and Matias, J. 1997. Angular momentum transport by internal waves in the solar interior. *A&A*, **322**(June), 320–328.

Zahn, J.-P., Brun, A. S., and Mathis, S. 2007. On magnetic instabilities and dynamo action in stellar radiation zones. *A&A*, **474**(Oct.), 145–154.

Zanni, C., and Ferreira, J. 2013. MHD simulations of accretion onto a dipolar magnetosphere: II. Magnetospheric ejections and stellar spin-down. *A&A*, **550**(Feb.), A99.

Zhao, H., and Chou, D.-Y. 2013. Mode conversion between different radial orders for solar acoustic waves scattered by sunspots. *ApJ*, **778**(Nov.), 4.

Zhao, J. 2007. Time-distance imaging of solar far-side active regions. *ApJ*, **664**(Aug.), L139–L142.

Zhao, J., Kosovichev, A. G., and Duvall, Jr., T. L. 2001. Investigation of mass flows beneath a sunspot by time-distance helioseismology. *ApJ*, **557**(Aug.), 384–388.

Zhao, J., Georgobiani, D., Kosovichev, A. G., *et al.* 2007. Validation of time-distance helioseismology by use of realistic simulations of solar convection. *ApJ*, **659**(Apr.), 848–857.

Zhao, J., Parchevsky, K. V., and Kosovichev, A. G. 2012a (Sept.). Interaction of helioseismic waves with sunspots: Observations and numerical MHD simulations. Page 277 of Shibahashi, H., Takata, M., and Lynas-Gray, A. E. (eds), *Progress in Solar/Stellar Physics with Helio- and Asteroseismology*. Astronomical Society of the Pacific Conference Series, vol. 462.

Zhao, J., Nagashima, K., Bogart, R. S., Kosovichev, A. G., and Duvall, Jr., T. L. 2012b. Systematic center-to-limb variation in measured helioseismic travel times and its effect on inferences of solar interior meridional flows. *ApJ*, **749**(Apr.), L5.

Zhao, J., Couvidat, S., Bogart, R. S., *et al.* 2012c. Time-distance helioseismology data-analysis pipeline for Helioseismic and Magnetic Imager onboard Solar Dynamics Observatory (SDO/HMI) and its initial results. *Sol. Phys.*, **275**(Jan.), 375–390.

Zhao, J., Bogart, R. S., Kosovichev, A. G., Duvall, Jr., T. L., and Hartlep, T. 2013. Detection of equatorward meridional flow and evidence of double-cell meridional circulation inside the Sun. *ApJ*, **774**(Sept.), L29.

Zhou, C. X., Cai, W., Luo, Y., Schuster, G. T., and Hassanzadeh, S. 1995. Acoustic wave-equation travel-time and wave-form inversion of crosshole seismic data. *Geophysics*, **60**, 765–773.

Zharkov, S., Zharkova, V. V., and Matthews, S. A. 2011. Comparison of seismic signatures of flares obtained by SOHO/Michelson Doppler Imager and GONG instruments. *ApJ*, **739**(Oct.), 70.

Zharkov, V. N., and Trubitsyn, V. P. 1978. *Physics of Planetary Interiors*. Tucson, AZ: Pachart.

Zumberge, M., Berger, J., Otero, J., and Wielandt, E. 2010. An optical seismometer without force. *Bull. Seismol. Soc. Am.*, **100**, 598–605.

Index